TOPOLOGICAL AND ALGEBRAIC STRUCTURES IN FUZZY SETS

AF181568

TRENDS IN LOGIC
Studia Logica Library

VOLUME 20

Managing Editor
Ryszard Wójcicki, *Institute of Philosophy and Sociology, Polish Academy of Sciences, Warsaw, Poland*

Editors
Daniele Mundici, *Department of Mathematics "Ulisse Dini", University of Florence, Italy*
Ewa Orłowska, *National Institute of Telecommunications, Warsaw, Poland*
Graham Priest, *Department of Philosophy, University of Queensland, Brisbane, Australia*
Krister Segerberg, *Department of Philosophy, Uppsala University, Sweden*
Alasdair Urquhart, *Department of Philosophy, University of Toronto, Canada*
Heinrich Wansing, *Institute of Philosophy, Dresden University of Technology, Germany*

SCOPE OF THE SERIES

Trends in Logic is a bookseries covering essentially the same area as the journal *Studia Logica* – that is, contemporary formal logic and its applications and relations to other disciplines. These include artificial intelligence, informatics, cognitive science, philosophy of science, and the philosophy of language. However, this list is not exhaustive, moreover, the range of applications, comparisons and sources of inspiration is open and evolves over time.

Volume Editor
Daniele Mundici

The titles published in this series are listed at the end of this volume.

TOPOLOGICAL AND ALGEBRAIC STRUCTURES IN FUZZY SETS

A Handbook of Recent Developments in the Mathematics of Fuzzy Sets

Edited by

STEPHEN ERNEST RODABAUGH
Youngstown State University, Youngstown, U.S.A.

and

ERICH PETER KLEMENT
Johannes Kepler Universität, Linz, Austria

KLUWER ACADEMIC PUBLISHERS
DORDRECHT / BOSTON / LONDON

A C.I.P. Catalogue record for this book is available from the Library of Congress.

ISBN 978-90-481-6378-6

Published by Kluwer Academic Publishers,
P.O. Box 17, 3300 AA Dordrecht, The Netherlands.

Sold and distributed in North, Central and South America
by Kluwer Academic Publishers,
101 Philip Drive, Norwell, MA 02061, U.S.A.

In all other countries, sold and distributed
by Kluwer Academic Publishers,
P.O. Box 322, 3300 AH Dordrecht, The Netherlands.

Printed on acid-free paper

All Rights Reserved
© 2003 Kluwer Academic Publishers
Softcover reprint of the hardcover 1st edition 2003
No part of this work may be reproduced, stored in a retrieval system, or transmitted
in any form or by any means, electronic, mechanical, photocopying, microfilming, recording
or otherwise, without written permission from the Publisher, with the exception
of any material supplied specifically for the purpose of being entered
and executed on a computer system, for exclusive use by the purchaser of the work.

TABLE OF CONTENTS

PART II: ALGEBRAIC STRUCTURES IN FUZZY SETS

PART III: RELATED TOPICS IN TOPOLOGICAL AND ALGEBRAIC STRUCTURES

VOLUME CONTRIBUTORS

IAN W. ALDERTON
Department of Mathematics,
Applied Mathematics & Astronomy
University of South Africa (UNISA)
0003, Pretoria, South Africa
E-mail: alderiw@risc5.unisa.ac.za
alderiw@alpha.unisa.ac.za

BERNARD DE BAETS
Department of Applied Mathematics,
Biometrics and Process Control
Ghent University, Coupure links 653
B-9000 Gent, Belgium
E-mail: bernard.debaets@rug.ac.be

BERNHARD BANASCHEWSKI
Department of Mathematics and Statistics
McMaster University
Hamilton, Ontario, L8S 4K1, Canada
E-mail: penny@mcmail.cis.mcmaster.ca
crag@maths.uct.ac.za
moira@maths.uct.ac.za

WERNER GÄHLER
Scheibenbergstraße 37
12685 Berlin, Germany
E-mail: gaehler@rz.uni-potsdam.de

MAI GEHRKE
Department of Mathematical Sciences
New Mexico State University
Las Cruces, NM 88003, USA
E-mail: mgehrke@nmsu.edu

COSIMO GUIDO
Department of Mathematics
University of Lecce
Via Arnesano, c.p. 193
73100 Lecce, Italy
E-mail: guido@ultra5.unile.it

JAVIER GUTIÉRREZ GARCÍA
Matematika Saila
Euskal Herriko Unibertsitatea
Apartado 644
48080 Bilbao, Spain
E-mail: mtpgugaj@lg.ehu.es

ULRICH HÖHLE
Bergische Universität
Gesamthochschule Wuppertal
Fachbereich 7, Mathematik
Gaußstraße 20
42097 Wuppertal, Germany
E-mail: uhoehle@uni-wuppertal.de
hoehle@wmfa2.math.uni-wuppertal.de

SÁNDOR JENEI
Institute of Mathematics and Informatics
University of Pécs
Ifjúság u. 6
7624 Pécs, Hungary
E-mail: jenei@ttk.pte.hu

ERICH PETER KLEMENT
Institut für Algebra, Stochastik und
wissensbasierte mathematische Systeme
Johannes Kepler Universität
4040 Linz, Austria
E-mail: ep.klement@jku.at

WESLSY J. KOTZÉ
Department of Mathematics (Pure & Applied)
Rhodes University
6140 Grahamstown, South Africa
E-mail: w.kotze@ru.ac.za

TOMASZ KUBIAK
Wydzial Matematyki i Informatyki
Uniwersytet im. Adama Mickiewicza
Umultowska 87
61-614 Poznań, Poland
E-mail: tkubiak@amu.edu.pl

Radko Mesiar
Department of Mathematics
Slovak University of Technology
Radlinského 11
81368 Bratislava, Slovak Republic
E-mail: mesiar@math.sk
Institute of Information Theory and Automation
Academy of Sciences of the Czech Republic
P.O. Box 18
182 08 Praha 8, Czech Republic

Christopher J. Mulvey
School of Mathematical Sciences
University of Sussex
Falmer, Brighton,
BN1 9QH, United Kingdom
E-mail: c.j.mulvey@sussex.ac.uk

Maria A. de Prada Vicente
Matematika Saila
Euskal Herriko Unibertsitatea
Apartado 644
48080 Bilbao, Spain
E-mail: mtpprvia@lg.ehu.es

Aleš Pultr
Department of Applied Mathematics and ITI
(Institute of Theoretical Informatics)
MFF
Charles University
Malostranske nam. 25
11800 Praha 1, Czech Republic
E-mail: pultr@kam.mff.cuni.cz
pultr@kam.ms.mff.cuni.cz

Stephen Ernest Rodabaugh
Department of Mathematics and Statistics
Youngstown State University
Youngstown, Ohio 44555-3609, USA
E-mail: rodabaug@math.ysu.edu
rodabaugh@as.ysu.edu, serod@concentric.net

ALEXANDER P. ŠOSTAK
Department of Mathematics
University of Latvia
1586 Riga, Latvia
E-mail: sostaks@latnet.lv
sostaks@lanet.lv

LAWRENCE NEFF STOUT
Department of Mathematics
Illinois Wesleyan University
Bloomington, Illinois 61702-2900, USA
E-mail: lstout@sun.iwu.edu

PETER VICENÍK
Department of Mathematics
Slovak University of Technology
Radlinského 11
813 68 Bratislava, Slovak Republic
E-mail: vicenik@math.sk

CAROL WALKER
Department of Mathematical Sciences
New Mexico State University
Las Cruces, NM 88003, USA
E-mail: hardy@nmsu.edu

ELBERT A. WALKER
Department of Mathematical Sciences
New Mexico State University
Las Cruces, NM 88003, USA
E-mail: elbert@nmsu.edu

PREFACE

This volume summarizes recent developments in the topological and algebraic structures in fuzzy sets and may be rightly viewed as a continuation of the standardization of the mathematics of fuzzy sets established in the "Handbook", namely the *Mathematics of Fuzzy Sets: Logic, Topology, and Measure Theory*, Volume 3 of *The Handbooks of Fuzzy Sets Series* (Kluwer Academic Publishers, 1999). Many of the topological chapters of the present work are not only based upon the foundations and notation for topology laid down in the Handbook, but also upon Handbook developments in convergence, uniform spaces, compactness, separation axioms, and canonical examples; and thus this work is, with respect to topology, a continuation of the standardization of the Handbook. At the same time, this work significantly complements the Handbook in regard to algebraic structures. Thus the present volume is an extension of the content and role of the Handbook as a reference work.

On the other hand, this volume, even as the Handbook, is a culmination of mathematical developments motivated by the renowned *International Seminar on Fuzzy Set Theory*, also known as the *Linz Seminar*, held annually in Linz, Austria. Much of the material of this volume is related to the Twentieth Seminar held in February 1999, material for which the Seminar played a crucial and stimulating role, especially in providing feedback, connections, and the necessary screening of ideas. Some chapters are mature developments and significant extensions of ideas presented at the Seminar; however, other chapters are wholly new works growing out of the informal discussions and roundtables of the Seminar. Though this volume is not a proceedings of the Twentieth Seminar, its content is in large measure a culmination of work motivated by that Seminar and the historical role of the Linz Seminar in the systematization of the mathematics of fuzzy sets.

The editors commend the authors for their willingness to extend, update, and rewrite their chapters within a formal peer-review procedure overseen by the editors; and appreciation is expressed to the reviewers for improving these chapters. A special thanks for generous support goes to the *Bundesministerium für Wissenshaft und Forschung in Vienna*, the *Linzer Hochschulfonds*, and the *Fuzzy Logic Laboratorium Linz* in Hagensberg; and a special thanks for logistical support goes to the *Bildungshaus St. Magdalena* and its staff for facilitating the intimate setting of the Twentieth Seminar. Finally, gratitude is expressed to Kluwer Academic Publishers editors for their cooperation and incredible patience with the volume editors during the long period of preparation.

The Editors

INTRODUCTION

As stated in the introduction of [45], "the mathematics of fuzzy sets is the mathematics of lattice-valued maps"; and consequently, the mathematics of fuzzy sets lends itself to a variety of structures, both topological and algebraic. Closely related to the standardization of the mathematics of fuzzy sets begun in [45], this volume continues the work of [45] in topology as well as gives several important developments in algebraic structures not available when [45] was published. At the same time, the chapters of the present work are motivated in significant measure by the presentations, informal discussions, and roundtables of the *Twentieth International Seminar on Fuzzy Set Theory*, or *Twentieth Linz Seminar*, held in Linz, Austria, February 1999.

The chapters in this volume are arranged into three main parts: Part I, topological structures; Part II, algebraic structures; and Part III, shorter chapters on topics related to Parts I and II. However, the chapters of this volume make so many substantive contributions across such a broad range of interrelated mathematical disciplines, that we have chosen to arrange our introductory remarks by the major themes of this volume in the order of their first appearance by chapter. We therefore discuss the chapters of this book as grouped by, and in the order of, the following major themes:

1. Uniformities and convergence structures
2. Fundamental examples in fuzzy topology
3. Modifications and extensions of sobriety
4. Categorical aspects
5. Logic and foundations of mathematics
6. Triangular norms and associated structures

Many chapters touch on more than one theme; however, most chapters have a primary emphasis which determines their place in the following remarks.

1 Uniformities and convergence structures

The first theme of this volume is uniformities and convergence structures, the former treated within both point-free and point-set contexts. To place the uniformity and later topology chapters into an overall context, we recall notions of fixed-basis *vis-a-vis* variable-basis topology. Frameworks for fixed-basis topology [47], e.g. L-**Top** or L-**Ftop**, are based on the same underlying lattice L of membership values, yet allow change of the underlying set for spaces. Frameworks for variable-basis topology [97], e.g. **C-Top** or **C-Ftop**, allow both underlying lattices and underlying sets to change from space to space (where **C** is some category of lattice-theoretic structures), thus providing a common framework for both point-set and point-free contexts, the latter including locales [60],

S.E. Rodabaugh and E.P. Klement (eds.), Topological and Algebraic Structures in Fuzzy Sets, 1–16.
© 2003 *Kluwer Academic Publishers.*

Hutton spaces [53], and topological molecular lattices [117]. It is known [89, 97] each point-free context is isomorphic to a subcategory of **C-Top** of *singleton* spaces (for prescribed **C** and singleton): e.g., **C** = **Loc** is isomorphic to the full subcategory of all singleton spaces of **Loc-Top** using a prescribed singleton. We note:

1. Point-free contexts, i.e. singleton spaces, have weaker requirements in foundations (see Chapters 1,11); cf. the localic Tihonov theorem [58].
2. Neither **C** nor L-**Top** (for $L \in |\mathbf{C}|$, which includes **Top** as **2-Top**) generalize the other and both are generalized by **C-Top**.
3. Each morphism in **C-Top** is a morphism embedded from **C** followed by a morphism embedded from some L-**Top**, and conversely [20, 21, 22].

Representing the point-free approach to uniformities is Chapter 1 of Prof. Banaschewski, a systematic tutorial which builds upon [12] and concerns uniform completions in the point-free context. This chapter deals with nearness and metric frames [10, 4, 11], a generalized sense of (regular) Cauchy filters, completeness of the natural uniformity on the localic reals [3], compactification [5, 6, 9], and other topics in which completeness plays a role. Needed specific foundations are pinpointed, including the role of Countable Dependent Choice. Noting completeness in spaces is equivalent to dense-embeddings always being isomorphisms, completeness is defined for nearness frames using strict surjections and it is proved that each nearness and metric frame has a completion.

Notions of filter and neighborhood play key roles in the chapters on uniformities, as well as in the theory of convergence developed in Chapter 2 of Prof. Gähler using partially-ordered monads of [24] and the L-filters of [43]. This theory includes probabilistic convergence structures, limit towers, and graded fuzzy convergence structures, and uses characterizations of neighborhood and closure operators to construct a scheme of separation axioms for convergence spaces [25, 26, 27].

Representing the point-set approach to uniformities are two chapters, Chapter 3 of Profs. Gutiérrez García, de Prada Vicente, and Šostak and Chapter 7 of Prof. Rodabaugh.

Chapter 3 categorically unifies in the fixed-basis case the entourage approach of [76], the T-uniformities approach of [38], and the uniform operator approach of [52] by extending the original approach of [52] to include the other approaches. The super category unifying these approaches is a topological construct. This is a significant advance in the unification of uniform spaces in fuzzy sets—the first such advance since [39]—and relies on a generalized notion of L-filter, cf. [23, 35, 36, 42, 47], which incorporates the notions of filter implicit in previous approaches to uniformities.

Developing the axiomatic foundations of a generalized Hutton approach to quasi-uniformities, Chapter 7 weakens Hutton's intersection operator to allow tensor products [47, 97] other than binary meets—e.g. cross products of t-norms, removes any distributivity of the underlying quasi-monoidal lattice, and constructs new classes of fundamental examples captured by such an axiomatization, including $\mathbb{R}$ and $\mathbb{I}$ for certain co-quantales L and L-**2** soberifications of $\mathbb{R}$ and $\mathbb{I}$ for certain complete quasi-monoidal lattices and quasi-uniform monoidal lattices L [98, 99].

It is an open question how much of Chapter 7 can be recovered within (a modification of) the unification of Chapter 3.

2 Fundamental examples in fuzzy topology

The importance and role of fundamental examples in fuzzy topology occur throughout many chapters; but three chapters are specifically dedicated to this topic, namely Chapter 4 of Prof. Höhle, Chapter 5 of Prof. Kubiak, and Chapter 17 of Profs. Pultr and Rodabaugh. Chapter 17 is is a trailer for Chapter 6 by the same authors and will be discussed later.

Chapter 4 proposes a new intersection axiom for $\mathbb{I}$-valued spaces using the arithmetic mean (monoidal mean operator derived from Łukasiewicz conjunction [41]) and constructs $\mathbb{I}$-valued rigid topological spaces, a setting in which Boolean negation $\neg : \{0,1\} \to \mathbb{I}$ has Łukasiewicz negation $' : \mathbb{I} \to \mathbb{I}$ as its *unique* continuous extension, uniqueness not being possible in traditional or standard $\mathbb{I}$-topology [47]. This remarkable result comes from the construction of new classes of examples by considering τ-smooth Borel probability measures on ordinary spaces and Radon measures on ordinary compact Hausdorff spaces, examples which include $\mathbb{I}$ interpreted as the $\mathbb{I}$-rigid topological space comprising all Radon measures on $\{0,1\}$, a space in which $\{0,1\}$ is densely embedded—thus providing $' : \mathbb{I} \to \mathbb{I}$ in this setting as the unique continuous extension of $\neg : \{0,1\} \to \mathbb{I}$.

The technique of constructing lattice-valued spaces by considering all measures of some kind on an ordinary topological space has also been considered in [77, 78] where saturated $\mathbb{I}$-topological spaces of probability measures on the Borel subsets of ordinary separable metric spaces are constructed. Such constructions are part of the broader phenomenon of fuzzification and fuzzy duals [91]. Perhaps the most studied fuzzifications are the L-fuzzy real line $\mathbb{R}(L)$ and the L-fuzzy unit interval $\mathbb{I}(L)$, the well-known fuzzy duals of $\mathbb{R}$ and $\mathbb{I}$ constructed using probability distributions on $\mathbb{R}$ and $\mathbb{I}$.

Chapter 5 is a tutorial dedicated to the study of $\mathbb{R}(L)$ and $\mathbb{I}(L)$ as codomains of particular families of L-continuous mappings, especially to the role played by these families in Urysohn Lemmas, insertion theorems, Tietze and Urysohn extension theorems, separation axioms, and the L-Tychonoff cube $\prod_\gamma \mathbb{I}(L)_\gamma$ [66, 67, 69, 70, 71, 72, 74]. Striking consequences of this approach include the following: the Urysohn lemma and the insertion theorem for normality characterize normality (for L a complete deMorgan algebra), but the extension theorem does not—normality guarantees extension (the "hard" direction in traditional topology) but the converse fails; and the L-Tychonoff cube $\prod_\gamma \mathbb{I}(L)_\gamma$ is shown to have a Brouwer fixed point theorem if L is completely distributive.

3 Modifications and extensions of sobriety

Sobriety has been an active area in the topology of fuzzy sets since [40] and [90, 92, 93]. Chapters 6 and 17 of Profs. Pultr and Rodabaugh, the Appendix to Chapter 6 of Profs. Höhle and Rodabaugh, and Chapter 16 of Prof. Kotzé develop modifications or extensions of sobriety as originally defined for L-topological spaces.

Chapter 6 introduces lattice-valued frames or L-frames, related to traditional frames analogously to how L-topological spaces are related to traditional topological spaces, motivated by the insight that level sets and level mappings associated with L-topologies are special systems of collection-wise monomorphic and collection-wise extremally epimorphic frame morphisms. The notion of an L-frame allows new descriptions of previously known classes sober spaces as well as generating a new class of sober spaces—which surprisingly turn out to be the ultra-sober or ι-sober spaces (in the notation of [68]), justifying examples for which are extensively documented in Chapter 17 primarily using results of [68] applied to various well-known spaces or their modifications.

Chapter 6 requires a complete chain L for much of its development. The Appendix to Chapter 6 outlines how this requirement may be relaxed to a spatial frame L for many constructions of Chapter 6 by focusing on the meet-irreducibles of L, which in the case of a complete chain are dually isomorphic to $L_\top \equiv \{\alpha \in L : \alpha < \top\}$.

Chapter 16 extends the semi-sobriety for L-topological spaces given in [64]—defined using irreducible L-closed subsets and the concept of β-closure of singletons—to (L, M)-topological spaces in the sense of [65, 102, 103, 73, 105]. Viewing (L, M)-topologies as towers of L-topologies ordered by the way below relation, where M is a continuous lattice, this chapter is the first extension of sobriety-like concepts to (L, M)-topologies.

It should be noted that traditional sobriety is equivalent to T_0 + semi-sobriety (restricted to the crisp case). On one hand, [64] shows that in L-topology—with L a complete De Morgan algebra—the standard L-sobriety [40, 62, 90, 92, 93, 94, 98] implies T_0 + semi-sobriety; but on the other hand, Chapter 17 shows that the converse fails for a rather large class of simple spaces. Thus for L a complete De Morgan algebra, [64] and Chapter 16 are implicitly proposing a new sobriety axiom for L-topological spaces—T_0 + semi-sobriety—which is more general than standard L-sobriety.

To summarize, Chapters 6, 16, 17 propose several new axioms of sobriety. Many questions concerning their interrelationships are resolved by the examples of Chapter 17. But open questions remain: e.g., it is an open question how the new sobriety of Chapter 6—ι-sobriety—is related to the new sobriety of Chapter 16—T_0 + semi-sobriety.

4 Categorical aspects

Although categorical aspects of fuzzy sets, topology, and related areas play a critical role in many chapters of this volume—see chapters 1,2,3,4,6,11, developments in this area are a primary emphasis of Chapter 8 of Prof. Stout, Chapter 12 of Prof. Šostak, Chapter 13 of Prof. Alderton, and Chapter 15 of Prof. Guido.

Chapter 8 develops a fully fuzzy topological point of view by considering topological space objects [108, 109] in Goguen's category SET(L) [33], where L is $[0, 1]$ equipped with a left-continuous t-norm. Since L has a closed structure, SET(L) has a closed structure [83, 84, 85], and remarkably, using its unbalanced subobjects, has sufficient structure to internize the operations needed for topology [48, 110, 111] despite the fact SET(L) is not a topos (since L is not **2**). The resulting category FFTOP(L) of fully fuzzy topological spaces is topological over SET(L). If the t-norm is $\wedge$, then fully topological spaces on a crisp set are saturated $\mathbb{I}$-topological spaces (cf. [40]). Interior operators, neighborhoods, and convergence are developed in this setting. Extensive computational examples are provided with the Likert sublattice of L used in opinion research.

Motivated by the crisp classification—induced by the standard forgetful functor V : L-**Top** $\to$ **Set**—of functions between L-topological spaces as being either L-continuous or not L-continuous, Chapter 12 applies the notion of fuzzy category [104, 106, 107] to construct the degree to which such a function is L-continuous, thereby constructing the L-fuzzy category $\mathcal{F}L$-**Top**, where L is a GL-monoid [46]. The issue of degree of objecthood is also addressed, including the degree to which an L-preinterior space is an L-topological space and the degree to which L-kernel space is an L-preinterior space, the concept of fuzzy functor is given, several L-fuzzy categories are constructed related to L-interior, L-preinterior, and L-kernel spaces, and initial and final structures in the fuzzy categories of L-kernel and L-preinterior spaces are obtained.

Categorical definitions of compactness rooted in certain categorical closure operators [16, 2] are applied in Chapter 13 to the fixed-basis categories **S$\mathbb{I}$-TOP** and **$\mathbb{I}$-TOP** [47] to generate corresponding compactness notions for $\mathbb{I}$-topology, namely the well-known α-compactness and α^*-compactness axioms of [28], as well as a new axiom termed semi-α-compactness arising via the closure spaces of Čech. The α-compactness axiom is also known to be categorically generated via factorization structures [1, 2] in the sense of [37].

Taking a different approach than Chapter 8, Chapter 15 constructs a category of spaces in which both the carrier and the topology are fuzzy sets. Building fuzzy functions using traditional Zadeh powerset operators [96, 95], generalized adjoint powerset operators between powersets of fuzzy subsets are then defined [34, 19]. A topology is a certain fuzzy subset on a powerset of a fuzzy set, and continuity is defined using the generalized preimage powerset operator. The resulting category is topological over its ground category, and products and initial subobjects are explicitly identified.

5 Logic and foundations of mathematics

Logic and foundations of mathematics are fundamentally related to the mathematics of fuzzy sets in part through the lattice-theoretic structures important in fuzzy sets as truth value or membership lattices, including De Morgan algebras and frames (or locales), structures dealt with respectively by Chapter 9 of Profs. Gehrke, Walker, and Walker and Chapter 11 of Prof. Mulvey. It is also of interest to note that common supercategories of these structures are important in the lattice-theoretic foundations of fuzzy topology, including semiframes (and semilocales) and complete quasi-monoidal lattices [47, 95, 96, 97].

Motivated by fuzzy logic applications using distinctly different logics, Chapter 9 examines truth value algebras from which these logics arise. The approach is to construct an algebraic model of a given logic in which the set of truth values is $[0,1]$ equipped with algebraic operations answering to logical connectives. In the case of idempotent conjunction, three logics arise from strict De Morgan systems [30]: classical Boolean logic, that having multiplication as conjunction, and that having minimum as conjunction [15]. The main focus is on the question of whether two choices of strict De Morgan systems—$[0,1]$ equipped with the usual order, a strict Archimedean t-norm, and a negation—can yield the same or even comparable logics. Using [29], the surprising answer is two such logics are comparable iff the strict De Morgan systems are isomorphic. The uncountably many resulting fuzzy logics are studied using equational properties and common extensions.

Chapter 11 is a tutorial examining the extent to which the Hahn-Banach Theorem of functional analysis can be proved constructively, without recourse to the Axiom of Choice (AC) or even the Axiom of Countable Choice. The traditional statement of the Hahn-Banach Theorem is reformulated so that it is both constructively proved without AC—the proof being valid in any topos with a natural numbers object—and logically equivalent to the classically formulated Hahn-Banach Theorem with AC. Drawing on the work of several authors [7, 8, 13, 14, 57, 59, 80, 81, 82, 112, 113], the geometric content of the classical theorem and its proof is linked to the constructive geometry of locales comprising weak* topologies of spaces of linear functionals bounded by 1 on seminormed linear spaces. The classical theorem is then recovered by applying the PT functor of [60] and the spatiality of these locales (which requires the Maximal Ideal Theorem and hence AC). The known applications of this approach—using the geometry of constructive logic to constructivize important mathematics, which also include the Tihonov Theorem and Stone-Čech compactification—together advance the relevant and important message that the means of contructivizing much mathematics is at hand and that doing so yields insights into the mathematics otherwise glossed over by the usual applicaions of AC.

Stone-Čech compactification reflectors for both fixed-basis and variable-basis topology (Section 1 preliminary remarks) have already been constructed [94, 98] using localic compactifications [60], the LPT functor, and the fact that **2**-spatiality implies L-spatiality (for each $L \hookleftarrow \mathbf{2}$). The work of Chapter 11 suggests that (with AC) there could be Hahn-Banach Theorems in fuzzy sets by combining the localic approach with the LPT, PT_{left}, PT_{right} functors of [94, 98] or the new spectrum functor $\mathbf{\Sigma}$ of Chapter 6 and its Appendix. Chapter 11's message implies AC is concerned with points—actually with having at least two points (Section 1 preliminary remarks). But AC gives something extra back as applied in *fuzzy sets*: compactifications in fuzzy sets have more limit points than traditional counterparts—e.g., [44] and Sections 2,6,8 of [98] show for $L \in |\mathbf{CBOOL}|$ that the L-**2** Stone-Čech compactification of $[0,1]$ is the L-fuzzy unit interval $\mathbb{I}(L)$; and the point-set variable-basis compactification from Section 9 of [98] is a strict generalization of localic compactification. Fuzzy topologists should examine what can be done constructively and then see what the investment of AC gives back, noting that point-set lattice-theoretic (poslat) topology in the form of variable-basis topology **C-Top**—when **C** = **Loc**—has a singleton ("constructive") part and a point-set ("AC") part (Section 1 preliminary remarks).

6 Triangular norms and associated structures

The most frequently occurring theme at the Twentieth Linz Seminar was triangular norms (t-norms) [79, 101, 61] and it is true of this volume as well. A t-norm is a binary operation on $[0,1]$ or other poset which is commutative, associative, isotone in each argument, and has the top element as identity. T-norms play a fundamental role in Chapters 3, 4, 7, 8, 9 discussed above and are the direct objects of study in the following chapters: Chapter 10 of Prof. Jenei, Chapter 14 of Profs. De Baets and Mesiar, Chapter 18 of Prof. Viceník, and Chapter 19 of Profs. Walker and Walker.

Chapter 10 is a systematic tutorial of Girard monoids on $[0,1]$. A Girard monoid is a commutative, residuated, zero-closed, lattice-ordered integral monoids, and when the underlying set is $[0,1]$, such a structure coincides with the concept of a left-continuous t-norm with strong induced negation [54, 55, 56], "strong" meaning the induced negation is an involution. Fundamental to nonclassical logics, such t-norms are geometrically described using properties of quasi-inverse and rotation invariance with respect to strong negation, one consequence being that the boundary condition of such a t-norm (1 is the identity) follows from its other axioms. Left-continuous t-norms with strong induced negation are constructed by rotation from any left-continuous t-norm without zero divisors and by rotation-annihilation from a t-norm with zero divisors in concert with a left-continuous, rotation-invariant t-subnorm. Decomposition theorems and informative three-dimensional plots are given.

Concerning discrete t-norms defined on a finite chain C [17, 32], Chapter 14 shows if $C \subset [0,1]$ with $\bigwedge C = 0$ and $\bigvee C = 1$, then each discrete t-norm on C admits a right-continuous extension; and if C is also divisible, then the t-norm admits a continuous extension having the same idempotent elements. Characterizing discrete t-norms having continuous extensions is an open problem. Conditions are given for restrictng t-norms from $[0,1]$ to discrete t-norms on finite subchains with $\bigwedge C = 0$ and $\bigvee C = 1$ or from complete chains to principal ideals or filters of idempotent elements.

Focusing on the generation of t-norms from generators (single-place functions) [61, 75, 101], Chapter 18 examines the question of associativity of binary operations generated by single-place conjunctive additive generators and presents necessary and sufficient conditions for such binary operations to be t-norms, as well as necessary and sufficient conditions for generated t-norms to be left-continuous or continuous. The latter include the result that a generated t-norm is left-continuous iff it is continuous and Archimedean.

Finally, Chapter 19 studies the group-theoretic aspects of several well known families of t-norms on $[0,1]$ [88]. Letting M denote the group of auto-order-isomorphisms and auto-anti-order isomorphisms on $[0,1]$, most of these families of t-norms are multiplicatively generated from subgroups of M or from cosets or conjugates of these subgroups. Important subgroups of M in this regard are A the subgroup of auto-order-isomorphisms, $\mathbb{R}^+$ embedded as a subgroup of A, the normalizer N of $\mathbb{R}^+$ in M, and S, where $N = \mathbb{R}^+ \times S$. Some families of strict t-norms are generated by conjugates of $\mathbb{R}^+$, while others are generated by S or its cosets or conjugates. The set of all nilpotent t-norms is bijective with A [31, 75] giving this set a group structure with Łukasiewicz conjunction as identity; and some nilpotent families are generated by or related to $\mathbb{R}^+$ or its conjugates by negation or to the centralizer of negation [31].

References

[1] I. W. Alderton, *α-Compactness of fuzzy topological spaces: a categorical approach*, in [63], 269–282.

[2] I. W. Alderton, *Compactness in topological categories*, in B. Banaschewski, C. R. A. Gilmour, H. Herrlich, eds, *Festschrift (On the Occasion of the 60th Birthday of Guillaume Brümmer) incorporating Proceedings, Symposium on Categorical Topology*, University of Cape Town, 1994, 9–16, Department of Mathematics and Applied Mathematics, University of Cape Town, 1999.

[3] B. Banaschewski, *The real numbers in pointfree topology*, Textos de Matemática Série B No. 12(1997), Departamento de Matemática da Universidade de Coimbra.

[4] B. Banaschewski, S. S. Hong, A. Pultr, *On the completion of nearness frames*, Quaestiones Mathematicae **21**(1998), 19–37.

[5] B. Banaschewski, C. J. Mulvey, *Stone-Čech compactification of locales I*, Houston J. Math. **6**(1980), 301–312.

[6] B. Banaschewski, C. J. Mulvey, *Stone-Čech compactification of locales II*, J. Pure Appl. Alg. **33**(1984), 107–122.

[7] B. Banaschewski, C. J. Mulvey, *The spectral theory of commutative C*-algebras: the constructive spectrum*, Quaestiones Mathematicae **25**(2000), 425–464.

[8] B. Banaschewski, C. J. Mulvey, *A globalisation of the Gelfand duality theorem*, to appear.

[9] B. Banaschewski, A. Pultr, *Samuel compactification and completion of uniform frames*, Math. Proc. Cambridge Phil. Soc. **108**(1990), 63–78.

[10] B. Banaschewski, A. Pultr, *Cauchy points of uniform and nearness frames*, Quaestiones Mathematicae **19**(1996), 107–127.

[11] B. Banaschewski, A. Pultr, *A new look at pointfree metrization theorems*, Comment. Math. Univ. Carolinae **39** (1998), 167–175.

[12] B. Banaschewski, A. Pultr, Uniformity In Pointfree Topology, *Chapter IV: Completion* (unpublished book manuscript).

[13] C. W. Burden, *The Hahn-Banach Theorem in a category of sheaves*, J. Pure Applied Algebra **17**(1980), 25–34.

[14] C. W. Burden, C. J. Mulvey, *Banach spaces in categories of sheaves, in Applications of Sheaves: Proceedings of the Durham Symposium* (July 1977), Lecture Notes in Mathematics **753**, 169–196, Springer-Verlag (Berlin/Heidelberg/New York), 1979.

[15] C. L. Chang, R. C. T. Lee, *Some properties of fuzzy logic*, Information and Control **19**(1971), 417–431.

[16] M. M. Clementino, E. Giuli, W. Tholen, *Topology in a category*, Portugal. Math. **53**(1996), 397–433.

[17] G. De Cooman, E. Kerre, *Order norms on bounded partially ordered sets*, J. Fuzzy Math. **2**(1994), 281–310.

[18] B. De Baets, R. Mesiar, *Triangular norms on product lattices*, Fuzzy Sets and Systems, Special Issue "Triangular norms" (B. De Baets and R. Mesiar, eds.), **104**(1999), 61–75.

[19] C. De Mitri, C. Guido, *Some remarks on fuzzy powerset operators*, Fuzzy Sets and Systems **126**(2002), 241–251.

[20] P. Eklund, *Category theoretic properties of fuzzy topological spaces*, Fuzzy Sets and Systems **13**(1984), 303–310.

[21] P. Eklund, *A comparison of lattice-theoretic approaches to fuzzy topology*, Fuzzy Sets and Systems **19**(1986), 81–87.

[22] P. Eklund, *Categorical fuzzy topology*, Doctoral Dissertation, (1986), Åbo Akademi (Turku, Finland).

[23] P. Eklund, W. Gähler, *Fuzzy filters, functors, and convergence*, Chapter 4 in [100], 109–136.

[24] W. Gähler, *Monadic topology—a new concept of generalized topology*, in Recent Developments of General Topology and its Applications, International Conference in Memory of Felix Hausdorff (1868–1942), Math. Research **67**, Akademie Verlag Berlin, 1992, 118–123.

[25] W. Gähler, A. S. Abd-Allah, A. Kandil, *Extended fuzzy topologies, Part I, On fuzzy stacks*, J. Fuzzy Mathematics **6**(1998) 223–261.

[26] W. Gähler, A. S. Abd-Allah, A. Kandil, *Extended fuzzy topologies, Part II, Characterizing notions*, J. Fuzzy Mathematics **6**(1998) 539–574.

[27] W. Gähler, A. S. Abd-Allah, A. Kandil, *Extended fuzzy topologies, Part III, Closedness and compactness*, J. Fuzzy Mathematics **6**(1998) 575–608.

[28] T. E. Gantner, R. C. Steinlage, R. H. Warren, *Compactness in fuzzy topological spaces*, J. Math. Anal. Appl. **62**(1978), 547–562.

[29] M. Gehrke, C. Walker, E. Walker, *De Morgan systems on the unit interval*, International Journal of Intelligent Systems, **11**(1996), 733–750.

[30] M. Gehrke, C. Walker, E. Walker, *A mathematical setting for fuzzy logics*, International Journal of Uncertainty, Fuzziness and Knowledge-Based Systems **5**(1997), 223–238.

[31] M. Gehrke, C. Walker, E. Walker, *A note on negations and nilpotent t-norms*, International Journal of Approximate Reasoning **21**(1999), 137–155.

[32] L. Godo, C. Sierra, *A new approach to connective generation in the framework of expert systems using fuzzy logic*, Proc. Eighteenth Internat. Symposium on Multiple-Valued Logic (Palma de Mallorca, Spain), IEEE Computer Society Press, 1988, pp. 157–162.

[33] J. A. Goguen, *Concept realizations in natural and artificial languages: axioms, extensions, applications for fuzzy sets*, Internat. J. Man-Machine Studies **6**(1974), 513–561.

[34] C. Guido, *The subspace problem in the traditional point set context of fuzzy topology*, in [63], 351–372.

[35] J. Gutiérrez García, M. A. de Prada Vicente, *Super uniform spaces*, in [63], 291–310.

[36] J. Gutiérrez García, M. A. de Prada Vicente, M. H. Burton, *Embeddings into the category of super uniform spaces*, in [63], 311–326.

[37] H. Herrlich, G. Salicrup, G. E, Strecker, *Factorizations, denseness, separation, and relatively compact objects*, Topology Appl. **27**(1987), 157–169.

[38] U. Höhle, *Probabilistic uniformization of fuzzy topologies*, Fuzzy Sets and Systems **1**(1978), 311–332.

[39] U. Höhle, *Probabilistic metrization of fuzzy topologies*, Fuzzy Sets and Systems **8**(1982), 63–69.

[40] U. Höhle, *Fuzzy topologies and topological space objects in a topos*, Fuzzy Sets and Systems **19**(1986), 299–304.

[41] U. Höhle, *Commutative, residuated ℓ–monoids*, Chapter IV in: U. Höhle, E. P. Klement, eds, Non-Classical Logics And Their Applications To Fuzzy Subsets—A Handbook of the Mathematical Foundations of Fuzzy Set Theory, Theory and Decision Library—Series B: Mathematical and Statistical Methods, Volume **32**, Kluwer Academic Publishers (Boston/Dordrecht/London), 1995, pp. 53–106.

[42] U. Höhle, *MV-algebra valued filter theory*, Quaestiones Mathematicae **19**(1996), 23–46.

[43] U. Höhle, *Locales and L-topologies*, in Mathematik-Arbeitspapiere 48, Univ. Bremen (1997) 223–250.

[44] U. Höhle, Many Valued Topology And Its Applications, Kluwer Academic Publishers (Boston/Dordrecht/London), 2001.

[45] U. Höhle, S. E. Rodabaugh, eds, Mathematics Of Fuzzy Sets: Logic, Topology, And Measure Theory, The Handbooks of Fuzzy Sets Series, Volume **3**(1999), Kluwer Academic Publishers (Boston/Dordrecht/London).

[46] U. Höhle, A. P. Šostak, *A general theory of fuzzy topological spaces*, Fuzzy Sets and Systems **73**(1995), 131–149.

[47] U. Höhle, A. Šostak, *Axiomatic foundations of fixed-basis fuzzy topology*, Chapter 3 in [45], 123–272.

[48] U. Höhle, L. N. Stout, *Foundations of fuzzy sets*, Fuzzy Sets and Systems, **40**(2)(1991), 257–296.

[49] R.-E. Hoffmann, *Irreducible filters and sober spaces*, Manuscripta Math. **22**(1977), 365-380.

[50] R.-E. Hoffmann, *Sobrification of partially ordered sets*, Semigroup Forum **17**(1979), 123-138.

[51] R.-E. Hoffmann, *On the soberification remainder* $sX - X$, Pacific J. Math. **83**(1979), 145–156.

[52] B. Hutton, *Uniformities on fuzzy topological spaces*, J. Math. Anal. Appl. **58**(1977), 559–571.

[53] B. Hutton, *Products of fuzzy topological spaces*, Topology Appl. **11**(1980), 59–67.

[54] S. Jenei, *New family of triangular norms via contrapositive symmetrization of residuated implications*, Fuzzy Sets and Systems **110**(2000), 157–174.

[55] S. Jenei, *Structure of left-continuous t-norms with strong induced negations. (I) Rotation construction*, Journal of Applied Non-Classical Logics **10**(2000), 83–92.

[56] S. Jenei, *Continuity of left-continuous triangular norms with strong induced negations and their boundary condition*, Fuzzy Sets and Systems **124**(2001), 35-41.

[57] P. T. Johnstone, Topos Theory, Academic Press (London), 1977.

[58] P.T. Johnstone, *Tychonoff's theorem without the axiom of choice,* Fund. Math. **113**(1981), 31-35.

[59] P. T. Johnstone, *The Gleason cover of a topos II*, J. Pure and Appl. Alg. **22**(1981), 229–247.

[60] P. T. Johnstone, Stone Spaces, Cambridge University Press (Cambridge), 1982.

[61] E. P. Klement, R. Mesiar, E. Pap, Triangular Norms, Trends in Logic, Studia Logica Library, Volume **8**(2000), Kluwer Academic Publishers (Dordrecht/Boston/London).

[62] W. Kotzé, *Lattice morphisms, sobriety, and Urysohn lemmas*, Chapter 10 in [100], 257-274.

[63] W. Kotzé, ed., Special Issue, Quaestiones Mathematicae **20(3)**(1997).

[64] W. Kotzé, *Sobriety and semi-sobriety of L-topological spaces*, Quaestiones Mathematicae **24**(2001), 549–554.

[65] T. Kubiak, *On Fuzzy Topologies*, Doctoral Dissertation, Adam Mickiewicz University (Poznan, Poland), 1985.

[66] T. Kubiak, *Extending continuous L-real functions*, Math. Japon. **31**(1986), 875–887.

[67] T. Kubiak, *L-fuzzy normal spaces and Tietze extension theorem*, J. Math. Anal. Appl. **125**(1987), 141–153.

[68] T. Kubiak, *The topological modification of the L-fuzzy unit interval*, Chapter 11 in [100], pp. 275–305.

[69] T. Kubiak, *A strengthening of the Katětov-Tong insertion theorem*, Comment. Math. Univ. Carolinae **34**(1993), 357–362.

[70] T. Kubiak, *The fuzzy Brouwer fixed-point theorem*, J. Math. Anal. Appl. **222**(1998), 62–66.

[71] T. Kubiak, *Separation Axioms: Extensions of Mappings and Embedding of Spaces*, Chapter 6 in [45], 433–479.

[72] T. Kubiak, M.A. de Prada Vicente, *On internal characterizations of completely L-regular spaces*, J. Math. Anal. Appl. **216**(1997), 581–592.

[73] T. Kubiak, A. P. Šostak, *Lower set-valued fuzzy topologies*, in [63], 423-429.

[74] T. Kubiak, D. Zhang, *On the L-fuzzy Brouwer fixed point theorem*, Fuzzy Sets and Systems **105**(1999), 287–292.

[75] C. M. Ling, *Representation of associative functions,* Publ. Math. Debrecen **12**(1965), 189–212.

[76] R. Lowen, *Fuzzy Uniform Spaces*, J. Math. Anal. Appl. **82**(1981), 370–385.

[77] R. Lowen, *On the existence of natural fuzzy topologies on spaces of probability measures*, Math. Nachr. **115**(1984), 33–57.

[78] R. Lowen, *The order aspect of the fuzzy real line*, Manuscripta Math. **49**(1985), 293–309.

[79] K. Menger, *Statistical metrics*, Proceedings of the National Academy of Sciences **28**(1942), 535–537.

[80] C. J. Mulvey, *Intuitionistic algebra and representation of rings*, Memoirs Amer. Math. Soc. **148**(1974), 3–57.

[81] C. J. Mulvey, J. W. Pelletier, *A globalisation of the Hahn-Banach theorem*, Advances in Mathematics **89**(1991), 1–59.

[82] C. J. Mulvey, J. J. C. Vermeulen, *A constructive proof of the Hahn-Banach theorem*, to appear.

[83] A. Pultr, *Closed categories of L-fuzzy sets*, Vorträge zur Automaten- und Algorithmentheorie, Heft **16/76**(1975), 60–68, TU Dresden.

[84] A. Pultr, *Remarks on dispersed (fuzzy) morphisms*, Vorträge aus dem Problemseminar Algebraische Methoden und ihre Auswendungen in der Automatentheorie, (1976) 26–37, TU Dresden.

[85] A. Pultr, *Fuzzy mappings and fuzzy sets*, Comment. Math. Univ. Carolinae **17(3)**(1976), 441–459.

[86] A. Pultr, A. Tozzi, *Separation facts and frame representation of some topological facts*, Applied Categorical Structures **2**(1994), 107–118.

[87] A. Pultr, A. Tozzi, *A note on reconstruction of spaces and maps from lattice data*, Quaestiones Mathematicae **24**(2001), 55–63.

[88] F. Richman, E. Walker, *Some group theoretic aspects of t-norms*, International Journal of Uncertainty, Fuzziness, and Knowledge-based Systems **8**(2000), 1–6.

[89] S. E. Rodabaugh, *A categorical accommodation of various notions of fuzzy topology*, Fuzzy Sets and Systems **9**(1983), 241–265.

[90] S. E. Rodabaugh, *A point set lattice-theoretic framework* $\mathbb{T}$ *which contains LOC as a subcategory of singleton spaces and in which there are general classes of Stone representation and compactification theorems*, first draft February 1986 / second draft April 1987, Youngstown State University Central Printing Office (Youngstown, Ohio).

[91] S. E. Rodabaugh, *Dynamic topologies and their applications to crisp topologies, fuzzifications of crisp topologies, and fuzzy topologies on the crisp real line*, J. Math. Anal. Appl. **131**(1988), 25–66.

[92] S. E. Rodabaugh, *Point-set lattice-theoretic topology*, Fuzzy Sets and Systems **40**(1991), 297–345.

[93] S. E. Rodabaugh, *Categorical frameworks for Stone representation theorems*, Chapter 7 in [100], 178-231.

[94] S. E. Rodabaugh, *Applications of localic separation axioms, compactness axioms, representations, and compactifications to poslat topological spaces*, Fuzzy Sets and Systems **73**(1995), 55–87.

[95] S. E. Rodabaugh, *Powerset operator based foundation for point-set lattice-theoretic (poslat) fuzzy set theories and topologies*, in [63], 463–530.

[96] S. E. Rodabaugh, *Powerset operator foundations for poslat fuzzy set theories and topologies*, Chapter 2 in [45], 91–116.

[97] S. E. Rodabaugh, *Categorical foundations of variable-basis fuzzy topology*, Chapter 4 in [45], 273–388.

[98] S. E. Rodabaugh, *Separation axioms: representation theorems, compactness, and compactifications*, Chapter 7 in [45], 481–552.

[99] S. E. Rodabaugh, *Fuzzy real lines and dual real lines as poslat topological, uniform, and metric ordered semirings with unity*, Chapter 10 in [45], 607–632.

[100] S. E. Rodabaugh, E. P. Klement, U. Höhle, eds, Application Of Category To Fuzzy Sets, Theory and Decision Library—Series B: Mathematical and Statistical Methods, Volume **14**, Kluwer Academic Publishers (Boston/Dordrecht/London), 1992.

[101] B. Schweizer, A. Sklar, Probabilistic Metric Spaces, North-Holland (Amsterdam), 1983.

[102] A. P. Šostak, *On a fuzzy topological structure*, Suppl. Rend. Circ. Matem. Palermo, sr. II, **11**(1985), 89-103.

[103] A. Šostak, *On compactness and connectedness degrees of fuzzy sets in fuzzy topological spaces.* General Topology and related Modern Analysis and Algebra. Proceedings of the V-th Prague Topological Symposium, Prague 1986, 519-523; Heldermann Verlag (Berlin), 1988.

[104] A. Šostak, *Towards the concept of a fuzzy category*, Acta Univ. Latviensis (ser. Math.) **562**(1991), 85–94.

[105] A. Šostak, *On a fuzzy syntopogeneous structure*, in [63], 431–463.

[106] A. Šostak, *Fuzzy categories versus categories of fuzzily structured sets: Elements of the theory of fuzzy categories,* Mathematik-Arbeitspapiere **48**(1997), 407–438: Categorical Methods in Algebra and Topology (A collection of papers in honor of Horst Herrlich), Hans-E. Porst, ed., Bremen, August 1997.

[107] A. Šostak, *Fuzzy categories related to algebra and topology*, Tatra Mount. Math. Publ. **16**(1999), 159–186.

[108] L. N. Stout, *Topology in a topos II: $\mathcal{E}$-completeness and $\mathcal{E}$-cocompleteness*, Manuscripta Math. **17**(1975), 1–14.

[109] L. N. Stout, *Quels sont les espaces topologiques dans les topos?* Annales des sciences mathématiques de Québec **II**(1)(1978), 123–141.

[110] L. N. Stout, *Fuzzy logic from a second closed structure on a quasitopos,* in *Proceedings of the Third IFSA World Congress* (1989), 458–461, IFSA (August 1989, Seattle).

[111] L. N. Stout, *The logic of unbalanced subobjects in a category with two closed structures*, Chapter 3 in [100], 73–105.

[112] J. J. C. Vermeulen, *Constructive techniques in functional analysis*, Thesis, University of Sussex, 1986.

[113] J. J. C. Vermeulen, *Proper maps of locales*, J. Pure and Appl. Alg. **92**(1994), 79–107.

[114] P. Viceník, *A note on generators of t-norms,* Busefal **75**(1998), 33–38.

[115] P. Viceník, *Additive generators and discontinuity,* Busefal **76**(1998), 25–28.

[116] H. Nguyen, E. Walker, A First Course In Fuzzy Logic, Second Edition, Chapman & Hall/CRC Press (Boca Raton), 2000.

[117] G. J. Wang, *Topological molecular lattices*, Monthly J. Science **1**(1984), 19–23 (Beijing).

Part I:

Topological Structures In Fuzzy Sets

CHAPTER 1

Uniform Completion In Pointfree Topology

B. Banaschewski

Introduction

Pointfree topology deals with certain complete lattices, called frames, which may be viewed as abstractly defined lattices of open sets, sufficiently resembling the concrete lattices of this kind that arise from topological spaces to make the treatment of a variety of topological questions possible. It turns out that a remarkable number of topological facts derive from results in this pointfree setting while the proofs of the latter are often more suggestive and transparent than those of their classical counterparts. But there is a deeper aspect of frames which endows them with a very specific significance: various topological spaces classically associated with other entities (such as several types of rings, or Banach spaces, or lattices) are actually the spectra of appropriate frames which themselves require weaker logical foundations for the proofs of their basic properties than those needed for the actual spaces but which can still serve much the same purposes as the spaces in question. In this way, pointfree topology acquires an autonomous rôle and appears as more fundamental than classical topology.

The subject as such originated in the Séminaire Ehresman in 1957 with early results by Bénabou [19] and S. Papert and D. Papert [36] and was subsequently pursued by Dowker and D. Papert-Strauss in a sequence of joint papers. A further, particularly important step in its development was provided by Isbell [29] who put the precise relationship between frames and spaces into categorical perspective, introduced a wealth of new concepts, and established a range of remarkable results. Of particular note was his introduction of uniformities into this setting, leading among other things to the unique existence of completions.

S.E. Rodabaugh and E.P. Klement (eds.), Topological and Algebraic Structures in Fuzzy Sets, 19–56.
© 2003 *Kluwer Academic Publishers.*

The present chapter provides a systematic exposition of the basic aspects of completion as they have evolved over the years, in particular taking into consideration the generalization from uniformity to nearness as well as the notion of metric frames. The material is organized as follows. After reviewing the general concepts and results required (Section 1), we introduce the specific notions to be treated here: nearness, uniformity, and metric diameter (Section 2), and then establish the central results, the unique existence of the completion for the corresponding types of frames (Section 3). Next, we give an account of the rôle of (regular) Cauchy filters, in an appropriately generalized sense, in connection with the completion, using them to establish that completion is a coreflection under suitable hypotheses—though not in general, as examples show—and to provide a completeness criterion which is an abstract variant of the classical Cauchy condition (Section 4). Further, we give a characterization of the nearness frames with compact completion and derive several consequences regarding compactifications of frames from this (Section 5), to be followed by a number of completeness results including the completeness of the frame $\mathcal{L}(\mathbb{R})$ of reals in its natural uniformity (Section 6). The chapter concludes with a brief survey of a few further topics in which completeness plays an important rôle (Section 7).

We conclude with a comment on foundations. Although it might be desirable to provide a constructive treatment, in the sense of topos theory, of the subject considered here, this is not quite available at this stage. It therefore seemed natural to settle for the next best approach, using as basis classical set theory in its least stringent form, that is, Zermelo-Fraenkel set theory without the Axiom of Choice, treated in ordinary logic. There are, however, a few instances where even this seems to be too weak a foundation. In these cases, the principle which has to be invoked is that of Countable Dependent Choice

(CDC) *Given any relation R on a set E such that, for each $x \in E$, $(x, y) \in R$ for some $y \in E$, there exists a sequence $x_0, x_1, \dots$ in E for which $(x_n, x_{n+1}) \in R$ for all $n = 0, 1, \dots$.*

Whenever this is required, this will be explicitly stated as well as indicated by an asterisk.

1 Background

1.1. A *frame* is a complete lattice L in which

$$a \wedge \bigvee S = \bigvee \{a \wedge t \mid t \in S\}$$

for all $a \in L$ and $S \subseteq L$, and a frame *homomorphism* is a map $h : L \to M$ between frames which preserves finitary meets including the unit (= top) e and arbitrary joins including the zero (= bottom) 0.

As standard examples of frames we mention: the finite distributive lattices, the complete Boolean algebras, the complete totally ordered sets, and for any topological space X the lattice $\mathfrak{O}X$ of its open sets.

Frames isomorphic to some $\mathfrak{O}X$ are called *spatial*. They are otherwise characterized as those frames L for which the homomorphisms from L into the two-element chain **2** distinguish the elements of L.

As a general reference to frames we suggest Johnstone [31] or Vickers [42].

1.2. We list some properties of frames which will be relevant here. A frame L is

compact if $e = \bigvee S$ implies $e = \bigvee T$ for some finite $T \subseteq S$,

regular if $a = \bigvee\{x \in L \mid x \prec a\}$ for each $a \in L$, where $x \prec a$ ("rather below" or "well inside") means that $x \wedge y = 0$ and $a \vee y = e$ for some $y \in L$, alternatively expressed as $a \vee x^* = e$, using the pseudocomplement

$$x^* = \bigvee\{z \in L \mid x \wedge z = 0\}$$

completely regular if $a = \bigvee\{x \in L \mid x \prec\!\!\prec a\}$ for each $a \in L$, where $x \prec\!\!\prec a$ ("completely below" or "really inside") means there exists a sequence $(x_{ik})_{i=0,1,\dots;k=0,\dots,2^i}$ such that

$$x_{00} = x,\ x_{01} = a,\ x_{ik} = x_{i+1\,2k},\ x_{ik} \prec x_{ik+1}$$

and

normal if $a \vee b = e$ implies there exist $u, v \in L$ for which $a \vee u = e = b \vee v$ and $u \wedge v = 0$.

Concerning these properties we have

1.2.1. *Any compact regular frame is normal.*

***1.2.2.** *Any normal regular frame is completely regular.*

Regarding the first, if $a \vee b = e$, then, by regularity, compactness, and the fact that $\prec$ is stable under $\vee$, there exist $x \prec a$ and $y \prec b$ such that $x \vee y = e$, and $u = x^*$, $v = y^*$ are then the desired elements.

As to the second, given $x \prec y$ there exist v such that $x \prec v \prec y : y \vee x^* = e$ implies that $y \vee u = e = x^* \vee v$ for some u and v such that $u \wedge v = 0$, and then indeed $x \prec v \prec y$. Now, it follows by CDC that $x \prec\!\!\prec y$ whenever $x \prec y$, and this proves the result.

Note that, for any topological space X, the frame $\mathfrak{O}X$ has any of the above four properties iff X has the property in the usual sense of the term involved. For the less obvious case of complete regularity see Banaschewski [1] and Behrend [18]. On the other hand, none of these properties imply spatiality.

Another example of a compact frame is the frame $\mathfrak{J}L$ of all ideals of a frame L, that is, the $J \subseteq L$ for which

$$0 \in J \text{ and } \ a, b \in J \text{ iff } a \vee b \in J$$

partially ordered by inclusion. Further, the map $\mathfrak{J}L \to L$ taking each J to its join $\bigvee J$ in L is a homomorphism. Note that $\mathfrak{J}L$ is regular iff L is Boolean.

1.3. The following properties of frame homomorphisms will be important here. $h : M \to L$ is called

dense if $h(a) = 0$ implies $a = 0$,

codense if $h(a) = e$ implies $a = e$, and

a *compactification* of L if h is dense onto and M compact regular.

Of particular significance for our purposes is

1.3.1. *For homomorphisms between regular frames, $h : M \to L$ is monic if it is dense and one-one if it is codense.*

Let h be dense and $f, g : N \to M$ such that $hf = hg$. Then $x \prec a$ in N implies $a \vee x^* = e$ and hence $g(a) \vee g(x^*) = e$ while $f(x) \wedge g(x^*) = 0$ (act the dense h) so that $f(x) \leq g(a)$. By the regularity of N it follows that $f(a) \leq g(a)$, and by symmetry we conclude that $f = g$.

If h is codense and $h(a) = h(b)$, consider any $x \prec a$. Then $h(b \vee x^*) = e$ since $a \vee x^* = e$, hence $b \vee x^* = e$ by hypothesis, and therefore $x \leq b$; it follows that $a \leq b$ by the regularity of M, showing $a = b$ by symmetry.

For any frame homomorphism $h : M \to L$, its *right adjoint* is the map $h_* : L \to M$ such that

$$h(a) \leq b \text{ iff } a \leq h_*(b)$$

for all $a \in M$ and $b \in L$, explicitly given by

$$h_*(b) = \bigvee\{x \in M \mid h(x) \leq b\}$$

h_* preserves arbitrary meets and hence in particular $\wedge$ and e; it preserves 0 exactly if h is dense.

A frame homomorphism $h : M \to L$ is called *strict* whenever $h_*[L]$ generates M. Note that

1.3.2. *For regular M, any dense $h : M \to L$ is strict.*

If $x \prec a$ in M, then $x \leq h_*h(x) \leq a$, the non-trivial second inequality because $a \vee x^* = e$ and $x^* \wedge h_*h(x) = 0$ (act the dense h).

The following puts the terminology introduced here in perspective. For any subspace X of a space Y, the frame homomorphism $\mathfrak{O}Y \to \mathfrak{O}X$ taking U to $U \cap X$ is dense iff X is a dense subspace of Y in the usual sense, a compactification iff Y is a compactification of X, and strict iff the open sets

$$U^{\#} = \bigcup\{V \in \mathfrak{O}Y \mid V \cap Y = U\}$$

generate $\mathfrak{O}Y$. In particular, if X is dense in Y the latter means that Y is a strict extension of X in the familiar sense of Banaschewski [2].

1.4. For the minimal number of category theoretical concepts used here—no more than the notions of category, functor, and coreflective subcategory—we refer to Mac Lane [34].

Regarding the basic category involved here, we take this to be the category **Frm** of frames and frame homomorphisms—although it should not be overlooked that, for some authors, the proper subject of pointfree topology is not this but rather the category of *locales*, the formal dual of **Frm**, introduced by Isbell [29]. Whatever the reasons for this, it remains a fact that any actual construction in this area is carried out in terms of frames and their homomorphisms, which suggests there might be some merit in focussing on **Frm** itself where the real structure is embodied. Of course, the counterparts of topological facts then appear in dual form but that seems easy enough to adjust to without any loss of intuition.

2 Structured frames

In classical topology, nearnesses, uniformities, and metrics are entities given on a specified set which then determine a topology on that set. By way of contrast, in the present context the specified object is already the "topology", that is, a frame, and consequently nearnesses, uniformities, and metrics must appear here as additional structures on a frame which we shall describe in this section.

2.1. The following terminology and notation will be used concerning subsets $A, B, \ldots$ and elements $a, b, \ldots, x, y, \ldots$ of a frame L.

$A \leq B$ (A *refines*, is a *refinement* of B) if, for each $x \in A$, there exist $y \in B$ such that $x \leq y$.

$A \wedge B = \{x \wedge y \mid x \in A,\ y \in B\}$.

$Ax = \bigvee\{s \in A \mid s \wedge x \neq 0\}$ (the *A-star* of x).

$AB = \{Ax \mid x \in B\}$.

$A \leq^* B$ means $AA \leq B$ (A *star-refines*, is a *star-refinement* of B).

A is called a *cover* of L if $\bigvee A = e$, and Cov$L$ is the set of all these, quasi-ordered by the refinement relation $\leq$. Further, FCov$L$ consists of all covers of L refined by some finite cover.

For any $\mathfrak{W} \subseteq \mathrm{Cov}L$, $a \lhd_{\mathfrak{W}} b$ (a is *$\mathfrak{W}$-strongly below* b) means that $Ca \leq b$ for some $C \in \mathfrak{W}$. This relation is called the *strong inclusion* for $\mathfrak{W}$.

$\mathfrak{W} \subseteq \mathrm{Cov}L$ is called *admissible* if $a = \bigvee\{x \in L \mid x \lhd_{\mathfrak{W}} a\}$ for each $a \in L$. Note that any set of covers containing an admissible set is itself admissible.

A *nearness* on L is an admissible filter $\mathfrak{N}$ in Cov$L$, the latter meaning that, for any $A, B \in \mathfrak{N}$, $A \wedge B \in \mathfrak{N}$, and if $A \in \mathfrak{N}$ and $A \leq B$ in Cov$L$, then $B \in \mathfrak{N}$. $\mathfrak{N}$ is called *strong* if

$$\check{C} = \{x \in L \mid x \lhd_{\mathfrak{N}} s \text{ for some } s \in C\}$$

belongs to $\mathfrak{N}$ for each $C \in \mathfrak{N}$.

A *uniformity* on L is a nearness $\mathfrak{U}$ on L such that each $C \in \mathfrak{U}$ is star-refined by some $B \in \mathfrak{U}$. Clearly, $B \leq^* C$ implies $B \subseteq \check{C}$ and hence $\check{C} \in \mathfrak{U}$: any uniformity is a strong nearness (but not conversely, as will be seen later).

For the strong inclusion $\lhd = \lhd_{\mathfrak{N}}$ of any nearness $\mathfrak{N}$ on a frame L, $0 \lhd 0$ and $e \lhd e$ trivially while $x \lhd a$ and $y \lhd b$ implies $x \wedge y \ \lhd\ a \wedge b$ and $x \vee y \ \lhd\ a \vee b$ by simple calculations and observations such as: if $A \leq B$ in $\mathfrak{N}$, then $Ax \leq Bx$ for any $x \in L$. Further, if $\mathfrak{N}$ is a *uniformity* and $x \lhd y$, then there exist z such that $x \lhd z \lhd y$: $A \leq^* B$ in $\mathfrak{N}$ implies $A(Ax) \leq Bx$ for any $x \in L$.

Note that, for a space X, the uniformities on $\mathfrak{O}X$ correspond exactly to the uniformities of X in the usual sense, that is, the uniformities on the underlying set of X for which $\mathfrak{O}X$ is the associated topology. Regarding nearness, any nearness of $\mathfrak{O}X$ for a space X is a nearness in the sense of Herrlich [25] with associated topology $\mathfrak{O}X$ but the converse does not hold. A characterization of the nearness spaces X that correspond to the nearnesses on $\mathfrak{O}X$ is given by Hong-Kim [27].

Concerning the history of these notions: Uniformities on frames were introduced by Isbell [29] as the exact translation into the frame setting of the notion of the uniformity of a space, expressed in terms of open covers derived from Tukey's cover approach to uniformity. Nearness, which is just uniformity without the star-refinement condition, was introduced more recently by Banaschewski-Pultr [14]. Apart from the fact that they have much wider scope than uniformities, nearnesses arise quite naturally in situations which originally only involve uniformities, making them practically unavoidable even if one's primary interest is in uniformities.

With respect to the latter, it should be added that as in the case of spaces there are several different ways of describing them, such as the entourage uniformities of Fletcher-Hunsaker [21] and the Weil uniformities of Picado [37]. These are certainly of interest, especially since they permit passage to a "non-symmetrical" generalization, but they will play no part in the present discourse.

2.2. In order to put these concepts in perspective we first establish some results concerning their scope. As a step towards this we need the following

Lemma. *For any onto frame homomorphism $h : L \to M$, if $\mathfrak{N}$ is a nearness on L, then $\mathfrak{M} = \{h[C] \mid C \in \mathfrak{N}\}$ is a nearness on M; further, $\mathfrak{M}$ is strong or a uniformity whenever $\mathfrak{N}$ is strong or a uniformity, respectively.*

Proof. The $h[C]$, $C \in \mathfrak{N}$, form a filter in Cov$M$ since

$$h[C] \wedge h[D] = h[C \wedge D]$$

and $h[C] \leq B$ implies $C \leq h_*[B]$ so that $h_*[B] \in \mathfrak{N}$ while $B = hh_*[B]$ since h is onto. Further, $h[C]h(x) \leq h(Cx)$ because $h(s) \wedge h(x) \neq 0$ implies $s \wedge x \neq 0$, hence $s \leq Cx$, and consequently $h(s) \leq h(Cs)$. It follows that $x \lhd_{\mathfrak{N}} a$ in L implies $h(x) \lhd_{\mathfrak{M}} h(a)$ and hence the admissibility of $\mathfrak{N}$ implies that of $\mathfrak{M}$, given that h is onto. Further, $\mathfrak{M}$ is strong whenever $\mathfrak{N}$ is strong since $h[\check{C}] \subseteq h[C]\check{}$. Similarly, if $A \leq^* B$ in $\mathfrak{N}$, then $h[A] \leq^* h[B]$ since

$$\{h[A]h(x) \mid x \in A\} \leq \{h(Ax) \mid x \in A\} \leq h[B]$$

and consequently $\mathfrak{M}$ is a uniformity if $\mathfrak{N}$ is a uniformity. □

Now we have

Proposition.

(1) *A frame L has a nearness iff it is regular.*

(2) *For a compact regular frame L, Cov$L$ is its unique nearness, and this is a uniformity.*

(3) *A frame L has a uniformity iff it has a compactification.*

Proof. (1) ($\Rightarrow$) If $Cx \leq a$ for any cover C of L, then $x \prec a$: for $y = \bigvee\{s \in C \mid s \wedge x = 0\}$, $x \wedge y = 0$ while $a \vee y = e$ since $(Cx) \vee y = e$ trivially.

(1) ($\Leftarrow$) If $x \prec a$ in a frame, then $C = \{a, x^*\}$ is a cover and $Cx \leq a$. Hence regularity makes Cov$L$ admissible and consequently a nearness on L. Note that the argument actually shows that FCov$L$ is already a nearness on L.

(2) For uniqueness it will be enough to show that any nearness $\mathfrak{N}$ on L contains each finite cover C of L. Let B be any finite subcover of $\check{C}$, by compactness, and for each $s \in B$, $A_s \in \mathfrak{N}$ such that $A_s s \leq \tilde{s}$ for some $\tilde{s} \in C$. Then $AB \leq C$ for any common refinement $A \in \mathfrak{N}$ of these A_s, $s \in B$, and since B is a cover this shows $A \leq C$ so that $C \in \mathfrak{N}$.

As to the second part of the assertion, we know already by (1) that Cov$L$ is a nearness, and it remains to establish the start-refinement property; further, by compactness, it is sufficient to do this for the finite covers. Now, for any such cover $A = \{a_1, \ldots, a_n\}$, there exists a cover $B = \{b_1, \ldots, b_n\}$ such that $b_i \prec a_i$ by compactness and the fact that $x, y \prec a$ implies $x \vee y \prec a$. Then, consider the cover

$$C = \{a_1, b_1^*\} \wedge \cdots \wedge \{a_n, b_n^*\}$$

Here, for any $u \in C$, $u = u_1 \wedge \cdots \wedge u_n$ where $u_k \in \{a_k, b_k^*\}$ for each k, and consequently $u \wedge b_i \neq 0$ implies $u_i = a_i$ so that $u \leq a_i$. It follows that $CB \leq A$, and if $D = C \wedge B$, then $D \leq^* A$, as desired.

(3) If L has a compactification $h : M \to L$, then by (2) and the lemma it has a uniformity. The converse is rather more involved. Given any uniformity $\mathfrak{U}$ on a frame L, one obtains a compactification of L in the form of a regular subframe of the compact frame $\mathfrak{J}L$ of ideals of L, given by the ideals J of L which are *regular* with respect to $\mathfrak{U}$ in the sense that, for each $a \in J$, there exist $b \in J$ for which $a \lhd_{\mathfrak{U}} b$. We refer to 5.4 for more details. □

Remark. (1) FCov$L$ is a *uniformity* iff L is regular and normal. The above argument for the second part of (2), together with the observation that the existence of the cover B used there already follows from normality, provides the proof for the less obvious "if" part.

(2) It is considerably more involved to characterize those L for which Cov$L$ is a uniformity, that is, every cover of L has a star-refinement. One result is that this holds iff L is regular and paracompact, the latter meaning that every cover A of L has a refinement B for which there exists a cover W such that $\{x \in B \mid x \wedge s \neq 0\}$ is finite for each $s \in W$ ("every cover has a locally finite refinement") (Isbell [29], Pultr-Úlehla [40]).

(3) For any regular frame L, the nearness Cov$L$ is trivially strong. Hence by (2) any regular Hausdorff space which is not paracompact provides an example of a strong nearness which is not a uniformity; a familiar space of this type is the Tychonoff plank (Gillman-Jerison [22], 8.20).

*(4) An alternative to part (3) of the proposition is that *a frame L has a uniformity iff it is completely regular* (Pultr [38]). Here, ($\Leftarrow$) is proved without using any choice principle: calling an ideal J of L completely regular if, for any $a \in J$, there exist $b \in J$ such that $a \prec\!\!\prec b$ one shows, by arguments similar to those in 5.4, that they form a regular subframe of $\mathfrak{J}L$, providing a compactification of L and hence a uniformity. The compactification arising here was first considered in Banaschewski-Mulvey [10].

The converse can be obtained from part (3) of the proposition together with 1.1.1 and *1.2.2 and the fact that complete regularity is preserved by taking homomorphic images. An alternative, direct proof would be to use that, for any uniformity $\mathfrak{U}$ on a frame L, $x \triangleleft_{\mathfrak{U}} y$ implies there exist z such that $x \triangleleft_{\mathfrak{U}} z \triangleleft_{\mathfrak{U}} y$ (as noted in 2.1), invoking CDC to conclude that $x \prec\!\!\prec y$ whenever $x \triangleleft_{\mathfrak{U}} y$.

2.3. Lemma 2.2 says that a nearness may be transferred from a frame L to a frame M by an onto homomorphism $h : L \to M$. We now provide a counterpart to this, concerning transfer in the opposite direction which will play a particularly important rôle later on.

First, we have the following general observation about stars and the right adjoints of homomorphisms.

Lemma. *For any dense onto frame homomorphism $h : M \to L$, $C \in CovL$, and $x, a \in L : Cx \leq a$ iff $h_*[C]h_*(x) \leq h_*(a)$.*

Proof. Given $Cx \leq a$, if $h_*(s) \wedge h_*(x) \neq 0$ for $s \in C$, then $s \wedge x \neq 0$ since h is dense onto, hence $s \leq a$ and consequently $h_*(s) \leq h_*(a)$, showing that $h_*[C]h_*(x) \leq h_*(a)$. Conversely, the latter implies

$$Cx = h[h_*[C]]h(h_*(x)) \leq h(h_*[C]h_*(s)) \leq a$$

since h is onto. □

The desired transfer principle is now:

Proposition. *If $h : M \to L$ is any strict dense onto frame homomorphism and $\mathfrak{N}$ a nearness on L such that*

(1) *$h_*[C]$ is a cover for each $C \in \mathfrak{N}$, and*
(2) *$h_*(a) = \bigvee\{h_*(x) \mid x \triangleleft_{\mathfrak{N}} a\}$ for each $a \in L$,*

then $\{h_*[C] \mid C \in \mathfrak{N}\}$ generates a nearness $\mathfrak{M}$ on M; further, $\mathfrak{M}$ is strong or a uniformity whenever $\mathfrak{N}$ is strong or a uniformity, respectively.

Proof. The $h_*[C]$, $C \in \mathfrak{N}$, form a filter basis in Cov$M$ since $A \leq B$ in Cov$L$ trivially implies $h_*[A] \leq h_*[B]$. Regarding admissibility, since $h_*(x) \triangleleft_{h_*(\mathfrak{N})} h_*(a)$ whenever $x \triangleleft_{\mathfrak{N}} a$ by the lemma, this follows immediately from (2) and the given strictness of h. Further, the resulting nearness $\mathfrak{M}$ is strong whenever $\mathfrak{N}$ is because $h_*[\check{C}] \subseteq h_*[C]\check{}$, again by the lemma. Finally, for the same reason, $A \leq^* B$ in $\mathfrak{N}$ implies $h_*[A] \leq^* h_*[B]$ and hence $\mathfrak{M}$ is a uniformity whenever this holds for $\mathfrak{N}$. □

2.4. We now turn to the other kind of additional structure on frames to be considered here.

A *metric diameter* on a frame L is a map $d : L \to \overline{\mathbb{R}}_+$, the extended non-negative reals, such that

(M1) $d(0) = 0$.

(M2) $d(a) \leq d(b)$ whenever $a \leq b$.

(M3) $d(a \vee b) \leq d(a) + d(b)$ if $a \wedge b \neq 0$.

(M4) For any $\alpha < d(a)$ and $\varepsilon > 0$, there exist $b, c \leq a$ such that $d(b), d(c) < \varepsilon$ and $\alpha < d(b \vee c)$.

(M5) For each $\varepsilon > 0$, $D_\varepsilon = \{a \in L \mid d(a) < \varepsilon\}$ is a cover of L.

(M6) The set of these covers is admissible.

Metric diameters, along with some weaker variants, were introduced by Pultr [39]. The following are some of the fundamental results concerning this notion.

2.4.1. Since $\varepsilon \leq \delta$ implies $D_\varepsilon \subseteq D_\delta$, (M5) and (M6) say that the D_ε generate a nearness. Moreover, one proves that $d(D_\varepsilon a) \leq d(a) + 2\varepsilon$ for any $a \in L$ and $\varepsilon > 0$ and consequently $D_\varepsilon \leq^* D_{3\varepsilon}$, showing this nearness is actually a uniformity, evidently with the countable basis $D_{1/n}$, $n = 1, 2, \ldots$, called the uniformity *associated with d*.

2.4.2. Conversely, any uniformity $\mathfrak{U}$ on a frame L with a countable basis determines a metric diameter d on L such that the uniformity associated with d is $\mathfrak{U}$. Consequently, *a frame has a metric diameter iff it has a uniformity with a countable basis* (Pultr [39]).

For a comprehensive treatment of all known results on the metrization of frames see Banaschewski-Pultr [15].

2.4.3. If $L = \mathfrak{O}X$ for some T_0-space X, the metric diameters d on L correspond exactly to the usual metrizations ϱ of X by the relations

$$d(U) = \sup\{\varrho(x, y) \mid x, y \in U\}, \ \varrho(x, y) = \inf\{d(U) \mid x, y \in U\}$$

Hence the metric diameters are exactly the pointfree counterparts of metrics in classical topology.

2.4.4. The existence of a metric diameter on a frame does not force L to be spatial. Indeed, the usual metric diameter on $\mathfrak{O}\mathbb{R}$ induces a metric diameter on the Boolean frame of all *regular open* sets of $\mathbb{R}$, that is, the $U \in \mathfrak{O}\mathbb{R}$ such that $U = U^{**}$, and this frame is as far from being spatial as possible since it has *no homomorphisms into* **2** *at all.* Moreover, this example is generic: any metric space without isolated points gives rise to the same phenomenon.

2.5. From the point of view of completion, the case of nearness is fundamental because for both, uniformities and metric diameters, completion is derived from that in the nearness case—somewhat similar to the fact that, for a metric *space*, its metric completion may be obtained from its uniform completion. Hence we first concentrate on nearness. We begin with some terminology.

A *nearness frame* is a frame together with a specified nearness on it. We write L for a nearness frame, $\mathfrak{N}L$ for its nearness, $\triangleleft$ for the corresponding strong inclusion, and permit notational confusion between L and its underlying frame.

As special types of nearness frames we have those that are

fine: $\mathfrak{N}L$ consists of all covers of the underlying frame,
finitely fine: $\mathfrak{N}L$ consists of all covers of the underlying frame which are refined by some finite cover,
strong: $\mathfrak{N}L$ is strong, and
uniform: $\mathfrak{N}L$ is uniform.

Concerning special types of maps between nearness frames L and M, a frame homomorphism $h : L \to M$ is called

uniform if $h[C] \in \mathfrak{N}M$ for each $C \in \mathfrak{N}L$,
a *surjection* if it is onto and $\mathfrak{N}M = \{h[C], C \in \mathfrak{N}L\}$, and
a *strict surjection* if it is a dense surjection and the $h_*[C]$, $C \in \mathfrak{N}M$, which belong to $\mathfrak{N}L$ because h is a surjection, actually generate $\mathfrak{N}L$.

The following describes an important property of the latter.

Lemma. *For any strict surjection $h : M \to L$ of nearness frames:*

(1) $h_*(a) = \bigvee\{h_*(x) \mid x \triangleleft a\}$ *for each* $a \in L$, *and*
(2) *if L is uniform or strong, then M is uniform or strong, respectively.*

Proof. It is immediate by Lemma 2.3 that $x \triangleleft a$ in L iff $h_*(x) \triangleleft h_*(a)$ in M. Consequently, if $z \triangleleft h_*(a)$, then also $h_*h(z) \triangleleft h_*(a)$ because $h(z) \triangleleft a$, and therefore $z \leq h_*(x)$ where $x = h(z) \triangleleft a$. Since $h_*(a) = \bigvee\{z \in M \mid z \triangleleft h_*(a)\}$ in M this proves (1).

Regarding (2), this now follows from (1) and Proposition 2.3 by the definition of strict surjections. □

Remark. For a *strong* nearness frame M, any dense surjection $h : M \to L$ is a strict surjection: since h is dense $x \triangleleft a$ in M implies $h_*h(x) \leq a$, showing that $h_*h[\check{C}] \leq C$ for any $C \in \mathfrak{N}M$, where $h[\check{C}] \in \mathfrak{N}L$ since M is strong. On the other hand, there are nearness frames M which have a dense surjection $M \to L$ that is not a strict surjection.

3 Completion

3.1. The familiar, and probably most commonly used, definition of the completeness of uniform or metric spaces, namely that every Cauchy sequence (metric case) or Cauchy filter (either case) converges, refers specifically to the points of the space and consequently does not suggest any immediate translation into the pointfree context. However, there is an equivalent condition, often somewhat ignored, which does not make any overt reference to points: X is complete iff any dense embedding $X \to Y$ is an isomorphism. Indeed, one might argue that it would be more natural to accept this as the *definition* of completeness and then derive the Cauchy condition as a completeness *criterion* – an approach that would tally well with some general perception as to what a notion of completeness ought to mean. But be that as it may, the importance of the alternative definition for the present situation lies in the fact that it immediately transfers to our pointfree setting, providing a natural notion of completeness.

Indeed, any dense embedding $X \to Y$ of uniform spaces, say, determines a dense surjection $\mathfrak{O}Y \to \mathfrak{O}X$ of the associated uniform frames, and this will be an isomorphism iff $X \to Y$ is an isomorphism of uniform spaces. Guided by this, but taking the more general features of nearness frames into account, we call a nearness frame L *complete* if any strict surjection $M \to L$ is an isomorphism, and a *completion* of L is then a strict surjection $M \to L$ with complete M. Our aim in 3.2–3.4 is to establish the existence of a completion, unique up to isomorphism, for any nearness frame.

3.1.1. There is one class of nearness frames which are automatically complete:

Any fine nearness frame is complete.

For any strict surjection $h : M \to L$ with fine L, if $h(a) = e$, then $C = \{h(x) \mid x \lhd a\}$ is a cover of L so that $C \in \mathfrak{N}L$ by hypothesis and consequently $h_*[C] \in \mathfrak{N}M$. On the other hand, $x \lhd a$ implies $h_*h(x) \leq a$ since h is dense and hence $e = \bigvee h_*[C] \leq a$. Thus h is codense and therefore an isomorphism (1.3).

Note that, in particular, any compact nearness frame is complete because it is fine by Proposition 2.2—an argument which differs remarkably from the classical proof that a compact uniform space must be complete. In Section 6, we shall prove completeness in some other special cases.

By way of contrast, it should be added that a fine uniform frame (= the uniformity is the largest uniformity on the underlying frame) need not be complete; for more detail see 7.2.

3.1.2. There is a case in which the preceding considerations lead to an immediate description of a completion:

For any finitely fine Boolean nearness frame M, the map $\mathfrak{J}M \to M$ from its ideal lattice $\mathfrak{J}M$ by taking joins is a completion.

$\mathfrak{J}M$ is compact regular and hence complete in its unique nearness, and the map is a strict surjection because its right adjoint takes $a \in M$ to its principal ideal $\downarrow a$, and any cover of $\mathfrak{J}M$ is obviously refined by a finite cover of principal ideals.

3.2. For the construction of the completion we need a number of general tools.

3.2.1. For any frame L, $\mathfrak{D}L$ will be *the frame of all downsets* of L, that is, the $U \subseteq L$ such that $0 \in U$ and $x \in U$ whenever $x \leq y$ and $y \in U$. $\mathfrak{D}L$, partially ordered by inclusion, is obviously a frame, with meet = intersection and join = union. Further, the join map $\bigvee : \mathfrak{D}L \to L$ taking U to $\bigvee U$ is a frame homomorphism, with right adjoint $\downarrow : L \to \mathfrak{D}L$ sending $a \in L$ to $\downarrow a = \{x \in L \mid x \leq a\}$. Moreover, the latter is the universal $0\wedge e$-homomorphism (= map preserving $0, \wedge$, and e) from L to frames: it clearly is of this kind, and any $0\wedge e$-homomorphism $\varphi : L \to M$ into a frame determines a (necessarily unique) frame homomorphism $h : \mathfrak{D}L \to M$ such that $h\downarrow = \varphi$, given by

$$h(U) = \bigvee \varphi[U]$$

As to the homomorphism $\bigvee : \mathfrak{D}L \to L$ itself, it is obviously dense, onto, and strict; in addition, it factors through any other homomorphism $h : M \to L$ of this kind with an onto homomorphism $\bar{h} : \mathfrak{D}L \to M$: h being dense $h_* : L \to M$ is a $0\wedge e$-homomorphism which determines the frame homomorphism $\bar{h} : \mathfrak{D}L \to M$ such that $\bar{h}(U) = \bigvee h_*[U]$; further,

$$h\bar{h}(U) = \bigvee hh_*[U] = \bigvee U$$

for any $U \in \mathfrak{D}L$ because h is onto, and $\bar{h}$ is onto since $\bar{h}(\downarrow a) = h_*(a)$ and h is strict.

3.2.2. In many respects, frames behave like algebras in the sense of Universal Algebra. This applies in particular to the formation of *quotients.* A *congruence* on a frame L is an equivalence relation Θ which is a subframe of $L \times L$, and the corresponding quotient L/Θ then becomes a frame with quotient homomorphism $\nu : L \to L/\Theta$ taking each $a \in L$ to its Θ-block

$$\Theta[a] = \{x \in L \mid (a, x) \in \Theta\}$$

further, just as for algebras, all homomorphic images of L are of this type, up to isomorphism. On the other hand, by the special properties of frames, there is a particularly convenient way of representing these quotients: each Θ-block has a largest element

$$k(a) = \bigvee \Theta[a]$$

and the resulting map $k : L \to L$ is a closure operator for which $k(a \wedge b) = k(a) \wedge k(b)$. Any such k is called a *nucleus*, and for any nucleus k on L, the complete lattice

$$\mathrm{Fix}(k) = \{a \in L \mid k(a) = a\}$$

is a frame, with $\wedge$ as in L and join given by $k(\bigvee \cdot)$, and $k : L \to \mathrm{Fix}(k)$ a frame homomorphism. Further, if k was derived from a congruence Θ as above, then $L/\Theta \simeq \mathrm{Fix}(k)$ so that the homomorphic images of L are also described as the $\mathrm{Fix}(k)$ for the nuclei k on L. In particular, for any homomorphism $h : L \to N$, the associated nucleus is h_*h. In addition, if k and ℓ are nuclei on a frame L such that $k \leq \ell$ (pointwise), then simple calculation shows that the map $\mathrm{Fix}(k) \to \mathrm{Fix}(\ell)$ effected by ℓ is a frame homomorphism.

3.2.3. Given any set $S \subseteq L \times L$ for a frame L, it is clear, at least conceptually, how to obtain the congruence generated by S: repeatedly alternate the two procedures of forming subframes in $L \times L$ and equivalence relations until this process stabilizes. On the other hand, the corresponding *nucleus* can be described rather more explicitly, with the added bonus that this at the same time provides a description of the corresponding quotient of L. First, define $\ell_0 : L \to L$ such that

$$\ell_0(a) = \bigvee\{x \wedge u \mid x \wedge v \leq a \ \text{ for some } (u, v) \in S \text{ or } (v, u) \in S\}$$

Then the following conditions are easily checked:

$$a \leq \ell_0(a), \ a \leq b \text{ implies } \ell_0(a) \leq \ell_0(b), \ \ell_0(a) \wedge b \leq \ell_0(a \wedge b)$$

showing that ℓ_0 is what is called a *prenucleus* (Banaschewski [3]). Now, for any prenucleus on a frame L, the set of fixed elements is evidently a closure system (= closed under $\bigwedge$ in L) by the first two conditions, and the associated closure operator ℓ, given by

$$\ell(a) = \bigwedge\{x \in L \mid a \leq x = \ell_0(x)\}$$

is a nucleus, as one derives from the third condition. Further, one shows that $\ell(u) = \ell(v)$ whenever $(u, v) \in S$ and $\ell \leq h_*h$ for any homomorphism $h : L \to M$ such that $h(u) = h(v)$ for all $(u, v) \in S$ which proves that ℓ is indeed the desired nucleus. Of course, the corresponding quotient frame $\mathrm{Fix}(\ell) = \mathrm{Fix}(\ell_0)$ is then characterized in L by the condition

$$\text{if } x \wedge u \leq a \text{ for some } (u, v) \in S \text{ or } (v, u) \in S, \text{ then } x \wedge v \leq a$$

3.3. It is clear that the completion of a nearness frame L should be an appropriately universal strict surjection to L and in order to see how to define this we note the following properties of the frame homomorphism $\bar{h} : \mathfrak{D}L \to M$ arising by 3.2.1 from any strict surjection $h : M \to L$:

3.3.1. For any $C \in \mathfrak{N}L$, if $\tilde{C} = \{\downarrow s \mid s \in C\}$, then

$$\bar{h}[\tilde{C}] = h_*[C]$$

is a cover of M by definition, and

3.3.2. For any $a \in L$, if $k(a) = \{x \in L \mid x \triangleleft a\}$, then

$$\bar{h}(k(a)) = \bar{h}(\downarrow a)$$

since

$$\bar{h}(k(a)) = \bigvee h_*[k(a)] = \bigvee \{h_*(x) \mid x \triangleleft a\} = h_*(a) = \bar{h}(\downarrow a)$$

the crucial next to the last step by Lemma 2.5.

Consequently, it seems promising to consider the quotient of $\mathfrak{D}L$ modulo the congruence determined by the conditions

(1) $\bigcup \tilde{C}$ is congruent to $\downarrow e$, for each $C \in \mathfrak{N}L$, and

(2) $k(a)$ is congruent to $\downarrow a$, for each $a \in L$,

and by 3.2.3 the associated nucleus ℓ on $\mathfrak{D}L$ results from the prenucleus ℓ_0 for which

$$\ell_0(U) = \{a \in L \mid \{a\} \wedge C \subseteq U \text{ for some } C \in \mathfrak{N}L\} \cup \{b \in L \mid k(b) \subseteq U\}$$

The corresponding quotient of $\mathfrak{D}L$ is then the closure system in $\mathfrak{D}L$ consisting of the $U \in \mathfrak{D}L$ such that

(C) $\{a\} \wedge C \subseteq U$ implies $a \in U$, for all $C \in \mathfrak{N}L$ and

(R) $k(a) \subseteq U$ implies $a \in U$, for all $a \in L$.

Further, since $\ell_0(U) \subseteq \downarrow(\bigvee U)$ the join map $\bigvee : \mathrm{Fix}(\ell) \to L$ is again a homomorphism. On the other hand, the map $r : L \to \mathrm{Fix}(\ell)$ such that $r(a) = \downarrow a$ is the right adjoint of this since obviously $\downarrow a \in \mathrm{Fix}(\ell)$ for each $a \in L$.

3.4. By the definition of $\ell, r[C]$ is a cover of $\mathrm{Fix}(\ell)$ for each $C \in \mathfrak{N}L$ and $r(a) = \bigvee \{r(x) \mid x \triangleleft a\}$ in $\mathrm{Fix}(\ell)$ for each $a \in L$, implying by Proposition 2.3 that $\{r[C] \mid C \in \mathfrak{N}L\}$ is a nearness basis on $\mathrm{Fix}(\ell)$. We let CL be the corresponding nearness frame and $\gamma_L : CL \to L$ the homomorphism given by taking joins. Evidently, γ_L is a strict surjection.

The following is the crucial step towards establishing the desired properties of $\gamma_L : CL \to L$.

Lemma. *For any strict surjection $h: M \to L$ there exists a strict surjection $g: CL \to M$ such that $hg = \gamma_L$.*

Proof. That there is a frame homomorphism $g: CL \to M$ for which $\gamma_L = hg$ is immediate from the discussion in 3.3 motivating the definition of the nucleus ℓ on $\mathfrak{D}L$: we have a factorization

$$\mathfrak{D}L \xrightarrow{\bar{h}} M = \mathfrak{D}L \xrightarrow{\ell} CL \xrightarrow{g} M$$

and hence $hg\ell = h\bar{h} = \gamma_L\ell$ so that $hg = \gamma_L$. Furthermore, g is onto since $\bar{h}$ is (3.2.1) and dense since hg is. Finally, the $g_*[A]$, $A \in \mathfrak{N}M$, generate $\mathfrak{N}(CL)$ because the $h_*[C]$, $C \in \mathfrak{N}L$, generate $\mathfrak{N}M$ while the corresponding $g_*h_*[C]$ generate $\mathfrak{N}(CL)$ since $g_*h_* = (hg)_* = r$. □

Now we have our principal result:

Proposition. *For any nearness frame L, $\gamma_L: CL \to L$ is a completion of L, unique up to isomorphism.*

Proof. To see that CL is complete, let $h: M \to CL$ be any strict surjection. Then $\gamma_L h: M \to L$ is a strict surjection so that there exists a strict surjection $g: CL \to M$ for which $\gamma_L hg = \gamma_L$ by the lemma and hence $hg = \mathrm{id}_{CL}$ (1.3.1). It follows that h is an isomorphism since h is dense.

To prove uniqueness, consider any completion $h: M \to L$. Again, there exists a strict surjection $g: CL \to M$ such that $hg = \gamma_L$ by the lemma, and since M is complete g is an isomorphism. □

Note that, by Proposition 2.4, the completion of a uniform or strong nearness frame is again of the same kind.

Remark. The uniform case of this proposition is due to Isbell [29], obtained by a relatively opaque proof. Alternative constructions, each different, were given by Kříž [33] and Banaschewski-Pultr [12]. The present general result was obtained by Banaschewski-Pultr [14]; the proof given here is related to Kříž's description of the uniform completion. The other two approaches do not seem to work in the present, more general setting.

3.5. We now turn to the metric case. Here we are dealing with *metric frames*, that is, frames together with a specified metric diameter. We use the notation L, d_L, D_ε^L, and $\mathfrak{U}L$, respectively for a metric frame, its metric diameter, its metric covers, and its corresponding uniformity; also, as usual, we allow notational confusion between L and its underlying frame.

For metric frames, a *dense surjection* is a frame homomorphism $h : M \to L$ which is dense, onto, and such that $d_M(a) = d_L(h(a))$ for all $a \in M$. Further, a metric frame L is called *complete* if any dense surjection $M \to L$ of metric frames is an isomorphism, and a *completion* of a metric frame L is a dense surjection $M \to L$ with complete metric M.

The following establishes the connection between metric and nearness frames needed in the context of completion.

Lemma.

(1) *If $h : M \to L$ is a strict nearness surjection for a nearness frame M and a metric frame L, then $d_M(a) = d_L(h(a))$ defines a metric diameter on M.*

(2) *Any dense surjection $h : M \to L$ of metric frames is a nearness surjection.*

Proof. (1) We have to check the six conditions defining metric diameters (2.4).

(M1) and (M2) hold trivially by the corresponding properties of d_L.

(M3) If $a \wedge b \neq 0$ in M, then $h(a) \wedge h(b) \neq 0$ in L since h is dense, and the condition for d_L then implies that for d_M.

(M4) Let $\alpha < d_M(a)$ and $\varepsilon > 0$ be given. Then $\alpha < d_L(h(a))$ and hence there exist $s, t \leq h(a)$ for which $d_L(s), d_L(t) < \varepsilon$ and $\alpha < d_L(s \vee t)$. Now, take $b = a \wedge h_*(s)$ and $c = a \wedge h_*(t)$; it follows that $d_M(b) = d_L(s)$ and $d_M(c) = d_L(t)$ as well as $d_M(b \vee c) = d_L(s \vee t)$, and this proves the condition for d_M.

(M5) and (M6). By hypothesis, the $h_*[D^L_\varepsilon]$ generate $\mathfrak{N}M$, and as $h_*[D^L_\varepsilon] \subseteq D^M_\varepsilon$ by the definition of d_M while $D^M_\varepsilon \leq h_*[D^L_\varepsilon]$ since $s \leq h_*h(s)$ for any $s \in M$, so do the D^M_ε.

(2) Again, we have $h_*[D^L_\varepsilon] \subseteq D^M_\varepsilon \leq h_*[D^L_\varepsilon]$, and this proves the assertion. □

As an obvious consequence of this lemma we have:

Corollary. *A metric frame is complete iff it is complete as a nearness frame.*

Further, calling d_M given in part(1) of the lemma the metric diameter *lifted* from L, we have the following result concerning completions.

Proposition. *Every metric frame has a completion, unique up to isomorphism, given by its nearness completion equipped with the lifted metric diameter.*

Remark. It may be of interest to describe what the procedure inherent in this proposition amounts to when it is translated into the spatial setting: for any metric space X, take its uniform completion $\tilde{X} \supseteq X$ and lift the metric ϱ of X to $\tilde{X}$ by first lifting the metric diameter of d given by ϱ on $\mathfrak{O}X$ to $\mathfrak{O}\tilde{X}$ so that

$$\tilde{\varrho}(x,y) = \inf\{d(U \cap X) \mid x, y \in U \in \mathfrak{O}\tilde{X}\}$$

One should note how radically different this is from traditional procedures such as extending $\varrho : X \times X \to \mathbb{R}$ to $\tilde{\varrho} : \tilde{X} \times \tilde{X} \to \mathbb{R}$ by uniform continuity!

4 Cauchy filters and completions

4.1. A (proper) *filter* in a frame is any subset F such that $0 \notin F$, $e \in F$, $a \wedge b \in F$ for any $a, b \in F$, and $a \in F$ whenever $a \geq b$ and $b \in F$. Further, a filter F is called *completely prime* if $\bigvee S \in F$ implies $S \cap F \neq \emptyset$ for any subset S of the frame.

Next, if L is a nearness frame, then a filter F in L is called a *Cauchy filter* if $F \cap C \neq \emptyset$ for each $C \in \mathfrak{N}L$, and *regular* if, for each $a \in F$, there exist $b \in F$ such that $b \lhd a$.

Note that in other contexts regular filters are also called *round* filters.

Our present purposes require an appropriate generalization of these notions. As a guide to this, we note that, in terms of the characteristic functions φ of these F, the above conditions are expressed as follows: φ is, respectively,

(1) a $0\wedge e$-homomorphism,

(2) a frame homomorphism,

(3) a $0\wedge e$-homomorphism such that $\varphi[C]$ is a cover for each $C \in \mathfrak{N}L$, and

(4) a $0\wedge e$-homomorphism such that $\varphi(a) = \bigvee\{\varphi(x) \mid x \lhd a\}$.

Hence we define, for an arbitrary frame T ("of truth values") as T-*valued* filters, completely prime filters, Cauchy filters, and regular filters of a nearness frame L the maps $\varphi : L \to T$ which satisfy (1), (2), (3), or (4), respectively. If T is left unspecified we refer to *general* filters, completely prime filters, Cauchy filters, or regular filters.

In particular, a T-valued entity of this kind is called *universal* if, for any $\psi : L \to H$ of the same kind there is a unique frame homomorphism $h : T \to H$ such that $h\varphi = \psi$. Of course, such universal entities are unique. Note that, rather trivially, the universal completely prime filter of L is the identity map $L \to L$. On the other hand, in view of our discussion of the frame $\mathfrak{D}L$ of downsets of L in 3.2.1, the map $L \to \mathfrak{D}L$ taking a to $\downarrow a$ is the universal filter of L. Here we have, in the same spirit:

Proposition. *For any nearness frame L, $(\gamma_L)_* : L \to CL$ is the universal regular Cauchy filter of L.*

Proof. Recall that $(\gamma_L)_*(a) \dashv\!\!\!\downarrow a$ (3.3).

For any strict surjection $h : M \to L$, h_* is an M-valued regular Cauchy filter, by definition and Lemma 2.5. Hence we only have to show the universality. Let $\varphi : L \to T$ be any general regular Cauchy filter and $\bar{\varphi} : \mathfrak{D}L \to T$ the frame homomorphism such that $\bar{\varphi}(\downarrow a) = \varphi(a)$, for each $a \in L$, since φ is a $0\wedge e$-homomorphism. Then, for any $C \in \mathfrak{N}L$ and the corresponding $\tilde{C}$ (3.3.1), $\bar{\varphi}[\tilde{C}] = \varphi[C]$ is a cover since φ is Cauchy, and $\bar{\varphi}(k(a)) = \bar{\varphi}(\downarrow a)$ for each $a \in L$ because

$$\bar{\varphi}(k(a)) = \bigvee \varphi[k(a)] = \bigvee\{\varphi(x) \mid x \triangleleft a\} = \varphi(a) = \bar{\varphi}(\downarrow a)$$

the third step precisely because φ is regular. It follows that $\bar{\varphi}$ factors through $\ell : \mathfrak{D}L \to CL$, producing the desired frame homomorphism $\tilde{\varphi} : CL \to T$ such that $\tilde{\varphi}(\downarrow a) = \varphi(a)$, saying that $\tilde{\varphi}(\gamma_L)_* = \varphi$. Finally, $\tilde{\varphi}$ is unique because the $\downarrow a$, $a \in L$, generate CL. □

The following obvious consequence of this proposition is an important principle which plays a decisive part in various different contexts.

Corollary. *Any general regular Cauchy filter of a complete nearness frame is a frame homomorphism.*

4.2. As a first application of Proposition 4.1 we now establish an important property of strong nearness frames. To begin with, we need the following

Lemma. *For any T-valued Cauchy filter $\varphi : L \to T$ of a strong nearness frame L, $\varphi^o : L \to T$ defined such that*

$$\varphi^o(a) = \bigvee\{\varphi(x) \mid x \triangleleft a\}$$

for each $a \in L$ is a T-valued regular Cauchy filter.

Proof. We first show, for any T-valued Cauchy filters $\varphi, \psi : L \to T$, that $\psi \leq \varphi$ implies $\varphi^o \leq \psi$. For any $x \triangleleft a$ in L, $\{a, x^*\} \in \mathfrak{N}L$ because $C \leq \{a, x^*\}$ whenever $Cx \leq a$, for any cover C, and hence $\psi(a) \vee \psi(x^*) = e$. Now, $\varphi(x) \wedge \psi(x^*) = 0$ since $\psi(x^*) \leq \varphi(x^*)$ and $\varphi(x) \wedge \varphi(x^*) = 0$, implying that $\varphi(x) \leq \psi(a)$ for all $x \triangleleft a$ and consequently $\varphi^o(a) \leq \psi(a)$, as claimed.

Next, for any T-valued Cauchy filter $\varphi : L \to T$, φ^o is again a T-valued Cauchy filter. We trivially have $\varphi^o(0) = 0$ and $\varphi^o(e) = e$ while

$$\varphi^o(a) \wedge \varphi^o(b) = \bigvee\{\varphi(x \wedge y) \mid x \triangleleft a, y \triangleleft b\} \leq \varphi^o(a \wedge b)$$

by the properties of $\triangleleft$ (2.1), and hence equality. Further, for any $C \in \mathfrak{N}L$,

$$\bigvee \varphi^o[C] = \bigvee\{\varphi(x) \mid x \triangleleft s,\ s \in C\} = \bigvee \varphi[\check{C}] = e$$

because L is strong. As a result, φ^{oo} is also a T-valued Cauchy filter, and since $\varphi^{oo} \leq \varphi$ it follows that $\varphi^o \leq \varphi^{oo}$ by our initial observation. This shows that $(\varphi^o)^o = \varphi^o$ which is the regularity condition for φ^o. □

Recall from 3.4 that the completion of a strong nearness frame is again strong; the following augments this in a substantial way.

Proposition. *The complete strong nearness frames are coreflective in the category of all strong nearness frames, with coreflection map* $\gamma_L : CL \to L$.

Proof. Consider, more generally, any uniform $h : M \to L$ where M is a complete strong nearness frame and L an arbitrary nearness frame. Then $(\gamma_L)_*h : M \to CL$ is a general Cauchy filter, hence $((\gamma_L)_*h)^o : M \to CL$ is a general regular Cauchy filter by the lemma, and consequently a frame homomorphism by Corollary 4.1. Moreover, for any $a \in M$,

$$\begin{aligned} \gamma_L(((\gamma_L)_*h)^o(a) &= \gamma_L(\bigvee\{(\gamma_L)_*h(x) \mid x \triangleleft a\}) \\ &= \bigvee\{h(x) \mid x \triangleleft a\} = h(a), \end{aligned}$$

showing that $h = \gamma_L((\gamma_L)_*h)^o$, and the uniqueness part is obvious.

It remains to show that $((\gamma_L)_*h)^o$ is uniform. For any $C \in \mathfrak{N}M$,

$$((\gamma_L)_*h)^o[C] = \{\bigvee\{(\gamma_L)_*h(x) \mid x \triangleleft s\} \mid s \in C\} \geq (\gamma_L)_*h[\check{C}]$$

and since $\check{C} \in \mathfrak{N}M$ because M is strong this shows that the set on the left belongs to $\mathfrak{N}(CL)$. □

Note that it is inherent in the above proposition that completion defines a *functor* on the category of strong nearness frames which takes each L to CL and each uniform homomorphism $h : L \to M$ to the unique uniform homomorphism Ch which makes the square

$$\begin{array}{ccc} CL & \xrightarrow{Ch} & CM \\ \gamma_L \downarrow & & \downarrow \gamma_M \\ L & \xrightarrow[h]{} & M \end{array}$$

commute, that is, $Ch = ((\gamma_M)_*h\gamma_L)^o$.

Remark. Given that the completion of a *uniform* frame is again a *uniform* frame (3.4) and uniform frames are strong nearness frames (2.1), we have an entirely analogous result for uniform frames. This actually goes all the way back to the paper by Isbell [29] which originally introduced the notion of uniform frames. The current extension of this result, along with some more general variants, was more recently obtained by Banaschewski-Hong-Pultr [9].

4.3. Since the completion of a metric frame is obtained by means of the completion with respect to the associated uniformity (3.5) we have the obvious analogue of Proposition 4.2 for complete metric frames. Moreover, there are two refinements of this result, as follows. For metric frames, one considers two further, more restricted types of homomorphisms $h : L \to M$ besides the uniform ones: the *Lipschitz* and the *contractive* homomorphisms. The former are defined by the condition that there exists a fixed real λ such that $D_\varepsilon^M \leq h[D_{\lambda\varepsilon}^L]$ for each $\varepsilon > 0$ while the latter are required to satisfy this for $\lambda = 1$. We note that, for a continuous map $\varphi : X \to Y$ between metric spaces, the corresponding frame homomorphism $\mathfrak{O}Y \to \mathfrak{O}X$ taking U to $\varphi^{-1}[U]$ is Lipschitz or contractive with respect to the derived diameters on $\mathfrak{O}Y$ and $\mathfrak{O}X$ (2.4.3) iff φ is of the corresponding type in the usual sense.

Now, it is clear from 3.5 that, for any metric frame L, the completion map $\gamma_L : CL \to L$ is contractive. Furthermore, by analyzing the proof of Proposition 4.2 one sees that the homomorphism $Ch : CL \to CM$ corresponding to a Lipschitz homomorphism $h : L \to M$ is again Lipschitz, with the same parameter λ that occurs in the Lipschitz condition for h. As a consequence, we therefore have

Proposition. *The complete metric frames are coreflective in the category of all metric frames and their uniform, Lipschitz, or contractive homomorphisms, respectively, with coreflection maps* $\gamma_L : CL \to L$.

4.4. In general, the completion homomorphism is not the coreflection map from complete nearness frames; in fact, there exist uniform homomorphisms $N \to M$ with complete N which do not even factor through $\gamma_M : CM \to M$ as a mere frame homomorphism.

The argument for this is fairly elaborate, and we only give an outline; for more detail see Banaschewski-Hong-Pultr [9].

To begin with, the *Booleanization* of any frame L is

$$\mathfrak{B}L = \{a \in L \mid a = a^{**}\}$$

which is a complete Boolean algebra (Glivenko [24]) such that the map $\beta_L : L \to \mathfrak{B}L$ taking a to a^{**} is a dense frame homomorphism. Further, a frame L

is called *DeMorgan* whenever

$$(a \wedge b)^* = a^* \vee b^*$$

for all $a, b \in L$ or, equivalently,

$$a^* \vee a^{**} = e$$

for all $a \in L$. Note that, for any such L, the binary join in $\mathfrak{B}L$ is the same as that in L, by an easy calculation.

Now let L be regular DeMorgan and, viewing both L and $M = \mathfrak{B}L$ as finitely fine nearness frames, suppose we have a frame homomorphism $f : CL \rightarrow CM$ such that $\gamma_M f = \beta_L \gamma_L$. Further, let K be the uniform frame with the same underlying frame as L but with the uniformity given by just the *finite partitions* of L, that is, the finite covers A such that $s \wedge t = 0$ for distinct $s, t \in A$. Then two things hold.

(i) The composite

$$h : CK \underset{\gamma_K}{\longrightarrow} K \underset{i}{\longrightarrow} L \underset{\beta_L}{\longrightarrow} \mathfrak{B}L = M$$

where i maps identically, is a strict surjection because L is DeMorgan, and by the uniqueness of completions we have an isomorphism $\ell : CK \rightarrow CM$ such that $\gamma_M \ell = h$.

(ii) Since CK is uniform there is a uniform homomorphism $g : CK \rightarrow CL$ such that $\gamma_L g = i\gamma_K$ by the proof of Proposition 4.2.

It follows now that $fg = \ell$, and since f is dense this makes it an isomorphism. Further, as noted in 3.1.2, CM is the ideal lattice of M and hence normal so that CL is normal. On the other hand, given that L is finitely fine, $a \vee b = e$ in L implies $(\gamma_L)_*(a) \vee (\gamma_L)_*(b) = e$ in CL, and the normality of the latter then implies that *L is normal.*

As a consequence, any *regular DeMorgan frame L which is not normal* will provide the desired $N \rightarrow M$ with complete N which fails to factor through $\gamma_M : CM \rightarrow M$. Since a space X is extremally disconnected (open subsets have open closures) iff $\mathfrak{O}X$ is DeMorgan, such L are readily obtained from classical topological examples, such as the extremally disconnected non-normal subspace Π of the Stone-Čech compactification of the discrete space $\mathbb{N}$ of natural numbers given in 6.Q of Gillman-Jerison [22] or the Gleason cover (Gleason [23]) of the Tychonoff plank. We note that these examples require the use of the Axiom of Choice but there is an alternative choice-free construction based on the pointfree version of the Gleason cover (Banaschewski [4]).

4.5. Despite the absence of points in our setting, we can characterize the complete nearness frames by a "Cauchy Criterion" in terms of convergence, specifically the convergence of general regular Cauchy filters.

First the requisite definition. Noting that, for a space X, a filter $\mathfrak{F}$ in $\mathfrak{O}X$ converges, that is, contains some

$$\mathfrak{O}(x) = \{U \mid x \in U \in \mathfrak{O}X\}$$

iff $\mathfrak{F}$ meets every cover of $\mathfrak{O}X$, we call a T-valued filter $\varphi : L \to T$ *convergent* if $\varphi[C]$ is a cover of T for every cover C of L (see Hong [26] for the case $T = \mathbf{2}$).

Then we have:

Lemma. *For any nearness frame L, a general regular filter $\varphi : L \to T$ is convergent iff it is a frame homomorphism.*

Proof. For the non-trivial part, let $a = \bigvee S$ in L and take any $x \lhd a$. Then $S \cup \{x^*\}$ is a cover so that $e = \varphi(x^*) \vee \bigvee \varphi[S]$, hence $\varphi(x) \leq \bigvee \varphi[S]$ and consequently $\varphi(a) \leq \bigvee \varphi[S]$ by the regularity of φ. □

The desired characterization is now the following

Proposition. *A nearness frame L is complete iff every general regular Cauchy filter of L is convergent.*

Proof. ($\Rightarrow$) By Corollary 4.1 any general regular Cauchy filter of L is a frame homomorphism and consequently convergent.

($\Leftarrow$) Here, $(\gamma_L)_* : L \to CL$ is a frame homomorphism, and since $\gamma_L(\gamma_L)_* = \mathrm{id}_L$ this makes γ_L an isomorphism and then L complete. □

For strong nearness frames there is a variant of this proposition which comes even closer to the classical definition of complete uniform spaces. If L is a complete strong nearness frame and $\varphi : L \to T$ any general Cauchy filter of L, then $\varphi^o : L \to T$ is a general regular Cauchy filter by Lemma 4.2, hence convergent by the proposition, and since $\varphi^o \leq \varphi$ it follows that φ is also convergent. This proves the non-trivial part of

Corollary. *A strong nearness frame is complete iff every general Cauchy filter of L is convergent.*

4.6. The characterization of completeness in Proposition 4.5 suggests a weaker variant of this notion, limited to the case $T = \mathbf{2}$ and hence defined as follows.

A nearness frame L is called *Cauchy complete* if every regular Cauchy filter of L is convergent, that is, is a completely prime filter.

Obviously, a *uniform space* X is complete iff the associated uniform frame $\mathfrak{O}X$ is Cauchy complete—which lends the latter notion some added significance. The following is a brief summary of the results in this context. A more detailed account may be found in Banaschewski [6], Appendix 3.

4.6.1. *Cauchy completeness does not imply completeness.*

The uniform frame $\mathfrak{O}(\mathbb{R}^m)$ for $\mathbb{R}^m$ with its usual uniformity is Cauchy complete for any exponent m but not complete for *uncountable* m: if it were $\mathfrak{O}(\mathbb{R}^m)$ and hence $\mathbb{R}^m$ would be normal by a result of Isbell [29] which it is not (Engelking [20], 3.12.15).

4.6.2. *A nearness frame L is Cauchy complete iff any homomorphism $CL \to$ **2** factors through $\gamma_L : CL \to L$.*

4.6.3. *The Cauchy complete strong nearness frames are coreflective in the category of all strong nearness frames.*

This can be obtained either by describing the coreflection functor by means of a functorial intermediate quotient of $\gamma_L : CL \to L$ (Banaschewski [6]) or by a direct construction involving the regular Cauchy filters of L which is the frame analogue of adjoining the non-convergent regular Cauchy filters of $\mathfrak{O}X$ as new points to a uniform space X in order to make it complete (Hong-Kim [28]).

Furthermore, as a consequence of either description, the Cauchy complete coreflection of a uniform, or metric, frame is uniform, or metric, respectively, so that the corresponding result holds for uniform, or metric, frames and their uniform homomorphisms.

5 Compact completions

5.1. In view of the importance of compact regular frames, these being exactly the pointfree versions of compact Hausdorff spaces, it will be of interest to determine when a nearness frame has a compact completion.

In the following, a nearness frame L is called *totally bounded* if $\mathfrak{N}L$ is generated by its finite members, that is, for each $C \in \mathfrak{N}L$ there exist finite $D \leq C$ in $\mathfrak{N}L$. Of course, the finitely fine nearness frames (2.5) are trivial examples of such L. Further, it is obvious that a nearness frame is totally bounded iff its completion is totally bounded.

Proposition. *For any nearness frame L, CL is compact iff L is uniform and totally bounded.*

Proof. ($\Rightarrow$) CL is totally bounded and uniform by Proposition 1.1, and L inherits this.

($\Leftarrow$) It will be sufficient to show that any complete totally bounded uniform frame M is compact. For this, let $\mathfrak{J}M$ be the frame of ideals of M (1.2) and $\lambda : M \to \mathfrak{J}M$ the map such that $\lambda(a) = \downarrow a$. Since this is a bounded lattice homomorphism (-preserving $0, e, \wedge, \vee$) and $\mathfrak{N}M$ is generated by its finite members, λ is a general Cauchy filter and hence λ^o is a frame homomorphism by the given completeness of M. Further

$$\lambda^o(a) = \bigcup\{\downarrow x \mid x \triangleleft a\} = \{x \in L \mid x \triangleleft a\} = k(a)$$

as defined in 3.3.2, and k is one-one because $\bigvee k(a) = a$. It follows that M is isomorphic to a subframe of $\mathfrak{J}M$ and consequently compact since the latter is compact. □

As an immediate consequence we have:

Corollary 1. *A regular frame is compact iff it has a totally bounded uniformity in which it is complete.*

Further, by Remark(1) in 2.2 we conclude

Corollary 2. *For a finitely fine nearness frame L, CL is compact iff L is normal.*

Remark 1. The argument for ($\Leftarrow$) in the above proof remains valid if M is only assumed to be strong. In particular, this implies that *any totally bounded strong nearness frame is actually uniform.*

Remark 2. Regarding total boundedness, this obviously implies that any $C \in \mathfrak{N}L$ has a finite subcover. Conversely, if a *strong* nearness frame has this property, then it is totally bounded. Given any $C \in \mathfrak{N}L$, let $B = \{b_1, \ldots, b_n\}$ be a finite subcover of $\check{C}$ and $b_i \triangleleft c_i$, $c_i \in C$, for each i. Then there exists $A \in \mathfrak{N}L$ such that $Ab_i \leq c_i$ for each i; and since $\bigvee B = e$, it follows that $A \leq \{c_1, \ldots, c_n\}$, showing the latter belongs to $\mathfrak{N}L$.

5.2. In the following, we are dealing with mere frames, specifically with compactifiable frames, that is, frames which have a compactification.

Proposition. *For any compactifiable frame, its compactifications are exactly the completions with respect to its totally bounded uniformities.*

Proof. That the completions referred to are compactifications is immediate by Proposition 5.1. For the converse, if $h : M \to L$ is any compactification we may view M as uniform with its unique nearness and take L as uniform with the uniformity transferred by h (Lemma 2.2). Then M is complete because it is fine, and h is a dense surjection by definition. It follows from this that it is a strict surjection (Remark 2.5) and consequently $h : M \to L$ is the completion of L. □

For any compactifiable frame L, the nearness generated by all its totally bounded uniformities is again a totally bounded uniformity: for any finite covers $C_1, \ldots, C_n$ of L, $C_1 \wedge \cdots \wedge C_n$ is obviously finite, and if $D_i \leq^* C_i$ for each i with covers D_i, then $D_1 \wedge \cdots \wedge D_n \leq^* C_1 \wedge \cdots \wedge C_n$ by a simple calculation. Hence any compactifiable frame has a finest totally bounded uniformity.

In the following, a compactification $K \to L$ of a frame L is called *universal* if any compactification $M \to L$ factors through it, meaning: there is a commuting triangle

$$\begin{array}{ccc} M & \longrightarrow & K \\ & \searrow \quad \swarrow & \\ & L & \end{array}$$

Then we have:

Corollary 1. *Any compactifiable frame has a universal compactification, given by the completion with respect to its finest totally bounded uniformity.*

Proof. Let $K \to L$ be the completion in question and $h : M \to L$ any compactification. Then h is uniform for the finest totally bounded uniformity on L and the desired factorization $M \to L = M \to K \to L$ then follows by the coreflectiveness of complete uniform frames (4.2, Remark 1). □

Note that the normal regular frames are exactly those compactifiable frames in which the finest totally bounded uniformity is just the finitely fine nearness. Hence:

Corollary 2. *For any normal regular frame, the universal compactification is the completion with respect to its finitely fine nearness.*

Remark. The universal compactification of a compactifiable frame is obviously the pointfree counterpart of the Stone-Čech compactification βX for a Tychonoff space X (Gillman-Jerison [22], Chapter 6). Moreover, βX can be derived from the present result as follows. For any such X, $\mathfrak{O}X$ is compactifiable because it has uniformities, and if $K \to \mathfrak{O}X$ is its universal compactification, then the corresponding continuous map $X \to \Sigma K$ for the spectrum of ΣK of K, that is, the space of all $\xi : K \to \mathbf{2}$ with open sets $\Sigma_a = \{\xi \in \Sigma K \mid \xi(a) = 1\}$, will be the Stone-Čech compactification *provided* we know that ΣK is compact. This will be the case for arbitrary compact regular frames iff the Boolean Prime Ideal Theorem (PIT) holds which says that every non-trivial Boolean algebra has a prime ideal, that is, a proper ideal P such that $a \wedge b \in P$ implies $a \in P$ or $b \in P$. It should be added that, conversely, PIT follows if βX exists for every Tychonoff space X. In all, this indicates that the result of Corollary 1 should be viewed as the more fundamental form of the Stone-Čech compactification. Alternative presentations of this result, not dependent on uniformities, are given in Johnstone [31] and Banaschewski-Mulvey [10,11].

5.3. A *partition* of a frame L is a cover C of L such that $s \wedge t = 0$ for any distinct $s, t \in C$. Since $C \leq^* C$ for such C any nearness which is generated by partitions is a uniformity, and we call these the *partition uniformities*, also sometimes referred to as the *transitive* uniformities.

It is clear that a frame L has partition uniformities iff it is *zero-dimensional*, that is, any $a \in L$ is the join of complemented $c \in L$, the latter meaning that $c \vee c^* = e$.

Note that, for any strict surjection $h : M \to L$ of nearness frames where $\mathfrak{N}L$ is a partition uniformity, the same holds for $\mathfrak{N}M$: the latter is generated by the $h_*[C]$, C any partition in $\mathfrak{N}L$, and if $h_*(s) \neq h_*(t)$ for $s, t \in C$, then $s \neq t$, hence $s \wedge t = 0$ and therefore also $h_*(s) \wedge h_*(t) = 0$ since h is dense.

Now we have the following analogue of Proposition 5.2:

Proposition. *For any zero-dimensional frame, its zero-dimensional compactifications are exactly the completions with respect to its totally bounded partition uniformities.*

Proof. ($\Rightarrow$) For any zero-dimensional compact frame (note this is automatically regular), its unique nearness is a partition uniformity. Hence the totally bounded uniformity induced on a frame by any zero-dimensional compactification is a partition uniformity, and the assertion follows from the proof of Proposition 5.2.

($\Leftarrow$) For any uniform frame L, if $\mathfrak{N}L$ is a partition uniformity, then so is $\mathfrak{N}(CL)$, as already noted, and hence CL is zero-dimensional. □

It is obvious that a zero-dimensional frame has a finest totally bounded partition uniformity, namely the uniformity generated by *all* finite partitions. Consequently, we have

Corollary. *Any zero-dimensional frame has a universal zero-dimensional compactification, given by the completion with respect to the uniformity generated by all its finite partitions.*

Note that this provides the pointfree counterpart of the universal zero-dimensional compactification ζX of a zero-dimensional Hausdorff space X, usually referred to as the *Banaschewski compactification* of X. Indeed, in complete analogy with the case of βX discussed in 5.2, if X is a zero-dimensional Hausdorff space and $M \to \mathfrak{O}X$ the universal zero-dimensional compactification of the zero-dimensional frame $\mathfrak{O}X$, then $\Sigma M \simeq \zeta X$ *provided* ΣM is compact. Again, this is the case for arbitrary compact zero-dimensional M iff PIT holds, and the latter in turn is equivalent to the general existence of ζX. For an alternative proof of the present result, not using uniformities, see Banaschewski [5].

5.4. The proof of Proposition 5.1 suggests an alternative description of the completion of a totally bounded uniform frame.

If an ideal J of L is called *regular* provided that, for each $a \in J$, there exist $b \in J$ such that $a \lhd b$, then

(i) the zero and the unit ideal are regular,

(ii) for any regular ideals I and J, $I \cap J$ and $I \vee J$ are regular,

and

(iii) any directed union of regular ideals is regular,

where (i) and (ii) are obvious consequences of the properties of $\lhd$ (2.1) and (iii) is trivial. It follows that the set $\mathfrak{R}L$ of regular ideals is a subframe of the frame $\mathfrak{J}L$ of all ideals of L.

Now, the $k(a) = \{x \in L \mid x \lhd a\}$ introduced in 3.3.2 are regular ideals, by the general properties of $\lhd$ and the specific fact that $\lhd$ interpolates in uniform frames (2.1) and consequently, they provide a map $k : L \to \mathfrak{R}L$. Further, by the proof of Proposition 5.1, k is a general regular Cauchy filter, and hence we have a frame homomorphism $h : CL \to \mathfrak{R}L$ such that $h(\downarrow a) = k(a)$. Finally, h is onto because

$$J = \bigcup\{k(a) \mid a \in J\}$$

precisely by the definition of regular ideals, and since h is clearly dense this makes it an isomorphism, obviously such that $\varrho_L = h^{-1}\gamma_L : \mathfrak{R}L \to L$ is the homomorphism given by taking joins in L. This proves the following

Proposition. *For any totally bounded uniform frame L, $\varrho_L : \mathfrak{R}L \to L$ is the completion of L.*

Remark 1. The regularity of $\mathfrak{R}L$ can easily enough be derived directly from its definition, by showing that $k(a) \prec k(b)$ in $\mathfrak{R}L$ whenever $a \lhd b$ in L, and one then readily shows that $\varrho_L : \mathfrak{R}L \to L$ is a dense surjection, and hence the completion, for any totally bounded uniform L, *without prior knowledge of its existence.* This approach was used by Banaschewski-Pultr [12].

Remark 2. It is clear that the notion of regular ideal depends only on $\lhd$ and hence makes sense for arbitrary uniform frames L, with the same properties as in the totally bounded case; consequently one may wonder about the significance of the homomorphism $\varrho_L : \mathfrak{R}L \to L$ in general. It is easy to check that this is always uniform and that the correspondence $L \mapsto \mathfrak{R}L$ is functorial such that the square

$$\begin{array}{ccc} \mathfrak{R}M & \xrightarrow{\mathfrak{R}h} & \mathfrak{R}L \\ \varrho_M \downarrow & & \downarrow \varrho_L \\ M & \xrightarrow[h]{} & L \end{array}$$

commutes for any uniform homomorphism h between uniform frames. Since ρ_L is an isomorphism iff L is compact it follows that it is the coreflection map from compact to arbitrary uniform frames and as such the pointfree counterpart of the universal uniform compactification of a uniform space X, called the *Samuel compactification* of X (Banaschewski-Pultr [12]).

Remark 3. There is an alternative approach to the general $\varrho_L : \mathfrak{R}L \to L$. For this, one first shows that the nearness generated by the *finite* uniform covers of a uniform frame L is actually a uniformity, that is, any such cover has a *finite* uniform star-refinement, and then concludes that $\varrho_L : \mathfrak{R}L \to L$ is just the completion relative to this totally bounded uniformity.

6 Further completeness results

We already know that any fine nearness frame is complete (3.1.1). Here we establish the completeness of some further nearness frames.

6.1. For any element a of a nearness frame L, the *closed quotient* $\uparrow a$ is the closed frame quotient of L, consisting of all $x \geq a$ in L, equipped with the image of the nearness $\mathfrak{N}L$ by the quotient homomorphism ν taking x to $x \vee a$.

Proposition. *Any closed quotient of a complete strong nearness frame is complete.*

Proof. L being strong we have a commuting triangle

$$\begin{array}{ccc} & & L \\ & {}^{h}\swarrow & \downarrow \nu \\ C(\uparrow a) & \underset{\gamma_{\uparrow a}}{\longrightarrow} & \uparrow a \end{array}$$

with a uniform homomorphism h by Proposition 4.2, and since $\nu(a) = a$, the zero of $\uparrow a$, it follows that $h(a) = 0$ because $\gamma_{\uparrow a}$ is dense. Consequently, h factors through ν, and if $h = g\nu$, then $\gamma_{\uparrow a} g = \mathrm{id}_{\uparrow a}$, showing that $\gamma_{\uparrow a}$ is an isomorphism. □

Corollary. *For strong nearness frames, if $h : L \to M$ is a surjection, then $Ch : CL \to CM$ is a surjection.*

Proof. Consider the commuting square

$$\begin{array}{ccc} CL & \overset{\nu}{\longrightarrow} & \uparrow a \\ \gamma_L \downarrow & & \downarrow f \\ L & \underset{h}{\longrightarrow} & M \end{array}$$

where $a = (h\gamma_L)_*(0)$ and $f\nu = h\gamma_L$ accordingly. Then, $\uparrow a$ is complete by the proposition while f is dense surjection, and by Remark 2.5 this makes $f : \uparrow a \to M$ the completion of M such that ν corresponds to Ch. □

6.2. Proposition 6.1 shows how new complete nearness frames arise from given ones. Here we have another result of this type.

Proposition. *If a regular frame L is complete relative to some strong nearness $\mathfrak{N}$ of L, then it is complete relative to any nearness $\mathfrak{M}$ finer than $\mathfrak{N}$.*

Proof. Let N and M be the nearness frames given by $\mathfrak{N}$ and $\mathfrak{M}$, respectively, and $h : N \to M$ the uniform homomorphism acting identically. N being strong we then have a commuting triangle

$$\begin{array}{ccc} & & N \\ & {}^{g}\swarrow & \downarrow h \\ CM & \underset{\gamma_M}{\longrightarrow} & M \end{array}$$

by the proof of Proposition 4.2, and since $\gamma_M g$ acts identically it follows that γ_M is an isomorphism. □

6.3. In the following, an element a of a frame L will be called *small* if $e = \bigvee S$ for any $S \subseteq L$ implies $a \leq \bigvee T$ for some finite subset T of S. Note that, in a regular frame L, $a \in L$ is small iff the closed quotient $\uparrow(a^*)$ is compact.

Proposition. *Any nearness frame L for which some $C \in \mathfrak{N}L$ consists of small elements is complete.*

Proof. We show that any strict surjection $h : M \to L$ is an isomorphism by proving that it is codense. Consider then any $a \in M$ such that $h(a) = e$. This implies that

$$e = \bigvee\{h(x) \mid x \triangleleft a\}$$

and since the join here is updirected by the properties of $\triangleleft$ (2.1) each $s \in C$ is below $h(x)$ for some $x \triangleleft a$. Hence $h_*(s) \leq h_*h(x) \leq a$, the second inequality because h is dense, and since $h_*[C]$ is a cover of M by the properties of h it follows that $a = e$. □

Remark. For any regular Hausdorff space X, $U \in \mathfrak{O}X$ is small iff its closure is compact. Hence this proposition embodies such classical results as the completeness of a locally compact group in its usual uniformities.

6.4. Our next result concerns the frame $\mathfrak{L}(\mathbb{R})$ of reals. This is defined as the frame generated by all pairs (p, q) for $p, q \in \mathbb{Q}$, subject to the following relations.

(R1) $(p, q) \wedge (r, s) = (p \vee r, q \wedge s)$
(R2) $(p, q) \vee (r, s) = (p, s)$ whenever $p \leq r < q \leq s$
(R3) $(p, q) = \bigvee\{(r, s) \mid p < r < s < q\}$
(R4) $e = \bigvee\{(p, q) \mid p, q \in \mathbb{Q}\}$

This definition, used by Banaschewski-Mulvey [11] and originally given by Joyal [32], is somewhat different from that in Johnstone [31] but readily seen to be equivalent to it. Note that $(p, q) = 0$ whenever $p \geq q$ as a consequence of (R3).

$\mathfrak{L}(\mathbb{R})$ is taken as a uniform frame with basic uniform covers

$$C_n = \{(p, q) \mid 0 < q - p < \tfrac{1}{n}\}$$

These are indeed covers: by successive application of (R2), any (r, s) is the join of $(p_1, q_1), \ldots, (p_k, q_k)$ where

$$p_1 = r < p_2 < q_1 < p_3 < q_2 < \cdots < p_k < q_{k-1} < q_k = s, \; q_i - p_i < \tfrac{1}{n}$$

and consequently $\bigvee C_n = e$ by (R4). Of course, $C_{n+1} \leq C_n$ trivially so that these C_n form a filter basis of covers. Further, it is clear from (R2) that $C_{3n} \leq^*$

C_n, and admissibility is an immediate consequence of (R3) by the observation that

$$(r,s) \triangleleft (p,q) \quad \text{iff} \quad p < r < s < q$$

because $C_n(r,s) \leq (p,q)$ iff $\frac{1}{n} < r-p, q-s$.

Note that this uniformity corresponds to the usual metric uniformity on $\mathbb{R}$.

Proposition. $\mathfrak{L}(\mathbb{R})$ *is complete.*

Proof. We show this by using Proposition 4.5. Consider, then, any T-valued regular Cauchy filter $\varphi : \mathfrak{L}(\mathbb{R}) \to T$ of $\mathfrak{L}(\mathbb{R})$. It is clear from the definition of $\mathfrak{L}(\mathbb{R})$ that φ will be homomorphism if it takes the defining relations (R1)–(R4) to identities in T. Now, this is trivial for (R1) and (R4) while it follows immediately for (R3) from our earlier remark about admissibility. For the remaining (R2), let $p \leq r < q \leq s$ and take $\frac{1}{n} < q - r$. Then

$$\varphi(p,s) = \varphi(p,s) \wedge \bigvee \varphi[C_n] = \bigvee\{\varphi(p \vee u, s \wedge v) \mid 0 < v - u < \tfrac{1}{n}\}$$

since $\varphi[C_n]$ is a cover. Here, for any of the $(p \vee u, s \wedge v)$, if $q < s \wedge v$, then $r \leq p \vee u$ by the choice of n, showing that

$$(p \vee u, s \wedge v) \leq (p,q) \quad \text{or} \quad (p \vee u, s \wedge v) \leq (r,s).$$

It follows that $\varphi(p,s) \leq \varphi(p,q) \vee \varphi(r,s)$, the non-trivial part of the desired inequality. □

The map taking each generator (p,q) of $\mathfrak{L}(\mathbb{R})$ to the rational open interval

$$\{r \in \mathbb{Q} \mid p < r < q\}$$

obviously turns (R1)–(R4) into identities in $\mathfrak{O}\mathbb{Q}$ and consequently defines a homomorphism $\mathfrak{L}(\mathbb{R}) \to \mathfrak{O}\mathbb{Q}$. Furthermore, this is dense onto and maps each of the covers C_n of $\mathfrak{L}(\mathbb{R})$ to the analogous covers of $\mathfrak{O}\mathbb{Q}$ which in turn evidently form a basis for the usual metric uniformity of $\mathfrak{O}\mathbb{Q}$. Hence we have:

Corollary 1. $\mathfrak{L}(\mathbb{R}) \to \mathfrak{O}\mathbb{Q}$ *is the completion of* $\mathfrak{O}\mathbb{Q}$ *in its metric uniformity.*

In a different direction, if

$$\mathfrak{L}([0,1]) = \uparrow((0,1)^*) \text{ for } (0,1)^* = \bigvee\{(p,0) \vee (0,q) \mid p < 0 < q\},$$

the frame version of the closed unit interval, we have

Corollary 2. $\mathfrak{L}([0,1])$ *is compact.*

Proof. It is complete by Proposition 6.1 and totally bounded by Remark 2 of 5.1 and the obvious fact that the quotient map $\mathfrak{L}(\mathbb{R}) \to \mathfrak{L}([0,1])$ takes the finitely many

$$(-\tfrac{1}{2n}\,,\,\tfrac{1}{2n}),\ (0,\,\tfrac{1}{n}),\ (\tfrac{1}{2n}\,,\,\tfrac{3}{2n}),\ldots,(1-\tfrac{1}{2n}\,,\,1+\tfrac{1}{2n})$$

to a cover for each n. □

Remark. An alternative approach to the above results would be to show *first* that each (p,q) in $\mathfrak{L}(\mathbb{R})$ is small, that is $\uparrow((p,q)^*)$ is compact, for which it is enough to consider the case $p=0$, $q=1$, and then conclude that $\mathfrak{L}(\mathbb{R})$ is complete by Proposition 6.2. For a direct proof of the compactness of $\mathfrak{L}([0,1])$ see Johnstone [31], IV, 1.2.

7 Concluding remarks

There are several other topics which a more extensive exposition of this kind ought to include. Here we list some of them.

Completion and coproducts

For any family $(L_\alpha)_{\alpha\in I}$ of nearness frames, there is a uniform isomorphism

$$h : \bigoplus CL_\alpha \to CL,\ L = \bigoplus L_\alpha .$$

This is obtained by first showing that each coproduct map $i_\alpha : L_\alpha \to L$ lifts to a uniform homomorphism $h_\alpha : CL_\alpha \to CL$ which is done by proving that $(\gamma_L)_* i_\alpha$ is a general regular Cauchy filter. The desired h is then the uniform homomorphism provided by these h_α (Banaschewski-Pultr [16]).

Paracompactness

Recall from Remark(2) of 2.2 that a frame L is called *paracompact* if every cover C of L has a locally finite refinement, that is, there exists a cover $D \leq C$ and a cover W such that $\{t \in D \mid t \wedge s \neq 0\}$ is finite for each $s \in W$. What makes this relevant in the present context is the result of Isbell [29], also treated in Banaschewski-Pultr [13], that a regular frame is paracompact iff it has a complete uniformity, and that the paracompact regular frames are coreflective in the category of all frames, the coreflection resulting from a suitable uniform completion.

Pseudocompactness

Following the terminology in the case of spaces, a frame L is called *pseudocompact* if any homomorphism $\mathfrak{L}(\mathbb{R}) \to L$ is bounded, that is, takes some (p, q) to $e \in L$, where $\mathfrak{L}(\mathbb{R})$ is the frame of reals discussed in 6.4. In the treatment of paracompactness by Banaschewski-Pultr [13], it is then shown that, for any frame L which has uniformities, L is pseudocompact iff all its uniformities are totally bounded iff its paracompact regular coreflection is compact.

Localic groups

The groups in the dual of the category of frames, which are the pointfree counterparts of the topological groups, come equipped with uniformities which arise in the same way as the familiar uniformities of topological groups (provided the latter are described without reference to points). A recent result in this area, solving a problem left open when this subject was first introduced, is that any localic group is complete in its two-sided uniformity (Banaschewski-Vermeulen [17]). The basic step in the proof of this involves lifting the given multiplication $L \to L \oplus L$ to a multiplication $CL \to CL \oplus CL$ for any of the completions CL of L, using a suitable general Cauchy filter $L \to CL \oplus CL$. In the case of the two-sided uniformity this then produces a localic group structure on CL for which $\gamma_L : CL \to L$ is a dense homomorphism—and any such is an isomorphism by a basic result of Isbell-Kříž-Pultr-Rosický [30].

Realcompactness

Here one considers completely regular frames L and their rings $\mathfrak{R}(L)$ of real-valued functions, consisting of the elements $\mathfrak{L}(\mathbb{R}) \to L$, with operations suitably induced by the operations of $\mathbb{Q}$ as lattice-ordered ring.

L is called *strongly realcompact* (or just: *realcompact*) if every unital ring homomorphism $\mathfrak{R}(L) \to \mathfrak{R}(M)$ (or: $\mathfrak{R}(L) \to \mathbb{R}$) arises from a frame homomorphism $L \to M$ (or: $L \to \mathbf{2}$). Strong realcompactness has been considered by several authors (Reynolds [41], Madden-Vermeer [35]) but always treated with arguments involving some form of the Axiom of Choice and leaving aside the puzzle concerning the relation between it and the weaker notion. A way to avoid the former as well as to resolve the puzzle is to make use of the *real uniformity* of the frames involved, that is, the uniformity induced by all $\mathfrak{L}(\mathbb{R}) \to L$. The result then is that L is strongly realcompact (or: realcompact) iff it is complete (or: Cauchy complete) with respect to this (Banaschewski [6], [7]). A good proof of the more subtle "if" part in the strong case (though not the one given in [6] or [7]) uses a suitable general Cauchy filter $L \to M$ which is derived from the given ring homomorphism, leading to a simple explicit formula for the desired frame homomorphism.

We note in addition that these results have their exact counterparts, as yet unpublished, for the rings of integer-valued functions on zero-dimensional frames, that is, homomorphisms $\mathfrak{P}\mathbb{Z} \to L$, with operations induced by those of $\mathbb{Z}$ (Banaschewski [8]).

Acknowledgements. This chapter grew out of a series of lectures given first at Sogang University, Seoul, Korea in May-June 1994 at the invitation of Hong Sung Sa and subsequently at the University of Cape Town on the occasion of SoCaT94, a symposium on categorical topology in honour of Guillaume Brümmer on his 60th birthday, and owes its inclusion in this volume to discussions that took place at the 20th Linz Seminar on Fuzzy Set Theory in February 1999. It is an extensively revised and augmented form of an earlier version which appeared in the Lecture Notes series of the Department of Mathematics and Applied Mathematics of the University of Cape Town in 1996.

Thanks for financial assistance go to the TGRC of Korea, the Categorical Topology Research Group at the University of Cape Town, the organizers of the Linz Seminar on Fuzzy Set Theory, and, as always, the Natural Sciences and Engineering Research Council of Canada.

References

[1] B. Banaschewski, *Untersuchungen über Filterräume,* Doctoral Dissertation, Universität Hamburg 1953.

[2] ——————, *Extensions of topological spaces,* Can. Math. Bull. **7**(1964), 1–22.

[3] ——————, *Another look at the localic Tychonoff Theorem,* Comment. Math. Univ. Carolinae **29**(1988), 647–656.

[4] ——————, *Compact regular frames and the Sikorski Theorem,* Kyungpook Math. J. **28**(1988), 1–14.

[5] ——————, *Universal zero-dimensional compactifications,* Categorical Topology and its Relations to Analysis, Algebra, and Combinatorics (Prague, August 1988), World Scientific Singapore 1989, pp. 257–269.

[6] ——————, *The real numbers in pointfree topology,* Textos de Matemática Série B, No. 12(1997), Departamento de Matemática da Universidade de Coimbra.

[7] ——————, *A uniform view of localic realcompactness,* J. Pure Appl. Alg. **143**(1999), 49–68.

[8] ______________, *Integer-valued functions in pointfree topology,* Unpublished notes, University of Cape Town 1997.

[9] B. Banaschewski, S. S. Hong, A. Pultr, *On the completion of nearness frames,* Quaest. Math. **21**(1998), 19–37.

[10] B. Banaschewski, C. J. Mulvey, *Stone-Čech compactification of locales I,* Houston J. Math. **6**(1980), 301–312.

[11] ______________, *Stone-Čech compactification of locales II,* J. Pure Appl. Alg. **33**(1984), 107–122.

[12] B. Banaschewski, A. Pultr, *Samuel compactification and completion of uniform frames,* Math. Proc. Cambridge Phil. Soc. **108**(1990), 63–78.

[13] ______________, *Paracompactness revisited,* Appl. Cat. Struct. **1**(1993), 181–190.

[14] ______________, *Cauchy points of uniform and nearness frames,* Quaest. Math. **19**(1996), 107–127.

[15] ______________, *A new look at pointfree metrization theorems,* Comment. Math. Univ. Carolinae **39**(1998), 167–175.

[16] ______________, Uniformity In Pointfree Topology, Chapter IV: Completion, (unpublished book manuscript).

[17] B. Banaschewski, J. J. C. Vermeulen, *On the completeness of localic groups,* Comment. Math. Univ. Carolinae **40**(1999), 273–307.

[18] F. A. Behrend, *Note on the compactification of separated uniform spaces,* Indag. Math. **18**(1956), 269–270.

[19] J. Bénabou, *Treillis locaux et paratopologie,* Séminaire C. Ehresman 1957-58, Fac. de Sciences de Paris.

[20] R. Engelking, General Topology, Sigma Series in Pure Mathematics Vol. 6, Heldermann Verlag, Berlin 1989.

[21] P. Fletcher, W. Hunsaker, *Entourage uniformities for frames,* Mh. Math. **112**(1991), 271–279.

[22] L. Gillman, M. Jerison, Rings Of Continuous Functions, Van Nostrand (Princeton), 1960.

[23] A. M. Gleason, *Projective topological spaces,* Ill. J. Math. **2**(1958), 482–489.

[24] V. Glivenko, *Sur quelques points de la logique de M. Brouwer,* Bull Acad. des Sci. de Belgique **15**(1929), 183–188.

[25] H. Herrlich, *A concept of nearness,* Gen. Top. Appl. **4**(1974), 191-212.

[26] S. S. Hong, *Convergence in frames,* Kyungpook Math. J. **35**(1995), 85–91.

[27] S. S. Hong, Y. K. Kim, *Nearness spaces and nearness frames,* Proceedings SoCaT94. University of Cape Town 1999; pp. 141–146.

[28] ——————, *Cauchy completions of nearness frames,* Appl. Cat. Struct. **3**(1995), 371–377.

[29] J. R. Isbell, *Atomless parts of spaces,* Math. Scand. **31**(1972), 5–32.

[30] J. R. Isbell, I. Kříž, A. Pultr, J. Rosický, *Remarks on localic groups,* Springer LNM 1348, Categorical Algebra and its Applications. Proceedings Lovain-la-Neuve 1987. Springer-Verlag, 1988; pp. 154–172.

[31] P. T. Johnstone, Stone Spaces, Cambridge University Press (Cambridge), 1982.

[32] A. Joyal, *Theorie des topos et le théorème de Barr,* Tagungsbericht, Category Theory Meeting Oberwolfach 1977.

[33] I. Kříž, *A direct description of uniform completion and a characterization of LT groups,* Cah. Top. Géom. Diff. Cat. **27**(1986), 19–34.

[34] S. Mac Lane, Categories For The Working Mathematician, Graduate Texts in Mathematics 5, Springer-Verlag (Berlin/New York), 1971.

[35] J. Madden, J. Vermeer, *Lindelöf locales and realcompactness,* Math. Proc. Cambridge Phil. Soc. **99**(1986), 473–480.

[36] D. Papert, S. Papert, *Sur les treillis des ouverts et la paratopologies,* Seminaire C. Ehresman 1957-58. Fac. de Sci. de Paris.

[37] J. Picado, *Weil uniformities for frames,* Comment. Math. Univ. Carolinae **36**(1995), 357–370.

[38] A. Pultr, *Pointless uniformities I: Complete regularity.* Comment. Math. Univ. Carolinae **25**(1984), 91–104.

[39] ——————, *Pointless uniformities II: (Dia)metrization.* Comment. Math. Univ. Carolinae **25**(1984), 105–120.

[40] A. Pultr, J. Úlehla, *Notes on characterization of paracompact frames.* Comment. Math. Univ. Carolinae **30**(1989), 377–384.

[41] G. Reynolds, *On the spectrum of a real representable ring,* Applications of Sheaves: Proceedings Durham 1977, Springer LNM 753, Springer-Verlag 1979, pp. 595–611.

[42] S. Vickers, Topology Via Logic, Cambridge Tracts in Theor. Comp. Sci. No. 5, Cambridge University Press (Cambridge), 1985.

CHAPTER 2

Monadic Convergence Structures

W. GÄHLER

Introduction

A basic notion for monadic convergence structures is that of partially ordered monad $\Phi = (\varphi, \leq, \eta, \mu)$, where (φ, η, μ) is a monad over **SET** and $(\varphi, \leq)$ is a functor from **SET** to the category **acSLAT** of almost complete semilattices. We introduce the notion of sup-inverse of an **acSLAT**-morphism and present some of its properties which are important in our theory. After defining the notion of partially ordered monad, there are given some examples, e.g. that of partially ordered fuzzy filter monad and of partially ordered fuzzy stack monad. Because of the generality considered here, microobjects may appear. For each partially ordered monad the notion of stratification is introduced.

One of the aims of this chapter is to demonstrate that a lot of different types of convergence structures can be included in our general theory, for instance probabilistic convergence structures, limit towers and graded fuzzy convergence structures. For showing this, the notion of partially ordered product monad is introduced.

Fuzzy pretopologies and extended fuzzy pretopologies are characterized by their interior operators. The related open fuzzy sets fulfill different conditions, that is $(\mathrm{O_f})$ and $(\mathrm{O_s})$, whereas in every case of a monadic convergence structure the open φ-objects fulfill one and the same condition, that is (O).

There is given a characterization of the neighborhood operator by sup-inverses which leads to a characterization of monadic topologies by a diagonal axiom. In some sense dually by means of the closure operator regularity is characterized. The last part of the chapter is devoted to separation axioms, applying previous results.

S.E. Rodabaugh and E.P. Klement (eds.), Topological and Algebraic Structures in Fuzzy Sets, 57–79.
© 2003 *Kluwer Academic Publishers.*

1 Sup-inverses and Galois connections

In our theory we are interested in posets $(X, \leq)$ in which all non-empty suprema exist. They are called *almost complete semilattices.* It is important that infima need not exist.

Let **acSLAT** denote the category of almost complete semilattices, where morphisms are the mappings between almost complete semilattices which preserve non-empty suprema.

In the following fix an **acSLAT**-morphism $f : (X, \leq) \to (Y, \leq)$. We assign to each element y of $D = \{y \in Y \mid \exists x \in X\ f(x) \leq y\}$ the greatest element x of X for which $f(x) \leq y$ holds. Let $g : D \to X$ denote this mapping. D equipped with the induced partial ordering of $(Y, \leq)$ is an almost complete subsemilattice of $(Y, \leq)$ and $g : (D, \leq) \to (X, \leq)$ is an **acSLAT**-morphism, called the *sup-inverse* of f.

Since $f[X] \subseteq D$ holds, f has the range restriction $f' : (X, \leq) \to (D, \leq)$.

For each $x \in X$ and each $y \in Y$ we have that $f(x) \leq y$ is equivalent to $y \in D$ and $x \leq g(y)$. Hence, f' together with g define a *Galois connection.* If f is surjective, then $D = Y$ and $g \circ f = 1_X$.

In the theory of monadic convergence structures a lot of results on sup-inverses appear. Most of these results are specializations of well-known results on Galois connections.

2 Partially ordered monads

There is a series of examples of partially ordered monads which are important in general topology.

By a *partially ordered monad* (over **SET**) (cf. [6]) we mean a quadrupel $\Phi = (\varphi, \leq, \eta, \mu)$ with the following properties.

Φ consists at first of a covariant functor $(\varphi, \leq) :$ **SET** $\to$ **acSLAT**, $X \mapsto (\varphi X, \leq)$ with $\varphi :$ **SET** $\to$ **SET** the underlying set functor.

Moreover, Φ consists of two natural transformations $\eta = (\eta_X)_{X \in \mathrm{Ob}\mathbf{SET}}$ and $\mu = (\mu_X)_{X \in \mathrm{Ob}\mathbf{SET}}$ of mappings $\eta_X : X \to \varphi X$ and $\mu_X : \varphi\varphi X \to \varphi X$, respectively.

We assume that the triple (φ, η, μ) is a monad over **SET** and that the following conditions are fulfilled:

(M0) φX is empty in case X is empty.

(M1) For any set X and each pair of different elements x and y of X, the infimum of $\eta_X(x)$ and $\eta_X(y)$ does not exist.

(M2) For all mappings $f, g : Y \to \varphi X$, $f \leq g$ implies $\mu_X \circ \varphi f \leq \mu_X \circ \varphi g$, where $\leq$ is defined argumentwise with respect to the partial ordering of φX.

(M3) For each set X, $\mu_X : (\varphi\varphi X, \leq) \to (\varphi X, \leq)$ preserves non-empty suprema.

Condition (M1) appears as a useful separation condition.

The partial orderings $\leq$ of the sets φX are considered as *finer relations*. For each set X, the elements of φX are called *φ-objects* on X, and the minimal elements of φX also *ultra objects*.

A φ-object $\mathcal{M}$ on a set X for which $\mathcal{M} < \eta_X(x)$ holds for some $x \in X$, will be called a *microobject* at x. Because of condition (M1), x is uniquely associated to $\mathcal{M}$. If there is a microobject at an element x of X, then there is a microobject at any element y of X, which follows by means of a bijection $f : X \to X$ for which $f(x) = y$.

Microobjects are in some sense properly finer than points. They may exist or do not. The conditions (M1) and that Φ has not any microobjects can be formulated together as follows:

(M1′) For each set X, $\eta_X : X \to \varphi X$ is an injection and all values $\eta_X(x)$ are ultra objects on X.

Let $\Phi = (\varphi, \leq, \eta, \mu)$ be a partially ordered monad. By a *partially ordered submonad* of Φ we mean a partially ordered monad $\Psi = (\varphi', \leq, \eta', \mu')$ such that

(1) $(\varphi', \leq) :$ **SET** $\to$ **acSLAT** is a subfunctor of $(\varphi, \leq)$ (in particular, for each set X and each non-empty set $A \subseteq \varphi' X$ the suprema of A with respect to $(\varphi' X, \leq)$ and to $(\varphi X, \leq)$ coincide) and

(2) (φ', η', μ') is a submonad of (φ, η, μ).

Instead of Ψ being a partially ordered submonad of Φ we also say that Φ is an *extension* of Ψ.

In the next sections we will give some examples of partially ordered monads and of their partially ordered submonads.

3 Filter case

Many classical topological structures can be completely described by means of the *partially ordered filter monad* $(\mathsf{F}, \leq, \eta, \mu)$. F is the *filter functor*, which assigns to each set X the set $\mathsf{F}X$ of all filters on X.

$\leq$ indicates that the sets $\mathsf{F}X$ are equipped with the finer relations of filters, that is, the inversion of the inclusion.

η and μ are natural transformations consisting of all mappings $\eta_X : X \to \mathsf{F}X$ and $\mu_X : \mathsf{FF}X \to \mathsf{F}X$ respectively, where for each $x \in X$, $\eta_X(x) = \dot{x}$, and for each filter $\mathcal{L}$ on $\mathsf{F}X$, $\mu_X(\mathcal{L}) = \bigcup_{A \in \mathcal{L}} \bigcap_{\mathcal{M} \in A} \mathcal{M}$. In this example microobjects do not exist.

4 General fuzzy filter case and related bounded case

The notion of partially ordered fuzzy filter monad is important in general fuzzy topology. This notion depends on a fixed *non-degenerate frame* L, that is, on an infinitely distributive complete lattice L with different smallest element 0 and largest element 1.

In case of $L = \{0, 1\}$ this partially ordered monad coincides with the partially ordered filter monad.

The mappings $f \in L^X$ are called *fuzzy subsets* of X. For each $\alpha \in L$ let $\bar{\alpha}$ denote the constant fuzzy subset of X with value α.

A *fuzzy filter* on X is a mapping $\mathcal{M} : L^X \to L$ such that

(F1) $\mathcal{M}(\bar{0}) = 0$ and $\mathcal{M}(\bar{1}) = 1$,

(F2) $\mathcal{M}(f \wedge g) = \mathcal{M}(f) \wedge \mathcal{M}(g)$ for all $f, g \in L^X$.

This general notion of fuzzy filter was proposed by U. Höhle ([16]). In [8] a notion of fuzzy filter was introduced which only differs from the notion given here in replacing in (F1) the condition that $\mathcal{M}(\bar{0}) = 0$, by the following stronger condition that

$$\mathcal{M}(\bar{\alpha}) \leq \alpha \tag{1}$$

holds for all $\alpha \in L$. Fuzzy filters which fulfill this stronger condition, will be called *bounded.*

In the general case, the *partially ordered fuzzy filter monad* $(\mathcal{F}_L, \leq, \eta, \mu)$ is defined as follows:

For each set X, $\mathcal{F}_L X$ consists of all fuzzy filters on X. The partial orderings on the sets $\mathcal{F}_L X$ are defined by $\mathcal{M} \leq \mathcal{N} \iff \mathcal{M}(f) \geq \mathcal{N}(f)$ for all $f \in L^X$.

For each mapping $f : X \to Y$, each $\mathcal{M} \in \mathcal{F}_L X$ and each $g \in L^X$ let

$$\mathcal{F}_L f(\mathcal{M})(g) = \mathcal{M}(g \circ f).$$

For each $x \in X$ and $f \in L^X$ let $\eta_X(x)(f) = f(x)$, and for each $\mathcal{L} \in \mathcal{F}_L \mathcal{F}_L X$ and $f \in L^X$ let $\mu_X(\mathcal{L})(f) = \mathcal{L}(e^f)$, where $e^f : \mathcal{F}_L X \to L$ is the mapping $\mathcal{M} \mapsto \mathcal{M}(f)$.

Proposition 1 *Let X be a set. For the supremum of a non-empty subset A of $\mathcal{F}_L X$ we have*

$$(\bigvee_{\mathcal{M}\in A} \mathcal{M})(f) = \bigwedge_{\mathcal{M}\in A} \mathcal{M}(f) \tag{2}$$

for all $f \in L^X$. If the infimum of a subset A of $\mathcal{F}_L X$ exists, then $(\bigwedge_{\mathcal{M}\in A} \mathcal{M})(f) = \bigvee_{\substack{f_1\wedge\cdots\wedge f_n \leq f \\ \mathcal{M}_{i_1},\ldots,\mathcal{M}_{i_n}}} (\mathcal{M}_{i_1}(f_1) \wedge \cdots \wedge \mathcal{M}_{i_n}(f_n))$ for all $f \in L^X$.

In the following a useful property of microobjects is presented.

Proposition 2 *Let $\mathcal{M}$ be a fuzzy filter on a set X which is a microobject at some $x \in X$. Then there is a mapping $f \in L^X$ such that $\mathcal{M}(f) > f(x)$ holds and $f(y) = 0$ for all $y \neq x$.*

Proof. There is a mapping $g \in L^X$ with $\mathcal{M}(g) > g(x)$. Define $h \in L^X$ by $h(x) = \mathcal{M}(g)$ and $h(y) = 0$ for all $y \neq x$ and take $f = g \wedge h$. We have $\mathcal{M}(h) \geq h(x) = \mathcal{M}(g)$ and $f(x) = g(x)$, hence $\mathcal{M}(f) = \mathcal{M}(g) \wedge \mathcal{M}(h) = \mathcal{M}(g) > f(x)$ and $f(y) = 0$ for $y \neq x$. □

Proposition 3 *In the general fuzzy filter case microobjects exist except for L being a complete Boolean algebra. If microobjects exist, then they even exist on each non-empty set.*

Proof. Follows immediately by examining the characterization of ultra fuzzy filters given in Höhle [16] on $\eta_X(x)$. □

Complete chains with at least three elements and in particular, the closed unit interval $[0, 1]$, are examples of L which are not complete Boolean algebras. In each of these cases, a simple example of a microobject $\mathcal{M}_x$ at an element x of a set X is given by defining $\mathcal{M}_x(f) = 0$ if $f(x) = 0$ and $\mathcal{M}_x(f) = 1$ if $f(x) > 0$ holds. The supremum $\bigvee_{x\in A} \mathcal{M}_x$ with A a subset of X of at least two elements is neither a microobject nor it is a bounded fuzzy filter.

Proposition 4 *(cf. [8]) The bounded fuzzy filters define a partially ordered submonad $(\mathcal{F}_L^b, \leq, \eta', \mu')$ of $(\mathcal{F}_L, \leq, \eta, \mu)$, called the* bounded partially ordered fuzzy filter monad. *In $(\mathcal{F}_L^b, \leq, \eta', \mu')$ microobjects do not exist.*

The bounded partially ordered fuzzy filter monad and the related monadic topologies are treated in [8, 9]. Proposition 1 holds analogously in the bounded case.

5 General fuzzy stack case and related bounded case

Let a non-degenerate frame L be fixed. The notion of *partially ordered fuzzy stack monad* $\Phi = (\mathcal{S}_L, \leq, \eta, \mu)$ is an extension of the notion of partially ordered fuzzy filter monad and is defined analogously. The only difference is that for each set X, the related set $\mathcal{S}_L X$ consist of all *fuzzy stacks* on X, which are the mappings $\mathcal{M} : L^X \to L$, for which both the following conditions are fulfilled:

(S1) (*Finite intersection property*) For each non-empty finite set of fuzzy subsets $f_1, \ldots, f_n$ of X, $f_1 \wedge \cdots \wedge f_n = \bar{0}$ implies $\mathcal{M}(f_1) \wedge \cdots \wedge \mathcal{M}(f_n) = 0$.

(S2) $\mathcal{M}$ is isotone.

Proposition 5 *Let X be a set. The supremum of a non-empty set $A \subseteq \mathcal{S}_L X$ fulfills, as in the fuzzy filter case, the equations (2). For each set $A \subseteq \mathcal{S}_L X$, for which the infimum exists, we have $(\bigwedge_{\mathcal{M} \in A} \mathcal{M})(f) = \bigvee_{\mathcal{M} \in A} \mathcal{M}(f)$ for all $f \in L^X$.*

A fuzzy stack $\mathcal{M}$ is called *bounded* provided that instead of (S1) the following stronger condition is fulfilled:

(S1′) (*Finite intersection property in the bounded case*) For each non-empty finite set of fuzzy sets $f_1, \ldots, f_n \in L^X$ we have $\mathcal{M}(f_1) \wedge \cdots \wedge \mathcal{M}(f_n) \leq \sup(f_1 \wedge \cdots \wedge f_n)$.

The notion of bounded fuzzy stack was introduced first in the common paper [10] of the author, A.S. Abd-Allah and A. Kandil. Clearly, for each bounded fuzzy stack $\mathcal{M}$ we have $\mathcal{M}(\bar{\alpha}) \leq \alpha$ for all $\alpha \in L$. The coarsest fuzzy stack $\mathcal{M}$ on a non-empty set X, defined by $\mathcal{M}(f) = 0$ for all $f \in L^X$, is bounded. Notice, that for a fuzzy stack $\mathcal{M}(\bar{1}) = 1$ need not hold.

Fuzzy filters are special fuzzy stacks. Moreover, a fuzzy filter is bounded if and only if it is bounded as a fuzzy stack.

According to the following proposition, for each fuzzy stack $\mathcal{M}$ a fuzzy filter, canonically associated to $\mathcal{M}$, exists.

Proposition 6 *For each fuzzy stack $\mathcal{M}$ on a set X, $\mathcal{M}^\wedge : L^X \to L$, defined by*

$$\mathcal{M}^\wedge(f) = \begin{cases} \bigvee_{f_1 \wedge \cdots \wedge f_n \leq f} (\mathcal{M}(f_1) \wedge \cdots \wedge \mathcal{M}(f_n) & \text{if } f \neq 1 \\ 1 & \text{if } f = 1, \end{cases}$$

is the coarsest fuzzy filter finer than $\mathcal{M}$.

If $\mathcal{M}$ is bounded, then $\mathcal{M}^\wedge$ also is.

Proposition 7 *In the general fuzzy stack case microobjects exist except for L being a complete Boolean algebra. If microobjects exist, then they even exist on each non-empty set.*

Proof. Immediate from Propositions 3 and 6. □

Proposition 8 *(cf. [10]) The bounded fuzzy stacks define a partially ordered submonad $(\mathcal{S}_L^b, \leq, \eta', \mu')$ of $(\mathcal{S}_L, \leq, \eta, \mu)$, called the* bounded partially ordered fuzzy stack monad. *In $(\mathcal{S}_L^b, \leq, \eta', \mu')$ microobjects do not exist.*

The bounded partially ordered fuzzy stack monad and the related monadic topologies are treated in detail in [10-13]. In particular, in [10] it is shown that Proposition 5 analogously holds in the bounded case.

For $L = \{0, 1\}$, the fuzzy stacks on X are the *stacks* on X which have the *finite intersection property*, that is, they are the sets $\mathcal{M}$ of subsets of X such that the following conditions are fulfilled:

(1) $M_1, \ldots, M_n \in \mathcal{M}$ implies $M_1 \cap \ldots \cap M_n \neq \emptyset$.

(2) $M \subseteq N \subseteq X$ and $M \in \mathcal{M}$ imply $N \in \mathcal{M}$.

On each non-empty set the coarsest fuzzy stack is empty.

Slightly modified notions of stacks appeared already in Grimeisen [14] and Netzer [21].

6 Some types of sup-inverses

In the following there are some remarks on three special types of sup-inverses. Let any partially ordered monad $\Phi = (\varphi, \leq, \eta, \mu)$ be fixed.

First type. This type is related to any mapping $f : X \to Y$.

$$\varphi f : (\varphi X, \leq) \to (\varphi Y, \leq)$$

is an acSLAT-morphism. The sup-inverse of φf is an acSLAT-morphism

$$\varphi^- f : (\mathcal{D}, \leq) \to (\varphi X, \leq) \text{ with } \mathcal{D} = \{\mathcal{N} \in \varphi Y \mid \exists \mathcal{M} \in \varphi X \ \varphi f(\mathcal{M}) \leq \mathcal{N}\}$$

Proposition 9 *If f is surjective, then $\mathcal{D} = \varphi Y$ and $\varphi^- f = \mu_X \circ \varphi(\varphi^- f \circ \eta_Y)$.*

Second type. This type of sup-inverses is related to the natural transformation μ. Because of condition (M3) and a monadic property, for each set X, $\mu_X : (\varphi\varphi X, \leq) \to (\varphi X, \leq)$ is a surjective acSLAT-morphism. Hence, its sup-inverse is an acSLAT-morphism $\mu_X^- : (\varphi X, \leq) \to (\varphi\varphi X, \leq)$.

Third type. For describing this type of sup-inverses let a mapping $f : X \to \varphi Y$ be fixed, where X and Y are any sets. Let $\varphi_\mu^- f$ denote the sup-inverse of the composition $\mu_Y \circ \varphi f : (\varphi X, \le) \to (\varphi Y, \le)$.

In the following proposition appears a useful condition which ensures that $\mu_Y \circ \varphi f$ is surjective without assuming that f is surjective.

Proposition 10 *Assume that there exists a mapping $e : Y \to X$ such that $f \circ e = \eta_Y$. Then $\mu_Y \circ \varphi f$ is a surjection and φY is the domain of the sup-inverse $\varphi_\mu^- f$. Moreover, then $\eta_X \le \varphi_\mu^- f \circ f$ holds.*

Proof. $f \circ e = \eta_X$ implies $\mu_Y \circ \varphi f \circ \varphi e = 1_{\varphi X}$ and therefore that $\mu_Y \circ \varphi f$ is surjective, hence φY is the domain of $\varphi_\mu^- f$. Since $1_{\varphi X} \le \varphi_\mu^- f \circ (\mu_Y \circ \varphi f)$ holds, it follows $\eta_X \le \varphi_\mu^- f \circ (\mu_Y \circ \eta_{\varphi Y}) \circ f = \varphi_\mu^- f \circ f$. □

Proposition 11 *In the fuzzy filter case, the fuzzy stack case and the related bounded cases under the assumption of Proposition 10, for each $\mathcal{M} \in \varphi Y$ and $g \in L^X$ we have $\varphi_\mu^- f(\mathcal{M})(g) = \bigvee_{f(.)h \le g} \mathcal{M}(h)$, where $f(.)h : X \to L$ is the mapping $x \mapsto f(x)(h)$.*

7 Stratified φ-objects

Let $\Phi = (\varphi, \le, \eta, \mu)$ be any partially ordered monad. In the following we introduce a special partially ordered submonad of Φ.

A φ-object $\mathcal{M}$ on a set X will be called *stratified* provided that $\mathcal{M} \le \bigvee_{x \in X} \eta_X(x)$ holds.

We obtain a subfunctor $\varphi' : \mathbf{SET} \to \mathbf{SET}$ of φ by assigning to each set X the set $\varphi' X = \{\, \mathcal{M} \in \varphi X \mid \mathcal{M} \text{ stratified} \,\}$ and to each mapping $f : X \to Y$ the domain-codomain restriction $\varphi' f : \varphi' X \to \varphi' Y$ of φf.

Each set $\varphi' X$ we equip with the partial ordering induced by that of φX. In this way we even get a subfunctor $(\varphi', \le) : \mathbf{SET} \to \mathbf{acSLAT}$ of $(\varphi, \le)$.

Since the φ-objects $\eta_X(x)$ are stratified, for each set X the codomain restriction $\eta'_X : X \to \varphi' X$ of η_X exists. $\eta' = (\eta'_X)_{X \in \mathrm{Ob}\mathbf{SET}}$ is a natural transformation of the identity set functor to φ'.

Let ι_X be the inclusion mapping of $\varphi' X$ into φX.

For each set X there is a mapping $\mu'_X : \varphi' \varphi' X \to \varphi' X$ which is the codomain restriction of $\mu_X \circ \varphi \iota_X \circ \iota_{\varphi' X} : \varphi' \varphi' X \to \varphi X$ to $\varphi' X$.

$\mu' = (\mu'_X)_{X \in \mathrm{Ob}\mathbf{SET}}$ is a natural transformation of $\varphi' \varphi'$ to φ'. Moreover, we have the following.

Proposition 12 $\Phi' = (\varphi', \le, \eta', \mu')$ *is a partially ordered submonad of Φ, called the stratification of Φ.*

A fuzzy filter and, more general, a fuzzy stack $\mathcal{M}$ is stratified if and only if $\mathcal{M}(\bar{\alpha}) \geq \alpha$ holds for all $\alpha \in L$. Bounded fuzzy filters and bounded fuzzy stacks which are stratified, are called *homogeneous.* All filters are homogeneous.

A fuzzy filter $\mathcal{M}$ is homogeneous if and only if

$$\mathcal{M}(\bar{\alpha}) = \alpha \tag{3}$$

for all $\alpha \in L$. The notion of homogeneous fuzzy filter was first introduced in the common paper [3] with Patrik Eklund.

Homogeneous fuzzy stacks fulfill also the inequality (3). The notion of homogeneous fuzzy stack was first introduced in [10].

The homogeneous fuzzy filters define a partially ordered submonad of $(\mathcal{F}_L^b, \leq, \eta', \mu')$ ([4,6]). Analogously, the homogeneous fuzzy stacks define a partially ordered submonad of $(\mathcal{S}_L^b, \leq, \eta', \mu')$ ([10]).

8 Products of partially ordered monads

There are a lot of methods to generate new partially ordered monads by given ones. One method is that of stratification. This section is devoted to the notion of product of partially ordered monads.

Let K be a non-empty set and for each $\alpha \in K$ let $\Phi_\alpha = (\varphi_\alpha, \leq, \eta_\alpha, \mu_\alpha)$ be a partially ordered monad.

We define a monad $\Psi = (\psi, \leq, \eta', \mu')$, called the *partially ordered product monad* of the family $(\Phi_\alpha)_{\alpha \in K}$, as follows:

For each set X, let $\psi X = \prod\limits_{\alpha \in K} \varphi_\alpha X$. We equip each of these cartesian products with the componentwise defined partial ordering $\leq$, that is, $\mathcal{M} \leq \mathcal{N}$ means that for all $\alpha \in K$ we have $\mathcal{M}_\alpha \leq \mathcal{N}_\alpha$.

For each mapping $f : X \to Y$ and each $\mathcal{M} \in \psi X$, let

$$\psi f(\mathcal{M}) = \big(\varphi_\alpha f(\mathcal{M}_\alpha)\big)_{\alpha \in K}$$

Moreover, for each $x \in X$ let $\eta'_X(x) = \big(\eta_{\alpha\, X}(x)\big)_{\alpha \in K}$.

Finally, for each $\mathcal{L} \in \psi\psi X$ let $\mu'_X(\mathcal{L}) = \big(\mu_{\alpha\, X}(\varphi_\alpha \pi_\alpha(\mathcal{L}_\alpha))\big)_{\alpha \in K}$, where $\pi_\alpha : \psi X \to \varphi_\alpha X$ are the related projections.

The partially ordered product monad Ψ will also be denoted by $\prod\limits_{\alpha \in K} \Phi_\alpha$.

We are especially interested in partially ordered product monads in which all factors Φ_α are one and the same partially ordered monad Φ. In these cases, Ψ will also be denoted by Φ^K and will be called the *K-power* of Φ.

9 Monadic convergence structures

Let $\Phi = (\varphi, \leq, \eta, \mu)$ be a partially orderd monad.

A *Φ-convergence structure* on a set X is a set $T \subseteq \varphi X \times X$ such that, writing $\mathcal{M} \to x$ instead of $(\mathcal{M}, x) \in T$, we have

(C1) $\eta_X(x) \to x$ for all $x \in X$,

(C2) $\mathcal{M} \to x$, $\mathcal{N} \leq \mathcal{M}$ imply $\mathcal{N} \to x$, and

(C3) $\mathcal{M} \to x$ implies $\mathcal{M} \vee \eta_X(x) \to x$.

For $\mathcal{M} \to x$ we say that $\mathcal{M}$ *converges to* x.

In the filter case the Φ-convergence structures are the *convergence structures* in sense of D. C. Kent, introduced first in [17].

If T and T' are Φ-convergence structures on a set X and $T \subseteq T'$ holds, then T is said to be *finer* than T' and T' is said to be *coarser* than T.

A Φ-convergence structure T is called a *Φ-limit structure* if instead of condition (C3) even the following condition is fulfilled:

(C3′) $\mathcal{M} \to x$, $\mathcal{N} \to x$ imply $\mathcal{M} \vee \mathcal{N} \to x$.

In the filter case, a Φ-limit structure is a *limit structure* in the usual sense, proposed first by H.-J. Kowalsky in [19].

A *Φ-pretopology* is a mapping $p : X \to \varphi X$ for which $\eta_X \leq p$ holds.

Proposition 13 *In the fuzzy filter case and the fuzzy stack case the values $p(x)$ of each Φ-pretopology p are bounded fuzzy filters and bounded fuzzy stacks respectively.*

Proof. Follows immediately from the inequality $\eta_X \leq p$. □

Each Φ-pretopology p will be identified with the Φ-limit structure $T = \{(\mathcal{M}, x) \mid \mathcal{M} \leq p(x)\}$.

For each Φ-convergence structure T, the mapping $p : X \to \varphi X$, given by $p(x) = \bigvee_{\mathcal{M} \to x} \mathcal{M}$ for all $x \in X$, is the finest Φ-pretopology coarser than T. It is called the *associated Φ-pretopology*.

By a *Φ-topology* we mean a Φ-pretopology p such that $\mu_X \circ \varphi p \circ p = p$.

If Φ is not specified, then instead of a Φ-convergence structure we also speak on a *monadic convergence structure*. Analogously the denotations *monadic limit structure*, *monadic pretopology* and *monadic topology* are used.

Clearly, a set X equipped with a Φ-convergence structure T on X is called a *Φ-convergence space*. Analogously the denotations *Φ-limit space*, *Φ-pretopological space*, *Φ-topological space* are used.

A mapping $f : (X, T_1) \to (Y, T_2)$ between Φ-convergence spaces is said to be *continuous* provided that $(\mathcal{M}, x) \in T_1$ implies $(\varphi f(\mathcal{M}), f(x)) \in T_2$. Thus, continuity of f means to preserve convergence.

A Φ-convergence space (X, T) and also its Φ-convergence structure T will be called *stratified* provided that all converging φ-objects are stratified.

Proposition 14 *A Φ-convergence structure T is stratified, if and only if the associated Φ-pretopology p is stratified, that is, all φ-objects $p(x)$ are stratified.*

Proposition 15 *Each constant mapping of a stratified Φ-convergence space into a Φ-convergence space is continuous.*

10 Fuzzy convergence structures

Let L be a non-degenerate frame and let Φ be the related partially ordered fuzzy filter monad. Then the Φ-convergence structures, Φ-limit structures, Φ-pretopologies and Φ-topologies are respectively called *fuzzy convergence structures*, *fuzzy limit structures fuzzy pretopologies*, and *fuzzy topologies*.

Clearly, if $L = \{0, 1\}$, then we have the filter case. In particular, then (up to an identification) the fuzzy pretopologies and fuzzy topologies are the classical pretopologies and topologies, respectively.

Each fuzzy pretopology $p : X \to \mathcal{F}_L X$ can be identified with the *interior operator* int : $L^X \to L^X$ of p defined by

$$(\mathrm{int} f)(x) = p(x)(f) \tag{4}$$

for all $f \in L^X$ and $x \in X$.

According to the following proposition, the interior operators of fuzzy pretopologies can be characterized independently on the related fuzzy pretopologies.

Proposition 16 ([9]) *The interior operator* int *of a fuzzy pretopology fulfills the following conditions:*

(1) $\mathrm{int} f \leq f$ *holds for all* $f \in L^X$ *and* $\mathrm{int}\bar{1} = \bar{1}$.

(2) $\mathrm{int}(f \wedge g) = \mathrm{int} f \wedge \mathrm{int} g$ *for all* $f, g \in L^X$.

On the other hand, each mapping int : $L^X \to L^X$ *which fulfills these conditions, is the interior operator of the fuzzy pretopology defined by equation* (4). int *characterizes a fuzzy topology if and only if additionally* $\mathrm{int} \circ \mathrm{int} = \mathrm{int}$.

f is called *open* if $\mathrm{int} f = f$.

Proposition 17 *The set τ of all open fuzzy sets $f \in L^X$ of a fuzzy pretopology fulfills the condition*

(O_f) *τ is closed with respect to all suprema and all finite infima.*

Each fuzzy topology p can be characterized as a subset τ of L^X which fulfills condition (O_f) by taking $\mathrm{int} f = \bigvee_{g \leq f,\, g \in \tau} g$ *for all $f \in L^X$.*

Corollary 1 *The set τ of all open fuzzy sets of a fuzzy topology equipped with the argumentwise defined partial ordering, is a frame.*

11 Extended fuzzy convergence structures

Let a non-degenerate frame L be fixed and let Φ be the related partially ordered fuzzy stack monad. Then the Φ-convergence structures, Φ-limit structures, Φ-pretopologies and Φ-topologies are respectively called *extended fuzzy convergence structures*, *extended fuzzy limit structures*, *extended fuzzy pretopologies*, and *extended fuzzy topologies.*

Analogously as the fuzzy pretopologies, each extended fuzzy pretopology p can be characterized by the *interior operator* $\mathrm{int} : L^X \to L^X$, defined by

$$(\mathrm{int} f)(x) = p(x)(f) \tag{5}$$

for all $f \in L^X$ and $x \in X$. However, in this case the properties of the interior operator are other ones.

Proposition 18 ([11]) *The interior operator int of an extended fuzzy pretopology fulfills the following conditions:*

(1) $\mathrm{int} f \leq f$ *holds for all $f \in L^X$.*

(2) $f \leq g$ *implies* $\mathrm{int} f \leq \mathrm{int} g$.

On the other hand, each mapping $\mathrm{int} : L^X \to L^X$ *which fulfills these conditions, is the interior operator of the extended fuzzy pretopology defined by equation* (5). int *characterizes an extended fuzzy topology if and only if additionally* $\mathrm{int} \circ \mathrm{int} = \mathrm{int}$.

f is called *open* if $\mathrm{int} f = f$.

As in the case of fuzzy topologies each extended fuzzy topology can be characterized by the set τ of open fuzzy sets. However, contrary to condition (O_f), in case of an extended fuzzy topology, a weaker condition holds.

Proposition 19 *The set τ of all open fuzzy sets $f \in L^X$ of an extended fuzzy topology fulfills the condition*

($\mathrm{O_s}$) *τ is closed with respect to all suprema.*

Each extended fuzzy topology p can be characterized as a subset τ of L^X which fulfills condition (O_s) by taking $\operatorname{int} f = \bigvee_{g \leq f,\, g \in \tau} g$ *for all $f \in L^X$.*

Corollary 2 *The set τ of all open fuzzy sets of an extended fuzzy topology equipped with the argumentwise defined partial ordering, is a complete lattice, which may not be distributive.*

If $L = \{0, 1\}$, then the extended fuzzy pretopologies and extended fuzzy topologies are called *extended pretopologies* and *extended topologies*, respectively.

An example of an extended topology is given in taking as X a vector space of a positive dimension and as the open subsets the complements of the convex subsets of X. In this case, τ is not distributive.

12 Gradation of monadic convergence

In our general theory, the graded monadic convergence appears as a special type of monadic convergence.

Let $\Phi = (\varphi, \leq, \eta, \mu)$ be a partially ordered monad, K a complete lattice and $\Phi^K = (\psi, \leq, \eta', \mu')$ the K-power of Φ.

By a *K-graded convergence structure with respect to Φ* we mean a family $(T_\alpha)_{\alpha \in K}$ of Φ-convergence structures on a set X with index set K.

Proposition 20 *The family $(T_\alpha)_{\alpha\in K}$ can be identified with the Φ^K-convergence structure* $T = \{\, (\mathcal{M}, x) \in \psi X \times X \mid x \in X,\ (\mathcal{M}_\alpha, x) \in T_\alpha \text{ for all } \alpha \in K \,\}$.

T is a Φ^K-limit structure, a Φ^K-pretopology and a Φ^K-topology if and only if for each $\alpha \in K$, T_α is a Φ-limit structure, Φ-pretopology and Φ-topology, respectively.

Proposition 21 *Each Φ^K-pretopology $p : X \to \psi X$ can be identified with the family $(p_\alpha)_{\alpha \in K}$ of Φ-pretopologies $p_\alpha : X \to \varphi X$, where $p_\alpha = \pi_\alpha \circ p$ with π_α the related projections $\mathcal{M} \mapsto \mathcal{M}_\alpha$.*

T is called *isotone* (resp. *antitone*), if the family $(T_\alpha)_{\alpha\in K}$ is isotone (resp. antitone) with respect to the finer relation of Φ-convergence structures.

If T is isotone, then for each $x \in X$ and $\mathcal{N} \in \varphi X$, $\bigwedge_{(\mathcal{N},x) \in T_\alpha} \alpha$ is called the *degree of convergence* of $\mathcal{N}$ to x.

If T is antitone, then we define the *convergence degree* in the same way but by replacing $\bigwedge$ by $\bigvee$.

13 Probabilistic convergence structures and limit towers

Let Φ be the partially ordered filter monad and K a complete lattice. Special Φ^K-convergence structures are the *probabistic convergence structures*, introduced by G. D. Richardson and D. C. Kent [23], and the *limit towers*, introduced by P. Brock and D. C. Kent [1]. Limit tower spaces are equivalent to the *convergence approach spaces*, introduced by E. Lowen and R. Lowen [20]. Special convergence approach spaces are R. Lowen's *approach spaces* [21].

In our terminology the definitions of probabilistic convergence structure and of limit tower are the following.

In case of $K = [0,1]$ a family $(T_\alpha)_{\alpha\in K}$ of Φ-convergence structures on a set X is called a *probabilistic convergence structure* on X, if the following conditions are fulfilled:

(P1) $(T_\alpha)_{\alpha\in K}$ is antitone.

(P2) T_0 is the indiscrete topology.

(P3) For each $\alpha \in K$ we have $T_\alpha = \bigwedge_{\beta<\alpha} T_\beta$.

In case of $K = [0,\infty]$ a family $(T_\alpha)_{\alpha\in K}$ of Φ-limit structures on a set X is called a *limit tower* on X, if the following conditions are fulfilled:

(L1) $(T_\alpha)_{\alpha\in K}$ is isotone.

(L2) T_∞ is the indiscrete topology.

(L3) $T_\alpha = \bigwedge_{\alpha<\beta} T_\beta$ for all $\alpha \in K$.

A set X which resp. is equipped with a probabilistic convergence structure and a limit tower is called a *probabilistic convergence space* and a *limit tower space*.

In an obvious manner both the notions of probabilistic convergence structure and limit tower can be generalized by taking as Φ any partially ordered monad and as K any complete lattice and assume that in the related conditions (P3) and (L3) only elements α of K are considered, for which respectively $\alpha = \bigvee_{\beta<\alpha} \beta$ and $\alpha = \bigvee_{\alpha<\beta} \beta$.

14 Graded fuzzy filter case

In the following let two non-degenerate frames L and K be fixed and let Φ be the partially ordered fuzzy filter monad $(\mathcal{F}_L, \leq, \eta, \mu)$.

Proposition 22 *Each Φ^K-pretopology $p = (p_\alpha)_{\alpha \in K}$ can be characterized by the* interior operator $\text{int} : L^X \times K \to L^X$ *of p defined by $\text{int}(f, \alpha)(x) = p_\alpha(x)(f)$ for all $f \in L^X$, $\alpha \in K$ and $x \in X$.*

The mappings $\text{int}_\alpha : L^X \to L^X$ with $\text{int}_\alpha(f) = \text{int}(f, \alpha)$ are the interior operators of the fuzzy pretopologies p_α.

If $\text{int}(f, \alpha) = f$, then f is called *α-open.*

In our case of Φ, each Φ^K-topology p can be characterized by the family $(\tau_\alpha)_{\alpha \in K}$, where τ_α are the sets of α-open fuzzy sets by taking $\text{int}(f, \alpha) = \bigvee\limits_{g \leq f,\, g\ \alpha\text{-open}} g$.

A Φ^K-topology $p = (p_\alpha)_{\alpha \in K}$ is called an *L, K-fuzzy topology*, if the following conditions are fulfilled:

(1) p is isotone, that is $\alpha \leq \beta$ implies $\tau_\alpha \supseteq \tau_\beta$.

(2) For each $f \in L^X$ there exists the greatest $\alpha \in K$ such that $f \in \tau_\alpha$.

For each L, K-fuzzy topology p we have that p_0 is the indiscrete L-fuzzy topology.

According to the following proposition, the L, K-fuzzy topologies can be characterized as special K-fuzzy subsets of L^X.

Proposition 23 *Each L, K-fuzzy topology p can be identified with the mapping $\mathcal{T} : L^X \to K$ which fulfills the following conditions:*

(1) $\mathcal{T}(\bar{0}) = \mathcal{T}(\bar{1}) = 1$.

(2) $\mathcal{T}(f) \wedge \mathcal{T}(g) \leq \mathcal{T}(f \wedge g)$ *for all* $f, g \in L^X$.

(3) $\bigwedge\limits_{f \in A} \mathcal{T}(f) \leq \mathcal{T}(\bigvee\limits_{f \in A} f)$ *for each subset A of L^X.*

The identification is given by $\mathcal{T}(f) = \bigvee\limits_{f \in \tau_\alpha} \alpha$ for all $f \in L^X$ and $\tau_\alpha = \{ f \in L^X \mid \alpha \leq \mathcal{T}(f) \}$ for all $\alpha \in K$.

First results on this type of fuzzy topologies are published in the eighties. In particular, this type was suggested and studied by A. P. Šostak (cf. [24,25]).

15 Neighborhood operator

Let $\Phi = (\varphi, \leq, \eta, \mu)$ be any partially ordered monad. Moreover, let T be a Φ-convergence structure on a set X and let p be the associated Φ-pretopology of T. The composition $\mathsf{nb} = \mu_X \circ \varphi p$ is called the *neighborhood operator* of T. For each $\mathcal{M} \in \varphi X$, the value $\mathsf{nb}\mathcal{M}$ is said to be the *neighborhood* of $\mathcal{M}$. $\mathsf{nb} : (\varphi X, \leq) \to (\varphi X, \leq)$ is an **acSLAT**-morphism.

Since $\mathsf{nb} \circ \eta_X = p$, for each $x \in X$, $p(x)$ is the neighborhood of $\eta_X(x)$, also called the *φ-neighborhood object* at x.

The property of p being a Φ-topology can be written $\mathsf{nb} \circ p = p$ and is equivalent to $\mathsf{nb} \circ \mathsf{nb} = \mathsf{nb}$ ([6]).

Proposition 24 ([6]) *nb is a hull operator, that is, $\mathcal{M} \leq \mathsf{nb}\mathcal{M}$ holds for all $\mathcal{M} \in \varphi X$.*

For each filter $\mathcal{M}$ on a topological space we have $\mathsf{nb}\mathcal{M} = \{M \subseteq X \mid \mathrm{int}M \in \mathcal{M}\}$, hence $\{M \in \mathcal{M} \mid M \text{ open}\}$ is a base of the filter $\mathsf{nb}\mathcal{M}$.

Proposition 25 ([6]) *Let $f : (X, p_1) \to (Y, p_2)$ be a mapping between Φ-pretopological spaces and let nb_1 and nb_2 be the neighborhood operators of (X, p_1) and (Y, p_2), respectively. Then f is continuous if and only if $\varphi f \circ p_1 \leq p_2 \circ f$ holds, and also, if and only if the inequality $\varphi f \circ \mathsf{nb}_1 \leq \mathsf{nb}_2 \circ \varphi f$ is fulfilled.*

16 Open φ-objects

Let $\Phi = (\varphi, \leq, \eta, \mu)$ be a partially ordered monad, T a Φ-convergence structure on a set X and nb the neighborhood operator of T.

A φ-object $\mathcal{M}$ on X is called *open* provided that $\mathcal{M} = \mathsf{nb}\mathcal{M}$. It is important that the set $\mathcal{O}$ of open φ-objects has one and the same characterization in all cases of partially ordered monads, even in the fuzzy filter cases and the fuzzy stack cases. Whereas in Propositions 17 and 19 opennes is meant for fuzzy sets, here openness is meant for φ-objects.

Proposition 26 ([6]) *The set $\mathcal{O}$ of all open φ-objects on X fulfills the following condition*

(O) *$\mathcal{O}$ is closed with respect to all non-empty suprema and all infima, as far as these infima exist.*

Each Φ-topology p can be characterized by the set $\mathcal{O}$ of all open φ-objects on X by taking

$$p(x) = \bigwedge_{\mathcal{M} \in \mathcal{O},\, \eta_X(x) \leq \mathcal{M}} \mathcal{M} \tag{6}$$

for all $x \in X$.

For any subset $\mathcal{O}$ of φX which fulfills condition (O), the mapping p defined by (6) may not be a Φ-topology. In general, we only have that p is a Φ-pretopology and that

$$p(x) \leq \mathcal{N} \iff \eta_X(x) \leq \mathcal{N} \tag{7}$$

holds for all $x \in X$ and $\mathcal{N} \in \mathcal{O}$. However the following holds.

Proposition 27 *Let $\mathcal{O}$ be a subset of φX which fulfills condition (O) and let p be the mapping defined by (6). Then the following are equivalent:*

(a) *p is a Φ-topology with $\mathcal{O}$ the set of all open φ-objects on X.*

(b) *For each $\mathcal{M} \in \varphi X$ we have*

$$\mathsf{nb}\mathcal{M} = \bigwedge_{\mathcal{N}\in\mathcal{O},\, \mathcal{M}\leq\mathcal{N}} \mathcal{N}. \tag{8}$$

Proof. Let (a) be fulfilled. For each $\mathcal{M} \in \varphi X$, $\bigwedge_{\mathcal{N}\in\mathcal{O},\, \mathcal{M}\leq\mathcal{N}} \mathcal{N}$ is the finest open φ-object on X which is coarser than $\mathcal{M}$, that is, it is $\mathsf{nb}\mathcal{M}$. Hence, (8) holds and therefore (b) is fulfilled.

Assume now that (b) is satisfied. Taking $p(x)$ as $\mathcal{M}$ in (8), by means of (6) and (7) it follows that $\mathsf{nb} \circ p = p$ and hence p is a Φ-topology. (8) implies that $\mathcal{M} \in \mathcal{O}$ is equivalent to $\mathsf{nb}\mathcal{M} = \mathcal{M}$. Hence, $\mathcal{O}$ consists of all $\mathcal{M} \in \varphi X$ which are open with respect to p. □

Proposition 28 *A mapping $f : (X, p_1) \to (Y, p_2)$ between Φ-topological spaces is continuous if and only if for each open $\mathcal{N} \in \varphi Y$, for which the preimage $\varphi^- f(\mathcal{N})$ exists, this preimage is also open.*

17 Characterization of neighborhood operator by sup-inverses

In the following let T be a Φ-convergence structure. Let t_1 and t_2 be the first and second projection of this subset of $\varphi X \times X$ into φX and X respectively, that is, $t_1(\mathcal{M}, x) = \mathcal{M}$ and $t_2(\mathcal{M}, x) = x$ for all $(\mathcal{M}, x) \in T$.

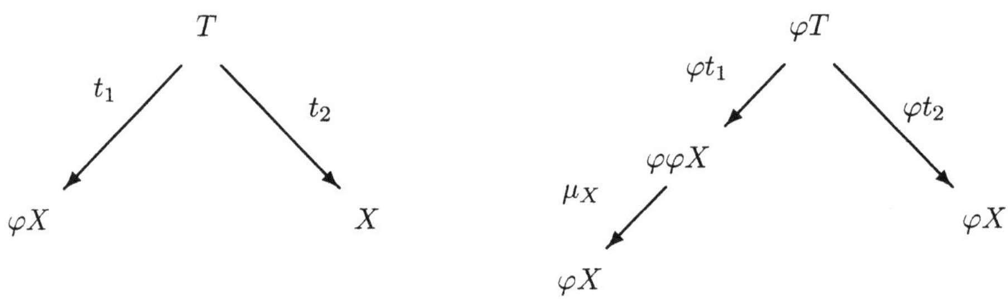

Since t_2 is surjective, the sup-inverse $\varphi^- t_2$ of φt_2 has φX as domain.

By definition, nb depends only on the associated Φ-pretopology p of T. In the following there is given a characterization of nb using T completely.

Proposition 29 *We have*

$$\mathsf{nb} = \mu_X \circ \varphi t_1 \circ \varphi^- t_2 \tag{9}$$

and, in particular,

$$p = \mu_X \circ \varphi t_1 \circ \varphi^- t_2 \circ \eta_X. \tag{10}$$

Proof. In this proof monadic properties are used several times. At first we prove (10). Applying $t_1 \leq p \circ t_2$, we get

$$\mu_X \circ \varphi t_1 \circ \varphi^- t_2 \circ \eta_X \ \leq \ \mu_X \circ \varphi p \circ \eta_X \ = \ p. \tag{11}$$

Fix any $x \in X$. Since $\varphi t_2 \circ \eta_T = \eta_X \circ t_2$, we have $(\varphi t_2 \circ \eta_T)(\mathcal{M}, x) = \eta_X(x)$ for all $\mathcal{M} \to x$, hence $\bigvee_{\mathcal{M}\to x} \eta_T(\mathcal{M}, x) \leq \varphi^- t_2(\eta_X(x))$. By means of $\eta_{\varphi X} \circ t_1 = \varphi t_1 \circ \eta_T$ it follows

$$\bigvee_{\mathcal{M}\to x} \eta_{\varphi X}(\mathcal{M}) \ = \ \varphi t_1(\bigvee_{\mathcal{M}\to x} \eta_X(\mathcal{M}, x) \ \leq \ (\varphi t_1 \circ \varphi^- t_2)(\eta_X(x))$$

and hence $p(x) \ = \ \mu_X(\bigvee_{\mathcal{M}\to x} \eta_{\varphi X}(\mathcal{M})) \ \leq \ (\mu_X \circ \varphi t_1 \circ \varphi^- t_2 \circ \eta_X)(x)$, because of (11) therefore (10). From (10) we obtain

$$\begin{aligned} \mu_X \circ \varphi p \ &= \ \mu_X \circ \mu_{\varphi X} \circ \varphi\varphi t_1 \circ \varphi(\varphi^- t_2) \circ \varphi \eta_X \\ &= \ \mu_X \circ \varphi t_1 \circ \mu_T \circ \varphi(\varphi^- t_2) \circ \varphi \eta_X) \\ &= \ \mu_X \circ \varphi t_1 \circ \varphi^- t_2 \quad \square \end{aligned}$$

18 Diagonal axiom (F)

The neighborhood operator is closely related to the *diagonal axiom* (F), which was introduced by Cook and Fischer [3] in the filter case. In the general case of a Φ-convergence structure T on a set X by this axiom we mean the following:

(F) Let Y be any set and let $f_1 : Y \to \varphi X$ and $f_2 : Y \to X$ be mappings such that $(f_1(y), f_2(y)) \in T$ for all $y \in Y$. For each $\mathcal{L} \in \varphi Y$ and $x \in X$, then $\varphi f_2(\mathcal{L}) \to x$ implies $(\mu_X \circ \varphi f_1)(\mathcal{L}) \to x$.

Because of the following proposition, axiom (F) is equivalent to the special case of this axiom in which $Y = T$ and f_1 and f_2 are the first and second projection of T respectively.

Proposition 30 *The following are equivalent:*

(1) *Axiom (F).*

(2) *Special case of axiom (F): For each $\mathcal{L} \in \varphi T$ and $x \in X$ from $\varphi t_2(\mathcal{L}) \to x$ it follows $(\mu_X \circ \varphi t_1)(\mathcal{L}) \to x$.*

(3) *For each $\mathcal{M} \in \varphi X$ and $x \in X$ with $\mathcal{M} \to x$ we have $\mathsf{nb}\mathcal{M} \to x$.*

(4) *T is a Φ-topology.*

In the filter case the equivalence of the conditions (1) and (4) was shown by D. C. Kent and G. D. Richardson in [18]. In the general case the proof of the proposition follows by means of Proposition 29.

19 Closure operator

Let T be a Φ-convergence structure on a set X,
let $t_1 : T \to \varphi X$ and $t_2 : T \to X$ be the first and second projection of T
and let $\sigma_X : X \to T$ be the mapping $x \mapsto (\eta_X(x), x)$.

Since $t_1 \circ \sigma_X = \eta_X$ holds, we have that the sup-inverse $\varphi_\mu^- t_1$ of the composition $\mu_X \circ \varphi t_1 : \varphi T \to \varphi X$ has φX as domain. The composition $\mathsf{cl} = \varphi t_2 \circ \varphi_\mu^- t_1$ of this sup-inverse with φt_2 is called the *closure operator* of T ([6]).

$$\mathsf{cl} : (\varphi X, \leq) \to (\varphi X, \leq)$$

is an **acSLAT**-morphism. Moreover, cl is a hull operator. For each $\mathcal{M} \in \varphi X$, $\mathsf{cl}\mathcal{M}$ is called the *closure* of $\mathcal{M}$.

The characterization of the neighborhood operator by means of sup-inverses and the definition of the closure operator show that both these notions are in some sense *dual.*

Proposition 31 *$\mathcal{M} \to x$ implies $\eta_X(x) \leq \mathsf{cl}\mathcal{M}$.*

Proof. Because of Proposition 10 we have $\eta_T \leq \varphi_\mu^- t_1 \circ t_1$, hence $\eta_X \circ t_2 = \varphi t_2 \circ \eta_T \leq \mathsf{cl} \circ t_1$. □

If $\mathcal{M} = \mathsf{cl}\mathcal{M}$, then $\mathcal{M}$ is called *closed*. Since cl is a hull operator, the infimum of any set of closed elements of φX is closed, as far as it exists.

By means of Proposition 11 we obtain the following.

Proposition 32 (cf. [9,13]) *In the fuzzy filter case, the fuzzy stack case and the related bounded cases we have* $(\mathsf{cl}\mathcal{M})(f) = \bigvee_{\substack{\mathcal{N}(g) \leq f(x) \\ \text{whenever } \mathcal{N} \to x}} \mathcal{M}(g)$ *for each φ-object $\mathcal{M}$ on a set X and each $f \in L^X$.*

In the cases considered in Proposition 32 each φ-object has a finer ultra object. Hence, by means of Proposition 32 it follows that in these cases the closure operator only depends only on the converging ultra objects.

For each filter $\mathcal{M}$ on a topological space we have $\mathsf{cl}\mathcal{M} = \{ N \subseteq X \mid \exists M \in \mathcal{M}\ \mathrm{cl}M \subseteq N\}$, hence $\{\mathrm{cl}M \mid M \in \mathcal{M}\}$ is a base of the filter $\mathsf{cl}\mathcal{M}$.

20 Regularity axiom (R)

The closure operator is closely related to the *regularity axiom* (R) introduced by Cook and Fischer [3] in the filter case. For a Φ-convergence structure T on a set X by this axiom we mean the following:

(R) Let Y be any set and let $f_1 : Y \to \varphi X$ and $f_2 : Y \to X$ be mappings such that $(f_1(y), f_2(y)) \in T$ for all $y \in Y$. For each $\mathcal{L} \in \varphi Y$ and $x \in X$, then $(\mu_X \circ \varphi f_1)(\mathcal{L}) \to x$ implies $\varphi f_2(\mathcal{L}) \to x$.

Because of the following proposition, axiom (R) is equivalent to the special case of this axiom in which $Y = T$ and f_1 and f_2 are the first and second projection of T respectively.

Proposition 33 (cf. [5,6]) *The following are equivalent:*

(1) *Axiom (R).*

(2) *Special case of axiom (R): For each $\mathcal{L} \in \varphi T$ and $x \in X$ from $(\mu_X \circ \varphi t_1)(\mathcal{L}) \to x$ it follows $\varphi t_2(\mathcal{L}) \to x$.*

(3) *For each $\mathcal{M} \in \varphi X$ and $x \in X$ with $\mathcal{M} \to x$ we have $\mathsf{cl}\mathcal{M} \to x$.*

In the filter case the equivalence of conditions (1) and (3) was already shown by D. C. Kent and G. D. Richardson in [18]. In the general case, the proof of the proposition is an easy consequence of our general definition of the closure operator.

21 Separation axioms

Let (X, T) be a Φ-convergence space.

(X, T) is said to be a *T_0-space* or to fulfill the *axiom* T_0 provided that $\eta_X(x) \to y$ and $\eta_X(y) \to x$ imply $x = y$.

(X, T) is said to be a *T_1-space* or to fulfill the *axiom* T_1 provided that $\eta_X(x) \to y$ implies $x = y$.

(X, T) is called a *T_1^+-space* or is said to fulfill the *axiom* T_1^+ if $\eta_X(x)$ is closed for all $x \in X$.

Moreover, (X, T) is said to be *separated* or Hausdorff or a *T_2-space* provided that $\mathcal{M} \to x$ and $\mathcal{M} \to y$ imply $x = y$.

Finally, (X, T) is said to be *regular* if $\mathcal{M} \to x$ implies $\mathsf{cl}\mathcal{M} \to x$ (that is, condition (3) of Proposition 33 is fulfilled), and to be *normal* provided that $\mathsf{cl}(\mathsf{nb}\mathcal{M}) \leq \mathsf{nb}(\mathsf{cl}\mathcal{M})$ holds for all $\mathcal{M} \in \varphi X$.

In case of a Φ-pretopological space with p the Φ-pretopology, separatedness means that $p(x) \wedge p(y)$ only exists if $x = y$, and regularity means that $p = \mathsf{cl} \circ p$.

Proposition 34 *For topological spaces both the axioms T_1^+ and T_1 coincide with the usual first separation axiom whereas separatedness, regularity and normality coincide with the related classical notions.*

Proof. Only the case of normality is not obvious. Let (X, T) be a topological space. For each filter $\mathcal{M}$ on X we have $\mathsf{cl}(\mathsf{nb}\mathcal{M}) = \{ N \subseteq X \mid \exists M \subseteq X \text{int}M \in \mathcal{M} \text{ and } \text{cl}M \subseteq N \}$ and $\mathsf{nb}(\mathsf{cl}\mathcal{M}) = \{ N \subseteq X \mid \exists M \in \mathcal{M} \ \text{cl}M \subseteq \text{int}N \}$.

The inequality $\mathsf{cl}(\mathsf{nb}\mathcal{M}) \leq \mathsf{nb}(\mathsf{cl}\mathcal{M})$ holds for a filter $\mathcal{M}$ on X if and only if for all $N \subseteq X$ and $M \in \mathcal{M}$ with $\text{cl}M \subseteq \text{int}N$ there is a $K \subseteq X$ with $\text{int}K \in \mathcal{M}$ and $\text{cl}K \subseteq N$. As is easily seen this inequality holds for all filters $\mathcal{M}$ on X if and only if it holds for all principal filters $[M]$ on X, that is, if and only if for all $N \subseteq X$ and $M \subseteq X$ with $\text{cl}M \subseteq \text{int}N$ there is a $K \subseteq X$ with $M \subseteq \text{int}K$ and $\text{cl}K \subseteq N$, hence if and only if (X, T) is normal in the classical sense. □

The following propositions presents some relations between the separation axioms.

Proposition 35 *For each Φ-convergence space (X, T), axiom T_1 implies T_0, axiom T_1^+ implies T_1, and separatedness implies T_1.*

For each Φ-pretopological space (X, T), regularity and T_1 imply separatedness, and normality and T_1^+ imply regularity.

Proof. Clearly, T_1 implies T_0. If T_1^+ is fulfilled and $\eta_X(x) \to y$, then by application of Proposition 31 it follows $\eta_X(y) \leq \mathsf{cl}\eta_X(x) = \eta(x)$, hence $x = y$. If T_2 is fulfilled and $\eta_X(x) \to y$, then because of $\eta_X(x) \to x$ we also have $x = y$.

In the following, let (X, T) be a Φ-pretopological space and p its Φ-pretopology.

Assume at first that (X,T) is a regular T_1-space and $\mathcal{M} \to x$ and $\mathcal{M} \to y$ holds. By applying Proposition 31 it follows $\eta_X(x) \leq \mathsf{cl}\, p(y) = p(y)$, hence $\eta_X(x) \to y$ and because of T_1 therefore $x = y$. Thus, (X,T) is separated.

Now assume that (X,T) is a normal T_1^+-space. We have $p \leq \mathsf{cl} \circ p \leq \mathsf{cl} \circ \mathsf{nb} \circ \eta_X \leq \mathsf{nb} \circ \mathsf{cl} \circ \eta_X = \mathsf{nb} \circ \eta_X = p$ and hence $p = \mathsf{cl} \circ p$. □

References

[1] P. Brock, D. C. Kent, *Approach spaces, limit tower spaces, and probabilistic convergence spaces*, Appl. Categorical Structures **5**(1997), 99–110.

[2] G. Choquet, *Convergence*, Ann. Univ. Grenoble, Sect. Sci., Paris, Ser. A **268**(1969), 1248–1250.

[3] C. H. Cook, H. R. Fischer, *Regular convergence spaces*, Math. Ann. **174**(1967), 1–7.

[4] P. Eklund, W. Gähler, *Fuzzy filter functors and convergence*, Chapter 4 in: S. E. Rodabaugh, E. P. Klement, U. Höhle, eds, Application Of Category To Fuzzy Sets, Theory and Decision Library—Series B: Mathematical and Statistical Methods, Volume **14**, Kluwer Academic Publishers (Boston/Dordrecht/London), 1992, pp. 109–136.

[5] W. Gähler, *Monads and convergence*, Proc. Conference Generalized Functions, Convergence Structures and Their Applications, Dubrovnik 1987, Plenum Press, New York (1988), 29–46.

[6] W. Gähler, *Monadic topology—a new concept of generalized topology*, in: Recent Developments of General Topology and its Applications, International Conference in Memory of Felix Hausdorff (1868–1942), Math. Research **67**(1992), Akademie Verlag Berlin, 118–123.

[7] W. Gähler, *Convergence*, Fuzzy Sets and Systems **73**(1995), 97–129.

[8] W. Gähler, *The general fuzzy filter approach to fuzzy topology, I*, Fuzzy Sets and Systems **76**(1995), 205–224.

[9] W. Gähler, *The general fuzzy filter approach to fuzzy topology, II*, Fuzzy Sets and Systems **76**(1995), 225–246.

[10] W. Gähler, A. S. Abd-Allah, A. Kandil, *Extended fuzzy topologies, Part I, On fuzzy stacks*, J. Fuzzy Mathematics **6**(1998), 223–261.

[11] W. Gähler, A. S. Abd-Allah, A. Kandil, *Extended fuzzy topologies, Part II, Characterizing notions*, J. Fuzzy Mathematics **6**(1998), 539–574.

[12] W. Gähler, A. S. Abd-Allah, A. Kandil, *Extended fuzzy topologies, Part III, Closedness and compactness*, J. Fuzzy Mathematics **6**(1998), 575–608.

[13] W. Gähler, A. S. Abd-Allah, A. Kandil, *On extended fuzzy topologies*, Fuzzy Sets and Systems, in press.

[14] G. Grimeisen, *Gefilterte Summation von Filtern und iterierte Grenzprozesse, I*, Math. Annal. **141**(1960), 318–342.

[15] P. C. Hammer, *Extended topology: set-valued set-functions*, Nieuw Archief voor Wiskunde **(3)**(1962), 55–77.

[16] U. Höhle, *Locales and L-topologies*, in: Mathematik-Arbeitspapiere 48, Univ. Bremen (1997), 223–250.

[17] D. C. Kent, *On convergence groups and convergence uniformities*, Fundamenta Math. **60**(1967), 213–222.

[18] D. C. Kent, G. D. Richardson, *Convergence spaces and diagonal conditions*, Topology and its Applications **70**(1996), 167–174.

[19] H.-J. Kowalsky, *Limesräume und Komplettierung*, Math. Nachr. **12**(1954), 301–340.

[20] E. Lowen, R. Lowen, *A quasi topos containing ONV and ET as full subcategories*, Int. J. Math. Math. Sci. **11**(1988), 417-438.

[21] R. Lowen, *Approach spaces—a common supercategory of OP and ET*, Math. Nachr. **141**(1989), 183–226.

[22] N. Netzer, *Verallgemeinerte topologische Strukturen*, Jahrbuch Überblicke Math. (1978), 87–106.

[23] G. D. Richardson, D. C. Kent, *Probabilistic convergence spaces*, J. Austral. Math. Soc. **61**(1996), 1–21.

[24] A. P. Šostak, *On a fuzzy topological structure*, Suppl. Rend. Circ. Matem. Palermo, Ser. II, **11**(1985), 89–103.

[25] A. P. Šostak, *On compactness and connectedness degrees of fuzzy sets in fuzzy topological spaces*, Proc. Sixth Prague Topological Symposium 1986, Heldermann Verlag Berlin 1988, 519–532.

CHAPTER 3

A Unified Approach To The Concept Of Fuzzy L-Uniform Space [1]

J. GUTIÉRREZ GARCÍA, M. A. DE PRADA VICENTE [2],

AND

A. P. ŠOSTAK

Introduction and preliminaries

The theory of uniform structures is an important area of topology which in a certain sense can be viewed as a bridge linking metrics as well as topological groups with general topological structures. In particular, uniformities form, the widest natural context where such concepts as uniform continuity of functions, completeness and precompactness can be extended from the metric case. Therefore, it is not surprising that the attention of mathematicians interested in fuzzy topology constantly addressed the problem to give an appropriate definition of a uniformity in fuzzy context and to develop the corresponding theory. Already by the late 1970's and early 1080's, this problem was studied (independently at the first stage) by three authors: B. Hutton [21], U. Höhle [11, 12], and R. Lowen [30]. Each of these authors used in the fuzzy context a different aspect of the filter theory of traditional uniformities as a starting point, related in part to the different approaches to traditional unformities as seen in [37, 2] *vis-a-vis* [36, 22]; and consequently, the applied techniques and the obtained results of these authors are essentially different. Therefore it seems natural and urgent to find a common context as broad as necessary for these theories and to develop a general approach containing the previously obtained results as special cases—it was probably S. E. Rodabaugh [31] who first stated this problem explicitly.

[1] An expanded version of this work (including proofs) will be published elsewhere.

[2] Partially supported by UPV127.310-EA018/99.

S.E. Rodabaugh and E.P. Klement (eds.), Topological and Algebraic Structures in Fuzzy Sets, 81–114.
© 2003 *Kluwer Academic Publishers.*

To present such a general approach is the aim of this chapter. Our strategy consists in considering the uniformity as a characteristic morphism on the power set $L^{X \times X}$ of all L-subsets of $X \times X$. As far as the lattice L is concerned, it will be enriched with an additional structure provided by binary operations $*$ and $\otimes$. The first one of these operations is assumed to distribute over arbitrary joins and as a result it gives rise to a Galois connection and the corresponding implication.

Apart from this introduction and preliminaries, which serve for describing and specifying the lattice context in which our work is accomplished, this chapter consists of seven sections. The first three are devoted to the study of different generalizations of the concept of a filter in an L-valued context: prefilters, L-filters, L-filters of ordinary subsets and $\top$-filters and the relations between them. In sections four to seven we introduce the new concept of a uniformity in "fuzzy context", called L-valued Hutton uniformity which is the central object of our work, study basic properties of the corresponding category **HL-UNIF** and establish the place of Hutton, Lowen, and Höhle approaches as special cases of **HL-UNIF**. Moreover, relations between the Hutton, Lowen and Höhle categories are obtained. This is done through the notion of L-uniformity which includes Lowen's and Höhle's uniformities. The notion of Hutton is then extended in such a way as to include L-uniformities. This extension of Hutton, for certain L, provides a fixed-basis unification of these diverse approaches, giving an affirmative answer to a fixed-basis analogue of the variable-basis question stated by S. E. Rodabaugh in [31]. More precisely, we can answer yes to this question: for appropriate L, can one find a common categorical framework which would include both the L-Lowen uniformities and the L-Hutton uniformities?

In what follows, we shall consider a quadruple $(L, \leq, \otimes, *)$, where $(L, \leq)$ is a complete lattice with $\top$ and $\bot$ respectively being the universal upper and lower lower bound and $*$ and $\otimes$ two operations defined on L such that:

(i) $(L, \leq, *)$ is a GL-monoid, that is:

- $(L, *)$ is a commutative semigroup;
- $*$ is distributive over *arbitrary* joins, i.e.

$$\left(\bigvee_{i \in J} \alpha_i\right) * \beta = \bigvee_{i \in J} (\alpha_i * \beta), \qquad \beta * \left(\bigvee_{i \in J} \alpha_i\right) = \bigvee_{i \in J} (\beta * \alpha_i)$$

  for all $\beta \in L$ and all $\{\alpha_i : i \in J\} \subseteq L$;
- $\top$ is the unit element in $(L, *)$;
- $\beta \leq \alpha \implies \exists\, \delta \in L$ such that $\beta = \alpha * \delta$.

(ii) $(L, \leq, \otimes)$ is a *cl*-quasi-monoid, that is:

- $\otimes$ is distributive over *non empty* joins, i.e.

$$\Big(\bigvee_{i\in J} \alpha_i\Big) \otimes \beta = \bigvee_{i\in J} (\alpha_i \otimes \beta), \qquad \beta \otimes \Big(\bigvee_{i\in J} \alpha_i\Big) = \bigvee_{i\in J} (\beta \otimes \alpha_i)$$

  for all $\beta \in L$ and all $\varnothing \neq \{\alpha_i : i \in J\} \subseteq L$.
- $\alpha \leq \alpha \otimes \top, \quad \alpha \leq \top \otimes \alpha, \quad \forall \alpha \in L$

(iii) $*$ is *dominated* by $\otimes$, that is:

- $(\alpha_1 \otimes \beta_1) * (\alpha_2 \otimes \beta_2) \leq (\alpha_1 * \alpha_2) \otimes (\beta_1 * \beta_2)$
  for all $\alpha_1, \alpha_2, \beta_1, \beta_2 \in L$.

The following conditions are also used and explicitly referenced in the text as necessary.

The quadruple $(L, \leq, \otimes, *)$ is said to be *pseudo-bisymmetric* if and only if

$$\begin{aligned}(\alpha_1 * \beta_1) \otimes (\alpha_2 * \beta_2) \;=\; & \Big((\alpha_1 \otimes \alpha_2) * (\beta_1 \otimes \beta_2)\Big) \vee \\ & \vee \Big((\alpha_1 \otimes \bot) * (\beta_1 \otimes \top)\Big) \vee \Big((\bot \otimes \alpha_2) * (\top \otimes \beta_2)\Big)\end{aligned}$$

for all $\alpha_1, \alpha_2 \in L$ and $\beta_1, \beta_2 \in L$.

In any GL-monoid $(L, \leq, *)$, $*$ is distributive over *arbitrary* joins, and hence for each $\alpha \in L$ the mapping $\_ \mapsto \alpha * \_$ has a right adjoint $\_ \mapsto \alpha \rightarrow \_$. The *implication* $\rightarrow$ is then determined by $\alpha \rightarrow \beta = \bigvee \{\lambda \in L : \alpha * \lambda \leq \beta\}$.

We recall that a GL-monoid is a complete MV-algebra if the following is satisfied:

(MV) $\quad (\alpha \rightarrow \bot) \rightarrow \bot = \alpha, \quad$ for each $\alpha \in L$.

If $L = [0,1]$ is equipped with the Łukasiewicz t-norm T_m defined by $T_m(\alpha, \beta) = \max\{\alpha + \beta - 1, 0\}$ for each $\alpha, \beta \in [0,1]$, then $([0,1], \leq, T_m)$ is a complete MV-algebra.

Finally, given a lattice L used in this chapter and a mapping $\varphi : X \mapsto Y$, the Zadeh image and preimage operators

$$\varphi^{\rightarrow} : L^X \mapsto L^Y, \quad \varphi^{\leftarrow} : L^Y \mapsto L^X$$

are given by

$$\varphi^{\rightarrow}(a)(y) = \bigvee_{f(x)=y} a(x), \quad \varphi^{\leftarrow}(b) = b \circ \varphi$$

It is well-known that $\varphi^{\rightarrow} \dashv \varphi^{\leftarrow}$; see [32].

1 L-filters

The notion of filter is the core of a generalized theory of convergence "valid" for all kind of topological spaces. But the importance of filters not only lies in a purely topological context but also plays an important role in the theory of uniform spaces. Therefore, any generalization of uniform structures to a fuzzy context will need the generalization and development of a filter theory.

Since the main objective of this work is to establish a generalized notion of uniformity which includes the most important among the fuzzy extensions of usual uniformities (precise reference to them will be given later) the first three sections are a systematic study of the different notions of filter involved in those extensions as well as the relations between them.

1.1 Definition. (L-filter) Let X be a set. A map $\mathfrak{F} : L^X \mapsto L$ is called an *L-filter* on X if $\mathfrak{F}$ satisfies the following properties:

($\mathfrak{F}$0) $\mathfrak{F}(1_X) = \top, \quad \mathfrak{F}(1_\emptyset) = \bot;$

($\mathfrak{F}$1) if $a_1 \leq a_2$, then $\mathfrak{F}(a_1) \leq \mathfrak{F}(a_2);$

($\mathfrak{F}$2) $\mathfrak{F}(a_1) \otimes \mathfrak{F}(a_2) \leq \mathfrak{F}(a_1 \otimes a_2)$ for all $a_1, a_2 \in L^X$.

1.2 Definitions. An L-filter is said to be *stratified* if:

(S) $\forall \alpha \in L, \quad \forall a \in L^X, \quad \alpha * \mathfrak{F}(a) \leq \mathfrak{F}(\alpha * a)\,.$

Sometimes we shall also need to consider the stronger axiom:

(S*) $\forall \alpha \in L, \quad \forall a \in L^X, \quad \alpha \to \mathfrak{F}(a) = \mathfrak{F}(\alpha \to a)\,.$

Finally, an L-filter is said to be *strongly stratified* if:

(SS) $\forall a \in L^X, \quad \mathfrak{F}(a) = \bigvee_{\alpha \in L} \alpha \otimes \mathfrak{F}\left(1_{[a \geq \alpha]}\right).$

The term strongly stratified is justified in the following:

1.3 Proposition. *If $\mathfrak{F}$ is a strongly stratified L-filter, then it is stratified.*

It will also be useful to consider the following notions:

1.4 Definition. A map $\mathfrak{B} : L^X \mapsto L$ is a *base* for an L-filter $\mathfrak{F} : L^X \mapsto L$ if $\mathfrak{F}(a) = \bigvee \{\mathfrak{B}(b) : b \leq a\}$ for each $a \in L^X$. A map $\mathfrak{B} : L^X \mapsto L$ is an *L-filter base* on X if it satisfies the following properties:

($\mathfrak{B}$0) $\bigvee \{\mathfrak{B}(a) : a \in L^X\} = \top$;

($\mathfrak{B}$1) $\mathfrak{B}(a_1) \otimes \mathfrak{B}(a_2) \leq \bigvee \{\mathfrak{B}(a) : a \leq a_1 \otimes a_2\}$ for all $a_1, a_2 \in L^X$;

($\mathfrak{B}$2) $\mathfrak{B}(1_\varnothing) = \bot$.

If $\mathfrak{B} : L^X \mapsto L$ is an L-filter base on X, the map $\mathfrak{F}_\mathfrak{B} : L^X \mapsto L$ defined by $\mathfrak{F}_\mathfrak{B}(a) = \bigvee \{\mathfrak{B}(b) : b \leq a\}$ for each $a \in L^X$, is an L-filter. The proof is based on the properties of $\otimes$, particularly that $\otimes$ distributes over arbitrary sups, is necessary for establishing ($\mathfrak{F}$2). Conversely, if $\mathfrak{B} : L^X \mapsto L$ is a base for an L-filter $\mathfrak{F}$, then it is an L-filter base.

1.5 Remark. Obviously, in the case $L = \mathbf{2}$ the notions of an L-filter base and an L-filter reduce respectively to the notions of a filter base and a filter on X in the sense of Bourbaki.

1.6 Examples.

(a) For each $x \in X$ the mapping $\mathfrak{F}_x : L^X \mapsto L$ defined for each $a \in L^X$ by $\mathfrak{F}_x(a) = a(x)$ is clearly a stratified L-filter on X satisfying (S*). Moreover, if $\top$ is the unit and $\bot$ the zero element with respect to $\otimes$, then

$$a(x) = \bigvee_{\alpha \in L} \alpha \otimes 1_{[a \geq \alpha]}(x)$$

and so $\mathfrak{F}_x$ is strongly stratified.

(b) The mapping $\mathfrak{F}_0 : L^X \mapsto L$ defined by $\mathfrak{F}_0(a) = \inf_x a(x)$ for each $a \in L^X$ is a stratified L-filter on X and since $\alpha \to \inf_x a(x) = \inf_x (\alpha \to a(x))$, it satisfies (S*). Moreover, if $\top$ is the unit and $\bot$ the zero element with respect to $\otimes$, then $\mathfrak{F}_0$ is strongly stratified.

A partial ordering, $\preceq$, can be introduced on the set of all L-filters on X by

$$\mathfrak{F}_1 \preceq \mathfrak{F}_2 \iff \forall a \in L^X, \quad \mathfrak{F}_1(a) \leq \mathfrak{F}_2(a).$$

Although the infimum of a collection of L-filters, $\{\mathfrak{F}_i\}_{i \in J}$, with respect to $\preceq$ always exists and it is defined for each $a \in L^X$ by $\left[\bigwedge_{i \in J} \mathfrak{F}_i\right](a) = \bigwedge_{i \in J} \mathfrak{F}_i(a)$, the supremum, $\mathfrak{F}_1 \vee \mathfrak{F}_2$, of two L-filters, does not always exist.

We are mainly interested in sufficient conditions for the existence of an upper bound for two given L-filters. The following propositions provide us with this kind of conditions.

1.7 Proposition. *Let $(L, \leq, \otimes, *)$ be pseudo-bisymmetric. Further let $\mathfrak{F}_1$ and $\mathfrak{F}_2$ be two L-filters on X. If $\mathfrak{F}_1(a_1) * \mathfrak{F}_2(a_2) = \bot$ for all $a_1, a_2 \in L^X$ such that $a_1 * a_2 = 1_\varnothing$, then there exists an upper bound for both $\mathfrak{F}_1$ and $\mathfrak{F}_2$. Moreover, if either $\mathfrak{F}_1$ or $\mathfrak{F}_2$ are stratified there exists a stratified L-filter finer than both $\mathfrak{F}_1$ and $\mathfrak{F}_2$.*

1.8 Proposition. *In case $\otimes = *$, we can conclude that given two (stratified) L-filters $\mathfrak{F}_1$ and $\mathfrak{F}_2$ on X, the supremum $\mathfrak{F}_1 \vee \mathfrak{F}_2$ exists if and only if $\mathfrak{F}_1(a_1) * \mathfrak{F}_2(a_2) = \bot$ for all $a_1, a_2 \in L^X$ such that $a_1 * a_2 = 1_\varnothing$. In particular the supremum is the (stratified) L-filter defined for each $a \in L^X$ by*

$$\left[\mathfrak{F}_1 \vee \mathfrak{F}_2\right](a) = \bigvee \left\{\mathfrak{F}_1(a_1) * \mathfrak{F}_2(a_2) \,:\, a_1, a_2 \in L^X \text{ and } a_1 * a_2 \leq a\right\}$$

Historical Comments. The notion of an L-filter appeared for the first time with this formulation in [20, Definition 6.1.4]. Similar notions in slightly different form already appeared in [6], [17], and in [8, Section 2].

A similar result to Proposition 1.8 was proved in a different lattice context in [20, Corollary 6.2.7].

2 L-filters of ordinary subsets

In this section we first present the definition and some related results and then give results showing that there exists a bijection between the collection of all L-filters of ordinary subsets and the collection of all strongly stratified L-filters.

In the following, we shall assume that $(L, \leq, \otimes, *)$ is pseudo-bisymmetric and that $\top$ is the unit and $\bot$ the zero element with respect to $\otimes$. We note that in this case for any $A_1, A_2 \subseteq X$ we have $1_{A_1} \otimes 1_{A_2} = 1_{A_1 \cap A_2}$.

2.1 Definition. **(L-filter of ordinary subsets)** A mapping $\mathfrak{f} : 2^X \mapsto L$ is said to be an *L-filter of ordinary subsets* on X if $\mathfrak{f}$ satisfies:

(f0) $\mathfrak{f}(X) = \top, \quad \mathfrak{f}(\varnothing) = \bot;$

(f1) if $A_1 \subseteq A_2 \subseteq X$, then $\mathfrak{f}(A_1) \leq \mathfrak{f}(A_2)$;

(f2) $\mathfrak{f}(A_1) \otimes \mathfrak{f}(A_2) \leq \mathfrak{f}(A_1 \cap A_2)$ for all $A_1, A_2 \subseteq X$.

If $\mathfrak{f}$ is an L-filter of ordinary subsets, then the value $\mathfrak{f}(A)$ can be interpreted as the *degree* to which A *belongs to* $\mathfrak{f}$.

2.2 Definition. A map $\mathfrak{b} : 2^X \mapsto L$ is a *base* for an L-filter of ordinary subsets $\mathfrak{f} : 2^X \mapsto L$ if $\mathfrak{f}(A) = \bigvee \{\mathfrak{b}(B) : B \subseteq A\}$ for each $A \in 2^X$. A map $\mathfrak{b} : 2^X \mapsto L$ is a *base* of some L-filter of ordinary subsets on X if it satisfies the following properties:

($\mathfrak{b}$0) $\quad \bigvee_{A \subseteq X} \mathfrak{b}(A) = \top;$

($\mathfrak{b}$1) $\quad \mathfrak{b}(A_1) \otimes \mathfrak{b}(A_2) \leq \bigvee \{\mathfrak{b}(B) : B \subseteq A_1 \cap A_2\}$ for all $A_1, A_2 \subseteq X$;

($\mathfrak{b}$2) $\quad \mathfrak{b}(\varnothing) = \bot.$

If $\mathfrak{b} : 2^X \mapsto L$ is a base of an L-filter of ordinary subsets on X, the map $\mathfrak{f}_{\mathfrak{b}} : 2^X \mapsto L$ defined for each $A \in 2^X$ by $\mathfrak{f}_{\mathfrak{b}}(A) = \bigvee \{\mathfrak{b}(B) : B \subseteq A\}$ is an L-filter of ordinary subsets.

2.3 Examples.

(a) Given $x \in X$ the mapping $\mathfrak{f}_x : 2^X \mapsto L$ defined for each $A \subseteq X$ by

$$\mathfrak{f}_x(A) = \begin{cases} \top, & x \in A \\ \bot, & x \notin A \end{cases}$$

is an L-filter of ordinary subsets on X.

(b) The mapping $\mathfrak{f}_0 : 2^X \mapsto L$ defined for each $A \subseteq X$ by

$$\mathfrak{f}_0(A) = \begin{cases} \top, & A = X \\ \bot, & A \neq X \end{cases}$$

is an L-filter of ordinary subsets on X.

A partial ordering, $\preceq$, can be introduced on the set of all L-filters of ordinary subsets on X by

$$\mathfrak{f}_1 \preceq \mathfrak{f}_2 \iff \forall A \subseteq X, \quad \mathfrak{f}_1(A) \leq \mathfrak{f}_2(A)$$

The infimum of two L-filters of ordinary subsets, $\mathfrak{f}_1$ and $\mathfrak{f}_2$, with respect to $\preceq$ always exists and it is defined for each $A \subseteq X$ by $\left[\mathfrak{f}_1 \wedge \mathfrak{f}_2\right](A) = \mathfrak{f}_1(A) \wedge \mathfrak{f}_2(A)$. However, the supremum, $\mathfrak{f}_1 \vee \mathfrak{f}_2$, of two L-filters of ordinary subsets, does not always exist. In particular, if there exist disjoint $A_1, A_2 \subseteq X$ such that $\mathfrak{f}_1(A_1) \otimes \mathfrak{f}_2(A_2) \neq \bot$, then there is no upper bound for $\mathfrak{f}_1$ and $\mathfrak{f}_2$, as ($\mathfrak{f}$2) shows.

2.4 Proposition. *Let* $\mathfrak{f}_1$ *and* $\mathfrak{f}_2$ *be two* L*-filters of ordinary subsets on* X*. If* $\mathfrak{f}_1(A_1) \otimes \mathfrak{f}_2(A_2) = \bot$ *for all disjoint* $A_1, A_2 \subseteq X$*, then the supremum is defined for each* $A \subseteq X$ *by*

$$\left[\mathfrak{f}_1 \vee \mathfrak{f}_2\right](A) = \bigvee \{\mathfrak{f}_1(A_1) \otimes \mathfrak{f}_2(A_2) : A_1 \cap A_2 \subseteq A\}$$

2.5 Comment. The notion of an L-filter of ordinary subsets as a mapping from 2^X into L is a particular case of an (L, M)-filter, that is a mapping from L^X into the (enriched) lattice M satisfying certain properties. Such (L, M)-filters are of particular interest when studying (L, M)-topologies (see [27]).

Relation between L-filters of ordinary subsets and L-filters

We shall now give results showing that there exists a bijection between the collection of all L-filters of ordinary subsets and the collection of all strongly stratified L-filters.

2.6 Theorem. [10]. *Let* $\mathfrak{f}$ *be an* L*-filter of ordinary subsets on* X*. Then the mapping* $\mathfrak{F}^{\mathfrak{f}} : L^X \mapsto L$ *defined for each* $a \in L^X$ *by*

$$\mathfrak{F}^{\mathfrak{f}}(a) = \bigvee_{\alpha \in L} \alpha \otimes \mathfrak{f}([a \geq \alpha])$$

is a strongly stratified L*-filter.*

Conversely, the restriction of any strongly stratified L*-filter to* 2^X *is an an* L*-filter of ordinary subsets.*

Moreover, given an L*-filter of ordinary subsets* $\mathfrak{f}$*, then* $\mathfrak{f}^{\mathfrak{F}^{\mathfrak{f}}} = \mathfrak{f}$*, and given a strongly stratified* L*-filter* $\mathfrak{F}$*, then* $\mathfrak{F}^{\mathfrak{f}^{\mathfrak{F}}} = \mathfrak{F}$*. Consequently, there exists a bijection between the collection of* L*-filters of ordinary subsets on* X *and the collection of strongly stratified* L*-filters on* X*.*

It is clear that the bijection defined in this way is an order isomorphism.

2.7 Remark. For any L-filter of ordinary subsets $\mathfrak{f}$ we have the following equality for each $a \in L^X$:

$$\bigvee_{\alpha \in L} \alpha \wedge \mathfrak{f}([a \geq \alpha]) = \bigvee_{A \subseteq X} \mathfrak{f}(A) \wedge \left(\bigwedge_{x \in A} a(x)\right)$$

Historical Comments. The notion of an L-filter of ordinary subsets, in the case $L = [0, 1]$, was implicitly used and studied for the first time in the context of uniformities in [3] and then explicitly defined and studied in [4] under the name of *generalized filter*.

Later on, in [10], the notion was extended for any enriched *cl*-premonoid $(L, \leq, \otimes, *)$ such that the universal lower bound $\perp$ is the zero element with respect to $\otimes$. In the mentioned paper this notion plays a central role in the subject of the paper, which is the relation between the different filter notions considered in Fuzzy Topology.

Independently, U. Höhle in [18] considers the restriction of L-filters to $\mathcal{P}(X)$ (when the lattice L is completely distributive and $\otimes = \wedge$) and calls them *L-filters of ordinary subsets* which is the terminology used in this work.

3 $\top$-Filters and characteristic values

The first generalization of the notion of filter to an "L–valued" context is that of prefilter [28] or filter in the lattice $[0,1]^X$. $\top$–filters, introduced in [12, Definition 1.5] are a generalization of prefilters and were used by Höhle in the context of probabilistic uniformities.

We present here an alternative but equivalent definition based on the notion of characteristic value, a notion crucial in the rest of this work and used in particular to establish the relation between $\top$-filters and L-filters and L-filters of ordinary subsets.

In this section we shall assume that $(L, \leq, *)$ is a complete MV-algebra.

3.1 Definition. ($\top$-Filter) A $\top$*-filter* is a non-empty subset $\mathbf{F}$ of L^X provided with the following properties:

($\mathbb{F}$1) If $a \in L^X$ and there exists $\varkappa : \mathbf{F} \mapsto L$ such that

(i) $\bigvee \{\varkappa(b) : b \in \mathbf{F}\} = \top$ and
(ii) $b * \varkappa(b) \leq a$ for all $b \in \mathbf{F}$,

then $a \in \mathbf{F}$. ($\varkappa$-condition).

($\mathbb{F}$2) If $a_1, a_2 \in \mathbf{F}$, then $a_1 \otimes a_2 \in \mathbf{F}$.

($\mathbb{F}$3) $\bigvee_{x \in X} a(x) = \top$ for all $a \in \mathbf{F}$.

3.2 Definition. ($\top$-Filter base) A non-empty subset $\mathbf{B} \subseteq L^X$ is called a $\top$*-filter base* if it satisfies the following conditions:

($\mathbb{B}$1) for all $a_1, a_2 \in \mathbf{B}$ there exists $\varkappa : \mathbf{B} \mapsto L$ such that
(i) $\bigvee \{\varkappa(b) : b \in \mathbf{F}b\} = \top$ and
(ii) $b * \varkappa(b) \leq a_1 \otimes a_2$ for all $b \in \mathbf{B}$;

($\mathbb{B}$2) $\bigvee_{x \in X} a(x) = \top$ for all $a \in \mathbf{B}$.

We introduce now important notions of a characteristic set and a characteristic value of a collection of L-sets.

3.3 Definition. (Characteristic value) Let $\mathbf{B} \subseteq L^X$ and $a \in L^X$. The *characteristic set* $\mathcal{C}^a(\mathbf{B})$ of $\mathbf{B}$ with respect to a is:

$$\mathcal{C}^a(\mathbf{B}) \quad = \quad \{\alpha \in L : \forall b \in \mathbf{B} \quad b * (\alpha \to \bot) \nleq a\}$$

In the same way, the *characteristic value* $\mathrm{c}^a(\mathbf{B})$ of $\mathbf{B}$ with respect to a is:

$$\begin{aligned} \mathrm{c}^a(\mathbf{B}) &= \bigwedge (L \setminus \mathcal{C}^a(\mathbf{B})) \\ &= \bigwedge \{\alpha \in L : \exists b \in \mathbf{B} \text{ such that } b \leq (\alpha \to \bot) \to a\} \end{aligned}$$

Obviously, in case $\mathbf{B}$ is an upper set (in particular if $\mathbf{B}$ is a $\top$-filter), then

$$\mathrm{c}^a(\mathbf{B}) \quad = \quad \bigwedge \{\alpha \in L : (\alpha \to \bot) \to a \in \mathbf{B}\}$$

Moreover, if $A \subseteq X$ we write $\mathrm{c}^A(\mathbf{B})$ instead of the more cumbersome $\mathrm{c}^{1_A}(\mathbf{B})$ and in this case we have

$$\begin{aligned} \mathrm{c}^A(\mathbf{B}) &= \bigwedge \{\alpha \in L : (\alpha \to \bot) \to 1_A \in \mathbf{B}\} \\ &= \bigwedge \{\alpha \in L : (\alpha \cdot 1_X) \vee 1_A \in \mathbf{B}\} \end{aligned}$$

where $\alpha \cdot 1_A(x) = \begin{cases} \alpha, & \text{if } x \in A \\ \bot, & \text{if } x \notin A \end{cases}$

3.4 Remark.

(1) In the case $L = \mathbf{2}$ the interpretation of the characteristic value of $\mathbf{B}$ with respect to a reduces to the following

$$\mathrm{c}^a(\mathbf{B}) = \bot \quad \iff \quad \exists b \in \mathbf{B} \text{ such that } b \leq a$$

(2) In the case $(L, \leq, *) = ([0,1], \leq, T_m)$ we have

$$\mathrm{c}^a(\mathbf{B}) \quad = \quad \bigwedge \{\alpha \in [0,1] : \exists b \in \mathbf{B} \text{ such that } b \leq a + \alpha\}$$

and the characteristic value of $\mathbf{B}$ with respect to a tells "how much $\mathbf{B}$ is above a".

We collect here some properties of the characteristic value. In particular we want to emphasize that (CV1) shows the equivalence between the notions of $\varkappa$-condition and the characteristic value of a $\top$-filter base.

3.5 Lemma. (Properties of the characteristic value) *Let* $\mathbf{B}$, $\mathbf{B}_1$, $\mathbf{B}_2 \subseteq L^X$ *and* $a \in L^X$*. Then the following statements hold:*

(CV0) $c^X(\mathbf{B}) = \bot$;

(CV1) $c^a(\mathbf{B}) = \bot$ *if and only if there exists* $\varkappa : \mathbf{B} \mapsto L$ *such that*

(i) $\bigvee \{\varkappa(b) : b \in \mathbf{B}\} = \top$ *and* (ii) $b * \varkappa(b) \leq a$ *for all* $b \in \mathbf{B}$.

(*That is,* $c^a(\mathbf{B}) = \bot$ *if and only if* $\mathbf{B}$ *satisfies the* $\varkappa$*-condition*).

(CV2) $c^{\varnothing}(\mathbf{B}) = \top$ *if and only if* $\bigvee_{x \in X} a(x) = \top$ *for all* $a \in \mathbf{F}$.

(CV3) *If* $a \in \mathbf{B}$, *then* $c^a(\mathbf{B}) = \bot$.

(CV4) *If* $\mathbf{B}_1 \subseteq \mathbf{B}_2$, *then* $c^a(\mathbf{B}_2) \leq c^a(\mathbf{B}_1)$.

As a consequence of (CV1) and (CV2), we have alternative definitions for $\top$-filters and $\top$-filter bases, as the propositions below show:

3.6 Proposition. *A non-empty subset* $\mathbf{B} \subseteq L^X$ *is a* $\top$*-filter base if and only if it satisfies the following conditions:*

($\mathbb{B}$1') *if* $a_1, a_2 \in \mathbf{B}$, *then* $c^{a_1 \otimes a_2}(\mathbf{B}) = \bot$

($\mathbb{B}$2') $c^{\varnothing}(\mathbf{B}) = \top$.

3.7 Proposition. *A* $\top$*-filter is a non-empty subset* $\mathbf{F}$ *of* L^X *provided with the following properties:*

($\mathbb{F}$1') *if* $c^a(\mathbf{F}) = \bot$, *then* $a \in \mathbf{F}$.

($\mathbb{F}$2) *if* $a_1, a_2 \in \mathbf{F}$, *then* $a_1 \otimes a_2 \in \mathbf{F}$.

($\mathbb{F}$3') $c^{\varnothing}(\mathbf{F}) = \top$.

Every $\top$-filter base $\mathbf{B}$ generates a $\top$-filter $\mathbf{F}_\mathbf{B}$ defined by

$$\mathbf{F}_\mathbf{B} = \{a \in L^X : \mathrm{c}^a(\mathbf{B}) = \bot\}$$

Conversely every non-empty subset $\mathbf{B}$ of a $\top$-filter $\mathbf{F}$ equipped with the property:

$$\forall a \in \mathbf{F}, \quad \mathrm{c}^a(\mathbf{B}) = \bot$$

is a $\top$-filter base.

The set of all $\top$-filters on X can be ordered with respect to the inclusion and in this case the infimum of two $\top$-filters, $\mathbf{F}_1$ and $\mathbf{F}_2$, with respect to $\subseteq$ always exists and it is defined by $\mathbf{F}_1 \wedge \mathbf{F}_2 = \mathbf{F}_1 \cap \mathbf{F}_2$. However, the supremum,

$\mathbf{F}_1 \vee \mathbf{F}_2$, of two $\top$-filters, does not always exist. In particular, if there exist $a_1 \in \mathbf{F}_1$ and $a_2 \in \mathbf{F}_2$ such that $\bigvee_{x\in X} [a_1 \otimes a_2](x) \neq \top$, then there is no $\top$-filter $\mathbf{F}$ satisfying $\mathbf{F}_1 \subseteq \mathbf{F}$ and $\mathbf{F}_2 \subseteq \mathbf{F}$.

Once again we are mainly interested in sufficient conditions for the existence of an upper bound for two given $\top$-filters.

3.8 Proposition. *Let $\mathbf{F}_1$ and $\mathbf{F}_2$ be two $\top$-filters on X. If for each $a_1 \in \mathbf{F}_1$ and each $a_2 \in \mathbf{F}_2$ we have $\bigvee_{x\in X} [a_1 * a_2] = \top$, then there exists an upper bound for both $\mathbf{F}_1$ and $\mathbf{F}_2$.*

Relation between $\top$-filters and L-filters

First of all we state some more properties of the characteristic value that will be necessary in subsequent development:

3.9 Lemma. (Properties of the characteristic value) *Let $\alpha \in L$ and $a_1, a_2 \in L^X$. Let $\mathbf{B}$ be a $\top$-filter base and $\mathbf{F_B} = \{a \in L^X : c^a(\mathbf{B}) = \bot\}$. Then the following statements hold:*

(CV5) *if $a_1 \leq a_2$, then $c^{a_2}(\mathbf{B}) \leq c^{a_1}(\mathbf{B})$;*

(CV6) $c^{\alpha * a}(\mathbf{B}) \leq \alpha \rightarrow c^a(\mathbf{B})$;

(CV7) $\alpha * c^a(\mathbf{B}) = c^{\alpha \rightarrow a}(\mathbf{B})$;

(CV8) $c^a(\mathbf{B}) = c^a(\mathbf{F_B})$, *that is, the characteristic value of a $\top$-filter does not depend on the selected base;*

(CV9) $c^{a_1 \otimes a_2}(\mathbf{B}) \leq c^{a_1}(\mathbf{B}) \boxtimes c^{a_2}(\mathbf{B})$, *where* $\alpha \boxtimes \beta = ((\alpha \rightarrow \bot) \otimes (\beta \rightarrow \bot)) \rightarrow \bot$.

The notion of characteristic value is now used to associate L-filters to $\top$-filters. The next theorem establishes such association and characterizes those L-filters which are arising from $\top$-filters.

3.10 Theorem. *Let $\mathbf{B} \subseteq L^X$ be a $\top$-filter base on X. Then the mapping $\mathfrak{F}^{\mathbf{B}} : L^X \mapsto L$ defined for each $a \in L^X$ by*

$$\mathfrak{F}^{\mathbf{B}}(a) = c^a(\mathbf{B}) \rightarrow \bot$$

is an L-filter satisfying the additional axiom (S*). *Conversely, to any L-filter on X satisfying* (S*) *we can associate a $\top$-filter*

$$\mathbf{F}^{\mathfrak{F}} = \{a \in L^X : \mathfrak{F}(a) = \top\}$$

Moreover, given a $\top$*-filter base* $\mathbf{B} \subseteq L^X$, *then* $\mathbf{F}^{\mathfrak{F}^{\mathbf{B}}} = \mathbf{F}_{\mathbf{B}}$ *and given an L-filter* $\mathfrak{F}$ *satisfying* (S*), *then* $\mathfrak{F}^{\mathbf{F}^{\mathfrak{F}}} = \mathfrak{F}$. *Consequently, there exists a bijection between the collection of* $\top$*-filters on* X *and the collection of* L*-filters on* X *satisfying axiom* (S*).

3.11 Remark. It is easy to check that

$$c^a(\mathbf{B}) \to \bot \quad = \quad \bigvee_{b \in \mathbf{B}} \bigwedge_{x \in X} \Big(b(x) \to a(x) \Big)$$

Relation between $\top$-filters and L-filters of ordinary subsets

Our intention now is to relate $\top$-filters and L-filters of ordinary subsets. First of all, as a consequence of Theorem 3.10 one can prove the following two results:

3.12 Theorem. *Let* $\mathbf{B} \subseteq L^X$ *be a* $\top$*-filter base, then the mapping* $\mathfrak{f}^{\mathbf{B}} : 2^X \mapsto L$ *defined for each* $A \subseteq X$ *by*

$$\mathfrak{f}^{\mathbf{B}}(A) \quad = \quad c^A(\mathbf{B}) \to \bot$$

is an L*-filter of ordinary subsets.*

3.13 Remark. Since $\alpha \to 1_A = 1_A \vee (\alpha \to \bot) \cdot 1_X$, it follows that

$$\mathfrak{f}^{\mathbf{B}}(A) \quad = \quad \bigvee \{\alpha \in L : \exists a \in \mathbf{F}b \text{ such that } a \leq 1_A \vee (\alpha \to \bot) \cdot 1_X\}$$

3.14 Corollary. *Let* $\mathbf{B} \subseteq L^X$ *be a* $\top$*-filter base on* X *and* $\mathbf{F_B}$ *the generated* $\top$*-filter. Then for each* $A \subseteq X$ *and* $\alpha \in L$ *the following are equivalent:*

(i) $1_A \vee (\alpha \to \bot) \cdot 1_X \in \mathbf{F_B}$.

(ii) $\mathfrak{f}^{\mathbf{B}}(A) \geq \alpha$.

It can now be proved that under certain restrictions this mapping $\mathbf{B} \mapsto \mathfrak{f}^{\mathbf{B}}$ has an inverse. This is done by first proving that if $\otimes = \wedge$ and $(L, \leq)$ is a lattice in which PRIME(L) is order generating, then any L-filter of ordinary subsets generates a $\top$-filter.

3.15 Theorem. *Let* $(L, \leq, \wedge, *)$ *be such that primes are order generating. For any* L*-filter of ordinary subsets* $\mathfrak{f}$, *the following collection is a* $\top$*-filter:*

$$\mathbf{F}^{\mathfrak{f}} \quad = \quad \Big\{ b \in L^X : \mathfrak{f}([b \not\leq \alpha]) \geq \alpha \to \bot, \quad \forall \alpha \in \text{PRIME}(L) \Big\}$$

3.16 Remark. We can rewrite $\mathbf{F}^{\mathfrak{f}}$ in the following way

$$\mathbf{F}^{\mathfrak{f}} = \Big\{ a \in L^X : \mathfrak{f}(A) \geq \bigwedge_{x \notin A} \big(a(x) \to \bot \big), \quad \forall A \subseteq X \Big\}$$

As a consequence of the previous remark we have the following:

3.17 Corollary. *Let $(L, \leq)$ be such that* PRIME(L) *is order generating. Further, let $\mathfrak{f}$ be an L-filter of ordinary subsets. Then for each $A \subseteq X$ and $\alpha \in L$ the following are equivalent:*

(i) $1_A \vee (\alpha \to \bot) \cdot 1_X \in \mathbf{F}^{\mathfrak{f}}$.

(ii) $\mathfrak{f}(A) \geq \alpha$.

Finally, in the case when $\otimes = \wedge$ and $(L, \leq)$ is completely distributive, we have also the following result:

3.18 Theorem. *Let $(L, \leq, \wedge, *)$ be such that the underlying lattice is completely distributive. Then there exists a bijection from the set of all $\top$-filters on X onto the set of all L-filters of ordinary subsets on X.*

Historical Comments. The first generalization of the notion of filter to an "L-valued" context is that of prefilter [28] or filter in the lattice $[0,1]^X$. $\top$-filters, introduced in [12] are a generalization of prefilters and were used by Höhle in the context of probabilistic uniformities.

In the case $(L, \leq, \otimes, *) = ([0,1], \leq, \wedge, T_m)$ the axioms $(\mathbb{F}1)$ $(\mathbb{F}1')$ and the saturation condition of prefilters (filters in the lattice $[0,1]^X$), considered by R. Lowen in the context of fuzzy neighborhood spaces (cf. [29]) and fuzzy uniform spaces (cf. [30]) coincide. Consequently $\top$-filters in this case are exactly saturated prefilters with characteristic value 1 (because of $(\mathbb{B}2')$).

It was already proved in [4] that in the case of the unit interval, there exists a bijection between saturated prefilters with characteristic value 1 and $[0,1]$-filters of ordinary subsets. This bijection was extended in [18] to a more general case $(L, \leq, \wedge, *)$ when $(L, \leq)$ is completely distributive and $(L, \leq, *)$ is a complete MV-algebra.

Taking into account Remark 3.11 we conclude that the construction considered in Theorem 3.10 coincides with [20, Remark 6.2.3] and [20, Lemma 2.12] (in the particular case $\odot = *$).

Theorem 3.12 and Corollary 3.14 extend [4, Theorem 4.3] and [18, see page 400] since complete distributivity is not needed in the proof and $\otimes$ is not necessarily equal to $\wedge$. We refer here to [4, Lemma 5.3] in the case of the unit interval.

For $(L, \leq)$ completely distributive, Theorem 3.15 has been already proved by Höhle in [18, Theorem 2.11] We want to emphasize here that the condition of PRIME(L) being order generating is sufficient. The first to use lattices with this condition was T. Kubiak [24].

4 L-uniformities

4.1 Definition. (L-uniform space) Let X be a non-empty set and $\mathfrak{U}$ be an L-filter on $X \times X$ in the sense of Definition 1.1. $\mathfrak{U}$ is called an *L-uniformity* on X if it satisfies the following axioms:

$$(\mathfrak{U}3) \quad \mathfrak{U}(d) \leq \bigwedge_{x \in X} d(x,x) \text{ for all } d \in L^{X \times X}.$$

$$(\mathfrak{U}4) \quad \mathfrak{U}(d) \leq \mathfrak{U}(d^{-1}) \text{ for all } d \in L^{X \times X} \quad (\text{where } d^{-1}(x,y) = d(y,x)).$$

$$(\mathfrak{U}5) \quad \mathfrak{U}(d) \leq \bigvee \Big\{\alpha * \mathfrak{U}(d_1) * \mathfrak{U}(d_2) : \\ : d_1, d_2 \in L^{X \times X},\ \alpha \in L \text{ and } \alpha * \Big(d_2 \circ d_1\Big) \leq d\Big\} \text{ for all } d \in L^{X \times X}.$$

We call $(X, \mathfrak{U})$ an *L-uniform space.* An L-uniformity will be said to be stratified (resp. strongly stratified) if it is stratified (resp. strongly stratified) as an L-filter (see Definitions 1.2).

4.2 Remark. One can prove that if $\mathfrak{U}$ is stratified, then $(\mathfrak{U}5)$ is equivalent to

$$(\mathfrak{U}5') \quad \mathfrak{U}(d) \leq \bigvee \{\mathfrak{U}(d_1) * \mathfrak{U}(d_2) : \\ : d_1, d_2 \in L^{X \times X} \text{ and } d_2 \circ d_1 \leq d\} \text{ for all } d \in L^{X \times X}.$$

4.3 Definition. Let $(X, \mathfrak{U})$ and $(Y, \mathfrak{U}')$ be two L-uniform spaces. A mapping $\varphi : X \mapsto Y$ is said to be *L-uniformly continuous* if $\mathfrak{U}((\varphi \times \varphi)^{\leftarrow}(e)) \geq \mathfrak{U}'(e)$ for all $e \in L^{Y \times Y}$.

We shall denote by L-**UNIF** the category corresponding to the above definitions. Note that in case $L = \mathbf{2}$ an L-uniform space is just a diagonal uniform space in the usual sense.

L-topologies associated with L-uniformities

To any stratified L-uniformity an L-topology can be associated as follows:

4.4 Definition. (L-interior operator) [20] A map $\mathcal{K} : L^X \mapsto L^X$ is called an *L-interior operator* on X if $\mathcal{K}$ satisfies the following axioms:

($\mathcal{K}$0) $\mathcal{K}(1_X) = 1_X$;

($\mathcal{K}$1) if $a_1 \le a_2 \in L^X$, then $\mathcal{K}(a_1) \le \mathcal{K}(a_2)$;

($\mathcal{K}$2) $\mathcal{K}(a_1) \otimes \mathcal{K}(a_2) \le \mathcal{K}(a_1 \otimes a_2)$ for all $a_1, a_2 \in L^X$;

($\mathcal{K}$3) $\mathcal{K}(a) \le a$ for all $a \in L^X$;

($\mathcal{K}$4) $\mathcal{K}(a) \le \mathcal{K}(\mathcal{K}(a))$ for all $a \in L^X$.

4.5 Definition. (L-neighborhood system) [20] Let $\mathcal{U} : X \mapsto L^{(L^X)}$ be a map and denote $\mathcal{U}(x) = \mathfrak{N}_x$. $\mathcal{U}$ is said to be an *L-neighborhood system* on X if $\mathfrak{N}_x$ is an L-filter on X for each $x \in X$ and satisfies the following axioms:

($\mathfrak{N}$3) $\mathfrak{N}_x(a) \le a(x)$ for all $a \in L^X$;

($\mathfrak{N}$4) $\mathfrak{N}_x(a) \le \bigvee \{\mathfrak{N}_x(b) : b(y) \le \mathfrak{N}_y(a) \quad \forall y \in X\}$ for all $a \in L^X$.

A set X equipped with an L-neighborhood system $\mathcal{U}$ is called an L-neighborhood space. $\mathcal{U}$ is called (strongly) stratified if $\mathfrak{N}_x$ is a (strongly) stratified L-filter for each $x \in X$.

4.6 Remark. In the presence of the axiom ($\mathfrak{N}$1), axiom ($\mathfrak{N}$4) can obviously be reformulated in the following form:

($\mathfrak{N}$4′) $\mathfrak{N}_x(a) \le \mathfrak{N}_x(\mathfrak{N}_-(a))$ for all $a \in L^X$.

where $\mathfrak{N}_-(a) \in L^X$ is defined by $\left[\mathfrak{N}_-(a)\right](y) = \mathfrak{N}_y(a)$ for each $y \in X$.

4.7 Definition. Let $\varphi : (X,\mathcal{U}_X) \mapsto (Y,\mathcal{U}_Y)$ be a mapping. φ is said to be *L-continuous at* $x \in X$ if $\mathfrak{N}^X_x(\varphi^{\leftarrow}(b)) \ge \mathfrak{N}^Y_{\varphi(x)}(b)$ for all $b \in L^Y$. A map $\varphi : X \mapsto Y$ is *L-continuous on* X if φ is L-continuous at each point $x \in X$. A bijective map $\varphi : X \mapsto Y$ is an *L-homeomorphism* if φ and φ^{-1} are L-continuous. L-neighborhood spaces and L-continuous maps form a category which will be denoted by L-**NS**.

In the following we always identify L-topologies τ with their corresponding L-interior operators. Also, L-interior operators and L-neighborhood systems are equivalent concepts.

Given an L-uniformity $\mathfrak{U}$ we consider the collection $\{\mathfrak{N}_x^{\mathfrak{U}}\}_{x\in X}$ defined for each $a \in L^X$ as follows:

$$\mathfrak{N}_x^{\mathfrak{U}}(a) \quad = \quad \bigvee \{\alpha * \mathfrak{U}(d) \,:\, \alpha \in L,\, d \in L^{X\times X} \text{ and } \alpha * d(x,-) \leq a\}$$

(where $\Big[d(x,-)\Big](y) = d(x,y)$ for each $y \in X$).

The next theorem shows how to each L-uniformity an L-topology can be assigned; moreover this assignment, as shown by Proposition 4.9, is categorical.

4.8 Theorem. *Given an L-uniformity $\mathfrak{U}$ on X, the mapping $\mathfrak{N}_x^{\mathfrak{U}}$ defined by:*

$$\mathfrak{N}_x^{\mathfrak{U}}(a) \quad = \quad \bigvee \{\alpha * \mathfrak{U}(d) \,:\, \alpha \in L,\, d \in L^{X\times X} \text{ and } \alpha * d(x,-) \leq a\}$$

for each $a \in L^X$ and $x \in X$ satisfies:

(i) *If $\mathfrak{U}$ is stratified, then $\{\mathfrak{N}_x^{\mathfrak{U}}\}_{x\in X}$ is a stratified L-neighborhood system.*

(ii) *If $(L, \leq, \otimes, *)$ is pseudo-bisymmetric, then $\{\mathfrak{N}_x^{\mathfrak{U}}\}_{x\in X}$ is a stratified L-neighborhood system.*

4.9 Proposition. *If a mapping $\varphi : (X, \mathfrak{U}) \mapsto (Y, \mathfrak{U}')$ is L-uniformly continuous, then the mapping $\varphi : (X, \mathcal{U}^{\mathfrak{U}}) \mapsto (Y, \mathcal{U}^{\mathfrak{U}'})$ is L-continuous.*

Historical Comments. We restate two longstanding questions from [31]:

Question F: Is there a categorical framework which includes both the category for [0, 1]-Lowen uniformites and the category for variable-basis Hutton quasi-uniformities given in [31]?

Question F′: Let $L = [0,1]$. Is there a common definition satisfied by the objects of Lowen and Hutton uniformities?

We add the following question to this list:

Question F″: For each L of an appropriate category, is there a categorical framework which includes both the category for L-Lowen uniformities and the category for L-Hutton uniformities? This question is approximately the fixed-basis analogue of Question F, and a positive answer would essentially answer the variable-basis Question F for objects, though not necessarily for morphisms.

In 1982, Höhle [12] obtained a significant partial answer to Question F′ (see also [31]), showing that Lowen uniformities and a special subclass of [0, 1]-Hutton uniformities were particular cases of T-uniformities for continuous t-norms min and Łukasiewicz, respectively.

In this section the category L-**UNIF** of L-uniform spaces has been presented which covers Höhle's approach to L-valued uniformities (cf. [12]) as well as Lowen's approach (viewed as generalized uniformities) (cf. [3, 30]), and which, together with the category **H**L-**UNIF** developed in Section 7 below, *answers Question F″ in the affirmative.*

5 L-uniformities of ordinary subsets

In this section we consider the case $(L, \leq, *, *)$ i. e. when $\otimes = *$.

5.1 Definition. **(L-uniformities of ordinary subsets)** Let X be a non-empty set and $\mathfrak{u}$ an L-filter of ordinary subsets on $X \times X$. $\mathfrak{u}$ is called an *L-uniformity of ordinary subsets* if $\mathfrak{u}$ satisfies the following axioms:

(u3) $\mathfrak{u}(A) \neq \bot \implies \Delta \subseteq A$;

(u4) $\mathfrak{u}(A) \leq \mathfrak{u}\left(A^{-1}\right)$, where $A^{-1} = \{(x,y) \in X \times X : (y,x) \in A\}$.

(u5) $\mathfrak{u}(A) \leq \bigvee \{\mathfrak{u}(B_1) \otimes \mathfrak{u}(B_2) : B_1, B_2 \subseteq X \times X \text{ and } B_2 \circ B_1 \subseteq A\}$ (where $B_2 \circ B_1 = \{(x,y) : \exists z \in X \text{ s.t. } (x,y) \in B_1 \text{ and } (z,y) \in B_2\}$).

5.2 Definition. Let $\mathfrak{u}$ and $\mathfrak{u}'$ be two L-uniformities of ordinary subsets on X and Y respectively. A mapping $\varphi : X \mapsto Y$ is said to be *L-uniformly continuous* if $\mathfrak{u}\left((\varphi \times \varphi)^{-1}(B)\right) \geq \mathfrak{u}'(B)$ for all $B \subseteq Y \times Y$.
We shall denote by **OS**L-**UNIF** the corresponding category.

5.3 Remark. Note that if $\otimes = * = \wedge$, axiom (u5) can be rewritten as follows:

(u5′) $\mathfrak{u}(A) \leq \bigvee \{\mathfrak{u}(B) : B \subseteq X \times X \text{ and } B \circ B \subseteq A\}$.

L-topologies associated with L-uniformities of ordinary subsets

For each $A \subseteq X \times X$ and $x \in X$, we denote $A[x] = \{y \in X : (x,y) \in A\}$.

5.4 Theorem. *Given an L-uniformity of ordinary subsets $\mathfrak{u}$, if we define*

$$\begin{aligned} \mathfrak{N}^{\mathfrak{u}}_x(a) &= \bigvee \left\{\alpha \otimes \mathfrak{u}(A) : \alpha \in L,\ A \subseteq X \times X \text{ and } \left[\alpha \cdot 1_A\right](x,-) \leq a\right\} \\ &= \bigvee \{\alpha \otimes \mathfrak{u}(A) : \alpha \in L,\ A \subseteq X \times X \text{ and } A[x] \subseteq [a \geq \alpha]\} \end{aligned}$$

for each $a \in L^X$ and $x \in X$, then $\{\mathfrak{N}^{\mathfrak{u}}_x\}_{x \in X}$ is a strongly stratified L-neighborhood system.

5.5 Proposition. *If a mapping $\varphi : (X, \mathfrak{u}) \mapsto (Y, \mathfrak{u}')$ is L-uniformly continuous, then the mapping $\varphi : (X, \mathcal{U}^{\mathfrak{u}}) \mapsto (Y, \mathcal{U}^{\mathfrak{u}'})$ is L-continuous.*

Relation between L-uniformities of ordinary subsets and L-uniformities

The relation between L-filters of ordinary subsets and strongly stratified L-filters studied in Section 2 can be extended to the context of uniformities.

We first state in the following theorem that there is a bijection between objects in both categories.

5.6 Theorem. *Every L-uniformity of ordinary subsets $\mathfrak{u}$ on X determines a strongly stratified L-uniformity $\mathfrak{U}^{\mathfrak{u}}$ by*

$$\mathfrak{U}^{\mathfrak{u}}(d) \;=\; \bigvee \{\alpha \otimes \mathfrak{u}([d \geq \alpha]) \,:\, \alpha \in L\} \qquad \textit{for all} \quad d \in L^{X\times X}$$

Conversely, if $\mathfrak{U} : L^{X\times X} \mapsto L$ is a strongly stratified L-uniformity, then its restriction $\mathfrak{u}^{\mathfrak{U}} : 2^{X\times X} \mapsto L$ defined by:

$$\mathfrak{u}^{\mathfrak{U}}(A) \;=\; \mathfrak{U}(1_A) \qquad \textit{for all} \quad A \subseteq X \times X$$

is an L-uniformity of ordinary subsets on X. Moreover $\mathfrak{u}^{\mathfrak{U}^{\mathfrak{u}}} = \mathfrak{u}$ and $\mathfrak{U}^{\mathfrak{u}^{\mathfrak{U}}} = \mathfrak{U}$. Consequently, there exists a bijection between the collection of L-uniformities of ordinary subsets and the collection of strongly stratified L-uniformities.

Based on Theorem 5.6, one can prove that the construction $\mathfrak{u} \mapsto \mathfrak{U}^{\mathfrak{u}}$ induces an isomorphism between the category of L-uniformities of ordinary subsets **OSL-UNIF** (cf. [3]) and the category of strongly stratified L-uniformities **SSL-UNIF**.

5.7 Theorem. *Let $\mathfrak{u}$ and $\mathfrak{u}'$ be two L-uniformities of ordinary subsets on X and Y respectively. A mapping $\varphi : (X, \mathfrak{u}) \to (Y, \mathfrak{u}')$ is L-uniformly continuous if and only if $\varphi : (X, \mathfrak{U}^{\mathfrak{u}}) \to (Y, \mathfrak{U}^{\mathfrak{u}'})$ is L-uniformly continuous.*

5.8 Proposition. *Let $\mathfrak{u}$ be an L-uniformity of ordinary subsets on X and $\mathfrak{U}$ be the strongly stratified L-uniformity induced by $\mathfrak{u}$. Further, let $\{\mathfrak{N}_x^{\mathfrak{u}}\}_{x\in X}$ and $\{\mathfrak{N}_x^{\mathfrak{U}}\}_{x\in X}$ be the L-neighborhood systems induced by $\mathfrak{u}$ and $\mathfrak{U}$ respectively. Then $\mathfrak{N}_x^{\mathfrak{u}} = \mathfrak{N}_x^{\mathfrak{U}}$ for each $x \in X$.*

Relation between L-uniformities of ordinary subsets and Lowen fuzzy uniformities

5.9 Definition. (Cf. [30]) Let X be a set, $(L, \leq, \otimes, *) = ([0,1], \leq, \wedge, \wedge)$ and $\mathbf{V}$ be a saturated prefilter on $X \times X$. $\mathbf{V}$ is called a *Lowen fuzzy uniformity* on X if it satisfies the following axioms:

(V3) $d(x,x) = 1$, for all $d \in \mathbf{V}$ and $x \in X$.

(V4) $d^{-1} \in \mathbf{V}$, for all $d \in \mathbf{V}$.

(V5) for all $d \in \mathbf{V}$ and $\epsilon > 0$ there exists $d_\epsilon \in \mathbf{V}$ such that $d_\epsilon \circ d_\epsilon \leq d+\epsilon$. (where $\left[d_\epsilon \circ d_\epsilon\right](x,y) = \bigvee \{d_\epsilon(x,z) \wedge d_\epsilon(z,y) : z \in X\}$).

We call $(X, \mathbf{V})$ a Lowen fuzzy uniform space. Let $(X, \mathbf{V})$ and $(Y, \mathbf{V}')$ be two Lowen fuzzy uniform spaces. A mapping $\varphi : X \mapsto Y$ is said to be *fuzzy uniformly continuous* if $(\varphi \times \varphi)^{\leftarrow}(e) \in \mathbf{V}$ for all $e \in \mathbf{V}'$. We shall denote by **Lo-UNIF** the corresponding category.

As shown in [30], every Lowen fuzzy uniformity $\mathbf{V}$ on X induces a $[0,1]$-closure operator on X defined for each $a \in L^X$ and each $x \in X$ by

$$\overline{a}(x) = \bigwedge_{d \in \mathbf{V}} \left(\bigvee_{y \in X} \left(d(x,y) \wedge a(y) \right) \right)$$

Consequently the corresponding L-neighborhood system $\{\mathfrak{N}_x^{\mathbf{V}}\}_{x \in X}$ on X is defined for each $a \in L^X$ and each $x \in X$ as follows:

$$\mathfrak{N}_x^{\mathbf{V}}(a) = \bigvee_{d \in \mathbf{V}} \left(\bigwedge_{y \in X} \left((1 - d(x,y)) \vee a(y) \right) \right)$$

The relation between $[0,1]$-filters of ordinary subsets and 1-filters (or saturated prefilters) can be extended to the context of uniformities:

5.10 Theorem. *Let* $(L, \leq, \otimes, \otimes) = ([0,1], \leq, \wedge, \wedge)$, *that is,* $[0,1]$ *considered as a Heyting algebra. Given a Lowen fuzzy uniformity* $\mathbf{V}$ *on* X *the mapping* $\mathfrak{u}_{\mathbf{V}} : \mathcal{P}(X \times X) \to [0,1]$ *defined for each* $A \subseteq X \times X$ *by*

$$\mathfrak{u}^{\mathbf{V}}(A) = 1 - c^{1_A}(\mathbf{V}) = \bigvee \{\alpha \in [0,1] : 1_A \vee (1-\alpha) \cdot 1_{X \times X} \in \mathbf{V}\}$$

is a $[0,1]$*-uniformity of ordinary subsets. Conversely, if* $\mathfrak{u}$ *is a* $[0,1]$*-uniformity of ordinary subsets, then the collection*

$$\mathbf{V}^{\mathfrak{u}} = \left\{ d \in [0,1]^{X \times X} : \mathfrak{u}([d > \alpha]) \geq 1 - \alpha \quad \forall \alpha \in [0,1) \right\}$$

is a Lowen fuzzy uniformity. Moreover, given a Lowen fuzzy uniformity $\mathbf{V}$ *on* X, *then* $\mathbf{V}^{\mathfrak{u}^{\mathbf{V}}} = \mathbf{V}$; *and given a* $[0,1]$*-uniformity of ordinary subsets, then* $\mathfrak{u}^{\mathbf{V}^{\mathfrak{u}}} = \mathfrak{u}$.

Besides one can prove that there exists an isomorphism between the category of $[0,1]$-uniform spaces of ordinary subsets **OS**$[0,1]$-**UNIF** and the category of Lowen fuzzy uniformities **Lo-UNIF**.

5.11 Theorem. *Let us consider* $(L, \leq, \otimes, \otimes) = ([0,1], \leq, \wedge, \wedge)$. *Let* (X, u) *and* (Y, u') *be two* $[0,1]$*-uniform spaces of ordinary subsets. A mapping* $\varphi : (X, \mathrm{u}) \to (Y, \mathrm{u}')$ *is uniformly continuous if and only if* $\varphi : (X, \mathbf{V}^{\mathrm{u}}) \to (Y, \mathbf{V}^{\mathrm{u}'})$ *is uniformly continuous.*

Moreover, if $\mathbf{V}^{\mathrm{u}}$ is the Lowen fuzzy uniformity induced by the L-uniformity of ordinary subsets u, then the $[0,1]$-neighborhood systems induced by u and $\mathbf{V}^{\mathrm{u}}$ coincide.

5.12 Proposition. *Let* u *be an* L*-uniformity of ordinary subsets on* X *and* $\mathbf{V}$ *be the Lowen fuzzy uniformity induced by* u. *Further, let* $\{\mathfrak{N}_x^{\mathrm{u}}\}_{x \in X}$ *and* $\{\mathfrak{N}_x^{\mathbf{V}}\}_{x \in X}$ *be the* $[0,1]$*-neighborhood systems induced by* u *and* $\mathbf{V}$ *respectively. Then* $\mathfrak{N}_x^{\mathrm{u}} = \mathfrak{N}_x^{\mathbf{V}}$ *for each* $x \in X$.

Taking into account that in the case $L = [0,1]$ and $\otimes = * = \wedge$, L-uniformities of ordinary subsets and Lowen's uniformities are equivalent concepts (cf. [3]) we can conclude that the category of Lowen uniform spaces is isomorphic to the category **SS**$[0,1]$-**UNIF** of strongly stratified $[0,1]$-uniform spaces.

6 Probabilistic uniformities

In this section we shall assume that $(L, \leq, *)$ is a complete MV-algebra and $(L, \leq, \otimes)$ is a commutative semigroup such that $\top$ acts as unit element.

6.1 Definition. (Probabilistic uniformities) (Cf. [12].) Let X be a set and $\mathbf{V}$ be a $\top$-filter on $X \times X$. $\mathbf{V}$ is called a *probabilistic uniformity* if $\mathbf{V}$ satisfies the following axioms:

(**V3**) $1_\Delta \leq d$ for all $d \in \mathbf{V}$.

(**V4**) $d^{-1} \in \mathbf{V}$ for all $d \in \mathbf{V}$.

(**V5**) $\forall d \in \mathbf{V}, \quad \exists \varkappa : \mathbf{V} \longrightarrow L$ such that:

(i) $\bigvee \left\{ \varkappa\left(\tilde{d}\right) : \tilde{d} \in \mathbf{V} \right\} = \top$, and

(ii) $\left(\tilde{d} \circ \tilde{d}\right) * \varkappa\left(\tilde{d}\right) \leq d$ for all $\tilde{d} \in \mathbf{V}$.

In this context $(X, \mathbf{V})$ is called a *probabilistic uniform space* and every element $d \in \mathbf{V}$ is called an *L-valued vicinity in* X. Let $(X, \mathbf{V})$ and $(Y, \mathbf{V}')$ be two probabilistic uniform spaces. A mapping $\varphi : X \mapsto Y$ is said to be *uniformly continuous* if $(\varphi \times \varphi)^{\leftarrow}(e) \in \mathbf{V}$ for all $e \in \mathbf{V}'$. We shall denote by **Prob-UNIF** the corresponding category.

6.2 Remark. One can prove that the axiom [(**V5**)] can be rewritten as:

(**V5′**) $\quad \forall d \in \mathbf{V} \quad c^d(\mathbf{V} \circ \mathbf{V}) = \bot,$

where $\mathbf{V} \circ \mathbf{V} = \{d_2 \circ d_1 \in L^{X \times X} : d_1, d_2 \in \mathbf{V}\}$.

6.3 Theorem. *Given a probabilistic uniformity* $\mathbf{V}$ *on* X*, if we define*

$$\mathfrak{N}_x^{\mathbf{V}}(a) = \bigvee_{d \in \mathbf{V}} \left(\bigwedge_{y \in X} \left(d(x,y) \to a(y) \right) \right)$$

for each $a \in L^X$ *and* $x \in X$*, then* $\{\mathfrak{N}_x^{\mathbf{V}}\}_{x \in X}$ *is an* L*-neighborhood system satisfying axiom:*

(S*) $\quad \forall \alpha \in L, \quad \forall a \in L^X, \quad \alpha \to \mathfrak{N}_x(a) = \mathfrak{N}_x(\alpha \to a).$

Relation between probabilistic uniformities and L-uniformities

The relation between L-filters and L-filters satisfying axiom (S*) can be extended to the context of uniformities. Here we shall give results showing that the category **Prob-UNIF** is isomorphic to a full subcategory of the category L-**UNIF**.

6.4 Theorem. *Let* $(L, \leq, *)$ *be a complete MV-algebra and* $\mathbf{V} \subseteq L^{X \times X}$ *a probabilistic uniformity on* X*. Then the mapping* $\mathfrak{U}^{\mathbf{V}} : L^{X \times X} \mapsto L$ *defined by*

$$\mathfrak{U}^{\mathbf{V}}(d) = c^d(\mathbf{V}) \to \bot = \bigvee \{\alpha \in L : \alpha \to d \in \mathbf{V}\}$$

for each $d \in L^{X \times X}$*, is an* L*-uniformity satisfying the additional axiom*

(S*) $\quad \forall \alpha \in L, \quad \forall d \in L^{X \times X}, \quad \alpha \to \mathfrak{U}(d) = \mathfrak{U}(\alpha \to d).$

Conversely, with any L*-uniformity on* X *satisfying* (S*) *we can associate a probabilistic uniformity*

$$\mathbf{V}^{\mathfrak{U}} = \{d \in L^{X \times X} : \mathfrak{U}(d) = \top\}$$

Moreover, given a probabilistic uniformity $\mathbf{V}$ *on* X*, then* $\mathbf{V}^{\mathfrak{U}^{\mathbf{V}}} = \mathbf{V}$*; and given an* L*-uniformity* $\mathfrak{U}$ *satisfying* (S*)*, then* $\mathfrak{U}^{\mathbf{V}^{\mathfrak{U}}} = \mathfrak{U}$.

Besides, one can prove that there exists an isomorphism between the two categories.

6.5 Theorem. *Let $(X,\mathbf{V})$ and $(Y,\mathbf{V}')$ be two probabilistic uniform spaces. A mapping $\varphi : (X,\mathbf{V}) \to (Y,\mathbf{V}')$ is uniformly continuous if and only if $\varphi : (X,\mathfrak{U}^{\mathbf{V}}) \to (Y,\mathfrak{U}^{\mathbf{V}'})$ is L-uniformly continuous.*

6.6 Proposition. *Let $(X,\mathbf{V})$ be a probabilistic uniformity and $\mathfrak{U}$ the L-uniformity induced by $\mathbf{V}$. Further, let $\{\mathfrak{N}_x^{\mathbf{V}}\}_{x\in X}$ and $\{\mathfrak{N}_x^{\mathfrak{U}}\}_{x\in X}$ the L-neighborhood systems induced by $\mathbf{V}$ and $\mathfrak{U}$ respectively. Then $\mathfrak{N}_x^{\mathbf{V}} = \mathfrak{N}_x^{\mathfrak{U}}$ for each $x \in X$.*

7 Hutton type uniformities

Join preserving mappings from L^X into L^X

Let $\mathcal{H}(L,X)$ denote the collection of all join preserving mappings ϕ from L^X into L^X. (In particular, since $1_\varnothing$ is the join of the empty set in L^X we have $\phi(1_\varnothing) = 1_\varnothing$).

Given $\phi_1, \phi_2 \in \mathcal{H}(L,X)$ we define $\phi_1 \preceq \phi_2 \iff \phi_1(a) \le \phi_2(a)$ for each $a \in L^X$. $(\mathcal{H}(L,X), \preceq)$ is a complete lattice in which $\phi_\perp$ and $\phi_\top$ are the bottom and the top elements respectively, where $\phi_\perp$ and $\phi_\top$ are defined by $\phi_\perp(a) = 1_\varnothing$ and $\phi_\top(a) = 1_X$ for each $a \in L^X$.

We observe that ϕ_1 and ϕ_2 can be join preserving without $\phi_1 \wedge \phi_2$ being order preserving; hence the infimum in $(\mathcal{H}(L,X), \preceq)$ does not coincide with the infimum in the lattice $\left(L^X\right)^{L^X}$. So for any $\phi_1, \phi_2 \in \mathcal{H}(L,X)$ let $\barwedge$ denote the infimum in $(\mathcal{H}(L,X), \preceq)$ of ϕ_1 and ϕ_2, that is,

$$\phi_1 \barwedge \phi_2 \quad = \quad \bigvee \{\phi \in \mathcal{H}(L,X) \,:\, \phi \preceq \phi_1 \text{ and } \phi \preceq \phi_2\}$$

or equivalently,

$$\phi_1 \barwedge \phi_2 = \bigvee \left\{\phi \in \mathcal{H}(L,X) \,:\, \phi(a_1 \wedge a_2) \le \phi_1(a_1) \wedge \phi_2(a_2),\ \forall a_1, a_2 \in L^X\right\}$$

We want to emphasize here that complete distributivity is not needed in this characterization—a fact previously noted in [33] and generalized in [34] using tensor products. On the other hand, if the lattice is completely distributive, Hutton proved (see [21]) that for each $a \in L^X$,

$$\left[\phi_1 \barwedge \phi_2\right](a) \quad = \quad \bigwedge \left\{\phi_1(a_1) \vee \phi_2(a_2) \,:\, a_1, a_2 \in L^X \text{ and } a_1 \vee a_2 = a\right\}$$

Consider a binary operation $\overline{\otimes}$ on $\mathcal{H}(L,X)$ induced by $\otimes$ and defined for all $\phi_1, \phi_2 \in \mathcal{H}(L,X)$ by

$$\phi_1 \overline{\otimes} \phi_2 = \bigvee \left\{\phi \in \mathcal{H}(L,X) \,:\, \phi(a_1 \otimes a_2) \le \phi_1(a_1) \otimes \phi_2(a_2) \quad \forall a_1, a_2 \in L^X\right\}$$

Given $\phi \in \mathcal{H}(L,X)$, let $\phi^* : L^X \mapsto L^X$ be the upper adjoint mapping of ϕ:

$$\phi^*(a) = \bigvee \left\{ b \in L^X : \phi(b) \leq a \right\} \text{ for each } a \in L^X$$

Then ϕ^* is meet preserving and so the mapping $\phi^{-1} : L^X \mapsto L^X$ defined by

$$\phi^{-1}(a) = \phi^*(a \to \bot) \to \bot = \bigwedge \left\{ b \in L^X : \phi(b \to \bot) \leq a \to \bot \right\}$$

for each $a \in L^X$, is join preserving, that is, ϕ^{-1} also belongs to $\mathcal{H}(L,X)$. We shall call ϕ^{-1} the *symmetric* of ϕ; moreover $\left(\phi^{-1}\right)^{-1} = \phi$.

Also it is easy to check that $\phi_1 \preceq \phi_2 \iff \phi_2^* \preceq \phi_1^* \iff \phi_1^{-1} \preceq \phi_2^{-1}$.

7.1 Remark. A join preserving mapping ϕ is completely determined by the collection of L-sets

$$\left\{ \phi\left(\alpha \cdot 1_{\{x\}}\right) : \alpha \in L, x \in X \right\}$$

Relation between $\mathcal{H}(L,X)$ and $L^{X\times X}$

We define a mapping $\Lambda : L^{X\times X} \longrightarrow \mathcal{H}(L,X)$ by

$$\left[\Lambda(d)\right](a) = \bigvee_{x\in X} a(x) * d(x,-) \text{ for each } d \in L^{X\times X} \text{ and each } a \in L^X$$

It can be easily seen that Λ is well-defined, injective, and preserves arbitrary sups. Consequently, Λ has an upper adjoint mapping Υ from $\mathcal{H}(L,X)$ into $L^{X\times X}$. Explicitly, Υ is given by

$$\left[\Upsilon(\phi)\right](x,y) = \bigwedge_{\alpha\in L} \alpha \to \left[\phi\left(\alpha \cdot 1_{\{x\}}\right)\right](y) \text{ for each } x, y \in X$$

As a consequence of (Λ, Υ) being a Galois connection we have

$$\Upsilon \circ \Lambda \geq 1_{L^{X\times X}} \quad \text{and} \quad \Lambda \circ \Upsilon \leq 1_{\mathcal{H}(L,X)}$$

and since Λ is injective we also have $\Upsilon \circ \Lambda = 1_{L^{X\times X}}$ (See e.g. [7]). Moreover, we have the following:

7.2 Proposition. *Let $\phi \in \mathcal{H}(L,X)$. Then $\phi \in \Lambda\left(L^{X\times X}\right)$ iff*

$(\bigstar)$ $\quad \phi\left(\alpha \cdot 1_{\{x\}}\right) = \alpha * \phi\left(1_{\{x\}}\right)$ *for each $\alpha \in L$ and $x \in X$*

As a consequence of the previous proposition we can conclude that L-valued binary relations $\phi \in L^{X\times X}$ and join preserving mappings $\phi : L^X \mapsto L^X$ satisfying condition $(\bigstar)$ are the same thing.

Historical Comments. We observe that this natural generalization already appeared in [11, Remark 2.2 (b)] and also in [13, Proposition 4.2] in the case $(L, \leq, *) = ([0,1], \leq, T_m)$.

This result was already used in [11] and in [12] to conclude that probabilistic uniformities are a particular type of Hutton uniformities.

The following propositions contain the most important properties of the mappings Λ and Υ.

7.3 Proposition. (Properties of Λ). *Let $d, d_1, d_2 \in L^{X\times X}$ and $\alpha \in L$. Then*

(1) $\Lambda(1_\varnothing) = \phi_\perp$ *and* $\Lambda(1_{X\times X}) = \phi_\top^\vee$;

(2) $\Lambda(d_1 \otimes d_2) \preceq \Lambda(d_1)\overline{\otimes}\Lambda(d_2)$;

(3) $\Lambda(1_\Delta) = 1_{L^X}$;

(4) $\left[\Lambda(d)\right]^{-1} = \Lambda(d^{-1})$;

(5) $\Lambda(d_2 \circ d_1) = \Lambda(d_2) \circ \Lambda(d_1)$;

(6) $\Lambda(\alpha * d) = \alpha * \Lambda(d)$.

7.4 Proposition. (Properties of Υ). *Let $\phi, \phi_1, \phi_2 \in \mathcal{H}(L, X)$ and $\alpha \in L$. Then*

(1) $\Upsilon(\phi_\top) = 1_{X\times X}$ *and* $\Upsilon(\phi_\perp) = 1_\varnothing$;

(2) $\Upsilon(\phi_1) \otimes \Upsilon(\phi_2) \leq \Upsilon(\phi_1 \overline{\otimes} \phi_2)$;

(3) $\Upsilon(1_{L^X}) = 1_\Delta$;

(4) $\left[\Upsilon(\phi)\right]^{-1} = \Upsilon(\phi^{-1})$;

(5) $\Upsilon(\phi_2) \circ \Upsilon(\phi_1) \leq \Upsilon(\phi_2 \circ \phi_1)$;

(6) $\alpha * \Upsilon(\phi) \leq \Upsilon(\alpha * \phi)$.

In both (2) and (5), for $\phi_1, \phi_2 \in \Lambda\left(L^{X\times X}\right)$, we have equalities.

Image and preimage operators between $\mathcal{H}(L,X)$ and $\mathcal{H}(L,Y)$

While Zadeh powerset operators suffice for our work with Lowen uniformities and Höhle's probabilistic uniformities above, our work with Hutton type uniformities below requires different powerset operators.

Following [21, 31, 32, 33, 38], we define for fixed L the image and preimage operators

$$\varphi^{\Rightarrow} : \left(L^X\right)^{\left(L^X\right)} \mapsto \left(L^Y\right)^{\left(L^Y\right)}, \quad \varphi^{\Leftarrow} : \left(L^Y\right)^{\left(L^Y\right)} \mapsto \left(L^X\right)^{\left(L^X\right)}$$

of a mapping $\varphi : X \mapsto Y$ defined for each $\phi \in \left(L^X\right)^{\left(L^X\right)}$, $\psi \in \left(L^Y\right)^{\left(L^Y\right)}$ by

$$\varphi^{\Rightarrow}(\phi) = \varphi^{\rightarrow} \circ \phi \circ \varphi^{\leftarrow}, \quad \varphi^{\Leftarrow}(\psi) = \varphi^{\leftarrow} \circ \psi \circ \varphi^{\rightarrow}$$

where $\varphi^{\rightarrow} : L^X \mapsto L^Y$ and $\varphi^{\leftarrow} : L^Y \mapsto L^X$ are the usual Zadeh powerset operators.

The following proposition [33] is a consequence of the adjunction $\varphi^{\rightarrow} \dashv \varphi^{\leftarrow}$.

7.5 Proposition. *The pair $(\varphi^{\Rightarrow}, \varphi^{\Leftarrow})$ is a Galois connection; i.e., $\varphi^{\Rightarrow} \dashv \varphi^{\Leftarrow}$.*

Hutton uniformities

In the original definition of Hutton, the involved lattice was a completely distributive one with an order reversing involution; $(L, \leq, ')$. We shall consider here the following definition which is slightly more general and in which complete distributivity is not needed (cf. [33, 34]).

7.6 Definition. (Hutton uniformities). (See [21]). A non-empty subset $\mathbf{D}$ of $\mathcal{H}(L,X)$ is called a *Hutton quasi-uniformity on* X if $\mathbf{D}$ satisfies the following conditions:

(D1) if $\phi_1 \in \mathbf{D}$, $\phi_2 \in \mathcal{H}(L,X)$, and $\phi_1 \leq \phi_2$, then $\phi_2 \in \mathbf{D}$;

(D2) if $\phi_1, \phi_2 \in \mathbf{D}$, then there exists $\phi \in \mathcal{H}(L,X)$ such that $\phi \preceq \phi_1$ and $\phi \preceq \phi_2$;

(D3) $\mathrm{Id} \preceq \phi$ for each $\phi \in \mathbf{D}$, i.e. $a \leq \phi(a)$ for each $a \in L^X$;

(D5) if $\phi \in \mathbf{D}$, then there exists $\varkappa : \mathbf{D} \longrightarrow L$ such that
(i) $\bigvee \left\{ \varkappa\left(\widetilde{\phi}\right) : \widetilde{\phi} \in \mathbf{D} \right\} = \top$ and
(ii) $\left(\widetilde{\phi} \circ \widetilde{\phi}\right) * \varkappa\left(\widetilde{\phi}\right) \leq \phi$ for all $\widetilde{\phi} \in \mathbf{D}$.

A Hutton quasi-uniformity is called a *Hutton uniformity* if:

(**D4**) $\phi \in \mathbf{D}$ implies $\phi^{-1} \in \mathbf{D}$.

The pair $(X, \mathbf{D})$ is called a Hutton (quasi-)uniform space.

7.7 Remark. Note that axioms (**D1**) and (**D2**) just say that $\mathbf{D}$ is a filter in the lattice $\mathcal{H}(L, X)$. The axiom (**D5**) is weaker than the original axiom (**D5***):

(**D5***) if $\phi \in \mathbf{D}$, then there exist $\psi_1, \psi_2 \in \mathbf{D}$ such that $\psi_2 \circ \psi_1 \preceq \phi$.

7.8 Definition. (Cf. [21].) Let $(X, \mathbf{D})$ and $(Y, \mathbf{D}')$ be two Hutton (quasi-)uniform spaces. A mapping $\varphi : X \mapsto Y$ is said to be *uniformly continuous* if and only if for every $\psi \in \mathbf{D}'$ there exists $\phi \in \mathbf{D}$ such that $\phi \preceq \varphi^{\Leftarrow}(\psi)$. Explicitly, this means that $\phi(a) \leq \varphi^{\leftarrow}(\psi(\varphi^{\rightarrow}(a)))$ for each $a \in L^X$.

We will denote by **H-UNIF** the corresponding category.

7.9 Comment. Given $\varphi : X \mapsto Y$ and $\psi \in \mathcal{H}(L, Y)$ it follows that $\varphi^{\Leftarrow}(\psi)$ belongs to $\mathcal{H}(L, X)$. Consequently, because of (**D1**) it follows that a mapping $\varphi : X \mapsto Y$ is uniformly continuous iff for every $\psi \in \mathbf{D}'$, $\varphi^{\Leftarrow}(\psi) \in \mathbf{D}$.

We can define now the notion of the characteristic set and the characteristic value in this situation as follows:

7.10 Definition. (Characteristic value) Let $\mathbf{D} \subseteq \mathcal{H}(L, X)$ be a family of join preserving mappings from L^X to L^X and $\phi \in \mathcal{H}(L, X)$. We define the *characteristic set* of $\mathbf{D}$ with respect to ϕ, and denote $\mathcal{C}^{\phi}(\mathbf{D})$, as follows:

$$\mathcal{C}^{\phi}(\mathbf{D}) = \{\alpha \in L : \forall \psi \in \mathbf{D} \quad (\alpha \to \bot) * \psi \not\preceq \phi\}$$

and the *characteristic value* of $\mathbf{D}$ with respect to ϕ as follows

$$\mathrm{c}^{\phi}(\mathbf{D}) = \bigwedge [L \setminus \mathcal{C}^{\phi}(\mathbf{D})] = \bigwedge \{\alpha \in L : \exists \psi \in \mathbf{D} \text{ such that } \psi \preceq (\alpha \to \bot) \to \phi\}$$

Obviously, in case $\mathbf{D}$ is an upper set (in particular if $\mathbf{D}$ is a filter), then

$$\mathrm{c}^{\phi}(\mathbf{D}) = \bigwedge \{\alpha \in L : (\alpha \to \bot) \to \phi \in \mathbf{D}\}$$

7.11 Lemma. (Properties of the characteristic value) *Let $(L, \leq, *)$ be a complete MV-algebra, $\mathbf{D}$ a filter in $\mathcal{H}(L, X)$, and $\phi, \phi_1, \phi_2 \in \mathcal{H}(L, X)$. Then:*

(CV0) $c^{\phi_\alpha}(\mathbf{D}) = \alpha \to \bot$ *for each* $\alpha \in L$;

(CV1) *if* $\phi_1 \preceq \phi_2$, *then* $c^{\phi_2}(\mathbf{D}) \leq c^{\phi_1}(\mathbf{D})$;

(CV2) $c^{\phi_1 \bar{\wedge} \phi_2}(\mathbf{D}) \leq c^{\phi_1}(\mathbf{D}) \vee c^{\phi_2}(\mathbf{D})$;

(CV3) $c^{\phi}(\mathbf{D}) \geq \bigvee \left\{\alpha * \left(\left[\phi\left(\alpha \cdot 1_{\{x\}}\right)\right](x) \to \bot\right) : \alpha \in L, x \in X\right\}$;

(CV4) $c^{\phi}(\mathbf{D}) \leq c^{\phi^{-1}}(\mathbf{D})$;

(CV5) $c^{\phi}(\mathbf{D}) \geq \bigwedge \left\{(c^{\phi_2}(\mathbf{D}) \to \bot) \to c^{\phi_1}(\mathbf{D}) : \phi_2 \circ \phi_1 \leq \phi\right\}$;

(CV6) $\alpha * c^{\phi}(\mathbf{D}) = c^{\alpha \to \phi}(\mathbf{D})$ *for each* $\alpha \in L$.

Now, if we define a mapping $\mathfrak{D}^{\mathbf{D}} : \mathcal{H}(L,X) \mapsto L$ for each $\phi \in \mathcal{H}(L,X)$ by $\mathfrak{D}^{\mathbf{D}}(\phi) = \mathrm{c}^{\phi}(\mathbf{D}) \to \bot$, then $\mathfrak{D}^{\mathbf{D}}$ satisfies the following:

($\mathfrak{D}$0) $\mathfrak{D}^{\mathbf{D}}(\phi_\alpha) = \alpha$ for each $\alpha \in L$;

($\mathfrak{D}$1) if $\phi_1 \preceq \phi_2$, then $\mathfrak{D}^{\mathbf{D}}(\phi_1) \leq \mathfrak{D}^{\mathbf{D}}(\phi_2)$;

($\mathfrak{D}$2) $\mathfrak{D}^{\mathbf{D}}(\phi_1) \wedge \mathfrak{D}^{\mathbf{D}}(\phi_2) \leq \mathfrak{D}^{\mathbf{D}}(\phi_1 \bar{\wedge} \phi_2)$;

($\mathfrak{D}$3) $\mathfrak{D}^{\mathbf{D}}(\phi) \leq \bigwedge \left\{\alpha \to \left[\phi\left(\alpha \cdot 1_{\{x\}}\right)\right](x) : \alpha \in L, x \in X\right\}$;

($\mathfrak{D}$4) $\mathfrak{D}^{\mathbf{D}}(\phi) \leq \mathfrak{D}^{\mathbf{D}}(\phi^{-1})$;

($\mathfrak{D}$5) $\mathfrak{D}^{\mathbf{D}}(\phi) \leq \bigvee \left\{\alpha * \mathfrak{D}^{\mathbf{D}}(\phi_2) * \mathfrak{D}^{\mathbf{D}}(\phi_1) : \alpha * (\phi_2 \circ \phi_1) \leq \phi\right\}$.

L-valued Hutton uniformities

7.12 Definition. **(L-valued Hutton uniformities)** Let X be a non-empty set and $\mathfrak{D}$ be a mapping $\mathfrak{D} : \mathcal{H}(L,X) \mapsto L$. $\mathfrak{D}$ is called a *L-valued Hutton uniformity* on X if it satisfies the following axioms:

($\mathfrak{D}$0) $\mathfrak{D}(\phi_\top) = \top$ and $\mathfrak{D}(\phi_\bot) = \bot$;

($\mathfrak{D}$1) if $\phi_1 \preceq \phi_2$, then $\mathfrak{D}(\phi_1) \leq \mathfrak{D}(\phi_2)$;

($\mathfrak{D}$2) $\mathfrak{D}(\phi_1) \otimes \mathfrak{D}(\phi_2) \leq \mathfrak{D}(\phi_1 \overline{\otimes} \phi_2)$ for all $\phi_1, \phi_2 \in \mathcal{H}(L,X)$;

($\mathfrak{D}$3) $\mathfrak{D}(\phi) \leq \bigwedge \left\{\alpha \to \left[\phi\left(\alpha \cdot 1_{\{x\}}\right)\right](x) : \alpha \in L, x \in X\right\}$ for all $\phi \in \mathcal{H}(L,X)$;

($\mathfrak{D}$4) $\mathfrak{D}(\phi) \leq \mathfrak{D}(\phi^{-1})$ for all $\phi \in \mathcal{H}(L,X)$;

($\mathfrak{D}$5) $\mathfrak{D}(\phi) \leq \bigvee \{\alpha * \mathfrak{D}(\phi_1) * \mathfrak{D}(\phi_2) : \alpha \in L,\ \phi_1, \phi_2 \in \mathcal{H}(L,X)$ and $\alpha * (\phi_2 \circ \phi_1) \preceq \phi\}$ for all $\phi \in \mathcal{H}(L,X)$.

$(X, \mathfrak{D})$ is called an *L-valued Hutton uniform space.*

7.13 Definition. Let $(X,\mathfrak{D})$ and $(Y,\mathfrak{D}')$ be two L-valued Hutton uniform spaces. A mapping $\varphi : X \mapsto Y$ is said to be *L-uniformly continuous* if $\mathfrak{D}\left(\varphi^{\Leftarrow}(\psi)\right) \geq \mathfrak{D}'(\psi)$ for all $\psi \in \mathcal{H}(L,Y)$.

We will denote by **HL-UNIF** the corresponding category.

L-topologies associated to L-valued Hutton uniformities

An l-valued Hutton uniformity $\mathfrak{D}$ induces an L-topology. In order to show it we first define a mapping $\mathcal{K}^{\mathfrak{D}} : L^X \to L^X$ as follows:

$$\mathcal{K}^{\mathfrak{D}}(a) = \bigvee \{\alpha \cdot 1_{\{x\}} * \beta * \mathfrak{D}(\phi) \,:\, \alpha,\beta \in L,\, x \in X,\, \phi \in \mathcal{H}(L,X) \text{ and } \beta * \phi\left(\alpha \cdot 1_{\{x\}}\right) \leq a\}$$

7.14 Comment. The previous definition is just an L-valued interpretation of the following (see [21]):

$$\mathcal{K}^{\mathbf{D}}(a) = \bigvee \{b \in L^X \,:\, \exists \phi \in \mathbf{D} \text{ such that } \phi(b) \leq a\}$$

which is equivalent to

$$\mathcal{K}^{\mathbf{D}}(a) = \bigvee \{\alpha \cdot 1_{\{x\}} \in L^X \,:\, \exists \phi \in \mathbf{D} \text{ such that } \phi\left(\alpha \cdot 1_{\{x\}}\right) \leq a\}$$

7.15 Theorem. *Let $(L,\leq,\otimes,*)$ be pseudo-bisymmetric. Given an l-valued Hutton uniformity $\mathfrak{D}$ on X, $a \in L^X$ and $x \in X$, the mapping:*

$$\mathfrak{N}_x^{\mathfrak{D}}(a) = \bigvee \{\alpha * \beta * \mathfrak{D}(\phi) : \alpha,\beta \in L,\, \phi \in \mathcal{H}(L,X) \text{ and } \beta * \phi\left(\alpha \cdot 1_{\{x\}}\right) \leq a\}$$

is a stratified L-neighborhood system.

7.16 Proposition. *Let $(L,\leq,\otimes,*)$ be pseudo-bisymmetric. If a mapping $\varphi : (X,\mathfrak{D}) \mapsto (Y,\mathfrak{D}')$ is L-uniformly continuous, then the mapping $\varphi : (X,\mathcal{U}^{\mathfrak{D}}) \mapsto (Y,\mathcal{U}^{\mathfrak{D}'})$ is L-continuous.*

Relation between Hutton uniformities and L-valued Hutton uniformities

We shall now give results showing that the category **H-UNIF** is isomorphic to a full subcategory of the category **HL-UNIF** when $(L,\leq,*)$ is a complete MV-algebra.

7.17 Theorem. *Let* $\mathbf{D} \subseteq \mathcal{H}(L,X)$ *be a Hutton uniformity on* X. *Then the mapping* $\mathfrak{D}^{\mathbf{D}} : \mathcal{H}(L,X) \mapsto L$ *defined for each* $\phi \in \mathcal{H}(L,X)$ *by*

$$\mathfrak{D}^{\mathbf{D}}(\phi) \quad = \quad c^{\phi}(\mathbf{D}) \to \bot \quad = \quad \bigvee \{\alpha \in L : \alpha \to \phi \in \mathbf{D}\}$$

is an l-valued Hutton uniformity satisfying the additional axiom

$$(\mathrm{S}^*) \qquad \forall \alpha \in L, \quad \forall \phi \in \mathcal{H}(L,X), \quad \alpha \to \mathfrak{D}(\phi) \;=\; \mathfrak{D}(\alpha \to \phi).$$

Conversely, to any L-valued Hutton uniformity on X *satisfying* (S^*) *we can associate a Hutton uniformity*

$$\mathbf{D}^{\mathfrak{D}} \quad = \quad \{\phi \in \mathcal{H}(L,X) : \mathfrak{D}(\phi) = \top\}$$

Moreover, $\mathbf{D}^{\mathfrak{D}^{\mathbf{D}}} = \mathbf{D}$; *and if* $\mathfrak{D}$ *satisfies* (S^*), *then* $\mathfrak{D}^{\mathbf{D}^{\mathfrak{D}}} = \mathfrak{D}$.

One can further prove that there exists an isomorphism between the two categories.

7.18 Theorem. *Let* $(X,\mathbf{D})$ *and* $(Y,\mathbf{D}')$ *be two Hutton uniform spaces. A mapping* $\varphi : (X,\mathbf{D}) \to (Y,\mathbf{D}')$ *is uniformly continuous if and only if* $\varphi : (X,\mathfrak{D}^{\mathbf{D}}) \to (Y,\mathfrak{D}^{\mathbf{D}'})$ *is L-uniformly continuous.*

Moreover, if $\mathfrak{D}$ is the L-valued Hutton uniformity induced by the Hutton uniformity $\mathbf{D}$, then the L-neighborhood systems induced by $\mathbf{D}$ and $\mathfrak{D}$ coincide.

7.19 Proposition. *Let* $(X,\mathbf{D})$ *be a Hutton uniformity and* $\mathfrak{D}$ *the the L-valued Hutton uniformity induced by* $\mathbf{D}$. *Further, let* $\{\mathfrak{N}_x^{\mathbf{D}}\}_{x\in X}$ *and* $\{\mathfrak{N}_x^{\mathfrak{D}}\}_{x\in X}$ *the L-neighborhood systems induced by* $\mathbf{D}$ *and* $\mathfrak{D}$ *respectively. Then* $\mathfrak{N}_x^{\mathbf{D}} = \mathfrak{N}_x^{\mathfrak{D}}$ *for each* $x \in X$.

Relation between L-valued Hutton uniformities and L-uniformities

We conclude this section by giving results showing that the category L-**UNIF** is isomorphic to a full subcategory of **H**L-**UNIF**.

We first establish in the following two propositions the equivalence between objects in both categories.

7.20 Theorem. *Let* $\mathfrak{U}$ *be an L-uniformity on* X. *Then the mapping* $\mathfrak{D}^{\mathfrak{U}} : \mathcal{H}(L,X) \mapsto L$ *defined by* $\mathfrak{D}^{\mathfrak{U}} = (\Upsilon)^{\leftarrow}(\mathfrak{U}) = \mathfrak{U} \circ \Upsilon$ *is an l-valued Hutton uniformity, which satisfies the additional axiom*

$$(\bigstar) \qquad \mathfrak{D}\left(\Lambda\left(\Upsilon\left(\phi\right)\right)\right) = \mathfrak{D}\left(\phi\right) \text{ for each } \phi \in \mathcal{H}(L,X).$$

7.21 Theorem. *Let $\mathfrak{D}$ be an l-valued Hutton uniformity on X satisfying axiom ($\bigstar$). Then the mapping $\mathfrak{U}^{\mathfrak{D}} : L^{X\times X} \mapsto L$ defined by $\mathfrak{U}^{\mathfrak{D}} = \Lambda^{\leftarrow}(\mathfrak{D}) = \mathfrak{D} \circ \Lambda$ is an L-uniformity.*

There exists an isomorphism between the category L-**UNIF** and the category of L-valued Hutton uniform spaces satisfying axiom ($\bigstar$).

7.22 Theorem. *Let $(X,\mathfrak{U})$ and $(Y,\mathfrak{U}')$ be two L-uniform spaces. A mapping $\varphi : (X,\mathfrak{U}) \to (Y,\mathfrak{U}')$ is L-uniformly continuous iff $\varphi : (X,\mathfrak{D}^{\mathfrak{U}}) \to (Y,\mathfrak{D}^{\mathfrak{U}'})$ is L-uniformly continuous.*

7.23 Proposition. *Let $(X,\mathfrak{U})$ be an L-uniformity and $\mathfrak{D}$ the L-valued Hutton uniformity induced by $\mathfrak{U}$. Further, let $\{\mathfrak{N}_x^{\mathfrak{U}}\}_{x\in X}$ and $\{\mathfrak{N}_x^{\mathfrak{D}}\}_{x\in X}$ be the L-neighborhood systems induced by $\mathfrak{U}$ and $\mathfrak{D}$ respectively. Then $\mathfrak{N}_x^{\mathfrak{U}} = \mathfrak{N}_x^{\mathfrak{D}}$ for each $x \in X$.*

Final Remark. It would be of interest to answer the following questions:

Q1. Can the solution of this chapter to Question F″ as stated at the end of Section 4 be given a variable-basis treatment which also answers the original Question F of [31] in the affirmative?

Q2. How much of the uniform operator approach to Hutton's uniformities as generalized by S. E. Rodabaugh in [34] can be accommodated within the lattice- and filter-theoretic framework used in this chapter?

References

[1] G. Birkhoff, Lattice Theory, Trans. Amer. Math. Soc. Colloquium Publications, third edition, Amer. Math. Soc. (Providence, RI), 1973.

[2] N. Bourbaki, General Topology: Part I, Section 6.4, Addison Wesley (Reading, MA), 1966.

[3] M. H. Burton, M. A. De Prada Vicente, J. Gutiérrez García, *Generalised uniform spaces*, J. Fuzzy Math. **4(2)**(1996), 363–380.

[4] M. H. Burton, M. Muraleetharan, J. Gutiérrez García, *Generalised filters 1, 2*, Fuzzy Sets and Systems **106**(1999), 275–284 and 393–400.

[5] H. Cartan, *Theorie des filters*, C. R. Acad. Paris **205**(1937), 777–779.

[6] P. Eklund, W. Gähler, *Fuzzy filters, functors, and convergence*, Chapter 4 in [35], 109–136.

[7] G. Gierz, K. H. Hofmann, K. Keimel, J. D. Lawson, M. Mislove, D. S. Scott, A Compendium Of Continuous Lattices, Springer-Verlag (Berlin/Heidelberg/New York), 1980.

[8] J. Gutiérrez García, M. A. de Prada Vicente, *Super uniform spaces*, Quaestiones Mathematicae **20(3)**(1997), 291–310.

[9] J. Gutiérrez García, M. A. de Prada Vicente, M. H. Burton, *Embeddings into the category of super uniform Spaces*, Quaestiones Mathematicae **20(3)**(1997), 311–326.

[10] J. Gutiérrez García, I. Mardones Pérez, M. H. Burton, *The relationship between various filter notions on a GL-Monoid*, J. Math. Anal. Appl. **230**(1999), 291–302.

[11] U. Höhle, *Probabilistic uniformization of fuzzy uniformities*, Fuzzy Sets and Systems **1**(1978), 311–332.

[12] U. Höhle, *Probabilistic topologies induced by L–fuzzy uniformities*, Manuscripta Math. **38**(1982), 289–323.

[13] U. Höhle, *Probabilistic metrization of fuzzy topologies*, Fuzzy Sets and Systems **8**(1982), 63–69.

[14] U. Höhle, *$\mathfrak{G}$-Fuzzy topologies on algebraic structures*, J. Math. Anal. Appl.**108**(1985), 113–150.

[15] U. Höhle, *M-valued sets and sheaves over integral commutative cl–monoids*, Chapter 2 in [35], 33–72.

[16] U. Höhle, *Commutative residuated L–monoids*, Chapter IV in: U. Höhle, E. P. Klement, eds, Non-Classical Logics And Their Applications To Fuzzy Subsets—A Handbook of the Mathematical Foundations of Fuzzy Set Theory, Theory and Decision Library—Series B: Mathematical and Statistical Methods, Volume **32**, Kluwer Academic Publishers (Boston/Dordrecht/London), 1995, pp. 53–106.

[17] U. Höhle, *MV-algebra valued filter theory*, Quaestiones Mathematicae **19**(1996), 23–46.

[18] U. Höhle, *Characterization of L–topologies by L–valued neighborhoods*, in [19], 389–432.

[19] U. Höhle, S. E. Rodabaugh, eds, Mathematics Of Fuzzy Sets: Logic, Topology, And Measure Theory, The Handbooks of Fuzzy Sets Series: Volume **3**, Kluwer Academic Publishers (Boston/Dordrecht/London), 1999.

[20] U. Höhle, A. Šostak, *Axiomatic foundations of fixed-basis fuzzy topology*, Chapter 3 in [19], 123–272.

[21] B. Hutton, *Uniformities on fuzzy topological spaces*, J. Math. Anal. Appl. **58**(1977), 559–571.

[22] J. R. Isbell, Uniform Spaces, Math. Surveys **12**(1964), American Mathematical Society (Providence, R.I.).

[23] W. Kotzé, *Uniform spaces*, Chapter 8 in [19], 553–580.

[24] T. Kubiak, *The topological modification of the L–fuzzy unit interval*, Chapter 11 in [35], 275–305.

[25] T. Kubiak, *Separation axioms: extensions of mappings and embedding of spaces*, Chapter 6 in [19], 433–479.

[26] T. Kubiak, A.P. Šostak, *Lower set valued topologies*, Quaestiones Mathematicae **20(3)**(1997), 415–422.

[27] T. Kubiak, A. P. Šostak, *On* (L, M)*-topological spaces*, (in preparation).

[28] R. Lowen, *Convergence in fuzzy topological spaces*, General Topology and its Applications **10**(1979), 147–160.

[29] R. Lowen, *Fuzzy neighbourhood spaces*, Fuzzy Sets and Systems **7**(1980), 165–189.

[30] R. Lowen, *Fuzzy uniform spaces*, J. Math. Anal. Appl. **82**(1981), 370–385.

[31] S. E. Rodabaugh, *A theory of fuzzy uniformities with applications to the fuzzy real lines*, J. Math. Anal. Appl.**129**(1988), 37–70.

[32] S. E. Rodabaugh, *Powerset operator based foundation for point-set lattice-theoretic (poslat) set theories and topologies*, Quaestiones Mathematicae **20(3)**(1997), 463–530.

[33] S. E. Rodabaugh, *Covering/Hutton fuzzy uniformities: preliminary remarks*, Twentieth Linz Seminar: Topological and Algebraic Structures (2–6 February 1999, Bildungshaus St. Magdalena, Linz, Austria).

[34] S. E. Rodabaugh, *Axiomatic foundations for uniform operator quasi-uniformities*, Chapter 7 in this Volume.

[35] S. E. Rodabaugh, E. P. Klement, U. Höhle, eds, Application Of Category To Fuzzy Sets, Theory and Decision Library—Series B: Mathematical and Statistical Methods, Volume **14**, Kluwer Academic Publishers (Boston/Dordrecht/London), 1992.

[36] J. W. Tukey, Convergence And Uniformity In Topology, Ann. Math. Studies, No. **1**, Princeton University Press, 1940.

[37] A. Weil, *Sur les Spaces à Structure Uniforme et sur la Topologie Générale*, Act. Sci. et Ind. **551**, Hermann Press (Paris), 1937.

[38] D. Zhang, *Stratified Hutton uniform spaces*, Fuzzy Sets and Systems (to appear).

CHAPTER 4

Many Valued Topologies And Borel Probability Measures

U. HÖHLE

Introduction

One of the fundamental problems of the discipline of fuzzy topology is the question of its applicability to those mathematical problems which cannot be solved by standard techniques from general topology. As a simple example we mention the problem to construct a unique continuous extension of the Boolean negation from $2 = \{0,1\}$ to the real unit interval $[0,1]$. Since 2 is not dense in $[0,1]$, it is clear that standard techniques from general topology fail! What is even more surprising is the fact that fuzzy topology[1] as originally defined in [7], also called $[0,1]$-topology, is also not able to solve this simple question. The reason for this situation has to do with the way how $[0,1]$-topology understands the "intersection axiom" of ordinary topologies. The interpretation of intersection of subsets by the minimum of $[0,1]$-valued functions entails the possibility to identify fuzzy topological with ordinary topological spaces. This insight is not new and traces back to an unpublished paper by E. Santos 1977 (cf. [17]) and a remark by R. Lowen 1978 (cf. [14]). Hence it is hopeless to look for problems which are solvable using $[0,1]$-topological spaces but unsolvable using ordinary topological spaces.

The aim of this chapter is to change Chang's original axioms of fuzzy topology and to propose a *new* intersection axiom such that Łukasiewicz' negation is the unique continuous extension of the Boolean negation from 2 to $[0,1]$

[1] For historical reasons we use here the terminology introduced by C. L. Chang 1968 [7].

S.E. Rodabaugh and E.P. Klement (eds.), Topological and Algebraic Structures in Fuzzy Sets, 115–135.
© 2003 *Kluwer Academic Publishers.*

(cf. Section 6). This approach leads to rigid I-valued topological spaces[2] whose axiom system has already been published in [12]. In this chapter we continue this investigation and present a detail analysis of the most important topological axioms for rigid I-valued topological spaces—e.g. density, Hausdorff's separation axiom, regularity, compactness. This study is accompanied by a sequence of examples showing the relevance of the proposed theory for τ-smooth Borel probability measures. A remarkable result is the fact that ordinary compact Hausdorff spaces X are dense subspaces of the corresponding spaces of Radon probability measures on X w.r.t. to a prominent rigid I-valued topology. Finally, we emphasize that rigid I-valued topological spaces form an enrichment of the theory of semitopological spaces[3] which traces back to the work of A. Appert and E. Čech in the mid-thirties of the last century (cf. [2, 3, 6]).

1 Axioms of rigid I-valued topologies

As lattice-theoretic basis we make use of the real unit interval $I = [0,1]$ viewed as a complete MV-algebra—i.e. on I we consider the usual partial ordering $\leq$ and *Lukasiewicz arithmetic conjunction* $*$ determined by the following relation:

$$\alpha * \beta \quad = \quad \max(\alpha + \beta - 1\,, 0)\;, \qquad \alpha, \beta \in I$$

Then $(I, \leq, *)$ is a quantale (cf. [16]) and 1 is the unit element in I w.r.t. $*$. In particular, the implication $\rightarrow$ associated with $*$

$$\alpha \rightarrow \beta \quad = \quad \sup\{\gamma \in I \mid \alpha * \gamma \leq \beta\}$$

coincides with *Lukasiewicz implication*—i.e. $\alpha \rightarrow \beta = \min(1 - \alpha + \beta, 1)$. Since $*$ is *not idempotent* there exists a non trivial formation of square roots w.r.t. $*$ (cf. [9]):

$$\alpha^{1/2} \quad = \quad \sup\{\beta \in I \mid \beta * \beta \leq \alpha\} \quad = \quad \frac{\alpha + 1}{2}$$

Therefore the monoidal mean operator $\circledast$ corresponding to $*$ (cf. [12]) coincides with the arithmetic mean in I—i.e.

$$\alpha \circledast \beta \quad = \quad (\alpha^{1/2}) * (\beta^{1/2}) \quad = \quad \frac{\alpha + \beta}{2}$$

The characteristic function of a subset A of a given set X is denoted by χ_A.

After these preparations we are now in the position to formulate the axioms of rigid I-valued topologies. A subset τ of I^X is said to be a **rigid I-valued topology** on X iff τ satisfies the following conditions:

[2] Here we take the freedom to modify the terminology proposed in [12]. In particular, I-topologies are also called I-valued topologies.

[3] See Section 3.

$$(\mathcal{O}1) \quad \chi_X\,,\,\chi_\varnothing \in \tau\,.$$

$$(\mathcal{O}2) \quad g_1, g_2 \in \tau \quad \Longrightarrow \quad g_1 \circledast g_2 \in \tau\,.$$

$$(\mathcal{O}3) \quad \{g_j \mid j \in J\} \subseteq \tau \quad \Longrightarrow \quad \sup_{j\in J} g_j \in \tau\,.$$

$$(\Sigma 1) \quad g \in \tau\,,\ \alpha \in I \quad \Longrightarrow \quad \alpha * g \in \tau\,.$$

$$(\Sigma 2) \quad g \in \tau \ \text{ with } \ g \le \tfrac{1}{2} \quad \Longrightarrow \quad 2 \cdot g \in \tau\,.$$

Two important properties of rigid I-valued topologies are given in the next Lemma.

Lemma 1.1 *Let τ be a rigid I-valued topology on X. Then τ fulfils the following properties:*

$$(\mathcal{O}2') \quad g_1\,,\,g_2 \in \tau \quad \Longrightarrow \quad g_1 * g_2 \in \tau\,.$$

$$(\mathcal{O}2'') \quad g_1\,,\,g_2 \in \tau \ \text{ with } \ g_1 + g_2 \le 1 \quad \Longrightarrow \quad g_1 + g_2 \in \tau\,.$$

Proof. Because of $\alpha * \beta = 2 \cdot ((\alpha \circledast \beta) * \frac{1}{2})$ the property $(\mathcal{O}2')$ follows immediately from $(\mathcal{O}2)$, $(\Sigma 1)$ and $(\Sigma 2)$. If $\alpha+\beta \le 1$, then $\alpha+\beta = 2\cdot(\alpha\circledast\beta)$; hence $(\mathcal{O}2)$ and $(\Sigma 2)$ imply $(\mathcal{O}2'')$. □

Let τ_1 and τ_2 be rigid I-valued topologies on X. τ_1 is said to be **coarser** than τ_2 (resp. τ_2 is said to be **finer** than τ_1) iff $\tau_1 \subseteq \tau_2$. Obvioulsy, I^X is the finest, rigid I-valued topology on X. Therefore, I^X is also called the *discrete*, rigid, I-valued topology on X and is denoted by τ_d. The axioms $(\mathcal{O}1)$—$(\mathcal{O}3)$ imply that every rigid I-valued topology contains all constant maps $\alpha\cdot\chi_X$ from X to I. Hence $\tau_i = \{\alpha \cdot \chi_X \mid \alpha \in I\}$ is the coarsest, rigid, I-valued topology on X and is called the *indiscrete* rigid I-valued topology. Since the axioms of rigid I-valued topologies are preserved under arbitaray intersections, it is easy to see that the set $\mathfrak{O}(X)$ of all rigid I-valued topologies on X forms a complete lattice with respect to the set inclusion $\subseteq$.

A pair (X,τ) is called a **rigid, I-valued topological space** iff X is a set and τ is a rigid, I-valued topology on X. Let (X_1,τ_1) and (X_2,τ_2) be rigid I-valued topological spaces. A map $\varphi : X_1 \longmapsto X_2$ is said to be **continuous** w.r.t. τ_1 and τ_2 iff the following relation holds:

$$\text{(C)} \quad \{g \circ \varphi \mid g \in \tau_2\} \subseteq \tau_1\,. \qquad \textit{(Continuity)}.$$

Obviously, rigid I-valued topological spaces and continuous maps form a category **R-TOP**. It is not difficult to show that initial and final structures do exist in **R-TOP**. Hence **R-TOP** is a *topological catgeory* over **SET** (cf. [1]).

In the following considerations we characterize rigid I-valued topologies by appropriate I-valued neighborhood systems. For every $p \in X$ let $\nu_p : I^X \longmapsto I$ be a map. Then $(\nu_p)_{p \in X}$ is called a *rigid I-valued neighborhood system* iff the following conditions are satisfied:

(U0) $[\nu_p](\chi_X) \ = \ 1 .$

(U1) $[\nu_p](f) \ \leq \ \nu_p(g)$ whenever $f \leq g \ (g, f \in L^X)$. (Isotonicity)

(U2) $\nu_p(f) \circledast \nu_p(g) \ \leq \ \nu_p(f \circledast g) .$

(U3) $\nu_p(f) \ \leq \ f(p) .$

(U4) $\nu_p(f) \ \leq \ \sup \{\nu_p(h) \mid h \in I^X, \, h(q) \leq \nu_q(f) \, , \, \forall q \in X\} .$

(U5) $\alpha * \nu(f) \ \leq \ \nu(\alpha * f) .$ (*Stratification*)

(U6) $\frac{1}{2} \cdot \nu_p(f) \ = \ \nu_p(\frac{1}{2} \cdot f) .$

It is not difficult to see that the axioms (U0)–(U4) represent an I-valued interpretation of the usual neighborhood axioms. Under the hypothesis of (U1) the stratification axiom (U5) is equivalent to the nonexpansivity of ν_p w.r.t. to the uniform norm on I^X and the usual metric on I—i.e.

$$|\nu_p(f) - \nu_p(g)| \ \leq \ \| f - g \|_\infty$$

Hence the stratification axiom can be viewed as a compatibility condition w.r.t. to the natural, underlying metrics in the domain and codomain of ν_p. Finally, the axiom (U6) is a homogeneity condition, and its role will become apparent when we are discussing maximal stratified I-valued filters.

Every rigid I-valued topology τ on X induces a rigid I-valued neighborhood system $(\nu_p)_{p \in X}$ by

$$\nu_p(f) \ = \ \sup \{g(p) \mid g \in \tau \, , \, g \leq f\} \, , \qquad f \in I^X \tag{1.1}$$

On the other hand, if $(\nu_p)_{p \in X}$ is a rigid neighborhood system, then $(\nu_p)_{p \in X}$ induces a rigid I-valued topology τ on X by:

$$\tau \ = \ \{g \in I^X \mid g(p) \ \leq \ \nu_p(g) \, , \, \forall p \in X\} \tag{1.2}$$

It is not difficult to prove that the relations (1.1) and (1.2) determine a bijective map between the set of all rigid I-valued topologies on X and the set of all rigid I-valued neighborhood systems. Hence rigid I-valued topologies and rigid I-valued neighborhood systems are *equivalent concepts.*

We close this section with an investigation concerning the interrelations between ordinary topologies and rigid I-valued topologies. First we observe that for every ordinary topology $\mathcal{O}$ on X the set $\tau_{\mathcal{O}}$ of all lower $\mathcal{O}$-semicontinues maps $g : X \longmapsto I$ forms a rigid I-valued topology. This situation gives rise to a functor ω : **TOP** $\longmapsto$ **R-TOP** as follows:

$$\omega(X, \mathcal{O}) \quad = \quad (X, \tau_{\mathcal{O}}) \,, \qquad \omega(\varphi) \,=\, \varphi \tag{1.3}$$

i.e. ω leaves maps invariant, but replaces ordinary topologies by their attached sets of lower semicontinuous, I-valued maps.

Theorem 1.2 ω : **TOP** $\longmapsto$ **R-TOP** *is a functorial embedding which has both a left adjoint and a right adjoint.*

Proof. It is easily seen that ω is faithful and one-to-one on objects. In order to show that ω has a right adjoint functor we first observe that for any non empty family $\{\mathcal{O}_i \mid i \in I\}$ of ordinary topologies on X the relation

$$\bigcap_{i \in I} \tau_{\mathcal{O}_i} \quad = \quad \tau_{\bigcap_{i \in I} \mathcal{O}_i}$$

holds. Hence the right adjoint functor ι : **R-TOP** $\longmapsto$ **TOP** leaves maps invariant and acts an objects as follows:

$$\iota(X, \tau) \,=\, (X, \mathcal{O}_\tau) \qquad \text{where} \quad \mathcal{O}_\tau \quad = \quad \bigcap \Big\{ \mathcal{O} \Big| \, \tau \subseteq \tau_{\mathcal{O}} \Big\}$$

Finally, the property $(\mathcal{O}2')$ implies that for every rigid I-valued topology τ on X the set

$$\mathcal{O}^{(\tau)} \quad = \quad \{G \subseteq X \mid \chi_G \in \tau\}$$

is an ordinary topology on X. This situation gives rise to a functor ℓ : **R-TOP** $\longmapsto$ **TOP** determined by:

$$\ell(X, \tau) \quad = \quad (X, \mathcal{O}^{(\tau)}) \,, \qquad \ell(\varphi) \quad = \quad \varphi$$

It is not difficult to show that ℓ is left adjoint to ω. □

Motivated by the previous theorem we will identify ordinary topological spaces $(X, \mathcal{O})$ with their rigid I-valued counterparts $\omega(X, \mathcal{O}) \,=\, (X, \tau_{\mathcal{O}})$. In particular, the functor ω traces back to R. Lowen 1976 and appeared for the first time in the connection with fuzzy topologies[4] (cf. [13]).

[4] Here the reader should be warned to establish any connection between fuzzy topologies (cf. [7]) and rigid I-valued topologies. One of the fundamental differences between fuzzy topologies and rigid I-valued topologies consists, e.g., in the formulation of the "intersection axiom" $(\mathcal{O}2)$.

2 Bases of rigid I-valued topologies

Let τ be a rigid I-valued topology on X. A subset $\mathcal{B}$ of τ is called a base of τ iff $\mathcal{B}$ satisfies the axioms $(\mathcal{O}1)$ and $(\mathcal{O}2)$, and every element $h \in \tau$ has the following representation: there exists a non empty index set J, and depending on J there exist three subsets $\{n_j \mid j \in J\} \subseteq \mathbb{N} \cup \{0\}$, $\{\alpha_j \mid j \in J\} \subseteq I$, and $\{g_j \mid j \in J\} \subseteq \mathcal{B}$ provided with the property $\alpha_j * g_j(x) \leq 2^{-n_j}$, $\forall x \in X$ such that the relation

$$h(x) \quad = \quad \sup_{j \in J} 2^{n_j} \cdot (\alpha_j * g_j(x)) \,, \qquad \forall x \in X$$

holds.

Proposition 2.1 *Let $\mathcal{B}$ be a subset of I^X satisfying the axioms $(\mathcal{O}1)$ and $(\mathcal{O}2)$. Further, let τ be the coarsest rigid I-valued topology on X containing $\mathcal{B}$. Then $\mathcal{B}$ is a base of τ.*

Proof. Obviously the axioms $(\mathcal{O}1)$ and $(\mathcal{O}2)$ entail the subsequent property:

$$g \in \mathcal{B} \quad \Longrightarrow \quad 2^{-n} \cdot g \in \mathcal{B} \,, \qquad \forall n \in \mathbb{N} \tag{2.1}$$

Further, let τ^* be the set of all $h \in I^X$ having the following representation:

$$h(x) \quad = \quad \sup_{j \in J} 2^{n_j} \cdot (\alpha_j * g_j(x)) \,, \qquad \forall x \in X$$

where $\alpha_j \in I$, $g_j \in \mathcal{B}$, $n_j \in \mathbb{N} \cup \{0\}$, $\alpha_j * g_j(x) \leq 2^{-n_j}$, $\forall x \in X$. It is sufficient to show that τ^* is a rigid I-valued topology. The axiom $(\mathcal{O}1)$ is evident. In order to verify $(\mathcal{O}2)$ we make the following observations:

$$2^{-m} \cdot (\alpha * g) \quad = \quad ((1 - 2^{-m} \cdot (1 - \alpha)) * (2^{-m} \cdot g) \tag{2.2}$$

$$\left. \begin{array}{l} (\alpha_1 * g_1) \circledast (\alpha_2 * g_2) \quad = \\ \max \left\{ (\alpha_1 \circledast \alpha_2) * (g_1 \circledast g_2) \,,\, \alpha_1^{1/2} * (2^{-1} \cdot g_1) \,,\, \alpha_2^{1/2} * (2^{-1} \cdot g_2) \right\} \end{array} \right\} \tag{2.3}$$

$$\left. \begin{array}{l} \left(2^{n_1} \cdot (\alpha_1 * g_1)\right) \circledast \left(2^{n_2} \cdot (\alpha_2 * g_2)\right) \quad = \\ \left(2^{n_0} \cdot \left(2^{-(n_0 - n_1)} \cdot (\alpha_1 * g_1)\right)\right) \circledast \left(2^{n_0} \cdot \left(2^{-(n_0 - n_2)} \cdot (\alpha_2 * g_2)\right)\right) \quad = \\ 2^{n_0} \cdot \left(\left(2^{-(n_0 - n_1)} \cdot (\alpha_1 * g_1)\right) \circledast \left(2^{-(n_0 - n_2)} \cdot (\alpha_2 * g_2)\right)\right) \end{array} \right\} \tag{2.4}$$

where $n_0 = \max(n_1, n_2)$. Now we choose $g_1, g_2 \in \mathcal{B}$ with $\alpha_1 * g_1 \leq 2^{-n_1}$, $\alpha_2 * g_2 \leq 2^{-n_2}$; then the relations (2.1), (2.2), (2.3) and (2.4) show that

$$\left(2^{n_1} \cdot (\alpha_1 * g_1)\right) \circledast \left(2^{n_2} \cdot (\alpha_2 * g_2)\right)$$

is an element of τ^*. Since $\circledast$ is distributive over non empty suprema, τ^* fulfils $(\mathcal{O}2)$. Further we note:

$$\beta * 2^n(\alpha * g) \quad = \quad 2^n \cdot \Big((1 - 2^{-n} \cdot (1-\beta)) * \alpha * g\Big)\,, \qquad \text{where} \quad \alpha * g \,\leq\, 2^{-n}$$

Therefore τ^* satisfies also $(\Sigma 1)$. Finally, the axioms $(\mathcal{O}3)$ and $(\Sigma 2)$ are trivial. □

An important application of Proposition 2.1 is given in the next example.

Example 2.2 (Borel probability measures) Let $(X, \mathcal{O})$ be an ordinary, topological space and $\mathcal{G}(X)$ be the set of all lower semicontinuous functions $g : X \longmapsto [0,1]$. A Borel probability measure π is called τ-smooth (cf. P13 in [19]) iff for every family $\mathcal{F}$ of closed sets filtering to the left[5] the following relation holds:

$$\pi\Big(\bigcap_{F \in \mathcal{F}} F\Big) \quad = \quad \inf_{F \in \mathcal{F}} \pi(F)$$

Let $\mathcal{P}_1(X, \mathcal{O})$ be the set of all τ*-smooth Borel probability measures* on X. Then every map $g \in \mathcal{G}(X)$ induces a map $\mathbb{A}_g : \mathcal{P}_1(X, \mathcal{O}) \longmapsto [0,1] = I$ by

$$\mathbb{A}_g(\pi) \quad = \quad \int\limits_X g \, d\pi \,, \qquad \forall \, \pi \in \mathcal{P}_1(X, \mathcal{O})$$

Then Proposition 2.1 implies that the set $\mathcal{B} = \{\mathbb{A}_g \mid g \in \mathcal{G}(X)\}$ forms a base of a unique rigid I-valued topology $\sigma_{\mathcal{O}}$ on $\mathcal{P}_1(X, \mathcal{O})$. Moreover the construction of image measures (cf. p. 33 and p. 34 in [15]) leads to a functor $\mathcal{F}_P$ from **TOP** to **R-TOP**. The details of this situation are as follows:

Let $\varphi : (X_1, \mathcal{O}_1) \longmapsto (X_2, \mathcal{O}_2)$ be a continuous map—i.e. a **TOP**-morphism. Since continuous images of τ-smooth measures are again τ-smooth, the formation of image measures induces a map $\Theta_\varphi : \mathcal{P}_1(X_1, \mathcal{O}_1) \longmapsto \mathcal{P}_1(X_2, \mathcal{O}_2)$ by

$$[\Theta_\varphi(\pi)](B) \quad = \quad \pi(\varphi^{-1}(B))\,, \qquad \pi \in \mathcal{P}_1(X_1, \mathcal{O}_1)$$

where B is a Borel subset of X_2. As an immediate consequence of the construction of image measures we obtain the important relation:

$$\mathbb{A}_g \cdot \Theta_\varphi \quad = \quad \mathbb{A}_{g \cdot \varphi}\,, \qquad \forall \, g \in \mathcal{G}(X_2, \mathcal{O}_2) \tag{2.5}$$

Hence Θ_φ is *continuous* w.r.t. to the rigid I-valued topologies $\sigma_{\mathcal{O}_1}$ and $\sigma_{\mathcal{O}_2}$ — i.e. Θ_φ is a **R-TOP**-morphism. Thus the functor $\mathcal{F}_P$: **TOP** $\longmapsto$ **R-TOP** is determined as follows:

$$\mathcal{F}_P(X, \mathcal{O}) \;=\; (\mathcal{P}_1(X, \mathcal{O}), \sigma_{\mathcal{O}})\,, \qquad \mathcal{F}_P(\varphi) \;=\; \Theta_\varphi$$

[5] A family $\mathcal{F}$ filters to the left iff for all $F_1, F_2 \in \mathcal{F}$ there exists $F_3 \in \mathcal{F}$ s.t. $F_3 \subseteq F_1 \cap F_2$.

In contrast to ω (cf. Theorem 1.2) the functor $\mathcal{F}_P$ has neither a left adjoint nor a right adjoint functor. The only interesting aspect of $\mathcal{F}_P$ consists of demonstrating that the construction of rigid I-valued topologies on spaces of Borel probability measures is fairly universal.

Finally, if for all $x \in X$ we attach the *Dirac measure* δ_x to x, then we obtain a well known map $\Delta : X \longmapsto \mathcal{P}_1(X, \mathcal{O})$. Because of $\mathbb{A}_g \circ \Delta = g$ it is easy to see that the rigid I-valued topology $\tau_{\mathcal{O}}$ (cf. Section 1) is the *initial* rigid I-valued topology on X w.r.t. $(\Delta, (\mathcal{P}_1(X, \mathcal{O}), \sigma_{\mathcal{O}}))$. Hence in the case of injectivity of Δ we can view $(X, \mathcal{O}) \cong \omega(X, \mathcal{O})$ as an I-valued topological subspace of $(\mathcal{P}_1(X, \mathcal{O}), \sigma_{\mathcal{O}}))$. □

3 Density, closures, and closed subsets

Let (X, τ) be a rigid I-valued topological space, and $\mathcal{P}(X)$ be the ordinary power set of X. Then τ induces a closure operator $cl_\tau : \mathcal{P}(X) \longmapsto \mathcal{P}(X)$ as follows:

$$cl_\tau(A) = \{p \in X \mid g(p) \leq \sup_{a \in A} g(a) \ , \ \forall g \in \tau\} \ , \qquad A \in \mathcal{P}(X) \quad (3.1)$$

In particular, cl_τ satisfies the following conditions (see also Definition 1.1.2 in [16]):

(CL1) $A \subseteq cl_\tau(A) \ , \quad cl_\tau(\varnothing) = \varnothing$.

(CL2) $A \subseteq B \implies cl_\tau(A) \subseteq cl_\tau(B)$.

(CL3) $cl_\tau(cl_\tau(A)) \subseteq cl_\tau(A)$.

The **closure** $\overline{A}$ of a subset A of X w.r.t. τ is defined by: $\overline{A} = cl_\tau(A)$. A subset A of X is **closed** w.r.t. τ iff $A = \overline{A}$. It follows immediately from (CL1) and (CL3) that the closure of a subset is a closed subset w.r.t. τ. Moreover, it is well known that closure operators can be characterized by their associated families of closed sets. In particular, the family of all complements of closed sets corresponding to a given closure operator forms a semitopology $\mathbb{O}$ on X—i.e. $\mathbb{O}$ is provided with the following properties[6] :

(OI) $\varnothing, X \in \mathcal{O}$.

(OII) $\mathcal{O}$ is closed under arbitrary unions .

In this sense rigid I-valued topological spaces can be viewed as an enrichment of the theory of semitopological spaces which appeared in the literature under

[6] The axioms (OI) and (OII) appeared for the first time in W. Sierpiński's book *Zarys Teorji Mnogości. Część Druga: Topologja Ogólna* (Warszawa 1928)).

various names—e.g. *F-spaces* (cf. [6]) or *espaces à topologie transitive* (cf. p. 7 in [3])[7]. We will return to this point in Section 6, and continue here our considerations with the definition of density.

Let (X, τ) be a rigid I-valued topological space and cl_τ be the associated closure operator. A subset A of X is said to be **dense** in (X, τ) iff $cl_\tau(A) = X$. First we show that in the sense of the functor ω (cf. Theorem 1.2) the proposed concept of density is a natural extension of the usual concept of density applied in the realm of ordinary topological spaces.

Proposition 3.1 *Let $(X, \mathcal{O})$ be an ordinary topological space, and A be a subset of X. Then the following assertions are equivalent:*

(i) *A is dense in $(X, \mathcal{O})$ (in the sense of **TOP**).*

(ii) *A is dense in $\omega(X, \mathcal{O}) = (X, \tau_{\mathcal{O}})$.*

Proof. The equivalence follows from the fact that every lower $\mathcal{O}$-semicontinuous function $g : X \longmapsto I$ can be represented as follows:

$$g = \sup_{j \in J} \alpha_j \cdot \chi_{G_j}, \qquad \text{where} \quad G_j \in \mathcal{O}, \quad \alpha_j \in I \qquad \square$$

Making use of the concept of bases of rigid I-valued topologies we can characterize density as follows:

Proposition 3.2 *Let (X, τ) be a rigid I-valued topological space, and A be a subset of X. If $\mathcal{B}$ is a base of τ, then the following assertions are equivalent:*

(i) *A is dense in (X, τ).*

(ii) *For all $p \in X$ and for all $g \in \mathcal{B}$ the following inequality holds:*

$$g(p) \leq \sup_{a \in A} g(a)$$

Proof. The equivalence follows immediately from the definition of density and the definition of bases of rigid I-valued topologies. $\square$

A simple application of the previous proposition is given in the next example.

---

[7]The term "semitopological space" has recently been proposed by H. Herrlich, M. Hušek, and G. Preuss (cf. [8]).

Example 3.3 We continue our considerations started in Example 2.2 and claim that the set $\{\delta_x \mid x \in X\}$ of all Dirac measures in X is a dense subset of the rigid I-valued topological space $(\mathcal{P}_1(X, \mathcal{O}), \sigma_{\mathcal{O}})$. We use the well known estimation

$$\int_X g \, d\pi \quad \leq \quad \| \, g \, \|_\infty \quad = \quad \sup_{x \in X} \delta_x(g) \,, \qquad g \in \mathcal{G}(X)$$

and observe that the base $\{\mathbb{A}_g \mid g \in \mathcal{G}(X)\}$ of $\sigma_{\mathcal{O}}$ satisfies the assertion (ii) of Proposition 3.2. □

4 Hausdorff's separation axiom and regularity

A rigid I-valued topological space (X, τ) is said to be Hausdorff separated iff for every pair $(p, q) \in X \times X$ with $p \neq q$ there exists a pair $(g_1, g_2) \in \tau \times \tau$ provided with the following property

$$g_1 * g_2 \quad = \quad \chi_\varnothing \,, \qquad g_1(p) * g_2(q) \quad \neq \quad 0 \tag{4.1}$$

The next proposition shows that the proposed Hausdorff separation axiom is a natural extension of the Hausdorff's original axiom for ordinary topological spaces.

Proposition 4.1 *Let $(X, \mathcal{O})$ be an ordinary topological space and $\omega(X, \mathcal{O})$ be its I-valued topological counterpart (cf. Theorem 1.2). Then the following assertions are equivalent:*

(i) *$(X, \mathcal{O})$ is Hausdorff separated in the usual sense.*

(ii) *$\omega(X, \mathcal{O})$ is Hausdorff separated.*

Proof. The implication (i) $\Longrightarrow$ (ii) is trivial. On the other hand, let (g_1, g_2) be a pair of lower $\mathcal{O}$-semicontinuous I-valued maps satisfying (4.1). Then there exists a natural number n with

$$0 \quad < \quad g_1(p) + g_2(q) - 1 - \frac{1}{n}$$

Then the open subsets $G_1 = \{x \in X \mid g_1(p) - \frac{1}{2n} < g_1(x)\}$ and $G_2 = \{x \in X \mid g_2(q) - \frac{1}{2n} < g_2(x)\}$ are disjoint and respective neighborhoods of p and q. Hence the implication (ii) $\Longrightarrow$ (i) follows. □

Example 4.2 (Borel probability measures) Let $(X, \mathcal{O})$ be an ordinary regular topological space, and $(\mathcal{P}_1(X, \mathcal{O}), \sigma_{\mathcal{O}})$ be the rigid I-valued topological space constructed in Example 2.2. Then $(\mathcal{P}_1(X, \mathcal{O}), \sigma_{\mathcal{O}})$ is Hausdorff separated. For this purpose we fix two different, τ-smooth Borel probability measures π_1 and π_2 on X. Since the lattice $\mathcal{O}$ generates the σ-algebra of all Borel subsets of X, there exists a set $G \in \mathcal{O}$ s.t. $\pi_1(G) \neq \pi_2(G)$—e.g.

$$\pi_1(G) \quad < \quad \pi_2(G)$$

Now we apply the regularity of $(X, \mathcal{O})$, and for every $x \in G$ we choose an open neighborhood V_x of x such that the closure $\overline{V_x}$ is contained in G—i.e. $x \in \overline{V_x} \subseteq G$. Since π_2 is τ-smooth, there exists a finite subset H of G provided with the following property:

$$\pi_1(G) \quad < \quad \pi_2\Big(\bigcup_{x \in H} V_x\Big)$$

We put

$$G_1 \quad := \quad \bigcap_{x \in H} X \cap \complement \overline{V_x} \,, \qquad G_2 \quad := \quad \bigcup_{x \in H} V_x$$

and show that the lower semicontinuous characteristic functions χ_{G_1} and χ_{G_2} fulfil the desired properties. Since G_1 and G_2 are disjoint, the additivity of probability measures implies:

$$\pi(G_1) + \pi(G_2) - 1 \quad \leq \quad 0 \,, \qquad \forall \pi \in \mathcal{P}_1(X, \mathcal{O})$$

i.e. $\mathbb{A}_{\chi_{G_1}} * \mathbb{A}_{\chi_{G_2}} \; = \; \chi_{\varnothing}$. Further, we conclude from the choice of the neighborhoods $V_x\, (x \in G)$: $X \cap \complement G \subseteq G_1$. We again apply the additivity of probability measures and obtain:

$$\begin{aligned} \pi_1(G_1) + \pi_2(G_2) - 1 \quad &\geq \quad \pi_1(X \cap \complement G) + \pi_2(G_2) - 1 \\ &= \quad \pi_2(G_2) - \pi_1(G) \quad > \quad 0 \end{aligned}$$

i.e. $\mathbb{A}_{\chi_{G_1}}(\pi_1) * \mathbb{A}_{\chi_{G_2}}(\pi_2) \; > \; 0$. Hence $(\mathcal{P}_1(X, \mathcal{O}), \tau_{\mathcal{O}})$ is Hausdorff separated. □

In the next proposition we show that continuous maps are uniquely determined on dense subsets, provided the range is Hausdorff separated.

Proposition 4.3 *Let* (X_1, τ_1) *and* (X_2, τ_2) *be rigid* I*-valued topological spaces. Further, let* A *be a dense subset in* (X_1, τ_1) *and* (X_2, τ_2) *be Hausdorff separated. If* $\varphi_1 : X_1 \longmapsto X_2$ *and* $\varphi_2 : X_1 \longmapsto X_2$ *are continuous maps, then the following assertions are equivalent:*

(i) $\varphi_1 \; = \; \varphi_2$; (ii) $\varphi_1(a) \; = \; \varphi_2(a)$, $\forall\, a \in A$.

Proof. The implication (i) $\Longrightarrow$ (ii) is trivial. In order to verify (ii) $\Longrightarrow$ (i) we assume that φ_1 and φ_2 do not coincide. Hence there exists an element $p \in X_1$ with $\varphi_1(p) \neq \varphi_2(p)$. Since (X_2, τ_2) is Hausdorff separated we can choose elements $g_1, g_2 \in \tau_2$ satisfying the following properties:

$$g_1 * g_2 \quad = \quad \chi_\varnothing \,, \qquad g_1(\varphi_1(p)) * g_2(\varphi_2(p)) \quad \neq \quad 0$$

Since φ_1 and φ_2 are continuous, we obtain: $g_1 \circ \varphi_1$, $g_2 \circ \varphi_2 \in \tau_1$. Finally we apply the axiom $(\mathcal{O}2')$ and use the density of A:

$$g_1(\varphi_1(p)) * g_2(\varphi_2(p)) \quad \leq \quad \sup_{a \in A} (g_1(\varphi_1(a)) * g_2(\varphi_2(a)) \quad = \quad 0$$

Hence we obtain a contradiction to $g_1(\varphi_1(p)) * g_2(\varphi_2(p)) \neq 0$—i.e. the assertion (i) holds. □

The formulation of an appropriate regularity axiom for rigid I-valued topological spaces requires nuclei on the partially ordered groupoid $(I^X, \leq, \circledast)$ [8]. Therefore, let (X, τ) be a rigid I-valued topological space. Then τ induces a nucleus $C_\tau : I^X \longmapsto I^X$ by

$$C_\tau(f) \quad = \quad 1 - \sup\{g \in \tau \mid g \leq 1 - f\} \,, \qquad f \in I^X \tag{4.2}$$

In particular, C_τ fufils the following properties (see also Definition 3.1.1 in [16]):

(C0) $C_\tau(\chi_\varnothing) \quad = \quad \chi_\varnothing$.

(C1) $f_1 \leq f_2 \quad \Longrightarrow \quad C_\tau(f_1) \leq C_\tau(f_2)$.

(C2) $C_\tau(f_1 \circledast f_2) \quad \leq \quad (C_\tau(f_1)) \circledast (C_\tau(f_2))$.

(C3) $f \leq C_\tau(f)$.

(C4) $C_\tau(f) \leq C_\tau(C_\tau(f))$.

(C5) $C_\tau(\alpha \to f) \quad \leq \quad \alpha \to C_\tau(f)$.

(C6) $C_\tau(f^{1/2}) \quad = \quad (C_\tau(f))^{1/2}$.

It can be shown that rigid I-valued topologies can be characterized by their associated nuclei. Let (X, τ) be a rigid I-valued topological space, and C_τ be its associated nucleus. Then (X, τ) is said to be **regular** iff every element $h \in \tau$ has the following representation:

$$h = \sup\{2^n \cdot (\alpha * g) \mid n \in \mathbb{N} \cup \{0\}, \alpha \in I, g \in \tau, \alpha * C_\tau(g) \leq 2^{-n} \cdot h\} \tag{4.3}$$

First we show that the proposed concept of regularity is a natural extension of the usual concept of regularity for ordinary topological spaces.

[8] The dual concept of nuclei is conuclei (cf. [10]).

Proposition 4.4 *Let $(X, \mathcal{O})$ be an ordinary topological space and $\omega(X, \mathcal{O})$ be its I-valued counterpart. Then the following assertions are equivalent:*

(i) $(X, \mathcal{O})$ *is regular in the usual sense.*

(ii) $\omega(X, \mathcal{O})$ *is regular.*

Proof. Let $(\mathbb{U}_p)_{p \in X}$ be the ordinary neighborhood system corresponding to $(X, \mathcal{O})$. We recall $\omega(X, \mathcal{O}) = (X, \tau_{\mathcal{O}})$ (cf. Section 1). Then $C_{\tau_{\mathcal{O}}}(f)$ coincides with the upper $\mathcal{O}$-semicontinuous regularization of f—i.e.

$$f(p) \quad = \quad \inf_{U \in \mathbb{U}_p} \big(\sup_{x \in U} f(x) \big)$$

If $(X, \mathcal{O})$ is regular, then for every lower $\mathcal{O}$-semicontinuous, I-valued function h the following relation holds:

$$h \quad = \quad \sup \{ g \in \tau \mid C_{\tau_{\mathcal{O}}}(g) \leq h \}$$

i.e. $\omega(X, \mathcal{O})$ is regular. On the other hand, let us assume the regularity of $\omega(X, \mathcal{O})$. Then we choose $G \in \mathcal{O}$ and $p \in G$. Because of (4.3) there exist a $n \in \mathbb{N} \cup \{0\}$, $\alpha \in I$, $g \in \tau_{\mathcal{O}}$ such that

$$0 \quad < \quad 2^n \cdot (\alpha * g(p)) \ , \qquad \alpha * C_{\tau_{\mathcal{O}}}(g) \quad \leq \quad 2^{-n} \cdot \chi_G \tag{4.4}$$

In particular, $1 - \alpha < g(p)$. Then we put $\varepsilon = \frac{1 - \alpha + g(p)}{2}$ and define an open neighborhood V of p by: $V = \{ x \in X \mid \varepsilon < g(x) \}$. Because of $\varepsilon \cdot \chi_V \leq g$ we obtain from (4.4):

$$(\alpha * \varepsilon) \cdot C_{\tau_{\mathcal{O}}}(\chi_V) \quad \leq \quad \alpha * C_{\tau_{\mathcal{O}}}(g) \quad \leq \quad \chi_G$$

Now we observe: $0 < \alpha * \varepsilon$; hence $\overline{V} \subseteq G$ follows—i.e. $(X, \mathcal{O})$ is regular. □

The next example is a continuation of Example 4.2.

Example 4.5 Let $(X, \mathcal{O})$ be an ordinary regular topological space, and $(\mathcal{P}_1(X, \mathcal{O}), \sigma_{\mathcal{O}})$ be the rigid I-valued topological space constructed in Example 2.2. Then $(\mathcal{P}_1(X, \mathcal{O}), \sigma_{\mathcal{O}})$ is regular.

Since $\{ \mathbb{A}_g \mid g \in \mathcal{G}(X) \}$ is a base of $\sigma_{\mathcal{O}}$, we conclude from (4.3) that it is sufficient to verify the following relation:

$$\mathbb{A}_g \quad = \quad \bigvee \Big\{ \mathbb{A}_k \Big| \ k \in \mathcal{G}(X), \ C_{\sigma_{\mathcal{O}}}(\mathbb{A}_k) \leq \mathbb{A}_g \Big\} \tag{4.5}$$

Since every lower semicontinuous function g is a uniform limit of a non decreasing sequence $(g_n)_{n \in \mathbb{N}}$ of lower semicontinuous, simple functions $g_n : X \longmapsto$

$[0,1]$, it is sufficient to verify (4.5) in the case of simple functions. For this purpose we fix a lower semicontinuous function g_n defined by

$$g_n \quad := \quad \frac{1}{2^n} \cdot \sum_{i=1}^{2^n-1} \chi_{\{x \in X \mid \frac{i}{2^n} < g(r)\}} \,, \qquad \text{where} \quad g \in \mathcal{G}(X)$$

Further, we choose a Borel probability measure $\pi \in \mathcal{P}_1(X, \mathcal{O})$ and a real number $\epsilon > 0$. Since π is τ-smooth and $(X, \mathcal{O})$ is regular, there exist open subsets $G_i \in \mathcal{O}\,(i = 1, \ldots, 2^n - 1)$ provided with the following properties:

$$G_i, \subseteq \overline{G_i} \subseteq \{x \in X \mid \tfrac{i}{2^n} < g(x)\} \quad , \quad \pi(\{\tfrac{i}{2^n} < g\} \cap \complement G_i) \; < \; 2^{-n} \cdot \epsilon$$

Then we put $k_n = 2^{-n} \cdot \sum_{i=1}^{2^n-1} \chi_{G_i}$ and $f_n = 2^{-n} \cdot \sum_{i=1}^{2^n-1} \chi_{\overline{G_i}}$. Obviously, k_n is lower semicontinuous, and f_n is upper semicontinuous. Hence there exists a lower semicontinuous function $h_n \in \mathcal{G}(X)$ with $1 - h_n = f_n$. In particular, the relation: $1 - \mathbb{A}_{f_n} = \mathbb{A}_{h_n} \in \sigma_{\mathcal{O}}$ holds—i.e. $C_{\sigma_{\mathcal{O}}}(\mathbb{A}_{f_n}) = \mathbb{A}_{f_n}$. Then we obtain:

$$\mathbb{A}_{k_n} \;\leq\; C_{\sigma_{\mathcal{O}}}(\mathbb{A}_{k_n}) \;\leq\; \mathbb{A}_{f_n} \;\leq\; \mathbb{A}_{g_n}$$

Because of $\mathbb{A}_{g_n}(\pi) \leq \mathbb{A}_{k_n}(\pi) + \epsilon$ we conclude from the previous arguments that (4.5) is valid in the case of g_n and a fortiori for any lower semicontinuous function of $\mathcal{G}(X)$. □

Finally, we emphasize that the *principle of continuous extension* remains valid for Hausdorff separated, regular, rigid I-valued topological spaces (cf. Theorem 7.3.10 in [12] or Subsection 6.3.2 in [11]).

5 Compactness axiom

It is our intention to base the compactness notion for rigid I-valued topological spaces on an appropriate I-valued filter theory.

Let X be a set. A map $\xi : I^X \longmapsto I$ is called a stratified I-valued filter on X iff ξ satisfies the following axioms:

(F0) $\quad \xi(\chi_X) = 1$.

(F1) $\quad \xi(f) \leq \xi(g)$ whenever $f \leq g \;\; (g, f \in L^X)$. (Isotonicity)

(F2) $\quad \xi(f)) \circledast \xi(g) \leq \xi(f \circledast g)$.

(F3) $\quad \xi(\chi_\varnothing) = 0$.

(F4) $\quad \alpha * \xi(f) < \xi(\alpha * f)$. (*Stratification*)

A rigid I-valued filter on X is a stratified I-valued filter ξ on X provided with the additional property:

(F5) $\frac{1}{2} \cdot \xi(f) \quad = \quad \xi(\frac{1}{2} \cdot f) \, , \qquad \forall \, f \in I^X \, .$

On the set $\mathfrak{F}(X)$ of all stratified I-valued filters on X we define a partial ordering by

$$\xi_1 \preceq \xi_2 \qquad \Longleftrightarrow \qquad \xi_1(f) \quad \leq \quad \xi_2(f) \, , \qquad \forall \, f \in I^X$$

An application of the *axiom of choice* shows that $(\mathfrak{F}(X), \preceq)$ contains maximal elements. In particular, every maximal, stratified I-valued filter is rigid, and every rigid filter is an infimum of an appropriate family of maximal, stratified I-valued filters (cf. Lemma 6.2.13 and Theorem 6.2.19 in [12], Lemma 4.3.7 and Corollary 4.3.14 in [11]). This situation is one of the motivations to introduce the axiom $(\Sigma 2)$ to the theory of I-valued topological spaces.

A characterization of maximal, stratified I-valued filters can be given as follows: First we note that every *finitely additive probability measure* Π on X (i.e. the ordinary power set $\mathcal{P}(X)$ of X is the domain of Π) induces a maximal, stratified I-valued filter ξ_Π by

$$\xi_\Pi(f) \quad = \quad \int\limits_X f \, d\Pi \, , \qquad \forall \, f \in I^X$$

Now we quote the following result from [11] (cf. Theorem 4.4.6 in [11]).

Theorem 5.1 *Let ξ be a stratified I-valued filter on X. Then the following assertions are equivalent:*

(i) *ξ is maximal.*

(ii) *There exists a (unique) finitely additive probability measure Π on X with $\xi = \xi_\Pi$.* □

After these preparations we are in the position to formulate a *convergence theory* for rigid I-valued topological spaces as follows: Let (X, τ) be a rigid I-valued topological space, and $(\nu_p)_{p \in X}$ be the corresponding I-valued neighborhood system (vgl. Section 1). A rigid I-valued filter ξ on X has a limit point $p \in X$ iff $\nu_p \preceq \xi$. Before we define the compactness notion for rigid I-valued topological spaces we first show that in the sense of the functor ω the proposed concept of limit points is a natural extension of the usual concept of limit points for ordinary topological spaces. For this purpose we note that every ordinary filter $\mathcal{F}$ on X induces a rigid I-valued filter $\xi_{\mathcal{F}}$ on X by

$$\xi_{\mathcal{F}}(f) \quad = \quad \sup_{F \in \mathcal{F}} \Big(\inf_{a \in F} f(a) \Big) \, , \qquad \forall \, f \in I^X \tag{5.1}$$

Proposition 5.2 *Let $(X, \mathcal{O})$ be an ordinary topological space, and $\omega(X, \mathcal{O})$ be its I-valued counterpart. Further, let p be an element of X and $\mathcal{F}$ be a filter on X. Then the following assertions are equivalent:*

(i) *p is a limit point of $\mathcal{F}$ w.r.t. $(X, \mathcal{O})$.*

(ii) *p is a limit point of $\xi_{\mathcal{F}}$ w.r.t. $\omega(X, \mathcal{O})$.*

Proof. Let $(\mathbb{U}_p)_{p \in X}$ be the neighborhood system corresponding to $\mathcal{O}$, and $(\nu_p)_{p \in X}$ be the I-valued neighborhood system corresponding to $\tau_{\mathcal{O}}$ (where $\omega(X, \mathcal{O}) = (X, \tau_{\mathcal{O}})$). Since $\tau_{\mathcal{O}}$ consists of all lower $\mathcal{O}$-semicontinuous, I-valued functions, we obtain:

$$\nu_p(f) \quad = \quad \sup_{U \in \mathbb{U}_p} \Big(\inf_{a \in U} f(a) \Big) \,, \quad f \in I^X \tag{5.2}$$

Then the equivalence between (i) and (ii) follows immediately from (5.1) and (5.2). □

It can be shown that a rigid I-valued topological space is Hausdorff separated iff every rigid I-valued filter has at most one limit point. This situation motivates the following compactness notion: a rigid I-valued topological space (X, τ) is said to be compact iff every maximal, stratified I-valued filter on X has at least one limit point. We quote the following results from [11]:

- A *Tychonov Theorem* is valid for Hausdorff separated I-valued topological spaces (cf. Theorem 3.5.11 and Corollary 3.5.12 in [11]).
- The Hausdorff separation axiom and compactness implies regularity (cf. Theorem 6.4.2.8 in [11]). Hence one of the fundamental properties of ordinary topological spaces (cf. [4]) has a non trivial translation to rigid I-valued topological spaces.

We close this section with a continuation of our sequence of examples.

Example 5.3 (Radon probability measures) Let $(X, \mathcal{O})$ be an ordinary compact Hausdorff space, and let $\mathcal{P}_1(X, \mathcal{O})$ be the set of all τ-smooth (i.e. tight) Borel probability measures on X[9]. Then we claim that the rigid I-valued topological space $(\mathcal{P}_1(X, \mathcal{O}), \tau_{\mathcal{O}})$ (constructed in Example 2.2) is Hausdorff separated and compact.

With regard to Example 4.2 we have only to verify the compactness of $(\mathcal{P}_1(X, \mathcal{O}), \tau_{\mathcal{O}})$. For this purpose let Υ be a maximal, stratified I-valued filter

[9] In the case of ordinary compact Hausdorff spaces, τ-tight Borel measures and Radon measures are synonymous notions (cf. p.13 in [19]).

on $\mathcal{P}_1(X,\mathcal{O})$. Then Υ can be identified with a finitely additive probability measure Π on $\mathcal{P}_1(X,\mathcal{O})$ (cf. Theorem 5.1), and the following relation holds:

$$\Upsilon(F) = \int_{\mathcal{P}_1(X,\mathcal{O})} F\,d\Pi\,, \qquad F \in [0,1]^{\mathcal{P}_1(X,\mathcal{O})}$$

Further, let $\mathcal{C}(X)$ be the vector space of all continuous, real valued functions defined on X. Obviously for all $f \in \mathcal{C}(X)$ the map $F_f : \mathcal{P}_1(X,\mathcal{O}) \longmapsto \mathbb{R}$ defined by

$$F_f(\xi) = \int_X f\,d\xi\,, \qquad \xi \in \mathcal{P}_1(X,\mathcal{O})$$

is a bounded function. In particular, the inequality $\| F_g \|_\infty \leq \| f \|_\infty$ holds. Hence Π induces a positive, linear functional L_Π on $\mathcal{C}(X)$ by:

$$L_\Pi(f) = \int_{\mathcal{P}_1(X,\mathcal{O})} F_f\,d\Pi\,, \qquad f \in \mathcal{C}(X)$$

The Radon measure on X corresponding to L_Π is denoted by ν_Π. Because of $F_{\chi_X} = \chi_{\mathcal{P}_1(X,\mathcal{O})}$ we obtain $L_\Pi(\chi_X) = 1$—i.e. ν_Π is a Radon probability measure on X. Hence ν_Π is an element of $\mathcal{P}_1(X,\mathcal{O})$.

We show that ν_Π is a limit point of Υ. Since every lower semicontinuous function $g : X \longmapsto [0,1]$ is an upper envelope of continuous functions — i.e.

$$g = \sup\{f \mid f \in \mathcal{C}(X)\,,\ 0 \leq f \leq g\}$$

we conclude from the definition of ν_Π and from the tightness of ν_Π (cf. Theorem 1 in Chapter IV, §1 in [5]) that for all lower semicontinuous maps $g \in \mathcal{G}(X)$ the relation

$$\begin{aligned} \mathbb{A}_g(\nu_\Pi) &= \int_X g\,d\nu_\Pi = \sup_{\substack{0 \leq f \leq g \\ f \in \mathcal{C}(X)}} \int_X f\,d\nu_\Pi \\ &= \sup_{\substack{0 \leq f \leq g \\ f \in \mathcal{C}(X)}} \int_{\mathcal{P}_1(X,\mathcal{O})} F_f\,d\Pi \leq \Upsilon(\mathbb{A}_g) \end{aligned} \tag{5.3}$$

holds. If $\alpha * \mathbb{A}_g(\nu_\Pi) \leq 2^{-n}$ and $2^n \cdot (\alpha * \mathbb{A}_g) \leq F$, then we obtain from (5.3) and the filter axioms (F1) and (F4):

$$\alpha * \mathbb{A}_g(\nu_\Pi) \leq \alpha * \Upsilon(\mathbb{A}_g) \leq \Upsilon(\alpha * \mathbb{A}_g) \leq \Upsilon(2^{-n} \cdot F)$$

Thus the inequality

$$2^n \cdot (\alpha * \mathbb{A}_g(\nu_\Pi)) \leq \Upsilon(F) \tag{5.4}$$

follows from the filter axiom (F5). Since $\{\mathbb{A}_g \mid g \in \mathcal{G}(X)\}$ is a base of $\sigma_{\mathcal{O}}$, we obtain from (5.4) that ν_Π is a limit point of Υ. □

6 Real unit interval in light of I-valued topologies

At the end of this chapter we direct our attention to the real unit interval $[0,1]$ viewed as a rigid I-valued topological space. Since every element $\alpha \in [0,1]$ can be viewed as a probability measure on the 2-valued discrete space $(\{0,1\}, \mathcal{O}_d)$, Example 2.2 shows that there exists an interesting rigid I-valued topology $\sigma_{\mathcal{O}_d}$ on $[0,1]$. Because of Example 5.3 $([0,1], \sigma_{\mathcal{O}_d})$ is a Hausdorff separated, compact, rigid I-valued topological space which contains $\{0,1\}$ as a dense subspace (cf. Example 2.2, Example 3.3). It is obvious that an *ordinary topology* with these properties does *not exist* on $[0,1]$. More precisely, there does not exist an ordinary topology on $[0,1]$ such that $\omega(X, \mathcal{O})$ is Hausdorff separated and contains $\{0,1\}$ as a dense subspace (cf. Proposition 3.1 and Proposition 4.1). We are convinced that this situation is a motivation for taking a closer look at the rigid I-valued topological space $([0,1], \sigma_{\mathcal{O}_d})$ and to clarify the question to what extent rigid I-valued topologies enrich the theory of semitopological spaces (cf. Section 3). First we observe that the set $\mathcal{B}$ of all function $f_{(\alpha,\beta)}$ of the following type:

$$f_{(\alpha,\beta)}(x) \quad = \quad (\alpha - \beta) \cdot x + \beta \,, \qquad x \in [0,1],\ \alpha, \beta \in [0,1] \qquad (6.1)$$

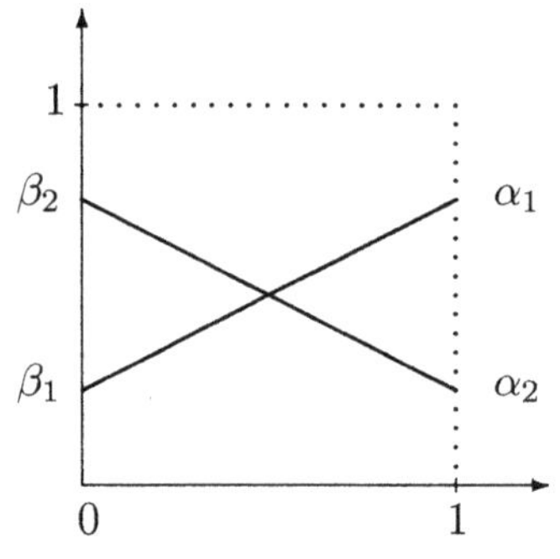

forms a base of $\sigma_{\mathcal{O}_d}$. Further, let $\mathcal{O}_0$ be the usual topology on $[0,1]$, and $\tau_{\mathcal{O}_0}$ be its I-valued counterpart (i.e. $\omega([0,1], \mathcal{O}_0) = ([0,1], \tau_{\mathcal{O}_0})$). Since all $f_{(\alpha,\beta)}$ are continuous w.r.t. the usual ordinary topology $\mathcal{O}_0$ on $[0,1]$, the rigid I-valued topology $\sigma_{\mathcal{O}_d}$ is strictly coarser than $\tau_{\mathcal{O}_0}$.

(a) Convergence Theory: Let $p \in [0,1]$ be a limit point of an ordinary filter $\mathcal{F}$ on $[0,1]$ w.r.t. $\sigma_{\mathcal{O}_d}$. With regard to (5.1) we obtain:

$$f_{(\alpha,\beta)}(p) \quad \leq \quad \sup_{F \in \mathcal{F}} \Big(\inf_{x \in F} f_{(\alpha,\beta)}(x) \Big) \,, \qquad \forall\, (\alpha,\beta) \in [0,1] \times [0,1] \qquad (6.2)$$

It is not difficult to show that (6.2) is equivalent to

$$\inf_{F\in\mathcal{F}}\Big(\sup_{x\in F} x\Big) \quad\leq\quad p \quad\leq\quad \sup_{F\in\mathcal{F}}\Big(\inf_{x\in F} x\Big)$$

Hence an ordinary filter $\mathcal{F}$ on $[0,1]$ converges in the sense of $\sigma_{\mathcal{O}_d}$ iff $\mathcal{F}$ converges in the sense of $\mathcal{O}_0$. In particular, the convergence theory for ordinary filters coincides with the usual order convergence in $[0,1]$. This is not surprising, because first $([0,1],\sigma_{\mathcal{O}_d})$ is Hausdorff separated, compact, rigid I-topological space, and secondly $\sigma_{\mathcal{O}_d}$ is coarser than $\tau_{\mathcal{O}_0}$. In particular, this situation entails the coincidence of the respective limit functions restricted to the set of all ordinary ultrafilters on $[0,1]$. On the other hand, there exist *finitely additive probability measures* which do not have a limit point w.r.t $\tau_{\mathcal{O}_0}$. A simple example can be constructed from two ultrafilters having different limit points w.r.t. $\mathcal{O}_0$ — e.g. let $\mathcal{U}_1$ and $\mathcal{U}_2$ be ultrafilters on $[0,1]$ with $p_1 = \lim\mathcal{U}_1 \neq \lim\mathcal{U}_2 = p_2$. Then

$$\Pi \quad=\quad \frac{1}{2}\cdot\chi_{\mathcal{U}_1} + \frac{1}{2}\cdot\chi_{\mathcal{U}_2}$$

is a finitely additive probability measure on $[0,1]$ which does not have any limit point w.r.t. $\tau_{\mathcal{O}_0}$. Hence the convergence theory in $([0,1],\sigma_d)$ is richer than in $\omega([0,1],\mathcal{O}_0)$.

(b) Density, closure operator and uniqueness of continuous maps: Referring to (3.1) and (6.1) it is easily seen that the closure operator *cl* associated with $\sigma_{\mathcal{O}_d}$ is determined by:

$$cl(A) \quad=\quad [\inf A\,,\,\sup A] \quad=\quad \{x\in[0,1] \mid \inf A \leq x \leq \sup A\}\,,\quad A\subseteq[0,1]$$

Obviously, the corresponding semitopology $\mathbb{O}$ is *not an ordinary topology* on $[0,1]$. Moreover, $([0,1],\mathbb{O})$ is Hausdorff separated, and $\{0,1\}$ is a dense subset of $[0,1]$ w.r.t. $\mathbb{O}$. Since $\mathbb{O}$ does not satisfy any type of intersection axiom, it is not difficult to see that $\mathbb{O}$-continuous self-mappings of $[0,1]$ are not uniquely determined on dense subsets. A simple example can be given as follows:

$$\varphi_1(x) \quad=\quad 1-x\,,\qquad \varphi_2(x) \quad=\quad 1-x^2\,,\qquad x\in[0,1]$$

Both functions are continuous w.r.t. the semitopology $\mathbb{O}$ and coincide with the *Boolean negation* on $\{0,1\}$.

In contrast to semitopologies rigid I-valued topologies satisfy a kind of intersection axiom (cf. $(\mathcal{O}2)$) which entails the *uniqueness* of the definition of continuous self-mappings on dense subsets of rigid I-valued topological spaces (cf. Proposition 4.3). In the previous situation the continuity of φ_1 follows from $f_{(\alpha,\beta)}\circ\varphi_1 = f_{(1-\alpha,1-\beta)}$; hence φ_1 is the *unique,* $\sigma_{\mathcal{O}_d}$*-continuous map* extending the Boolean negation. In particular, φ_1 coincides with *Łukasiewicz' negation.*

Finally, we can summarize the situation as follows: With regard to the problem of uniquely continuous extension rigid I-valued topological spaces are richer than their associated semitopological spaces. Moreover, the axioms $(\mathcal{O}2)$, $(\Sigma 1)$ and $(\Sigma 2)$ can be viewed as an enrichment of the theory of semitopological spaces—an enrichment which is not limited by ordinary topological spaces in the sense of F. Hausdorff or C. Kuratowski.

References

[1] J. Adámek, H. Herrlich, G. E. Strecker, Abstract and Concrete Categories: The Joy of Cats, Wiley Interscience Pure and Applied Mathematics, John Wiley & Sons (Brisbane/Chicester/New York/Singapore/Toronto), 1990.

[2] A. Appert, *Sur une condition jouant une rôle important dans la topologie des espaces abstraits*, C.R. Acad. des Sc. **194**(1932), 2277.

[3] A. Appert, Ky Fan, Espaces Topologiques Intermédiaires, Actualités Scientifiques et Industrielles **1121**, Hermann (Paris), 1951.

[4] N. Bourbaki, Topologie Générale, Éléments de Mathématique: Livre III, Chapitres 1–2, Actualités Scientifiques et Industrielles **858**, Hermann (Paris), 1940.

[5] N. Bourbaki, Intégration, Éléments de Mathématique: Livre VI, Chapitres 1–4, Actualités Scientifiques et Industrielles **858**, Hermann (Paris), 1940.

[6] E. Čech, *Topologické prostory*, Časopis pro pěstován ı matematiky a fysiky **66**(1937), 225–264 (English translation: E. Čech: Topological Papers, Prague 1968, 437–472).

[7] C.L. Chang, *Fuzzy topological spaces*, J. Math. Anal. Appl. **24** (1968), 182–190.

[8] H. Herrlich, M. Hušek, G. Preuss, *Zum Begriff des topologischen Raumes*, University of Bremen (Bremen, Germany), 1998.

[9] U. Höhle, *Commutative, residuated ℓ-monoids*, Chapter IV in: U. Höhle, E. P. Klement, eds, Non-Classical Logics And Their Applications To Fuzzy Subsets—A Handbook of the Mathematical Foundations of Fuzzy Set Theory, Theory and Decision Library—Series B: Mathematical and Statistical Methods, Volume **32**, Kluwer Academic Publishers (Boston/Dordrecht/London), 1995, pp. 53–106.

[10] U. Höhle, *Conuclei and many valued topology*, Acta Mathematica Hungarica **88**(2000), 259–257.

[11] U. Höhle, Many Valued Topology And Its Applications, Kluwer Academic Publishers (Boston/Dordrecht/London), 2001.

[12] U. Höhle, A.P. Šostak, *Axiomatic foundations of fixed-basis fuzzy topology*, Chapter 3 in: U. Höhle, S.E. Rodabaugh, eds, Mathematics Of Fuzzy Sets: Logic, Topology, And Measure Theory, 123–272, Kluwer Academic Publishers (Boston/Dordrecht/London), 1999.

[13] R. Lowen, *Fuzzy topological spaces and fuzzy compactness*, J. Math. Anal. App. **56** (1976), 621–633.

[14] R. Lowen, *A comparison of different compactness notions in fuzzy topological spaces*, J. Math. Anal. Appl. **64** (1978), 446–454.

[15] P.A. Meyer, Probabilités et Potentiel, Actualités Scientifiques et Industrielles **1318**, Hermann (Paris), 1966.

[16] K.I. Rosenthal, Quantales And Their Applications, Pitman Research Notes in Mathematics **234** (Longman/Burnt Mill/Harlow), 1990.

[17] E.S. Santos, *Topology versus fuzzy topology*, preprint, Youngstown State University (Youngstown, Ohio), 1977.

[18] W. Sierpiński, Introduction To General Topology, University of Toronto Press (Toronto), 1934. (English translation of the Polish edition: Zarys Teorji Mnogości, Część Druga: Topologja Ogólna (Warszawa), 1928).

[19] F. Topsøe, Topology And Measure, Lecture Notes in Mathematics **133**, Springer-Verlag (Berlin/Heidelberg/New York), 1970.

CHAPTER 5

Fuzzy Reals: Topological Results Surveyed, Brouwer Fixed Point Theorem, Open Questions

T. KUBIAK [1]

Introduction

There is no doubt that the L-reals $\mathbb{R}(L)$ and, in particular, the unit L-interval $\mathbb{I}(L)$ are among the most important and canonical examples of L-topological spaces. These two spaces have a particularly well-established theory within the framework of the, so to say, classical fuzzy topology (as originated by Chang [1], Goguen [3], and Hutton [7, 8]). It is presently a part of so-called fixed-basis fuzzy topology (cf. Höhle [4] and Höhle and Šostak [5]) and even more a part of variable-basis fuzzy topology (cf. Rodabaugh [31] and Section 7 of [32]).

This chapter is a survey of some of the *topological* results involving L-reals in the context of higher separation axioms and Brouwer-like fixed point theorems for L-cubes. Thus, other aspects such as, e.g., L-uniformities, L-proximities or algebraic structures of $\mathbb{R}(L)$ will not be treated here. For an account of these topics we refer to [22] and [34]; see also [36]. Some related questions are quoted from [24].

The range space for a fuzzy set will generally be a complete lattice L with an order-reversing involution. Some extra assumptions will occasionally be imposed, all of them being of a purely lattice-theoretic character. Most of the presented results (related to separation) are, nevertheless, complete lattice results. In the part of this chapter which is concerned with separation we discuss (as in [19] and [21]) only those results whose proofs are valid for the two-point lattice $\{0,1\}$. As a consequence of this, results which depend on the order-density of L (e.g., when $L = [0,1]$) may only be mentioned when making historical comments. Finally notice that all the discussed results (but those concerning the fixed point property of L-cubes) are free of dependence on their classical topological counterparts.

[1] Partially supported by UPV: 127.310-EA018/99.

S.E. Rodabaugh and E.P. Klement (eds.), Topological and Algebraic Structures in Fuzzy Sets, 137–151.
© 2003 *Kluwer Academic Publishers.*

1 Terminological background

We state here only those information which are necessary to understand the presented results. Their proofs may require more detailed preliminaries, as, e.g., in [21].

Let $(L,')$ be a complete lattice with an order-reversing involution, i.e., a unary operation $'$ such that $\alpha'' = \alpha$, and $\alpha' \le \beta'$ whenever $\beta \le \alpha$ for all $\alpha, \beta \in L$. The bounds in L are denoted by 0 and 1.

More specific assumptions about $(L,')$ include:

- **meet-continuity**: for each $\alpha \in L$ and every directed subset $D \subset L$ one has
$$\alpha \wedge \bigvee D = \bigvee\{\alpha \wedge \gamma : \gamma \in D\}$$
- **infinite distributivity** (= frame law): for each $\alpha \in L$ and every subset $B \subset L$ one has
$$\alpha \wedge \bigvee B = \bigvee\{\alpha \wedge \beta : \beta \in B\}$$

1.1 ***L*-sets.** Given a set X, L^X is the complete lattice (under the pointwise ordering and involution) of all mappings from X to L, which are called ***L*-sets** of X. Given $\mathcal{A} \subset L^X$ we have $(\bigvee \mathcal{A})(x) = \bigvee\{a(x) : a \in \mathcal{A}\}$ and $b'(x) = b(x)'$ for all $\mathcal{A} \subset L^X$, $b \in L^X$ and $x \in X$. Bounds of L^X are denoted by $1_\emptyset$ and 1_X (in general, 1_A is the characteristic function of $A \subset X$). For $A \subset X$ and $a \in L^X$, $a|A$ is the restriction of a to A. Constant member of L^X with value α is denoted α too.

Let $f\colon X \to Y$ be a map. We define (the powerset operators) $f^{\rightarrow}\colon L^X \to L^Y$ and $f^{\leftarrow}\colon L^Y \to L^X$ by $f^{\rightarrow}(a) = \bigvee_{x \in X} a(x) 1_{\{f(x)\}}$ (where $\beta 1_B = \beta \wedge 1_B$) and $f^{\leftarrow}(b) = b \circ f$ for all $a \in L^X$ and $b \in L^Y$ (the symbol $\circ$ stands for the usual composition of mappings). We note that all the properties of the standard powerset operators are inherited by $f^{\rightarrow}$ and $f^{\leftarrow}$. In most cases we write bf rather than $f^{\leftarrow}(b)$.

1.2 ***L*-topological spaces.** Following Goguen [3] and Hutton [7], a subfamily $\mathcal{T}$ of L^X is called an L-**topology** on X if:

(1) $1_\emptyset, 1_X \in \mathcal{T}$,

(2) $u, v \in \mathcal{T} \Rightarrow u \wedge v \in \mathcal{T}$,

(3) $\mathcal{U} \subset \mathcal{T} \Rightarrow \bigvee \mathcal{U} \in \mathcal{T}$

(the operations being formed pointwise in L^X). Members of $\mathcal{T}$ are called **open** (L)-**sets**, and k is **closed** if k' is open. The L-**topological space** $(X, \mathcal{T})$ will usually be denoted by just X. (The actual originator of this concept is Chang [1] who considered [0,1]-valued maps.) If $a \in L^X$, then $\mathrm{Cl}_X\, a = \overline{a} = \bigwedge\{k \in L^X :$

$a \leq k$ and k is closed$\}$ is the closure of a and $\mathrm{Int}_X a = (\mathrm{Cl}_X(a'))'$ is the interior of a. These operators satisfy the four well-known axioms of closure and interior. If $A \subset X$, the set $\mathcal{T}_A = \{u|A : u \in \mathcal{T}\}$ is called the **subspace L-topology** on (the subspace) A. A family $\mathcal{S} \subset L^X$ is said to **generate** $\mathcal{T}$ if $\mathcal{T} = \bigcap\{\mathcal{U} \supset \mathcal{S} : \mathcal{U}$ is an L-topology on $X\}$. With certain abuse of terminology, members of $\mathcal{S}$ will be called **subbasic** open.

Given two L-topological spaces X and Y, a function $f\colon X \to Y$ is **continuous** if uf is open in X whenever u is open in Y.

An L-topological space is called **stratified** if all the constant members of L^X are open (this category of spaces was introduced by Lowen [27]).

The product X^J is the usual Cartesian product L-topologized by the subbase $\{u\pi_j : u$ is open in $X, j \in J\}$ where π_j is the jth projection.

The topological modification $\imath_L X$ of X is the topological space with X as the underlying set and the topology generated by the subbase $\{u^{-1}(L \setminus \downarrow \alpha) : u$ is open in $X, \alpha \in L\}$. This assignment is functorial: a function continuous from X to Y is continuous from $\imath_L X$ to $\imath_L Y$. It is important to note that the functor $\imath_L$ preserves products (see [15] and [16]).

1.3 L-reals. Let $(L, ')$ be a complete lattice. Following [7] and [2], let $\mathbb{R}_L$ be the set of all order-reversing members of $L^{\mathbb{R}}$ and such that $\bigvee a(\mathbb{R}) = 1$ and $\bigwedge a(\mathbb{R}) = 0$. For $a \in \mathbb{R}_L$ and $t \in \mathbb{R}$, we let $a^+(t) = \bigvee a(t, \infty)$ and $a^-(t) = \bigwedge a(-\infty, t)$. Given $a, b \in \mathbb{R}_L$, let $a \sim b$ if and only if $a^+ = b^+$ (this is equivalent to the statement that $a^- = b^-$). The set $\mathbb{R}(L)$ of all the equivalence classes $[a]$ is called the **L-real line.** With $[a] \leq [b]$ if and only if $a^+ \leq b^+$ (iff $a^- \leq b^-$), $\mathbb{R}(L)$ becomes a poset. The **natural L-topology** on $\mathbb{R}(L)$ is generated by the L-sets $R_t, L_t \in L^{\mathbb{R}(L)}$ $(t \in \mathbb{R})$ defined by $R_t[a] = a^+(t)$ and $L_t[a] = a^-(t)'$. In what follows, $\mathbb{R}(L)$ carries its natural L-topology. The set $\mathbb{I}(L) = \{[a] \in \mathbb{R}(L) : a^-(0)' = a^+(1) = 0\}$ is called the (**unit**) **L-interval.** It has the subspace L-topology and the ordering induced from $\mathbb{R}(L)$. For a fuller account we refer to [21]. Note the notation $\mathbb{R}_L$ is used in Chapter 7 (of this Volume) of the usual real line with a particular L-topology.

1.4 Continuous L-real functions. We write $C(X, \mathbb{R}(L))$ and $C(X, \mathbb{I}(L))$ for the posets of all continuous functions from an L-topological space X to $\mathbb{R}(L)$ and $\mathbb{I}(L)$, respectively, with the partial ordering induced from $\mathbb{R}(L)$. An $a \in L^X$ is called an **L-zero-set** if $a = R_0' f$ for some $f \in C(X, \mathbb{I}(L))$. Also, a is an **L-cozero-set** if a' is an L-zero-set. We refer to [23] for a more detailed discussion.

Let X and Y be L-topological spaces, $A \subset X$, and let $f\colon A \to Y$ be continuous. A continuous $g\colon X \to Y$ is called a **continuous extension** of f over X if $g|A = f$ (i.e., $g(x) = f(x)$ for all $x \in A$).

There are two more topologies on $\mathbb{R}(L)$: $\mathcal{R}_L = \{R_t : t \in \mathbb{R}\} \cup \{1_\emptyset, 1_{\mathbb{R}(L)}\}$ and $\mathcal{L}_L = \{L_t : t \in \mathbb{R}\} \cup \{1_\emptyset, 1_{\mathbb{R}(L)}\}$. Given an L-topological space X, $LSC(X, \mathbb{R}(L))$ (resp. $USC(X, \mathbb{R}(L))$) denotes the family of all continuous mappings from

X into $(\mathbb{R}(L), \mathcal{R}_L)$ (resp. into $(\mathbb{R}(L), \mathcal{L}_L)$). The sets $LSC(X, \mathbb{I}(L))$ (resp. $USC(X, \mathbb{I}(L))$) are defined analogously. Members of $LSC(X, \mathbb{R}(L))$ are called **lower semicontinuous**, and members of $USC(X, \mathbb{R}(L))$ are called **upper semicontinuous**.

For a more detailed description we refer to [21]. Other terminology and notation is explained as it comes up.

2 Normality-type axioms and L-reals

The three normality-type axioms which follow have been introduced by Hutton [7] and Hutton and Reilly [10].

Let $(L, ')$ be a complete lattice. An L-topological space X is called:
L-normal if for any closed $k \in L^X$ and any open $u \in L^X$ such that $k \leq u$ there exists an open set $v \in L^X$ such that $k \leq v \leq \overline{v} \leq u$;
completely L-normal if given $a, b \in L^X$ with $\overline{a} \leq b$ and $a \leq \operatorname{Int} b$, there exists an open $u \in L^X$ such that $a \leq u \leq \overline{u} \leq b$;
perfectly L-normal if it is L-normal and every open L-set is a countable sup of closed L-sets (i.e., every open L-set is F_σ).

The first three theorems which follow have been originally proved for $(L, ')$ a frame, the reason being the wrong thinking (well, at least in [14]) that characterizing continuity in terms of the inverse images of subbasic open L-sets does require L to be a frame. Rodabaugh [31] has observed that completeness of L suffices. We thus have:

2.1 Theorem (Urysohn lemma; Hutton [7]). *Let $(L, ')$ be a complete lattice. For X an L-topological space, the following are equivalent:*

(1) *X is L-normal.*

(2) *If $k \in L^X$ is closed, $u \in L^X$ is open, and $k \leq u$, then there exists a continuous $f\colon X \to \mathbb{I}(L)$ such that $k \leq L_1' f \leq R_0 f \leq u$.*

The above theorem is a simple corollary of the following:

2.2 Theorem (Insertion theorem [14]). *Let $(L, ')$ be a complete lattice. For X an L-topological space, the following are equivalent:*

(1) *X is L-normal.*

(2) *If $g \in USC(X)$, $h \in LSC(X)$ and $g \leq h$, then there exists an $f \in C(X)$ such that $g \leq f \leq h$.*

Another corollary of 2.2 is the Tietze-Urysohn extension theorem. It is convenient to use the notion of suitability (for extending functions) introduced in [30]. An L-topological space is called **suitable** if there exists a nonempty proper subset A of X such that 1_A is closed. Then A is called **suitable closed subspace** of X.

2.3 Theorem (Tietze-Urysohn extension theorem [14]). *Let $(L,')$ be a complete lattice. Let X be L-normal and $A \subset X$ be suitable closed. Then every continuous function $f\colon A \to \mathbb{I}(L)$ has a continuous extension over X.*

A direct proof of 2.3 (avoiding 2.2) is given in [21]. This theorem yields several problems. First, unlike its $\{0,1\}$-counterpart, it does not characterize L-normality (if $|L| > 2$) simply because there are non-suitable L-normal spaces. We refer to [21] for an example of a non-$[0,1]$-normal suitable space for which the extension theorem holds true.

It will be convenient to have the following standard topological terminology. Given an L-topological space, a subset $A \subset X$ is said to be **C^*-embedded** (respectively, **C-embedded**) in X if every continuous function f from A to $\mathbb{I}(L)$ (respectively, $\mathbb{R}(L)$) can be extended continuously to the whole of X.

Our first question quoted from [24] is this:

2.4 Question. Under what conditions is L-normality equivalent to C^*-embedding of suitable closed subspaces?

Let us also recall the following long-standing question:

2.5 Question [30]. For which $(L,')$ with $|L| > 2$ does every continuous $f\colon A \to \mathbb{R}(L)$, where A is a suitable closed subspace of an L-normal space X, have a continuous extension to the whole of X?

More generally, no relation between C^*-embedding and C-embedding is known to hold for L-topological spaces. Clearly, for every complete $(L,')$, a subspace of X is C^*-embedded if it is C-embedded in X, but we do not know any necessary and sufficient condition in terms of L (when $|L| > 2$). We only know the following:

2.6 Proposition [21]. *Let X be an L-topological space with $(L,')$ a meet-continuous lattice. If $A \subset X$ is such that 1_A is an L-zero-set, then A is C^*-embedded in X if and only if it is C-embedded in X.*

Let us now turn to complete L-normality. The following extends a characterization of topological complete normality of [17] and [18] to L-topological setting. A proof is given in the very recent paper [28].

2.7 Theorem. *Let* $(L,')$ *be meet-continuous. For* X *an* L*-topological space, the following are equivalent:*

(1) X *is completely* L*-normal.*

(2) *If* $a, b\colon X \to \mathbb{R}(L)$ *are such that* $a \leq f \leq b$ *and* $a \leq g \leq b$ *with an* $f \in LSC(X, \mathbb{R}(L))$ *and* $g \in USC(X, \mathbb{R}(L))$*, then there exist* $l \in LSC(X, \mathbb{R}(L))$ *and* $h \in USC(X, \mathbb{R}(L))$ *such that* $a \leq l \leq h \leq b$.

For perfectly L-normal spaces, we have the following sharpened version of Urysohn's lemma:

2.8 Theorem [7]. *Let* $(L,')$ *be a meet-continuous lattice. For* X *an* L*-topological space, the following are equivalent:*

(1) X *is perfectly* L*-normal.*

(2) *If* $k \in L^X$ *is closed,* $u \in L^X$ *is open, and* $k \leq u$ *then there exists a continuous function* $f\colon X \to \mathbb{I}(L)$ *such that* $k = L'_1 f \leq R_0 f = u$.

Since in a perfectly L-normal space every closed L-set is an L-zero-set, we have on account of 2.6 the following extension theorem which provides a partial answer to the Question 2.5:

2.9 Theorem ([13], [21]). *Let* $(L,')$ *be meet-continuous. Let* X *be a perfectly* L*-normal space and* A *be a suitable closed subspace of* X*. Then every continuous* $f\colon A \to \mathbb{R}(L)$ *has a continuous extension over* X*.*

For $(L,')$ a meet-continuous lattice, there are abstract separation, insertion and extension theorems which have the presented results as corollaries. We cite from [21]:

2.10 Theorem (General insertion theorem). *Let* $(L,')$ *be a meet-continuous lattice. Let* X *be an* L*-topological space and let* $g, h\colon X \to \mathbb{R}(L)$ *be two arbitrary functions. The following are equivalent:*

(1) *There exists* $f \in C(X, \mathbb{R}(L))$ *such that* $g \leq f \leq h$.

(2) *If* $s < t$ *in* $\mathbb{R}$*, then* $R_s{}'h$ *and* $L_t{}'g$ *are completely separated.*

If $A \subset X$ and $a \in L^A$, then $(a)_0 \in L^X$ is defined as follows: $(a)_0 = a$ on A, and $(a)_0 = 0$ on $X - A$.

2.11 Theorem (General extension theorem). *Let $(L,')$ be a meet-continuous lattice. Let X be an L-topological space, $A \subset X$, and let $f \in C(A, \mathbb{I}(L))$. The following are equivalent:*

(1) *f extends continuously over X.*

(2) *$(R_s{}'f)_0$ and $(L_t{}'f)_0$ are completely separated in X for every $s < t$ in $[0,1]$.*

2.12 Theorem (Urysohn extension theorem). *Let $(L,')$ be a meet-continuous lattice. Let X be an L-topological space and $A \subset X$. The following are equivalent:*

(1) *Every $f \in C(A, \mathbb{I}(L))$ has a continuous extension over X.*

(2) *For all $a, b \in L^A$, if a and b are completely separated in A, then $(a)_0$ and $(b)_0$ are completely separated in X.*

3 Complete L-regularity and L-reals

Completely L-regular spaces have been introduced by Hutton [8] (independently, Katsaras [12] introduced another definition which is however equivalent to that of Hutton; cf. [19, 4.3 and 4.4]).

Let $(L,')$ be a complete lattice. An L-topological space is **completely L-regular** if for every open L-set $u \in L^X$, there exists a family $\mathcal{A} \subset L^X$ and a family $\{f_a : a \in \mathcal{A}\} \subset C(X, \mathbb{I}(L))$ such that

$$a \leq L_1{}'f_a \leq R_0 f_a \leq u \qquad (\star)$$

for all $a \in \mathcal{A}$.

If $a, b \in L^X$, we shall sometimes write $a \prec b$ if a and b satisfy condition $(\star)$ for some $f \in C(X, \mathbb{I}(L))$. Also, $a, b \in L^X$ are said to be **completely separated** in X if $a \prec b'$.

The following provides a list of various useful characterizations of completely L-regular spaces.

3.1 Theorem. *Let $(L,')$ be a complete lattice. For X an L-topological space, the following are equivalent:*

(1) *X is completely L-regular.*

(2) $u = \bigvee\{v \in L^X : v \prec u$ and v is open $\}$ *for every open $u \in L^X$.*

(3) $\mathrm{Int}_X\, a = \bigvee\{v \in L^X : v \prec a$ and v is open$\}$ *for every $a \in L^X$.*

(4) *Every lower semicontinuous function* $f\colon X \to \mathbb{I}(L)$ *is a pointwise sup of a subfamily of* $C(X, \mathbb{I}(L))$.

(5) *The family of all* L*-cozero-sets is a base for* X.

These conditions imply (6) $\iff$ (7) $\implies$ (8) *below, and if* $(L, ')$ *is a frame, then* (1)–(8) *are equivalent:*

(6) $k = \bigwedge\{f^{\leftarrow}(\overline{f^{\rightarrow}(k)}) : f \in C(X, \mathbb{I}(L))\}$ *for every closed* $k \in L^X$.

(7) $\mathrm{Cl}_X\, a = \bigwedge\{f^{\leftarrow}(\overline{f^{\rightarrow}(a)}) : f \in C(X, \mathbb{I}(L))\}$ *for every* $a \in L^X$.

(8) X *has the smallest* L*-topology with respect to which all members of* $C(X, \mathbb{I}(L))$ *are continuous.*

3.2 Comment. (1) The equivalences (1)–(7) come from [19] (also see [21]). When $L = \{0, 1\}$, the condition (6) says that $C(X, [0, 1])$ separates points from closed sets, i.e., given $x \notin K$ (a closed subset of the topological space X), there is a continuous function $f\colon X \to [0, 1]$ such that $f(x) \notin \overline{f(K)}$. Condition (6) has a pointed equivalent formulation in [26], as well as in [37] with $L = [0, 1]$ (cf. [19] and [21]). Condition (8) has been obtained by Katsaras in [12], and a completely distributive lattice argument involving L-uniformities for (1) $\Leftrightarrow$ (8) is in Liu [26].

(2) A frame argument for (8) $\Rightarrow$ (1) is given in [19]. The frame law has been used in proving that the open L-set u in the definition of complete L-regularity can be assumed to be subbasic open. This yields the following question:

3.3 Question ([19], [24]). Let $(L, ')$ be a complete lattice and let $\mathcal{S} \subset L^X$ generate the L-topology of X. If every member of $\mathcal{S}$ satisfies condition (2) of 3.1, must X then be completely L-regular?

There are also separation and extension theorems for completely L-regular spaces. After [9], an L-topological space X is H(utton)-**compact** if every closed L-set k is compact, i.e., whenever $k \le \bigvee \mathcal{U}$, $\mathcal{U}$ being a family of open L-sets, there is a finite subfamily $\mathcal{V} \subset \mathcal{U}$ with $k \le \bigvee \mathcal{V}$. Also, a complete lattice L is **algebraic** if every $\alpha \in L$ is a sup of compact elements of L, where $\beta \in L$ is **compact** if, whenever $a \le \bigvee A$ for $A \subset L$, there is a finite subset $B \subset A$ with $a \le \bigvee B$. Note that every algebraic L is meet-continuous.

3.4 Proposition [19]. *Let* $(L, ')$ *be an algebraic lattice. For an* L*-topological space* X, *the following are equivalent:*

(1) X *is completely* L*-regular.*

(2) *If* $k \in L^X$ *is compact,* $u \in L^X$ *is open, and* $k \le u$ *then there exists a continuous function* $f\colon X \to \mathbb{I}(L)$ *such that* $k \le L_1' f \le R_0 f \le u$.

Note. In Proposition 3.4, (1) implies (2) for any meet-continuous $(L,')$.

3.5 Theorem [19]. *Let* $(L,')$ *be meet-continuous. Let* A *be a suitable closed* H*-compact subspace of a completely* L*-regular space* X. *Then every continuous* $f\colon A \to \mathbb{I}(L)$ *has a continuous extension over* X.

The next theorem is an L-extension of the Nakano-Stone theorem on conditional completeness of the poset of all continuous real-valued functions.

We recall that an L-topological space X is **extremally disconnected** if $\overline{u}$ is open for every open $u \in L^X$.

3.6 Theorem [19]. *Let* $(L,')$ *be a complete lattice. For* X *a completely* L*-regular space, the following statements are equivalent:*

(1) X *is extremally disconnected.*

(2) *The poset* $C(X, \mathbb{R}(L))$ *is conditionally complete.*

The point-free condition (6) of 3.1 does not allow us to have an L-version of the Tychonoff embedding theorem with the amount of factors of the range space being controlled according to the weight of the embeddable space. For a general treatment of the problems of embedding L-topological spaces into products we refer to [19], [21], [26] and [37] (the last two papers have L a completely distributive lattice or $L = [0,1]$, respectively). Historically, the first embedding theorem comes from [12]. Here we discuss only those embedding theorems that are concerned with L-Tychonoff spaces.

Recall [26] that an L-topological space X is L-**Tychonoff** if X is completely L-regular and open L-sets separate points of X, i.e., whenever $x \neq y$ in X, there exists an open $u \in L^X$ such that $u(x) \neq u(y)$. The latter is the L-T_0 axiom introduced independently by a number of authors (see [19] for original source references).

Also, the **weight** $\omega(X)$ of an L-topological space is the smallest infinite cardinal such that X has a base of cardinality $\leq \omega(X)$, i.e., $\omega(X) = \min\{|\mathcal{B}| : \mathcal{B}$ is a base for $X\} + \aleph_0$.

3.7 Theorem. *For every complete* $(L,')$, *each* L*-Tychonoff space* X *can be embedded into a cube* $\mathbb{I}(L)^{\mathbf{m}}$ *for some* $\mathbf{m} \leq \omega(X)$.

Note. If we assume that X is stratified, then with $\mathbb{I}(L)^{\mathbf{c}}$ we can simply state that X is embeddable into $(\mathbb{I}(L)^{\mathbf{c}})^{\omega(X)}$. (Here $\mathbb{I}(L)^{\mathbf{c}}$ stands for $\mathbb{I}(L)$ with its natural L-topology enriched with all the constant L-sets.)

3.8 Comment. Theorems 3.7 and 3.1 are typical examples of the situation when completeness of $(L,')$ usually does suffice to establish of what is implied by complete L-regularity. Statements that imply complete L-regularity usually require $(L,')$ to be a frame. An example of this circumstance is provided by (8) $\Rightarrow$ (1) in Theorem 3.1 as well as in the theorem which follows. The reason is that we have an answer to Question 3.3 only for $(L,')$ being a frame.

3.9 Theorem. *Let $(L,')$ be a frame. The following hold:*

(1) *A space is L-Tychonoff if and only if it is homeomorphic to a subspace of an L-Tychonoff cube.*

(2) *A space is L-regular, second countable and L-T_0 if and only if it is homeomorphic to a subspace of an L-Hilbert cube.*

Note. An L-topological space is L-**regular** [10] if $u = \bigvee\{v \in L^X : \overline{v} \leq u$ and v is open$\}$ for every open $u \in L^X$.

Recall that a **universal** L-topological space for a given class of L-topological spaces is a space in which every space belonging to the class can be embedded. We have the following (see [19], [21]):

3.10 Theorem. *Let $(L,')$ be a frame. Then $(\mathbb{I}(L)^c)^{\max\{\mathbf{m},|L|\}}$ is universal for all stratified L-Tychonoff spaces of weight $\leq \max\{\mathbf{m},|L|\}$ where $\mathbf{m}$ is an infinite cardinal.*

We close this section with discussing the problem of an internal characterization of complete L-regularity. The discussion comes from [23]. First we recall that an internal characterization is one that depends only on the lattice of all open (or closed) L-sets and does not involve $\mathbb{R}(L)$ or $\mathbb{R}$ (even as an index set).

The internal characterization of topological complete regularity due to Johnson and Mandelker [11] extends to an L-topological setting with $(L,')$ a complete lattice. Given an L-topological space X, a **separating chain** in X is a family $\mathcal{U}$ of open L-sets such that: (1) $\mathcal{U}$ is a countable chain in L^X, (2) $\bigwedge\mathcal{U} = 1_\emptyset$ and $\bigvee\mathcal{U} = 1_X$, and (3) if $u, v \in \mathcal{U}$, $u \leq v$, and $u \not= v$, then there exists $w \in \mathcal{U}$ such that $\overline{u} \leq w \leq \overline{w} \leq v$. We then have:

3.11 Theorem [23]. *Let $(L,')$ be a complete lattice. For X an L-topological space, the following are equivalent:*

(1) *X is completely L-regular.*

(2) *For every open $u \in L^X$ there exists a family $\mathcal{V} \subset L^X$ of open L-sets such that $\bigvee\mathcal{V} = u$, and for every $v \in \mathcal{V}$ there exists a separating chain $\mathcal{U}$ such that $v \leq g \leq \overline{g} \leq h \leq u$ for some $g, h \in \mathcal{U}$.*

Another internal characterization of complete L-regularity due to Steiner [38]. The following definition is a point-free formulation of the concept of a normal and separating family of closed subsets of a topological space, which is written down in an L-topological setting.

Let $(L,')$ be a complete lattice. A family $\mathcal{K} \subset L^X$ of closed L-sets of an L-topological space is called:
separating, if for every open $u \in L^X$ there exist two families $\{a_\gamma : \gamma \in \Gamma\}$, $\{b_\gamma : \gamma \in \Gamma\} \subset \mathcal{K}$ such that $u = \bigvee_{\gamma \in \Gamma} a_\gamma$ and $a_\gamma \leq {b_\gamma}' \leq u$ for every $\gamma \in \Gamma$,
normal, if for every $a, b \in \mathcal{K}$ with $a \leq b'$ there exist $c, d \in \mathcal{K}$ such that $a \leq c' \leq d \leq b'$.

The following generalizes the hardest part of Steiner's characterization:

3.12 Theorem [23]. *Let $(L,')$ be a complete lattice. Every L-topological space with a normal and separating family of closed L-sets is completely L-regular.*

The necessity part of Steiner's characterization is easy and follows from the fact that in a completely regular topological space the family of all zero-sets is separating and normal while the family of all L-zero-sets in a completely L-regular space is separating for any complete $(L,')$ (see [23, 5.5]), we do not know if it is normal (note that in any topological space, the family of all zero-sets is obviously normal).

We thus have another open question which arises from the fact that if $|L| > 2$ then $C(X, \mathbb{R}(L))$ is not a ring (and merely a lattice provided $(L,')$ is meet-continuous, cf. [19]).

3.13 Question ([19], [23], [24]). For which complete lattices $(L,')$ with $|L| > 2$, are every two L-zero-sets a and b completely separated. Of course, this is the same question as to when is the family of all L-zero-sets a normal family.

An affirmative answer to 3.13 would also provide a positive answer to Question 2.4, among others.

4 Brouwer fixed point theorem for L-cubes

The Brouwer fixed point theorem is one of the most important existence theorems in mathematics.

The study of whether an arbitrary L-Tychonoff cube has the fixed point property has recently been initiated in [20]. We recall that an L-topological X space has the **fixed point property** (FPP for short) if every continuous function from X to X has a fixed point. It was shown in [20] that every L-cube with finitely or countably many factors has FPP provided $(L,')$ is a completely distributive lattice with a **countable base** (i.e., there is a countable subset $B \subset L$ such that $\alpha = \bigvee\{\beta \in B : \beta \leq \alpha\}$ for every $\alpha \in L$).

Then it has been shown in [25] that the latter assumption actually guarantees the FPP for arbitrary stratified L-cubes, while complete distributivity is all we need to have FPP for an arbitrary L-Tychonoff cube.

As already mentioned in the Introduction, the fuzzy Brouwer fixed point theorem depends more or less explicitly upon the Brouwer's theorem. In [20] the problem was reduced to a theorem of Papert Strauss [29] which asserts that every connected and metrizable completely distributive lattice with respect to its interval topology has FPP (a result depending on Brouwer's theorem) by using the product-preserving functor $\iota_L : L\text{-}\mathbf{TOP} \to \mathbf{TOP}$. Similarly, in [25] the proof depends upon a statement (proved in [25] by using the classical Brouwer's theorem) that if a completely distributive lattice with its interval topology has FPP, then its L-fuzzy modification $(M(L), \delta_L(M))$, as introduced in Zhang and Liu [39], has FPP. In the case $M = [0,1]$ we have Hutton's L-interval.

To summarize, the present status of Brouwer fixed point theorem in L-topological setting is as follows.

4.1 Theorem (Brouwer fixed point theorem [25]). *Let $(L, ')$ be a completely distributive lattice. Then:*

(1) *Every L-Tychonoff cube $\mathbb{I}(L)^m$ has the fixed point property.*

(2) *If L has a countable base, then every stratified L-Tychonoff cube has the fixed point property.*

4.2 Remark. For some related questions we refer to [25].

5 Final comment

It would be useful to know how much of the presented results go over to the new classes of fuzzy real lines defined in [33, 2.16.6–2.16.7 cf. 8.15.5(3)] as well as to the many valued real lines of [4, 5.5.1]. We note in particular (as suggested by S. E. Rodabaugh) that the Stone-Weierstraß Theorem obtained in [33, 8.15.5], using $\mathbb{R}(L)$ and requiring L to be a complete Boolean algebra, can be generalized to a Stone Weierstraß Theorem using the new $\mathbb{R}^*(L)$ of [33, 2.16.6–2.16.7], but only requiring L a distributive semiframe, i.e. a distributive complete lattice.

References

[1] C.L. Chang, *Fuzzy topological spaces*, J. Math. Anal. Appl. **24**(1968), 182–190.

[2] T.E. Gantner, R.C. Steinlage, R.H. Warren, *Compactness in fuzzy topological spaces*, J. Math. Anal. Appl. **62**(1978), 547–562.

[3] J.A. Goguen, *The fuzzy Tychonoff theorem*, J. Math. Anal. Appl. **43**(1973), 734–742.

[4] U. Höhle, Many Valued Topology And Its Applications, Kluwer Academic Publishers (Boston/Dordrecht/London), 2001.

[5] U. Höhle, A. P. Šostak, *Axiomatic foundations of fixed-basis fuzzy topology*, Chapter 3 in [6], 123–272.

[6] U. Höhle, S. E. Rodabaugh, eds, Mathematics Of Fuzzy Sets: Logic, Topology, And Measure Theory, The Handbooks of Fuzzy Sets Series: Volume **3**(1999), Kluwer Academic Publishers (Boston/Dordrecht/London).

[7] B. Hutton, *Normality in fuzzy topological spaces*, J. Math. Anal. Appl. **50**(1975), 74–79.

[8] B. Hutton, *Uniformities on fuzzy topological spaces*, J. Math. Anal. Appl. **58**(1977), 559–571.

[9] B. Hutton, *Products of fuzzy topological spaces*, Topology Appl. **11**(1980), 59–67.

[10] B. Hutton, I. L. Reilly, *Separation axioms in fuzzy topological spaces*, Fuzzy Sets Syst. **3**(1980), 93–104.

[11] D. G. Johnson, M. Mandelker, *Separating chains in topological spaces*, J. London Math. Soc. (2), **4**(1972), 510–512.

[12] A.K. Katsaras, *Fuzzy proximities and fuzzy completely regular spaces I*, Anal. Stiin. Univ. "Al. I. Cuza" Iasi **26**(1980), 31–41.

[13] T. Kubiak, *Extending continuous L-real functions*, Math. Japon. **31** (1986), 875–887.

[14] T. Kubiak, *L-fuzzy normal spaces and Tietze extension theorem*, J. Math. Anal. Appl. **125**(1987), 141–153.

[15] T. Kubiak, *Notes on the topological modification of the L-fuzzy unit interval*, Communications IFSA Math. Chapter **4**(1989), 26–30.

[16] T. Kubiak, *The topological modification of the L-fuzzy unit interval*, Chapter 11 in [35], 275–305.

[17] T. Kubiak, *A strengthening of the Katětov-Tong insertion theorem*, Comment. Math. Univ. Carolinae **34**(1993), 357–362.

[18] T. Kubiak, *Are higher fuzzy separation axioms good extensions?*, Quaest. Math. **16**(1993), 443–451.

[19] T. Kubiak, *On L-Tychonoff spaces*, Fuzzy Sets Syst. **73**(1995), 25–53.

[20] T. Kubiak, *The fuzzy Brouwer fixed-point theorem*, J. Math. Anal. Appl. **222** (1998), 62–66.

[21] T. Kubiak, *Separation axioms: extensions of mappings and embedding of spaces*, Chapter 6 in [6], 433–479.

[22] T. Kubiak, *L-fuzzy reals and topology*, Math. Japon. (Mathematical Plaza), to appear.

[23] T. Kubiak, M.A. de Prada Vicente, *On internal characterizations of completely L-regular spaces*, J. Math. Anal. Appl. **216**(1997), 581–592.

[24] T. Kubiak, M. A. de Prada Vicente, *Some questions in fuzzy topology*, Fuzzy Sets and Systems **105**(1999), 277–285.

[25] T. Kubiak, D. Zhang, *On the L-fuzzy Brouwer fixed point theorem*, Fuzzy Sets and Systems **105**(1999), 287–292.

[26] Liu Ying-Ming, *Pointwise characterization of complete regularity and imbedding theorem in fuzzy topological spaces*, Scientia Sinica, Series A, **26**(1983), 138–147.

[27] R. Lowen, *Fuzzy topological spaces and fuzzy compactness*, J. Math. Anal. Appl. **56**(1976), 621–633.

[28] I. Mardones Pérez, *Higher separation axioms in L-topological generated I(L)-topological spaces*, submitted.

[29] D. Papert Strauss, *Topological lattices*, Proc. London Math. Soc. **18**(1968), 217–230.

[30] S. E. Rodabaugh, *Suitability in fuzzy topological spaces*, J. Math. Anal. Appl. **79**(1981), 273–285.

[31] S. E. Rodabaugh, *A point-set lattice-theoretic framework $\mathbb{T}$ for topology which contains **LOC** as a subcategory of singleton spaces and in which there are general classes of Stone Representation and Compactification Theorems*, First printing February 1986, Second printing April 1987, Youngstown State University Printing Office (Youngstown, Ohio(USA)).

[32] S. E. Rodabaugh, *Categorical foundations of variable-basis fuzzy topology*, Chapter 4 in [6], 273–388.

[33] S. E. Rodabaugh, *Separation axioms: representation theorems, compactness, and compactifications*, Chapter 7 in [6], 481–552.

[34] S. E. Rodabaugh, *Fuzzy real lines and dual real lines as poslat topological, uniform, and metric ordered semirings with unity*, Chapter 10 in [6], 607–631.

[35] S. E. Rodabaugh, E.P. Klement, U. Höhle, eds, Applications Of Category Theory To Fuzzy Subsets, Theory and Decision Library—Series B: Mathematical and Statistical Methods, Volume **14**, Kluwer Academic Publishers (Boston/Dordrecht/London), 1992.

[36] A.P. Šostak, *Two decades of fuzzy topology: basic ideas, notions and results*, Russian Math. Surveys **4:6**(1989), 125–186.

[37] A.P. Šostak, *On the concept of E-regularity for fuzzy topology*, Mat. Vesnik **41:3**(1989), 189–203.

[38] E. F. Steiner, *Normal families and completely regular spaces*, Duke Math. J. **33**(1966), 743–745.

[39] Zhang Dexue, Liu Ying-Ming, *L-fuzzy modification of completely distributive lattices*, Math. Nachr. 168(1994), 79–95.

CHAPTER 6

Lattice-Valued Frames, Functor Categories, And Classes Of Sober Spaces

A. PULTR [1] AND S. E. RODABAUGH

Introduction and motivation

This chapter introduces *lattice-valued frames* or *L-frames*, related to traditional frames analogously to how L-topological spaces relate to traditional spaces, via level sets and level mappings viewed as *systems of frame morphisms* (Proposition 3.3.2). En route, the well-known Ω and $L\Omega$ functors [25, 10, 11, 12, 16, 17, 18, 29, 33, 43, 44, 48, 54, 55, 56, 57, 58, 61] relating traditional spaces and L-spaces, respectively, to their associated (semi)locales of open and L-open sets are analogized and modified. This study both gives new descriptions of classes of sober spaces extant in the literature and creates a new class of sober spaces, justifying examples for which are given in Chapter 17 [42] of this Volume.

To precisely state the main purpose of this chapter, first recall for L a complete lattice that (X, τ) is an **L-topological space** if $\tau \subset L^X$ is closed under arbitrary joins and finite meets. Note L a frame implies τ is a frame. Typically the lattice-theoretic structure of τ does not carry enough information to regenerate the space (X, τ); restated, if τ is order-isomorphic to a lattice A, then we may not be able to recover (X, τ) up to L-homeomorphism—in particular the *carrier set* X [25] up to bijection—from A using only the structure of A. Consider the following two examples—see [5, 19, 12], Subsections 2.15–2.16 of [61], and 2.3 below for documentation:

- If L is a complete Boolean algebra, then the standard L-topology $\tau(L)$ of the L-fuzzy real line $(\mathbb{R}(L), \tau(L))$ is order-isomorphic to the standard topology $\mathfrak{T}$ of $(\mathbb{R}, \mathfrak{T})$, and in this case we can recover the carrier set $\mathbb{R}(L)$ up to bijection from $\mathfrak{T}$ via (universality of) the L-soberification functor.

[1] Support of the Grant Agency of the Czech Republic under Grant 201/99/0310 is gratefully acknowledged.

S.E. Rodabaugh and E.P. Klement (eds.), Topological and Algebraic Structures in Fuzzy Sets, 153–187.
© 2003 *Kluwer Academic Publishers.*

- But if $L = \mathbb{I} = [0,1]$ with the usual ordering, then the L-topology $\tau(L)$ of $(\mathbb{R}(L), \tau(L))$ is still order-isomorphic to the topology $\mathfrak{T}$ of some traditional space $(Y, \mathfrak{T})$ (by 2.3 below), but in this case we cannot recover the carrier set $\mathbb{R}(L)$ up to bijection from $\mathfrak{T}$ via the L-soberification functor.

This chapter solves the conundrum of the second example above by enriching a frame A so that we can determine:

- what the underlying carrier set X is (up to bijection); and
- in what way the elements of A appear as L-subsets of X.

Such an enrichment brings us to the structure of an *L-frame*, related to frames in a way analogous to the relationship between traditional and L-topological spaces given by level sets and level mappings. Further, this enrichment allows us to extend the $L\Omega$ functor, from L-topology to (semi)locales, to a functor between L-spaces and *L-locales* (categorical dual of L-frames). The diagram

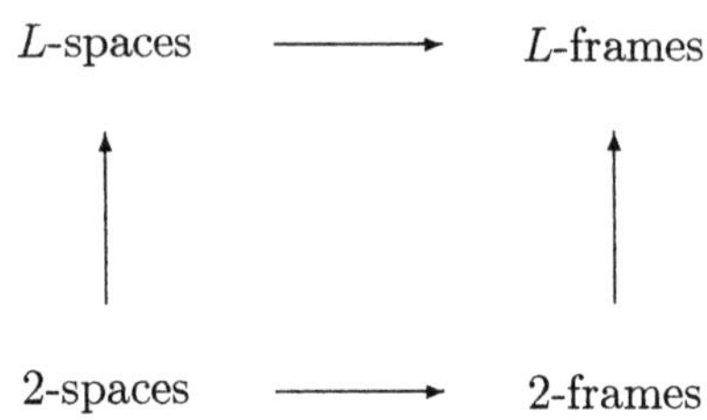

illustrates this situation, where vertical arrows indicate enrichment by fuzzification—replacing traditional subsets by lattice-valued maps—and horizontal arrows indicate the process of going point-free—essentially taking the lattice structure of a space and forgetting the given carrier set. It should be emphasized right away that the point is *not* to fuzzify a frame as an algebra by mechanically considering the associated fuzzy algebra: this would not make sense in the upper horizontal arrow.

We now discuss the goal of enrichment in the broader context of what is known about regenerating spaces from the topologies, both in traditional and lattice-valued topology. Strictly speaking, not every traditional space $(X, \mathfrak{T})$ can be recovered from the lattice $\mathfrak{T}$ of open sets. The class of spaces in which one looks for such recovery, or "representation", has to be restricted. For instance, if one restricts the class of traditional spaces to the subclass of *sober* spaces, then $(X, \mathfrak{T})$ can be recovered (up to homeomorphism) from $\mathfrak{T}$ by taking the *spectrum* of $\mathfrak{T}$ [16, 17, 18, 25, 43, 44]. For a fixed complete lattice L, an analogous restriction of the class of L-topological spaces to the subclass of *sober L-spaces* (or *L-sober spaces*) again leads to the recovery of each such (X, τ) (up to L-homeomorphism) from τ by taking the *L-spectrum* of τ introduced and studied by several authors [10, 11, 12, 29, 33, 48, 54, 55, 56, 57, 58, 61].

How rich are sober and L-sober spaces, are there important examples? It is well-known [25] that Hausdorff spaces, and hence compact T_0 (and so finite T_0 spaces), are sober spaces in the traditional setting. Furthermore, the soberification of a space—the spectrum of the topology of a space—is always sober; and if the original space is not Hausdorff, then its soberification is a sober space which is *not* Hausdorff. So there are many non-Hausdorff sober spaces as well. Thus the class of sober spaces restricts sufficiently to guarantee that each such space is recoverable (up to homeomorphism) from the lattice-theoretic structure of its topology, but does not restrict too much so as to preclude a rich inventory of important examples.

If we identify traditional spaces with **2**-topological spaces via the characteristic functor (Section 6.2 of [60])—**2** is the lattice $\{\bot, \top\}$—and further consider the **2**-spectrum, then the preceding paragraph may be restated to say that **2**-sober spaces allow both representation and rich examples. But if **2** is replaced by an arbitrary semiframe (or complete lattice) L, is there a comparably rich store of examples? For arbitrary L, the inventory of L-sober spaces is at least as rich as for traditional sober spaces, and in the cases of certain semiframes, the store of examples is strictly richer. This is a consequence of the L-**2** soberification functor $LPT \circ \Omega$ of Subsections 2.15–2.16 of [61] (cf. [12] and Subsections 6.3, 8.14–8.15 of [61]) which constructs the L-spectrum of the topology of a traditional space: for each semiframe L, the L-**2** soberification functor embeds (homeomorphism classes of) traditional sober spaces as a subclass of the class of (L-homeomorphism classes of) L-sober spaces, namely that subclass for which the L-topologies are **2**-spatial (i.e., traditionally spatial). This embedding thus gives L-topology all the examples of sober traditional spaces plus possibly more. Furthermore, it can be shown that L is **2**-spatial iff the class of L-sober spaces coincides precisely with its subclass of **2**-sober spaces (which are the embedding of the traditionally sober spaces). This means that if L is **2**-spatial, then L-topology has, up to embedding via the L-**2** soberification functor, the same inventory of sober spaces as traditional topology; and if L is not **2**-spatial—and canonical examples of such exist [4, 25]—then the class of L-sober spaces strictly contains the embedded class. We should also note that even for **2**-spatial L, an L-sober space can be much richer than the traditional sober corresponding to it [12, 61]; e.g. the L-2 soberification $\mathbb{I}^*(L)$ of $\mathbb{I} \equiv [0,1]$ is also its L-**2** compactification, adding many more points of closure to $\mathbb{I}$. Indeed, L-sobriety plays a critical and systematic role in separation axioms, Stone representation theorems, and Stone-Čech compactification reflectors for L-topological spaces [55, 56, 57, 61].

The L-sobriety of the preceding paragraph has been extended and modified in several directions. (1) Variable-basis topology, see [60] and its references, includes in one category all the distinct classes of L-sober spaces for all L in some specified category of semiframes, and the important role of these spaces in variable-basis compactification reflectors is in [61]. (2) Modifications of fixed basis or L-sobriety, allowing for L-spectra stratified to any desired degree, are given in [54, 55] and [33], including fully stratified spaces *à la* [37]; cf. Subsection 2.2 below. (3) Semi-sobriety for L-topological spaces is given in [32] and extended in [33] to (L, M)-fuzzy topological spaces in the sense of [66, 35] (cf. [15, 60]), spaces in which the topology is a lattice-valued mapping representing the predicate of openness.

To summarize, traditional topology has a significant inventory of important sober examples, included by fuzzy topology in many distinct ways, enriched (in some cases) with additional examples, and, at the same, significantly enriching many individual, traditional examples with additional points of closure.

The justification of lattice-valued frames, from the standpoint of the above overview of sobriety, is twofold. Lattice-valued frames categorically generate via a new spectrum adjunction (1) new characterizations of important examples of L-sober spaces and ω_L-generated L-spaces which are modified L-sober to degree L, and (2) important examples of a new sobriety not previously encountered in the literature. In particular, the L-spaces determined by the L-frames of this chapter include all L-spaces whose topological modification by the ι_L functor [38, 34] is a traditional sober space. Thus lattice-valued frames generate a new axiom of sobriety which, using the nomenclature standardized in [34], says an L-space (X, τ) is *ultra L-sober* or *ι_L-sober* if $(X, \iota_L(\tau))$ is traditionally sober. See Proposition 3.5.2 below and [42] for examples.

Future work based on this chapter focuses on two natural forgetful functors from the category of L-frames into the category of frames. One of them has a left adjoint determining for (traditional) frame A a universal L-frame with the "operating part" A; in particular, this adjoint for a given spatial frame A yields the "system" of all L-spaces (X, A^*) with A^* isomorphic to A. The other forgetful functor has a right adjoint yielding for a given frame A the universal L-frame carried by A; in particular, this adjoint for a given (traditional) topological space $(X, \mathfrak{T})$ yields the "system" of all subframes of $L^{\mathbb{X}}$ the topological modification of which system is $\mathfrak{T}$.

Throughout this chapter, the membership-value frame L is assumed to be a complete chain to streamline the exposition. It had been rightly pointed out to the authors by U. Höhle that the linear case essentially deals with the meet-irreducibles of L and that the meet-irreducibles may suffice for extensions to more general cases, say for spatial L's. As this chapter and book were going to press, Prof. Höhle kindly furnished comments to the authors making these suggestions concrete. His comments are given as an appendix [14] to this chapter.

On the other hand, we envisage potential applications of the notion of an L-frame in the linear case. One of the questions of interest in fuzzy topology concerns well-founded definitions of the structures of the uniformity type (see, e.g., [2, 7, 8, 9, 20, 31, 39, 50, 62]). The case of a complete chain L, important historically and still very important for applications, does not allow the direct uniformization of an L-topology τ as a frame: a uniformity on τ induces a uniformity on L (as observed by B. Banaschewski); and since the only linearly ordered frame admitting a uniformity is the two-point Boolean algebra $\{\bot, \top\}$ we would be left with the crisp case. One can, however, think of definitions based on the concept of an L-frame which would be more satisfactory.

Finally, extensions of this chapter involving a tensor product $\otimes$ [15, 60] or a quasi-complementation (order-reversing involution) on L do not seem to present difficulties and either could be built into the theory if one wished. Again, in order to streamline the exposition and constrain chapter size, we have not pursued this line here.

This chapter is organized as follows: Section 1 contains general facts from category theory, a review of traditional spatiality and sobriety, and basic information on frames; Section 2 discusses L-topological spaces, reviews L-spatiality, L-sobriety, level mappings, level sets, the ι functor, and presents a technical lemma repeatedly used throughout the text; Section 3 gives a point-free description of L-spaces by viewing the level mappings associated with an L-topology as a system of frame morphisms and, justified by this, presents L-frames and L-frame morphisms, along with relationships to traditional frames and to functor categories; and Section 4 is devoted to the ιL-spectrum of L-locales, its properties, its associated adjunction, and the categorical equivalence it induces between ιL-sober topological spaces and ιL-spatial L-locales. Thus the remaining sections of this chapter are as follows:

1. Preliminaries
2. L-topological spaces
3. L-frames
4. ιL-spectrum of L-locales

1 Preliminaries

1.1 Notation. A mapping associating elements a_i with elements i of an index set J will be sometimes called a **family** or a **system** and denoted by

$$(a_i \mid i \in J), \quad \text{or simply} \quad (a_i)_{i \in J}, \quad \text{or} \quad (a_i)_i$$

as opposed to $\{a_i \mid i \in J\}$, the set of the elements a_i where the order or repetition of the elements is irrelevant. Sometimes we will think of indexed sets as families even if the actual indexing is not important.

1.2. A family of morphisms $(\varphi_i : A \to B_i)_{i \in J}$ (resp. $(\varphi_i : A_i \to B)_{i \in J}$) in a category $\mathcal{C}$ is **collectionwise monomorphic** (resp., **collectionwise epimorphic**)—briefly, c.m. (resp., c.e.) if

$$(\forall i \in J,\ \varphi_i \alpha = \varphi_i \beta \ \ (\text{resp}, \alpha \varphi_i = \beta \varphi_i)) \quad \Rightarrow \quad \alpha = \beta$$

A c.m. (resp., c.e.) system is **collectionwise extremally monomorphic** (resp., **collectionwise extremally epimorphic**)—or c.e.m. (resp., c.e.e.)—if it additionally satisfies:

$$(\forall i \in J,\ \varphi_i = \varphi_i'\varepsilon)\ \&\ \varepsilon \text{ epimorphic} \quad\Rightarrow\quad \varepsilon \text{ is an isomorphism}$$
$$(\text{resp., } (\forall i \in J,\ \varphi_i = \mu\varphi_i')\ \&\ \mu \text{ monomorphic} \quad\Rightarrow\quad \mu \text{ is an isomorphism })$$

Note that

limits constitute c.e.m.systems, colimits constitute c.e.e. systems (1.2.1)

Also note c.m. is the condition of **mono-source** and c.e.e. is the condition of **epi-extremal sink** in [1].

We shall use the following simple fact:

Lemma Let $\mathcal{C}$ have coequalizers. Then the following are equivalent:

1. A system $(\varphi_i : A \to B_i)_{i\in J}$ is c.m.
2. Each $\gamma : A \to C$ satisfying $(\forall j \in J,\ \exists \psi_j : C \to B_j,\ \psi_j\gamma = \varphi_j)$ is a monomorphism.

Proof. For (1) $\Rightarrow$ (2), let $(\varphi_i)_i$ be c.m., $\gamma\alpha = \gamma\beta$, and $i \in J$. Then $\exists\, \psi_i$ satisfying $\psi_i\gamma = \varphi_i$. But $\varphi_i\alpha = \psi_i\gamma\alpha = \psi_i\gamma\beta = \varphi_i\beta$; hence $\alpha = \beta$. For (2) $\Rightarrow$ (1), assume (2) and that $(\forall i \in J,\ \varphi_i\alpha = \varphi_i\beta)$, and choose $\gamma = \text{Coequ}(\alpha, \beta)$. Then $\forall\, j \in J,\ \exists\,(!)\psi_j$ with the stated condition. Applying (2), γ is a monomorphism, hence an isomorphism; and hence $\alpha = \beta$ (using 7.70(3) of [1]). □

1.3. A **frame** is a complete lattice A satisfying the first infinite distributive law:

$$\forall\, S \subset A,\ \forall\, b \in A,\ (\bigvee S) \wedge b = \bigvee\{a \wedge b \mid a \in S\}$$

and a **frame morphism** $h : A \to M$ is a mapping preserving all joins (including the bottom $\bot$) and finite meets (including the top $\top$). The category of frames and frame morphisms will be denoted by **Frm**. Its dual category $\mathbf{Frm}^{op}$ is **Loc**, so that frames and **locales** as objects are the same; but the morphisms of **Loc** are those of **Frm** read in the opposite direction.

A **semiframe** is a complete lattice and a **semiframe** morphism preserves all joins and finite meets; and by analogy to the above, we have **semilocales** and the categories **SFrm** and $\mathbf{SLoc} \equiv \mathbf{SFrm}^{op}$.

The one-element frame $\{\bot = \top\}$ will be denoted by **1** and the two-element frame (Boolean algebra) $\{\bot < \top\}$ will be denoted by **2**. For details about frames see [25, 67]; semiframes and semilocales are given in [60].

The powerset $\mathfrak{P}(X)$ of a set X comprises all of its subsets; i.e. $\mathfrak{P}(X) = \{A : A \subset X\}$. Each function $f : X \to Y$ induces its **image** and **preimage** operators $f^{\to} : \mathfrak{P}(X) \to \mathfrak{P}(Y)$, $f^{\leftarrow} : \mathfrak{P}(X) \leftarrow \mathfrak{P}(Y)$, resp., by $f^{\to}(A) = \{f(x) \in Y : x \in A\}$, $f^{\leftarrow}(B) = \{x \in X : f(x) \in B\}$. The adjunction $f^{\to} \dashv f^{\leftarrow}$ is fundamental for much mathematics. See [40, 58, 59] for these operators and their generalizations.

Given that $\mathfrak{P}(X)$ is order-isomorphic to $\mathbf{2}^X$, we shall often use the notation $\mathbf{2}^X$ for $\mathfrak{P}(X)$.

Let **Top** denote the category of traditional topological spaces and continuous mappings. The functor $\Omega : \mathbf{Top} \to \mathbf{Loc}$ [25, 44] is defined as follows: $\Omega(X)$ is the topology of space X, i.e. the locale or frame of its open sets, and $\Omega(f : X \to Y)$, for continuous map f, is $\left[f^{\leftarrow}|_{\Omega(Y)}\right]^{op} : \Omega(X) \to \Omega(Y)$, i.e. the dual of the frame morphism $f^{\leftarrow}|_{\Omega(Y)} : \Omega(X) \leftarrow \Omega(Y)$. Note $\Omega(X) = \mathbf{2}^X$ if X is a discrete space.

For convenience, a traditional space may also be denoted $(X, \mathfrak{T})$, in which case we write $\Omega(X, \mathfrak{T}) = \mathfrak{T}$.

A semilocale A is **spatial** if A is order-isomorphic to $\Omega(X)$ for some space X; so all topologies are spatial locales. Equivalently, A is spatial iff for any two distinct $a, b \in A$, there is a frame morphism $h : A \to \mathbf{2}$ such that $h(a) \neq h(b)$. See [25, 67].

1.4. If A is a frame, then a **subframe** B is a subset of A having the property that each join [meet] in A of a subset [finite subset, resp.] of B is again in B. The family of all subframes of A, ordered by inclusion, is a complete lattice in which meets are given by intersection; i.e., arbitrary intersections of subframes are again subframes. It follows that for S any subset of A, there is a smallest subframe of A containing S. Such a subframe is said to be **generated** by S and will be denoted by $\langle S \rangle$.

Standard facts about **Frm** [25], together with Lemma 1.2 above, yield the following:

- **Frm** is complete and cocomplete (and so has equalizers and coequalizers).
- $(h_i : A \to B_i)_{i \in J}$ is c.m. in **Frm** iff $\forall\, a \neq b \in A$, $\exists\, i \in J$, $h_i(a) \neq h_i(b)$. Also, **Frm** monomorphisms are exactly the injective frame morphisms.
- $(h_i : A_i \to B)_{i \in J}$ is c.e.e. in **Frm** iff $\langle \bigcup_{i \in J} h_i^{\to}(A_i) \rangle = B$. Also, **Frm** extremal epimorphisms are precisely the surjective frame morphisms (and hence are precisely the **Frm** regular epimorphisms).

Lemma. The following hold:

I. Let $(\mu_i : A \to B_i)_i$ be c.m. in **Frm**.

(a) Let $h : C \to B$ be a mapping such that some $\mu_i h : C \to B_i$ is a frame morphism. Then h is a frame morphism.

(b) Let h, k be mappings such that $\mu_i h \le \mu_i k$ for some i. Then $h \le k$.

II. Let $(\varepsilon_i : A_i \to B)_i$ be c.e.e. in **Frm**.

(a) Let h, k be frame morphisms such that $h\varepsilon_i \le k\varepsilon_i$ for all i. Then $h \le k$.

In **Frm** we have the obvious (extremal epi, mono)-factorization of morphisms:

$$(h : A \to B) = \left(A \xrightarrow{a \mapsto h(a)} h^{\to}(A) \hookrightarrow B\right)$$

Due to the contravariant relation of spaces to frames indicated by Ω above, a surjective frame morphism $h : A \to B$—a regular epimorphism in **Frm**—concretely underlies the regular monomorphism $h^{op} : B \to A$ in **Loc**; and since the embeddings in **Top** are the regular monomorphisms, h^{op} may be viewed as an embedding of a point-free space representing B into a point-free space representing A; hence, B (with h^{op}) may be viewed as a **sublocale** of A (of course, not to be confused with subframes). Recall from above that the subframes of a frame constitute a complete lattice in which the meet is given by intersection and the join is given by taking the smallest subframe containing the union. If we regard a subframe as a frame with a mapping, then in **Frm** these meets and joins are given by limits and colimits, respectively. Analogously, if we have a system of sublocales $(h_i^{op} : B_i \to A)_{i \in J}$, then the meet and join of this system are given in **Loc** by taking limits and colimits. Further, the meet of $(h_i^{op} : B_i \to A)_{i \in J}$ in **Loc** can be obtained by taking the dual of the join of $(h_i : A \to B_i)_{i \in J}$ in **Frm**, i.e. by taking the colimit of the naturally ensuing diagram. This brings us to our next result.

1.5 Proposition. Let $(h_i : A \to B)_{i \in J}$ be a system of frame morphisms linearly ordered by the natural order. Then $h = \bigvee_{i \in J} h_i = (a \mapsto \bigvee_{i \in J} h_i(a))$ is a frame morphism.

Proof. Obviously $h(\bot) = \bot$, $h(\top) = \top$, $h(\bigvee a_j) = \bigvee h(a_j)$ and $h(a \wedge b) \le h(a) \wedge h(b)$. Finally,

$$h(a) \wedge h(b) = \bigvee h_i(a) \wedge \bigvee h_i(b) = \bigvee \{h_i(a) \wedge h_j(b) \mid i, j \in J\} = *$$

and denoting by h_k the larger of h_i, h_j we obtain

$$* \le \bigvee \{h_k(a) \wedge h_k(b) \mid i, j \in J\} = \bigvee \{h_k(a \wedge b) \mid k \in J\} = h(a \wedge b) \quad \square$$

1.6 Sobriety [16, 17, 18, 25, 44]. An element a of a frame is **meet-irreducible** if $a \neq \top$ and if the inequality $a \geq b \wedge c$ implies that either $a \geq b$ or $a \geq c$ (equivalently, $a = b \wedge c \Rightarrow a = b$ or $a = c$). Obviously, in the frame $\Omega(X)$, each $X \setminus \overline{\{x\}}$ is meet-irreducible. A space X is said to be **sober** if it is T_0 and each meet irreducible is of the form $X \setminus \overline{\{x\}}$.

Standard Ω Facts. The following hold:

- Ω is injective on objects.
- Ω is faithful if it is restricted to T_0 spaces, in which case Ω is an embedding.
- Ω is a full embedding if it is restricted to sober spaces.

Indeed, $f \mapsto \Omega(f)$ sets up a categorical equivalence between **SobTop** (the full subcategory of **Top** generated by the sober spaces) and **SpatLoc** (the full subcategory of **Loc** of spatial locales). **Loc** and **SpatLoc** may be replaced in these statements by **SLoc** and **SpatSLoc.**

1.7. The standard **spectrum** construction given in [16, 17, 18, 25, 44] for a semiframe or semilocale A may be summarized as follows:

$$Pt(A) \equiv \mathbf{SFrm}(A, \mathbf{2}) = \{p : A \to \mathbf{2} \mid p \in \mathbf{SFrm}\}$$

$$\Phi : A \to \mathbf{2}^{Pt(A)} \text{ by } \Phi(a) = \{p \in Pt(A) : p(a) = \top\}$$

$$PT(A) \equiv (Pt(A), \Phi^{\to}(A)) \in \mathbf{Top}$$

and given $f : A \to B$ in **SLoc**, then $PT(f) : PT(A) \to PT(B)$ by

$$PT(f)(p) = p \circ f^{op}$$

where $f^{op} : A \leftarrow B$ is the dual frame morphism. The **spectrum** or **point** functor

$$PT : \mathbf{SLoc} \to \mathbf{Top}$$

is thereby constructed. The following are well-known facts:

- $\Omega \dashv PT$.
- each $\Omega(X)$ is spatial.
- each $PT(A)$ is sober.
- the unit $\Psi : X \to PT(\Omega(X))$ of $\Omega \dashv PT$, given by $\Psi(x)(U) = \chi_U(x)$ [49, 51, 54, 55, 56], is a homeomorphism iff X is sober (cf. 2.2 below).
- the counit $[\Phi : A \to \Phi^{\to}(A)]^{op}$ of $\Omega \dashv PT$ is an isomorphism iff A is spatial.
- $\Omega \dashv PT$ restricts to the categorical equivalences **SpatSLoc** $\sim$ **SobTop** $\sim$ **SpatLoc** given in 1.6 above.

2 L-topological spaces

2.1. Let L be a semiframe. Recall that an L-**topological space** (or L-**space**) is a couple (X, τ), where $\tau \subset L^X$ is a sub-semiframe of the semiframe $L^{\mathbb{X}}$ of all mappings $X \to L$. See [3, 6] and cf. [15, 60]. If L is a frame, then τ is a frame or locale. The bottom and top constant maps are denoted $\underline{\perp}$ and $\underline{\top}$; and generally if $\alpha \in L$, the associated constant map is denoted $\underline{\alpha}$.

To discuss L-continuity, we need the appropriate powerset operators. Given a function $f : X \to Y$, the image and preimage operators are defined as follows:

$$\begin{aligned} f_L^{\to} &: L^X \to L^Y \text{ by } f_L^{\to}(a)(y) = \bigvee_{f(x)=y} a(x), \\ f_L^{\leftarrow} &: L^X \to L^Y \text{ by } f_L^{\leftarrow}(b) = b \circ f \end{aligned}$$

It is well-known that $f_L^{\to} \dashv f_L^{\leftarrow}$; and these operators have other fundamental properties as well, including $f_L^{\leftarrow}$ is a semiframe morphism. See [40, 58, 59, 70].

An L-**continuous map** $f : (X, \tau) \to (Y, \sigma)$ is a map $f : X \to Y$ such that $\forall\, v \in \sigma$, $f_L^{\leftarrow}(v) \in \tau$. Note that $(f_L^{\leftarrow})_{|\sigma} : \tau \leftarrow \sigma$ is a semiframe morphism, and hence that

$$\left[(f_L^{\leftarrow})_{|\sigma}\right]^{op} : \tau \to \sigma$$

is a semilocalic morphism. If L is a frame, then $(f_L^{\leftarrow})_{|\sigma}$ and $\left[(f_L^{\leftarrow})_{|\sigma}\right]^{op}$ are frame and localic morphisms, respectively.

The category of L-spaces and L-continuous mappings (with the usual composition and identity morphisms) is denoted L-**Top**, in keeping with [34, 15, 60].

2.2 Spatiality and Sobriety. The above remarks construct the functor $L\Omega$: L-**Top** $\to$ **SLoc** by stipulating

$$\begin{aligned} L\Omega(X, \tau) &= \tau, \\ L\Omega(f : (X, \tau) \to (Y, \sigma)) &= \left[(f_L^{\leftarrow})_{|\sigma}\right]^{op} : \tau \to \sigma \end{aligned}$$

If L is a frame, then $L\Omega$ is a functor from L-**Top** into **Loc**.

A semilocale A is L-**spatial** if A is order-isomorphic to τ for some L-space (X, τ). Each L-topology τ is L-spatial. We therefore have the full subcategory L-**SpatSLoc** of L-spatial objects of **SLoc**; and we note that $L\Omega$ maps L-**Top** into L-**SpatSLoc**.

The right adjoint of $L\Omega$, namely the extension of PT to the L-**spectrum** or L-**point** functor LPT, is constructed by the following sequence of definitions, beginning with a semiframe A:

$$Lpt(A) = \mathbf{SFrm}(A, L), \quad \Phi_L : A \to L^{Lpt(A)} \text{ by } \Phi_L(a)(p) = p(a)$$

$$LPT(A) = (Lpt(A), \Phi_L^{\to}(A))$$

$$LPT(g : A \to B) : Lpt(A) \to Lpt(B) \text{ by } p \mapsto p \circ g^{op}$$

It follows that Φ_L is a semiframe morphism, $\Phi_L^{\to}(A)$ is an L-topology on $Lpt(A)$, and $LPT(g)$ is L-continuous. Thus we have the functor $LPT : L\text{-}\mathbf{Top} \leftarrow \mathbf{SLoc}$.

If there is no confusion, we may drop the prefix L for $L\Omega$ and LPT. The following are well-known L-analogues for $L\Omega$ and LPT of the traditional facts given in 1.7 above:

- $\Omega \dashv PT$;
- each L-topology $\Omega(X, \tau) \equiv \tau$ is L-spatial;
- each $PT(A)$ is L-sober, where an L-space is L-**sober** if $\Psi_L : X \to Lpt(\tau)$, defined by
$$\Psi_L(x)(u) = u(x)$$
is a bijection;
- $\Psi_L : (X, \tau) \to (Lpt(\tau), \Phi_L^{\to}(\tau))$ is the unit of $\Omega \dashv PT$ and is an L-**homeomorphism** iff (X, τ) is L-sober, where an L-homeomorphism is a bijection such that both it and its inverse are L-continuous;
- $[\Phi_L : A \to \Phi_L^{\to}(A)]^{op}$ is the counit of $\Omega \dashv PT$ and is an isomorphism iff A is L-spatial; and
- $\Omega \dashv PT$ restricts to categorical equivalences L-**SpatSLoc** $\sim$ L-**SobTop**, where L-**SpatSLoc** [L-**SobTop**] is the full subcategory of **SLoc** [L-**Top**] of all L-spatial [L-sober] objects.

If L is a frame, then **Loc** may replace **SLoc** in the above constructions and statements. Further, if $L = \mathbf{2}$, then up to isomorphism of categories via the characteristic functor $G_\chi : \mathbf{Top} \to \mathbf{2}\text{-}\mathbf{Top}$ [54, 60], the traditional Ω [PT, Φ, Ψ, sobriety] becomes $\mathbf{2}\Omega$ [$\mathbf{2}PT, \Phi_\mathbf{2}, \Psi_\mathbf{2}$, **2**-sobriety, resp.]; and we shall not necessarily distinguish between a traditional concept and its **2**-analogue.

Modified L**-sobriety** and **modified** L**-sobriety to degree** D, where $D \hookrightarrow L$ is a subsemiframe of L, are respectively defined by making the following modifications:

$$\begin{aligned} Lpt_{\text{mod}}(\tau) &= \{p \in \mathbf{SFrm}(\tau, L) : \forall\, \underline{\alpha} \in \tau,\ p(\underline{\alpha}) = \alpha\}, \\ Lpt_D(\tau) &= \{p \in \mathbf{SFrm}(\tau, L) : \forall\, \alpha \in D,\ p(\underline{\alpha}) = \alpha\} \end{aligned}$$

and respectively replacing $Lpt(\tau)$ by $Lpt_{\text{mod}}(\tau)$ and $Lpt_D(\tau)$ in the definition of L-sobriety above (cf. [54, 55]). The following hold for any (X,τ): $\Psi_L : X \to Lpt_{\text{mod}}(\tau) \subset Lpt(\tau)$; the L-space $(Lpt_{\text{mod}}(\tau), \tau_{\text{mod}})$ is modified L-sober and the L-space $(Lpt_{\text{mod}}(\tau), \tau_{\text{mod}})$ is modified L-sober to degree D, where

$$\begin{aligned}\tau_{\text{mod}} &\equiv [\Phi_L^{\rightarrow}(\tau)]_{|Lpt_{\text{mod}}(\tau)} \equiv \left\{\Phi_L(u)_{|Lpt_{\text{mod}}(\tau)} : u \in \tau\right\},\\ \tau_D &\equiv [\Phi_L^{\rightarrow}(\tau)]_{|Lpt_D(\tau)} \equiv \left\{\Phi_L(u)_{|Lpt_D(\tau)} : u \in \tau\right\}\end{aligned}$$

are respectively the subspace topologies on $Lpt_{\text{mod}}(\tau)$ and $Lpt_D(\tau)$ from the L-sober space $(Lpt(\tau), \Phi_L^{\rightarrow}(\tau))$; $(Lpt_{\text{mod}}(\tau), \tau_{\text{mod}})$ has precisely the constant maps in its topology that (X,τ) has in its topology; and $(Lpt_{\text{mod}}(\tau), \tau_{\text{mod}})$ is the space $(Lpt_{D_\tau}(\tau), \tau_{D_\tau})$, where D_τ is the subsemiframe $\{\alpha \in L : \underline{\alpha} \in \tau\}$.

The preceding paragraphs are found in, or closely related to, [10, 11, 12, 29, 33, 48, 54, 55, 56, 57, 58, 61]. A significant advantage of these lattice-valued definitions is that Φ_L, Ψ_L are given the explicit form of evaluation maps, something not possible in traditional topology and which has nice consequences. Also note the notation **Sob-L-Top** of these references is rewritten as L-**SobTop**.

Examples justifying the above sobrieties, along with those introduced in this chapter, are found in [42].

2.3 L-T_0 and Properties of $L\Omega$. We say that an L-space (X,τ) is (L-)T_0 iff $\forall\, x \neq y \in X$, $\exists\, u \in \tau$, $u(x) \neq u(y)$.

2.3.1 Proposition. An L-space (X,τ) is T_0 iff the map $\Psi_L : (X,\tau) \to (Lpt(\tau), \Phi_L^{\rightarrow}(\tau))$, given in 2.2 above, is injective.

This proposition can be found in [48, 54, 55, 57], and the definition of L-T_0 first appears independently in [69], [36], and [48]. As above, we do not necessarily distinguish between the traditional concept of T_0 and its analogue **2**-T_0. We are now in a position to state the following analogue for $L\Omega$ of the standard facts for the traditional Ω given in 1.6 above:

2.3.2 Standard $L\Omega$ Facts. The following hold for $\Omega : L$-**Top** $\to$ **SLoc**:

1. Ω is injective on objects.
2. Ω is faithful if it is restricted to L-T_0 spaces, in which case Ω is an embedding.
3. Ω is a full embedding if it is restricted to L-sober spaces.

2.4 Proposition. Let L, M be semiframes. Then the following hold:

1. Semiframe A is L-spatial $\Leftrightarrow \forall a \neq b \in A, \exists p \in Lpt(A), p(a) \neq p(b)$.

2. L is M-spatial $\Leftrightarrow$ each L-topology on each set is M-spatial.

3. M is a subsemiframe of $L \Rightarrow M$ is L-spatial.

Proof. (1) follows from the evaluation map formulation of Φ_L.

For ($\Rightarrow$) of (2), let τ be an L-topology on a set X and let $u \neq v \in \tau$. Then $\exists x \in X, u(x) \neq v(x)$. But the evaluation map formulation of Ψ_L yields that $\Psi_L(x)(u) \neq \Psi_L(x)(v)$ in L. Also, L is M-spatial implies by (1) that $\exists p : L \to M, p(\Psi_L(x)(u)) \neq p(\Psi_L(x)(v))$. Since each of $\Psi_L(x) : \tau \to L$ and $p : L \to M$ are semiframe morphisms, then $p \circ \Psi_L(x)$ is the needed semiframe morphism separating u and v.

For ($\Leftarrow$) of (2), choose any non-empty set and choose the L-topology on that set consisting precisely of all constant maps into L. Then this topology is M-spatial and isomorphic to L. Hence L is M-spatial.

For (3), M is a subsemiframe of L implies that $\hookrightarrow : M \to L$ is a semiframe morphism, and this morphism separates all distinct elements of M. □

Corollary. If L is (traditionally) spatial, then each L-topology on each set is spatial. This holds if L is completely distributive, and hence if L is a complete chain.

2.5 Notation. If L is a semiframe, we let $L_\top = L - \{\top\}$.

2.6. We now consider the Lowen-Kubiak ι_L functor [37, 34] and (in the next section) the system of mappings packaged inside the fibre-map (also denoted ι) underlying the ι_L functor. Let L be a semiframe and (X, τ) an L-space. Fix $t \in L_\top$ and let $a \in L^X$,and consider, using the Halmos notation, the t-**level set**

$$[a \not\leq t] \equiv \{x \in X : a(x) \not\leq t\}$$

This defines the t-**level mapping** $\iota_t : L^X \to \mathbf{2}^X$ by

$$\iota_t(a) = [a \not\leq t] \tag{2.6.1}$$

This sets up the well-known *fibre-map* ι_L from the fibre of L-topologies on X to the fibre of traditional topologies on X by

$$\iota_L(\tau) = \langle\{\iota_t(u) : u \in \tau, t \in L_\top\}\rangle$$

the smallest subframe of $\mathbf{2}^X$ containing $\{\iota_t(u) : u \in \tau, t \in L_\top\}$ (cf. 1.4 above).

Now define the *functor* $\iota_L : L\text{-}\mathbf{Top} \to \mathbf{Top}$ by

$$\iota_L(X, \tau) = (X, \iota_L(\tau)), \quad \iota_L(f) = f \tag{2.6.2}$$

The space $\iota_L(X, \tau) = (X, \iota_L(\tau))$ in keeping with [34], will be called the **topological modification** of (X, τ).

We do not distinguish between level maps applied to the L-powerset or to the L-topology. In the sequel, we often drop the subscript L in ι_L if the L is understood. We also recall that both the fibre map ι and the functor ι admit adjoints, namely the well-known ω. In particular, $\iota \dashv \omega$ as fibre maps and $\omega \dashv \iota$ as functors.

2.7 Convention. For the remainder of this chapter L is a complete chain (but see the appendix [14] to this chapter by Profs. Höhle and Rodabaugh). This has two immediate consequences:

- In the definition of the ι fibre map, we may replace $\not\leq$ by $>$ and write $\iota_t(a) = [a > t]$.
- Let $a, b \in L^X$ such that $\forall t \in L_\top$, $\iota_t(a) = \iota_t(b)$. Then $a = b$.

2.8. We repeatedly use the following fact:

Lemma. Let L be a complete chain, let A be a semilocale, consider the traditional spectrum $Pt(A)$ as a poset with the natural order, and let $\nu : L_\top \to Pt(A)$ have the property that for all non-void $S \subset L_\top$, ν changes meets to joins, i.e.

$$\nu\left(\bigwedge S\right) = \bigvee \nu^{\to}(S) \equiv \bigvee \{v(s) : s \in S\}$$

Then the following statements hold:

1. ν is antitone; and
2. $\forall t \in L_\top$, $\forall a \in A$,

$$\nu(t)(a) = \top \;\Leftrightarrow\; t < \bigvee \{s \in L_\top : \forall r < s \text{ in } L_\top, \nu(r)(a) = \top\} \;\; (\in L)$$

Proof. For (1), let $a, b \in A$ with $a \leq b$ and put $S = \{a, b\}$. Then S is non-void and

$$\nu(a) = \nu(a \wedge b) = \nu\left(\bigwedge S\right) = \bigvee \nu^{\to}(S) = \nu(a) \vee \nu(b)$$

which implies $\nu(a) \geq \nu(b)$.

For $(\Leftarrow)$ of (2), let $t < \bigvee \{s \in L_\top : \forall r < s \text{ in } L_\top, \nu(r)(a) = \top\}$. Then there is $s \in L_\top$ such that $t < s$ and for all $r < s$, $\nu(r)(a) = \top$. Hence $\nu(t)(a) = \top$.

Now for ($\Rightarrow$) of (2), let $\nu(t)(a) = \top$. Set $S = \{s \in L_\top : t < s\}$ and $s_0 = \bigwedge S$. On one hand, suppose $t = s_0$. Then $S \neq \emptyset$, and $\nu(t)(a) = \bigvee_{s \in S} \nu(s)(a) = \top$; so then there is an $s > t$ such that $\nu(s)(a) = \top$, and consequently $\nu(r)(a) = \top$ for all $r \leq s$ and $t < s \leq \bigvee \{z \in L_\top : \forall r < z \text{ in } L_\top, \nu(r)(a) = \top\}$. On the other hand, if $t < s_0$, then for all $r < s_0$ we have $r \leq t$ and hence $\nu(r)(a) = \top$ and $t < s_0 \leq \bigvee \{s \in L_\top : \forall r < s \text{ in } L_\top, \nu(r)(a) = \top\}$. $\square$

3 L-frames

3.1. Recall L is a complete chain and let (X, τ) be a L-topological space. The formula (2.6.1) defines a family of mappings

$$\iota_t : \tau \to \iota(\tau), \quad t \in L_\top$$

from the L-topology τ to its topological modification $\iota(\tau)$. Using the linearity of L, we can easily check each of the following facts:

- Each ι_t is a frame morphism.
- For each non-empty $S \subset L_\top$,

$$\iota_{\bigwedge S} = \bigcup_{s \in S} \iota_s \tag{3.1.1}$$

  Hence the assignment $t \mapsto \iota_t$ is antitone.

- The family $(\iota_t : \tau \to \iota(\tau) \mid t \in L_\top)$ is a c.m. system (1.2 above) in **Frm**.
- The family $(\iota_t : \tau \to \iota(\tau) \mid t \in L_\top)$ is a c.e.e. system (1.2 above) in **Frm** (using the characterization of c.e.e. systems in 1.4).

We now show that the system $(\iota_t : \tau \to \iota(\tau) \mid t \in L_\top)$ contains all the information necessary to recover the structure of (X, τ). In particular, we have:

Proposition. Let $(X, \mathfrak{T})$ be a traditional topological space and let $(\varphi_t : A \to \mathfrak{T})_{t \in L_\top}$ be a c.m. and c.e.e. system of frame morphisms satisfying (3.1.1). Then there is a frame τ and an isomorphism $\kappa : A \to \tau$ such that following hold:

1. τ is a subframe of L^X, i.e. (X, τ) is an L-space;
2. $\mathfrak{T} = \iota(\tau)$, the topological modification of τ; and
3. for each $t \in L_\top$, $\iota_t \circ \kappa = \varphi_t$.

Proof. For $a \in A$, define $\kappa(a) : X \to L$ by setting

$$\kappa(a)(x) = \bigvee\{t : \forall\, s < t,\ x \in \varphi_s(a)\}\ (\in L)$$

and put

$$\tau = \kappa^{\rightarrow}(A) = \{\kappa(a) : a \in A\}$$

For $t \in L_\top$, define $\nu(t) : A \to \mathbf{2}$ by setting $\nu(t)(a) = \top$ iff $x \in \varphi_t(a)$.

We now check each $\nu(t)$ is a frame morphism by satisfying the hypotheses of Lemma 2.8. First note $\nu : L_\top \to Pt(A)$ is antitone from 3.3.1. Next, for non-empty $S \subset L_\top$ we have

$$\nu(\bigwedge S)(a) = \top \quad \text{iff} \quad x \in \varphi_{\wedge S}(a) = \bigcup \varphi_s(a) \quad \text{iff} \quad \bigvee_{s \in S} \nu(s)(a) = \top$$

Thus, Lemma 2.8 obtains that

$$t < \kappa(a)(x) \quad \text{iff} \quad x \in \varphi_t(a) \tag{*}$$

From (*) we infer $\kappa(\bot) = \underline{\bot}$ and $\kappa(\top) = \underline{\top}$; and additionally, (*), the assumption that each φ_t is a frame morphism, and the linearity of L imply $\kappa(a \wedge b) = \kappa(a) \wedge \kappa(b)$ and $\kappa(\bigvee a_i) = \bigvee \kappa(a_i)$. It follows:

- $\kappa : A \to L^X$ is a frame morphism; and
- $\tau = \kappa^{\rightarrow}(A)$ is a subframe of L^X, i.e. an L-topology on X.

Further, (*) says that $\iota_t(\kappa(a)) = [\kappa(a) > t] = \varphi_t(a)$, i.e.

$$\varphi_t = \iota_t \circ \kappa \quad \text{and} \quad \mathfrak{T} = \iota(\tau)$$

where the latter uses the fact that $(\varphi_t)_t$ is c.e.e. Now by Lemma 1.2, κ is injective, which implies it is an isomorphism onto τ. □

Comment. The above proposition captures an L-topology, up to frame isomorphism, by considering levels of fuzzy subsets. Describing L-topologies using levels and level topologies is an old theme in the fuzzy literature, e.g. [64, 38, 45, 46, 47, 54, 34, 65] with various level-dependent constructions. Some of these techniques have captured L-topologies (under certain conditions) by "integrating" together all the levels, examples including the hypergraph functor of [64, 38, 45, 54], the fundamental level-dependent constructions of [26, 27, 28], and the level-indexed nests of traditional topologies of [65]. But the above discussion and proposition appear to be a new approach *vis-a-vis* the previous literature in that it interprets the levels as a *system of frame morphisms* satisfying *categorical* conditions from which the original L-topology may be recaptured (up to frame isomorphism).

3.2. Combining 1.3 and 2.6, we have that if $f : (X, \tau) \to (Y, \sigma)$ is L-continuous, then $\iota(f) = f : (X, \iota(\tau)) \to (Y, \iota(\sigma))$ is (traditionally) continuous, in which case

$$f^{\leftarrow}_{|\iota(\sigma)} : \iota(\tau) \leftarrow \iota(\sigma)$$

is a frame morphism. It can be checked that the following identities in **Frm** hold when the traditional and Zadeh preimage operators are respectively restricted to the traditional topology $\iota(\sigma)$ and the L-topology σ:

$$\forall t \in L_{\top}, \quad f^{\leftarrow} \circ \iota_t^{\sigma} = \iota_t^{\tau} \circ f_L^{\leftarrow} \tag{3.2.1}$$

3.3. Subsection 3.1 and identities 3.2.1 motivate the following definition:

Definition (***L*-Frames and *L*-Frm**). An ***L*-frame** is a system of frame morphisms

$$A \equiv \left(\varphi_t^A : A^{\mathfrak{u}} \to A^{\mathfrak{l}}\right)_{t \in L_{\top}}$$

satisfying these axioms:

(F0) For each non-empty $S \subset L_{\top}$, $\varphi^A_{\bigwedge S} = \bigvee_{s \in S} \varphi_s^A$ (cf. 3.3.1).

(F1) $\left(\varphi_t^A\right)_t$ is c.e.e. in **Frm**, i.e. $A^{\mathfrak{l}} = \langle \bigcup_t \left(\varphi_t^A\right)^{\to} [A^{\mathfrak{u}}] \rangle \equiv \langle \bigcup_t \varphi_t^A [A^{\mathfrak{u}}] \rangle$ (cf. 1.4).

(F2) $\left(\varphi_t^A\right)_t$ is c.m. in **Frm**.

And an ***L*-frame morphism** $h : A \to B$ is an ordered pair of frame morphisms

$$\left(h^{\mathfrak{u}} : A^{\mathfrak{u}} \to B^{\mathfrak{u}},\ h^{\mathfrak{l}} : A^{\mathfrak{l}} \to B^{\mathfrak{l}}\right)$$

satisfying

$$\forall t \in L_{\top},\ h^{\mathfrak{l}} \cdot \varphi_t^A = \varphi_t^B \cdot h^{\mathfrak{u}}$$

The resulting category, with composition and identities done coordinate-wise from (squares in) **Frm**, is denoted L-**Frm**.

3.3.1 Note. From (F1) and (F2) we immediately see that in an L-frame morphism $h \equiv \left(h^{\mathfrak{u}}, h^{\mathfrak{l}}\right)$, each of the frame morphisms $h^{\mathfrak{u}}$, $h^{\mathfrak{l}}$ guarantees the uniqueness of the other.

3.3.2 Relationship to Frm and Functor Categories: Special Case.. If we choose $L = \mathbf{2}$ in Definition 3.3, then we have one frame morphism $\varphi_{\bot}^A$ which automatically satisfies (F0); and the imposition of (F1, F2) makes $\varphi_{\bot}^A$ an isomorphism. Thus a **2**-frame A stands for a pair of (possibly distinct) isomorphic frames $\left(A^{\mathfrak{u}}, A^{\mathfrak{l}}\right)$. Further, each L-frame morphism is a pair of frame morphisms $\left(h^{\mathfrak{u}}, h^{\mathfrak{l}}\right)$ such that each factors through the other via isomorphisms, i.e. $h^{\mathfrak{u}}, h^{\mathfrak{l}}$ are in the same *morphism class* (8.15.1 of [61]) of **Frm.**

It follows that **2-Frm** is not **Frm** nor isomorphic to **Frm.** Yet it is somehow "like" **Frm**. In anticipation of the discussion of functor categories in 3.4 below, let **2** also denote the category $\{\mathfrak{u}, \mathfrak{l}\}$ with 2 objects and a non-trivial arrow $\mathfrak{u} \to \mathfrak{l}$. Each **2**-frame is a functor

$$A : \mathbf{2} \to \mathbf{Frm}$$

It follows that any natural transformation between two such functors is a pair of frame morphisms from the same morphism class of **Frm**. The following is not difficult to check:

Proposition. The following hold:

1. $\mathbf{2\text{-}Frm} = \mathbf{Frm}^{\mathbf{2}}$.

2. **Frm** is categorically equivalent to $\mathbf{Frm}^{\mathbf{2}}$ via $F : \mathbf{Frm} \to \mathbf{Frm}^{\mathbf{2}}$ and $G : \mathbf{Frm} \leftarrow \mathbf{Frm}^{\mathbf{2}}$ defined by

$$F(A) = (A, A) \text{ with } \varphi_{\perp}^{A} = id_A, \; F(h) = (h, h)$$

$$G\left(A^{\mathfrak{u}}, A^{\mathfrak{l}}\right) = A^{\mathfrak{u}}, \; G\left(h^{\mathfrak{u}}, h^{\mathfrak{l}}\right) = h^{\mathfrak{u}}$$

Restated, each of F and G are functors and $F \dashv G \dashv F$.

Remark. By the proposition, we are assured that **Frm** and **2-Frm** are categorically equivalent and hence behave the same categorically. Up to categorical equivalence, we may say that **2-Frm** captures traditional frames—in fact F is an embedding. However, G is not an embedding (e.g., not injective on objects), and indeed **Frm** and **2-Frm** are not isomorphic, which contrasts to the topological situation wherein **Top** and **2-Top** are isomorphic via the characteristic functor. Thus **2-Frm** is a "bigger" category than **Frm** which nonetheless behaves as **Frm**. This justifies traditional frames being viewed as "crisp" frames and L-frames being viewed as "lattice-valued" or "fuzzy" frames.

3.4 Relationship to Functor Categories: General Case. In [53, 54, 55], it is pointed out that several approaches to topology in fuzzy sets yield categories that can be viewed, in a natural way, as functor categories over various base categories. Such approaches include Chang-Goguen (Section 2 above), Hutton [21, 22], Lowen and topologies with degrees of stratification [37, 53, 54, 55], and variable-basis topology and variable-basis compactifications [53, 54, 55, 61]. As an example, the Hutton approach has the exponent category **2**, with objects of the functor category being functors from **2** into **SLoc**; so that Hutton embeds into the functor category $\mathbf{SLoc}^{\mathbf{2}}$.

The notion of an L-frame can be viewed in a similar way; only, we have to consider enriched categories (see, e.g., [41]), that is, roughly speaking, categories with structured morphism sets as the codomain. Consider $\mathbf{Frm}_{\leq}$ to be **Frm** with the morphism sets endowed by the natural order. The "exponent" category **TL** has two objects $\mathfrak{u}$, $\mathfrak{l}$ and, besides the identities, morphisms $t : \mathfrak{u} \to \mathfrak{l}$ for each $t \in L_{\top}$, with $\mathbf{TL}(\mathfrak{u}, \mathfrak{l})$ ordered by L. Trivially, $(t)_{t \in L_{\top}}$ is both c.m. and c.e.e. in **TL**. Now, an L-frame can be viewed as a functor $A : \mathbf{TL} \to \mathbf{Frm}_{\leq}$ satisfying these axioms:

(F0') convert non-empty meets to joins;

(F1') preserve the c.e.e. property of $(t)_{t \in L_{\top}}$; and

(F2') preserve the c.m. property of $(t)_{t \in L_{\top}}$.

L-frame morphisms then coincide with natural transformations between such functors. Thus

$$L\text{-}\mathbf{Frm} \hookrightarrow \mathbf{Frm}_{\leq}^{\mathbf{TL}}$$

The reader can check that for $L = \mathbf{2}$,

$$\mathbf{Frm}^{\mathbf{2}} \hookrightarrow \mathbf{Frm}_{\leq}^{\mathbf{T2}}$$

thus reconciling this representation with 3.3.2 above.

As pointed out in [53, 54, 55], two point-set lattice-theoretic (poslat) approaches to topology may be seen as different if the functor categories representing their lattice-theoretic aspects are different. It would seem that L-**Frm** points to a new approach for poslat (or fuzzy) topology.

3.5 Functor $L\boldsymbol{\Omega}$. Recalling 3.1 and 3.2 and letting L-**Loc** $\equiv L$-**Frm**^{op}, we see that we have a functor

$$\mathfrak{l}L\boldsymbol{\Omega} : L\text{-}\mathbf{Top} \to L\text{-}\mathbf{Loc}$$

defined by

$$\mathfrak{l}L\boldsymbol{\Omega}\,(X, \tau) = (\iota_t : \tau \to \iota\,(\tau) \mid t \in L_{\top})$$
$$\mathfrak{l}L\boldsymbol{\Omega}\,(f : (X, \tau) \to (Y, \sigma)) = \left(\mathfrak{l}L\boldsymbol{\Omega}^{\mathfrak{u}}\,(f)\,, \mathfrak{l}L\boldsymbol{\Omega}^{\mathfrak{l}}\,(f)\right)$$

where

$$\mathfrak{l}L\boldsymbol{\Omega}^{\mathfrak{u}}\,(f) = \left[(f_L^{\leftarrow})_{|\sigma}\right]^{op}, \ \mathfrak{l}L\boldsymbol{\Omega}^{\mathfrak{l}}\,(f) = \left[(f^{\leftarrow})_{|\iota(\sigma)}\right]^{op}$$

and $f^{\leftarrow}$ is the traditional preimage operator of f (1.3) and $f_L^{\leftarrow}$ is the Zadeh L-preimage operator of f (2.1).

Note. The action of $\mathfrak{l}L\boldsymbol{\Omega}$ is precisely that of $L\Omega$ (2.2) in the first component and precisely that of traditional Ω (1.3) in the second component. This is a useful observation below.

Notation. If there is no confusion, we may drop the prefix ɩL for this new $L\Omega$ as well as for its component functions.

3.5.1 Definition.

1. An L-frame is **ɩL-spatial** if it is isomorphic (in L-**Frm**) to some $\mathbf{\Omega}(X,\tau)$. Note that each $\mathbf{\Omega}(X,\tau)$ is therefore ɩL-spatial.
2. An L-space (X,τ) is **ɩL-sober** if its topological modification $(X,\iota(\tau))$ is (traditionally) sober (1.6).

3.5.2 Discussion. Using the nomenclature scheme of [34], ɩ-sobriety is the same as **ultra** L-**sobriety** or ι-**sobriety**: an L-topological space (X,τ) has property **ultra** (L-)P or ι-P if $\iota(X,\tau) \equiv (X,\iota(\tau))$ has property P. Often in the literature, a property P for traditional topological spaces is generalized to a property L-P for L-topological spaces; and then one frequently has the result that "ultra" L-P is stronger than L-P (hence the label "ultra")—see [38, 34] and the references of the latter for examples. For L-spaces, is ultra-sobriety stronger than L-sobriety, is ultra-sobriety stronger than modified L-sobriety to degree D in the sense of 2.2 [54, 55], and which nomenclature of [34] should be applied to ɩ-sobriety? This brings us to the next proposition, in which ω is the adjoint of ι (2.6) and in which ι-T_0 (in accordance with [34]) means the topological modification $\iota(X,\tau)$ is T_0:

Proposition. Let L be a semiframe and (X,τ) be an L-topological space.

1. (X,τ) is T_0 implies (X,τ) is ι-T_0 (cf. [34]).
2. If L is a complete chain, (X,τ) is ι-T_0 implies (X,τ) is T_0.
3. If L is completely distributive and admits a semiframe morphism other than the identity morphism (e.g., if L is a complete chain), then each sober traditional space $(X,\mathfrak{T})$ generates via ω an ultra-sober L-space (X,τ) which is *not* L-sober.
4. If L is a complete chain, then each sober traditional space $(X,\mathfrak{T})$ generates via ω an ultra-sober L-space (X,τ) which is modified L-sober to degree L in the sense of 2.2 above.
5. If L is a frame, $D \hookrightarrow L$ is a subsemiframe, and (X,τ) is L-sober, then its stratification to degree D, namely

$$G_D(X,\tau) \equiv (X, G_D(\tau)) \equiv (X, \tau \vee \{\underline{\alpha} : \alpha \in D\})$$

with $G_D(\tau)$ the smallest L-topology containing $\tau \cup \{\underline{\alpha} : \alpha \in D\}$, is modified L-sober to degree D in the sense of 2.2 above (cf. the stratification functor in [34, 54] and Section 6 of [60]).

Proof. For (1), let $x \neq y$ in X. By 2.3, $\exists\, u \in \tau,\, u(x) \neq u(y)$. Then WLOG $u(x) \not\leq u(y)$; so that $x \in [u \not\leq u(y)]$ and $y \notin [u \not\leq u(y)]$.

For (2), let $x \neq y$ in X, and let $U \equiv \bigcup_i \bigcap_{j=1}^{n} [u_{ij} > \alpha_{ij}] \in \iota(\tau)$, such that WLOG $x \in U$ and $y \notin U$. Then $\exists\, i, j,\, x \in [u_{ij} > \alpha_{ij}]$ and $y \notin [u_{ij} > \alpha_{ij}]$. It follows $u_{ij}(x) \neq u_{ij}(y)$.

For (3), let $(X, \mathfrak{T})$ be a traditional sober space. Then put $\tau = \omega(\mathfrak{T})$, where ω is the fibre map right adjoint to ι_L [38, 34]. Then from Property 3.5(2) of [34], L is completely distributive implies that

$$\iota(X, \tau) = (X, \mathfrak{T})$$

and so (X, τ) is ultra-sober. But by Theorem 6.6 of [48] and the fact that (X, τ) is stratified (τ contains all the constant maps), (X, τ) cannot be L-sober in the sense of 2.2.

For (4), let $(X, \mathfrak{T})$ be a traditional sober space and put $\tau = \omega(\mathfrak{T})$ as in (3). Then as in (3), we have $\iota(X, \tau) = (X, \mathfrak{T})$, which implies (X, τ) is ultra sober and by (2) that (X, τ) is L-T_0. Recall $Lpt_{\mathrm{mod}}(\tau)$ (2.2) is the modified carrier set

$$Lpt_{\mathrm{mod}}(\tau) \equiv \{p \in \mathbf{SFrm}(\tau, L) : \forall\, \underline{\alpha} \in \tau,\, p(\underline{\alpha}) = \alpha\}$$

where $\underline{\alpha}$ is the constant map with value α, L-T_0 means the L-comparison map $\Psi_L : X \to Lpt_{\mathrm{mod}}(\tau)$ (2.2) is injective (Proposition 2.3 plus the observation that we always have $\Psi_L : X \to Lpt_{\mathrm{mod}}(\tau) \subset Lpt(\tau)$). To show that (X, τ) is modified sober, it remains to show Ψ_L is onto $Lpt_{\mathrm{mod}}(\tau)$, i.e. that $Lpt_{\mathrm{mod}}(\tau) \subset \{\Psi_L(x) : x \in X\}$. Let $p \in Lpt_{\mathrm{mod}}(\tau)$ and put $q : \mathfrak{T} \to \mathbf{2}$ by

$$q(U) = p(\chi_U)$$

It follows that $q \in Pt(\mathfrak{T})$. Since $(X, \mathfrak{T})$ is sober, the traditional comparison map $\Psi : X \to Pt(\mathfrak{T})$, given by $\Psi(x)(U) = \chi_U(x)$, is surjective (1.7); so $\exists\, x \in X,\, q = \Psi(x)$. We claim $p = \Psi_L(x)$ to finish the proof. By Theorem 3.3 of [52] (cf. Proposition 3.8 of [34]), L is a chain implies that $\omega_L = G_k \circ G_\chi$, where G_χ is the characteristic functor $G_\chi(\mathfrak{S}) = \{\chi_U : U \in \mathfrak{S}\}$ and G_k is the stratification functor (cf. statement (5) with $D = L$). Each $u \in \tau = \omega_L(\mathfrak{T})$ can be written in the form

$$u = \bigvee_i (\underline{\alpha_i} \wedge \chi_{U_i})$$

where $\alpha_i \in L$ and $U_i \in \mathfrak{T}$. We calculate:

$$\begin{aligned} p(u) &= p\left[\bigvee_i (\underline{\alpha_i} \wedge \chi_{U_i})\right] = \bigvee_i (\alpha_i \wedge q(U_i)) \\ &= \bigvee_i (\alpha_i \wedge \Psi(x)(U_i)) = \bigvee_i (\alpha_i \wedge \chi_{U_i}(x)) \end{aligned}$$

$$= \left[\bigvee_i (\underline{\alpha_i} \wedge \chi_{U_i})\right](x) = \Psi_L(x)\left[\bigvee_i (\underline{\alpha_i} \wedge \chi_{U_i})\right]$$
$$= \Psi_L(x)(u)$$

For (5), assume (X, τ) is L-sober. For clarity, we denote $\Psi_L : X \to Lpt(\tau)$ by Ψ_τ and denote $\Psi_L : X \to Lpt_{\text{mod}}(G_D(\tau))$ by $\Psi_{G_D(\tau)}$. Now (X, τ) is T_0 and $\tau \subset G_D(\tau)$ imply $(X, G_D(\tau))$ is T_0 and hence that $\Psi_{G_D(\tau)}$ is injective. For surjectivity, let $p \in Lpt_{\text{mod}}(G_D(\tau))$. Then p a semiframe morphism and τ a subsemiframe of $G_D(\tau)$ imply $p_{|\tau} : \tau \to L$ is a semiframe morphism, and so $p_{|\tau} \in Lpt(\tau)$. Now Ψ_τ surjective implies $\exists\, x \in X$, $\psi_\tau(x) = p_{|\tau}$. Using steps similar to those finishing the proof of (4), the reader can show $p = \Psi_{G_D(\tau)}(x)$, which completes the proof. □

The primary point of the proposition is that the term "ultra" is questionable in regard to both T_0 and ɩ-sobriety, even though ɩL-sober, ω generated spaces are modified L-sober to degree L. We shall primarily use the terms L-T_0 (or T_0) and ɩL-sober (or ɩ-sober) in the context of the spectrum adjunction being developed in the sequel; and in the context of ɩ-sobriety as a new separation axiom of interest in its own right, we shall primarily use the term ι-sobriety.

Justifying Examples. To finish this discussion, we refer the reader to Chapter 17 [42] of this Volume for examples of important ι-sober spaces which are not sober in senses previously defined in the literature or defined in 2.2 above. □

We may restate statement (4) of the above proposition as follows: let L be a complete chain and (X, τ) be an ω generated space; then (X, τ) ɩL-sober $\Rightarrow$ (X, τ) is modified L-sober. This leads to an *open question*: does $\Rightarrow$ reverse? A similar open question: does the implication of (5) reverse?

Denote by ɩL-**SobTop** [ɩL-**Spat**L-**Loc**] the full subcategory of L-**Top** [L-**Loc**, resp.] generated by ɩL-sober L-spaces [ɩLspatial L-locales, resp.]. In analogy with the standard facts of 1.6 for Ω and traditional spaces and with Proposition 2.3.2 for $L\Omega$ and L-spaces, we have the following proposition:

3.5.3 Standard ɩ$L\boldsymbol{\Omega}$ Facts. The following hold for $\boldsymbol{\Omega} : L$-**Top** $\to L$-**Loc**:

1. $\boldsymbol{\Omega}$ is injective on objects.

2. $\boldsymbol{\Omega}$ is faithful if it is restricted to L-T_0 spaces, in which case $\boldsymbol{\Omega}$ is an embedding.

3. $\boldsymbol{\Omega}$ is a full embedding if it is restricted to ɩL-sober spaces.

Proof. (1) is immediate. For (2), apply 3.5.2(1) to get that the topological modifications of the L-T_0 spaces are (traditionally) T_0. Now apply Proposition 2.3.2(2) in the first coordinate and Standard Fact (2) of 1.6 in the second coordinate to get that $\mathbf{\Omega}$ is faithful on hom-sets.

For (3), let $(h^{\mathfrak{u}}, h^{\mathfrak{l}})^{op} : \mathbf{\Omega}(X, \tau) \to \mathbf{\Omega}(Y, \sigma)$ be a morphism in L-**Loc**-. Then $[h^{\mathfrak{u}}]^{op} : \tau \to \sigma$, $[h^{\mathfrak{l}}]^{op} : \iota(\tau) \to \iota(\sigma)$ are localic morphisms, i.e. $h^{\mathfrak{u}} : \tau \leftarrow \sigma$, $h^{\mathfrak{l}} : \iota(\tau) \leftarrow \iota(\sigma)$ are frame morphisms. By sobriety—of the spaces $(X, \iota(\tau))$ and $(Y, \iota(\sigma))$—and Standard Fact (2) of 1.6, $\exists f : (X, \iota(\tau)) \to (Y, \iota(\sigma))$ such that $h^{\mathfrak{l}} = f^{\leftarrow}_{|\iota(\sigma)}$. It remains to check that $h^{\mathfrak{u}} = (f^{\leftarrow}_{L})_{|\sigma}$, which would both show that $f : (X, (\tau)) \to (Y, \iota(\sigma))$ is continuous and that $\mathbf{\Omega}(f) = (h^{\mathfrak{u}}, h^{\mathfrak{l}})^{op}$. Since $(h^{\mathfrak{u}}, h^{\mathfrak{l}}) \in L$-**Frm**, we have $\forall t \in L_{\top}$, $h^{\mathfrak{l}} \circ \iota_t^{\iota(\sigma)} = \iota_t^{\iota(\tau)} \circ h^{\mathfrak{u}}$, i.e.

$$f^{\leftarrow}_{|\iota(\sigma)} \circ \iota_t^{\iota(\sigma)} = \iota_t^{\iota(\tau)} \circ h^{\mathfrak{u}}$$

Let $v \in \sigma$. It follows

$$[(f^{\leftarrow}_{L})(v) > t] = f^{\leftarrow}[v > t] = [h^{\mathfrak{u}}(v) > t]$$

where the first equality is a consequence of the definition of $f^{\leftarrow}_{L}$ (2.1) and the second equality is a consequence of the previous display. Since L is a complete chain, we apply the third bullet of Convention 2.7 to the above display to conclude $h^{\mathfrak{u}} = (f^{\leftarrow}_{L})_{|\sigma}$. □

4 $\mathbb{L}$-spectrum of L-locales

4.1 First Comparison Map and $\mathbb{L}$-Spectrum of L-Locales. Recall the first comparison maps for previous spectral theories:

$$\begin{aligned} \Phi &: A \to \mathbf{2}^{Pt(A)}, \text{ counit}^{op} \text{ of } \Omega \dashv PT \text{ (1.7)}, \\ \Phi_L &: A \to L^{Lpt(A)}, \text{ counit}^{op} \text{ of } L\Omega \dashv LPT \text{ (2.2)}, \\ \Phi &: A(t) \to A(l)^{Pt(A)}, \text{ part of counit}^{op} \text{ of } \mathbf{\Lambda} \dashv \mathbf{\Sigma} \text{ [54, 55, 56, 57]} \end{aligned}$$

where in the first two maps A is a semilocale and for the third A is a functor from functor category $\mathbf{SLoc}^{\mathbf{2.5}}$. This subsection constructs an analogous first comparison map for L-locales, subsequently used in 4.4 below to construct an analogous first comparison "morphism" for L-locales.

Consider carrier set $Pt(A) \equiv \{p : A \to \mathbf{2} \mid p \in \mathbf{SFrm}\}$ (1.7) of points for traditional spectrum $PT(A)$ of semilocale (or semiframe) A. Now let $A = (\varphi_t^A : A^{\mathfrak{u}} \to A^{\mathfrak{l}} \mid t \in L_{\top})$ be an L-locale and let a be in $A^{\mathfrak{u}}$. Define the map

$$\Phi_{L\mathfrak{ul}} : A^{\mathfrak{u}} \to L^{Pt(A^{\mathfrak{l}})}$$

by

$$\Phi_{L\mathfrak{ul}}(a)(p) = \bigvee \{t \in L_{\top} : \forall s < t, p(\varphi_s^A(a)) = \top\}$$

The map Φ_{Lul} resembles Φ, Φ_L, Φ cited in the previous paragraph. Set $\nu : L_\top \to Pt\left(A^{u}\right)$ by $\nu(t) = p \circ \varphi_t^A$. Then Lemma 2.8 implies—*à la* the proof of (*) within the proof of Proposition 3.1—that

$$t < \Phi_{Lul}(a)(p) \quad \text{iff} \quad p(\varphi_t^A(a)) = \top \tag{4.1.1}$$

Proposition. Φ_{Lul} is a frame morphism, $\Phi_{Lul}^{\to}\left(A^{u}\right) \equiv \{\Phi_{Lul}(a) : a \in A^{u}\}$ is an L-topology on $Pt\left(A^{l}\right)$, and the **l*L*-spectrum** $\mathbf{\Sigma}(A) \equiv \left(Pt\left(A^{l}\right), \Phi_{Lul}^{\to}\left(A^{u}\right)\right)$ of the L-locale A is an L-topological space in L-**Top**.

Proof. From (4.1.1), we easily infer (*à la* proof of Proposition 3.1) that

$$\begin{aligned}
\Phi_{Lul}(\bot) &= \underline{\bot},\ \Phi_{Lul}(\top) = \underline{\top}, \\
\Phi_{Lul}(a \wedge b) &= \Phi_{Lul}(a) \wedge \Phi_{Lul}(b), \\
\Phi_{Lul}\left(\bigvee S\right) &= \bigvee\{\Phi_{Lul}(a) : a \in S\}
\end{aligned}$$

which establishes that Φ_{Lul} is a frame morphism. □

4.2 For an L-localic morphism $h = \left(h^{u}, h^{l}\right)^{op} : A \to B$, define map $\mathbf{\Sigma}(h) : Pt\left(A^{l}\right) \to Pt\left(B^{l}\right)$ by

$$\mathbf{\Sigma}(h)(p) = p \circ h^{l}$$

Lemma. $\mathbf{\Sigma}(h) : \mathbf{\Sigma}(A) \to \mathbf{\Sigma}(B)$ is L-continuous.

Proof. We show that $\forall\, \Phi_{Lul}(b) \in \Phi_{Lul}^{\to}\left(B^{u}\right), \mathbf{\Sigma}(h)_L^{\leftarrow}\left(\Phi_{Lul}(b)\right) \in \Phi_{Lul}^{\to}\left(A^{u}\right)$. Let $p \in Pt\left(A^{l}\right)$. Using condition (3.3.1) and the above definitions, we have:

$$\begin{aligned}
\Phi_{Lul}(b)\left[\mathbf{\Sigma}(h)(p)\right] &= \Phi_{Lul}(b)\left(p \circ h^{l}\right) = \bigvee\{t : \forall\, s < t,\ \ p(h^{l}\left(\varphi_s^A(b)\right)) = \top\} \\
&= \bigvee\{t : \forall\, s < t,\ \ p(\varphi_s^A(h^{u}(b))) = \top\} = \mathbf{\Sigma}\left(h^{u}(b)\right)(p)
\end{aligned}$$

Observing that $h^{u}(b) \in A^{u}$ completes the proof. □

4.3. From 4.1, 4.2, we have the **l*L*-spectrum functor** $\mathbf{\Sigma} : L$-**Top** $\leftarrow L$-**Loc**.

4.4 First Comparison (*L*-) Morphism (Counit) ϕ. Let $A \in |L\text{-}\mathbf{Loc}|$. The map $\Phi_{Lul} : A^{u} \to L^{Pt\left(A^{l}\right)}$ constructed in 4.1 is a frame morphism, and its restriction restricted with respect to codomain

$$\Phi_{Lul} : A^{u} \to \Phi_{Lul}^{\to}\left(A^{u}\right)$$

is also a (surjective) frame morphism by (the proof of) Proposition 4.1 and may be viewed as a map $\Phi_{Lul} : A^{u} \to L\Omega\left(\mathbf{\Sigma}(A)\right)$, where $L\Omega$ comes from 2.2 and $\mathbf{\Sigma}(A)$ is the L-space constructed by functor $\mathbf{\Sigma}$ from A.

Next, consider the counit Φ^{op} of traditional adjunction $\Omega \dashv PT$ (1.7). Then $\Phi : A^{\mathfrak{l}} \to \mathbf{2}^{Pt(A^{\mathfrak{l}})}$ is a frame morphism, and its restriction restricted with respect to codomain

$$\Phi : A^{\mathfrak{l}} \to \Phi^{\to}\left(A^{\mathfrak{l}}\right)$$

is also a (surjective) frame morphism.

Proposition. Let $L\mathbf{\Omega}$ and $\mathbf{\Sigma}$ be the functors constructed in 3.5 and 4.1–4.3, respectively. The pair $(\Phi_{L\mathfrak{u}\mathfrak{l}}, \Phi)^{op}$ is an L-localic morphism from $L\mathbf{\Omega}\left(\mathbf{\Sigma}\left(A\right)\right)$ to A, i.e.

$$\phi_A \equiv (\Phi_{L\mathfrak{u}\mathfrak{l}}, \Phi)^{op} : L\mathbf{\Omega}\left(\mathbf{\Sigma}\left(A\right)\right) \to A \text{ in } L\text{-}\mathbf{Loc}$$

where both $\Phi_{L\mathfrak{u}\mathfrak{l}}, \Phi$ are taken in the restricted sense as discussed above.

Proof. Given Lemma 4.2, the main point is to check that the commutative squares condition of Definition 3.3 for L-frame morphisms is satisfied, i.e. that $\forall t \in L_{\top}$,

$$\Phi \circ \varphi_t^A = \iota_t \circ \Phi_{L\mathfrak{u}\mathfrak{l}}$$

Let $a \in A^{\mathfrak{u}}$. Then using (4.1.1), we have

$$\begin{aligned}\Phi(\varphi_t^A(a)) &= \{p : p\left(\varphi_t^A(a)\right) = \top\} = \{p : \Phi_{L\mathfrak{u}\mathfrak{l}}\left(a\right)\left(p\right) > t\} \\ &= \iota_t(\Phi_{L\mathfrak{u}\mathfrak{l}}\left(a\right)) = \iota_t(\Phi_{L\mathfrak{u}\mathfrak{l}}(a))\end{aligned}$$

which completes the proof. □

Corollary. The proposition gives us a natural transformation

$$\phi : L\mathbf{\Omega} \circ \mathbf{\Sigma} \to Id_{L\text{-}\mathbf{Loc}}$$

Proof. This can be proved using Lemma 4.2 and the above proposition. □

We shall speak of ϕ as the **first comparison (L-)morphism.** It is built out of comparison map $\Phi_{L\mathfrak{u}\mathfrak{l}}$—a modification of Φ_L and the variable-basis Φ—in the first component and the traditional Φ in the second component. It will be seen below that ϕ will play the same categorical role as do these other comparison maps.

4.5. Second Comparison Morphism (Unit) ψ. To motivate this subsection, we remark in further detail on the units of the $\Omega \dashv PT$ and $L\Omega \dashv LPT$ adjunctions given in 1.7 and 2.2 above, respectively. The 1.7 formulation of Ψ is quite different from its original formulation in [25] (ultimately coming from the work of Marshall Stone) and is a consequence of the much later formulation of Ψ_L given in 2.2.

The original formulation [25] of $\Psi : (X, \mathfrak{T}) \to (Pt(\mathfrak{T}), \Phi^{\to}(\mathfrak{T}))$ is

$$\begin{array}{rcl} x & \longmapsto & \overline{\{x\}} \text{ (irreducible closed set)} \\ & \longmapsto & X - \overline{\{x\}} \text{ (meet-irreducible open set)} \\ & \longmapsto & \downarrow\left(X - \overline{\{x\}}\right) \text{ (prime principal ideal of } \mathfrak{T}) \\ & \longmapsto & \Psi(x) \in Pt(\mathfrak{T}) \text{ by } \ker(\Psi(x)) = \downarrow\left(X - \overline{\{x\}}\right), \end{array}$$

the formulation of $\Psi_L : (X, \tau) \to (Lpt(\tau), \Phi_L^{\to}(\tau))$ is $\Psi_L(x)(u) = u(x)$—an explicit evaluation map, and the characteristic functor G_χ : **Top** $\to$ **2-Top** is given on objects by $G_\chi(X, \mathfrak{T}) = (X, G_\chi(\mathfrak{T}) \equiv \{\chi_U : U \in \mathfrak{T}\})$, where the latter G_χ is the fibre map underlying functor G_χ. These facts now follow:

- The map $\Psi_{\mathbf{2}} : (X, G_\chi(\mathfrak{T})) \to (\mathbf{2}pt(G_\chi(\mathfrak{T})), \Phi_{\mathbf{2}}^{\to}(G_\chi(\mathfrak{T})))$ is given by $\Psi_{\mathbf{2}}(x)(\chi_U) = \chi_U(x)$;
- the correspondence F_χ, given by $p \mapsto p \circ G_\chi^{-1}$, is a bijection between $Pt(\mathfrak{T})$ and $\mathbf{2}pt(G_\chi(\mathfrak{T}))$; and
- $\Psi_{\mathbf{2}} \circ G_\chi = F_\chi \circ \Psi$, i.e. $\Psi = F_\chi^{-1} \circ \Psi_{\mathbf{2}} \circ G_\chi$.

From the last fact, i.e. from the evaluation map formulation of Ψ_L, we have the following simplified formulation of the traditional Ψ given in 2.2 and [49, 51] and implicit in [54, 55, 56, 57], namely:

$$\Psi(x)(U) = \chi_U(x)$$

With the above overview in mind, let $(X, \tau) \in |L\text{-}\mathbf{Top}|$ and consider its topological modification $(X, \iota(\tau)) \in |\mathbf{Top}|$. We define the *function* underlying

$$\begin{array}{rcl} \psi_{(X,\tau)} & : & (X, \tau) \to \mathbf{\Sigma}(L\mathbf{\Omega}(X, \tau)) \equiv \mathbf{\Sigma}\left[(\iota_t : \tau \to \iota(\tau))_{t \in L_\top}\right] \\ & \equiv & (Pt(\iota(\tau)), \Phi_{L\mathfrak{U}\mathfrak{l}}^{\to}(\tau)) \end{array}$$

to be the function underlying the traditional

$$\Psi_{(X,\iota(\tau))} : (X, \iota(\tau)) \to (Pt(\iota(\tau)), \Phi^{\to}(\iota(\tau)))$$

i.e., $\forall x \in X$, $\forall U \in \iota(\tau)$,

$$\psi_{(X,\tau)}(x)(U) = \chi_U(x)$$

If there is no confusion, we may drop the subscript (X, τ) in $\psi_{(X,\tau)}$.

The following lemma is useful here (via its Corollary) and in subsection 4.10 below. For this lemma we need the forgetful functor $\mathcal{U}^{\mathfrak{l}} : L\text{-}\mathbf{Frm} \to \mathbf{Frm}$ defined by $\mathcal{U}^{\mathfrak{l}}(A) = A^{\mathfrak{l}}$, $\mathcal{U}^{\mathfrak{l}}(h) = h^{\mathfrak{l}}$.

Lemma. The following hold:

1. $\forall A \in |L\text{-}\mathbf{Frm}|$, $\iota(\Phi^{\rightarrow}_{L\mathfrak{u}\mathfrak{l}}(A^{\mathfrak{u}})) = \Phi^{\rightarrow}(A^{\mathfrak{l}})$, in which case we have the functor equality $\iota \circ \mathbf{\Sigma} = PT \circ \mathcal{U}^{\mathfrak{l}}$ from L-**Frm** to **Top**.
2. $\forall (X, \tau) \in |L\text{-}\mathbf{Top}|$, $\iota(\Phi^{\rightarrow}_{L\mathfrak{u}\mathfrak{l}}(\tau)) = \Phi^{\rightarrow}(\iota(\tau))$, in which case we have the functor equality $\iota \circ \mathbf{\Sigma} \circ L\mathbf{\Omega} = PT \circ \Omega \circ \iota$ from L-**Top** to **Top**.

Proof. For (1), it is easy to check that both sides of the claimed functor equality hold for L-continuous mappings, and the functor equality holds for objects provided $\iota(\Phi^{\rightarrow}_{L\mathfrak{u}\mathfrak{l}}(A^{\mathfrak{u}})) = \Phi^{\rightarrow}(A^{\mathfrak{l}})$ holds. First note that for $a \in A^{\mathfrak{u}}$, $t \in L_{\top}$, and $p \in Pt(A^{\mathfrak{l}})$, applying (4.1.1) yields

$$\begin{aligned} p &\in [\Phi_{L\mathfrak{u}\mathfrak{l}}(a) > t] \Leftrightarrow \Phi_{L\mathfrak{u}\mathfrak{l}}(a)(p) > t \\ &\Leftrightarrow p(\varphi_t^A(a)) = \top \Leftrightarrow p \in \Phi(\varphi_t^A(a)) \end{aligned}$$

so that $[\Phi_{L\mathfrak{u}\mathfrak{l}}(u) > t] = \Phi(\varphi_t^A(a))$. It follows, since $[\Phi_{L\mathfrak{u}\mathfrak{l}}(a) > t]$ is a subbasic set of $\iota(\Phi^{\rightarrow}_{L\mathfrak{u}\mathfrak{l}}(A^{\mathfrak{u}}))$, that $\Phi^{\rightarrow}(A^{\mathfrak{l}})$ contains a subbasis of $\iota(\Phi^{\rightarrow}_{L\mathfrak{u}\mathfrak{l}}(A^{\mathfrak{u}}))$ and hence $\iota(\Phi^{\rightarrow}_{L\mathfrak{u}\mathfrak{l}}(A^{\mathfrak{u}}))$ itself. Now recall from Definition 3.3(F1), that

$$\{\varphi_t^A(a) : a \in A^{\mathfrak{u}}, t \in L_{\top}\}$$

is a subbasis for the frame $A^{\mathfrak{l}}$. Since Φ is a frame morphism, it follows

$$\{\Phi(\varphi_t^A(a)) : a \in A^{\mathfrak{u}}, t \in L_{\top}\}$$

is a subbasis for $\Phi^{\rightarrow}(A^{\mathfrak{l}})$, showing by the above computation that $\iota(\Phi^{\rightarrow}_{L\mathfrak{u}\mathfrak{l}}(A^{\mathfrak{u}}))$ contains a subbasis of $\Phi^{\rightarrow}(A^{\mathfrak{l}})$ and hence $\Phi^{\rightarrow}(A^{\mathfrak{l}})$ itself. Finally, (2) follows immediately from (1). □

Corollary. The traditional $\Psi_{(X,\iota(\tau))} : (X, \iota(\tau)) \to (Pt(\iota(\tau)), \Phi^{\rightarrow}(\iota(\tau)))$ may be rewritten $\Psi_{(X,\iota(\tau))} : (X, \iota(\tau)) \to (Pt(\iota(\tau)), \iota(\Phi^{\rightarrow}_{L\mathfrak{u}\mathfrak{l}}(\tau)))$, so that the domain [codomain] of $\Psi_{(X,\iota(\tau))}$ is the topological modification of the domain [codomain, resp.] of $\psi_{(X,\tau)}$.

Proposition. $\forall u \in \tau$, $\psi_L^{\leftarrow}(\Phi_{L\mathfrak{u}\mathfrak{l}}(u)) \equiv \Phi_{L\mathfrak{u}\mathfrak{l}}(u) \circ \psi = u$ on X, and hence each $\psi_{(X,\tau)}$ is L-continuous. Further, $\psi : Id \to \mathbf{\Sigma} \circ L\mathbf{\Omega}$ is a natural transformation.
Proof. Let $u \in \tau$, $x \in X$, .$s \in L_{\top}$. From 4.1.1 we have

$$s < \Phi_{L\mathfrak{u}\mathfrak{l}}(u)(\psi(x)) \quad \text{iff} \quad \psi(x)(\iota_s(u)) = \top \quad \text{iff} \quad x \in \iota_s(u) \quad \text{iff} \quad s < u(x)$$

which proves (using Convention 2.7) that ψ is subbasic continuous, which suffices by Theorem 3.2.6 of [60] to say that ψ is L-continuous. □

4.6 Proposition. $\forall$ complete chain L, $L\boldsymbol{\Omega} \dashv \boldsymbol{\Sigma}$ with unit ψ and counit ϕ.

Proof. We follow the lifting and naturality diagrams of [41] as given in [54, 55]. First, let $(X,\tau) \in |L\text{-}\mathbf{Top}|$, consider

$$\psi : (X,\tau) \to \boldsymbol{\Sigma}\left[L\boldsymbol{\Omega}(X,\tau)\right] \equiv \left(Pt\left(\iota(\tau)\right), \Phi^{\rightarrow}_{Lu\text{l}}(\tau)\right)$$

let $(Y,\sigma) \in |L\text{-}\mathbf{Top}|$, let $A \equiv \left(\varphi^A_t : A^{\text{u}} \to A^{\text{l}}\right)_{t\in L_\top} \in |L\text{-}\mathbf{Loc}|$, and let

$$f \in L\text{-}\mathbf{Top}\left((X,\tau), \boldsymbol{\Sigma}(A) \equiv \left(Pt\left(A^{\text{l}}\right), \Phi^{\rightarrow}_{Lu\text{l}}\left(A^{\text{u}}\right)\right)\right)$$

We must show $\exists!\, \overline{f} : L\boldsymbol{\Omega}(X,\tau) \to A$ in L-**Loc** such that

$$f = \boldsymbol{\Sigma}\left(\overline{f}\right) \circ \psi \tag{4.6.1}$$

Let us impose (4.6.1) in order to get both uniqueness as well as possibly existence. Suppose for now that we have $\overline{f} : L\boldsymbol{\Omega}(X,\tau) \to A$, with

$$f = \left(f^{\text{u}}, f^{\text{l}}\right)^{op}$$

Let $x \in X$ and $a^{\text{l}} \in A^{\text{l}}$. Then assuming (4.6.1), we have

$$\begin{aligned} f(x)\left(a^{\text{l}}\right) &= \boldsymbol{\Sigma}\left(\overline{f}\right)\left(\psi(x)\right)\left(a^{\text{l}}\right) = \psi(x)\left(f^{\text{l}}\left(a^{\text{l}}\right)\right) \\ &= \chi_{f^{\text{l}}(a^{\text{l}})}(x) \end{aligned}$$

This uniquely determines the function f^{l} by putting

$$f^{\text{l}}\left(a^{\text{l}}\right) = \left\{x \in X : f(x)\left(a^{\text{l}}\right) = \top\right\}$$

Further, it can be shown that since each $f(x)$ is a frame morphism, then f^{l} is a frame morphism. Now by Note 3.3.1, we have that f^{u} is uniquely determined provided we can show the existence a frame morphism f^{l} such that $\forall\, t \in L_\top$, $\iota_t \circ f^{\text{u}} = f^{\text{l}} \circ \phi^A_t$. Using the L-continuity of f, we have that $\left[f^{\leftarrow}_L\right] : \tau \leftarrow \Phi^{\rightarrow}_{Lu\text{l}}\left(A^{\text{u}}\right)$ is a frame morphism; so that

$$f^{\text{u}} \equiv f^{\leftarrow}_L \circ \Phi_{Lu\text{l}} : A^{\text{u}} \to \tau$$

is a frame morphism. Now the required commutivity can be shown from

$$\forall\, t \in L_\top, \quad \Phi_{Lu\text{l}}\left(a^{\text{u}}\right)\left(f(x)\right) > t \quad \text{iff} \quad f(x)\left(\phi^A_t\left(a^{\text{u}}\right)\right) = \top$$

which in turn is a consequence of (4.1.1) above.

The required naturality diagram follows from Corollary 4.4 and Proposition 4.5. Thus we have justified the adjunction $L\boldsymbol{\Omega} \dashv \boldsymbol{\Sigma}$ with unit ψ. It can also be shown that the dual lifting and naturality diagrams hold with ϕ as the counit. □

4.7 Characterization of $\mathfrak{l}L$-Spatial L-Locales. Since each frame morphism component of

$$\phi_A \equiv (\Phi_{L\mathfrak{u}\mathfrak{l}}, \Phi)^{op} : L\boldsymbol{\Omega}(\boldsymbol{\Sigma}(A)) \to A$$

is taken as restricted to its range (or image)—see 4.4, then $(\Phi_{L\mathfrak{u}\mathfrak{l}}, \Phi)$ is an epimorphism in L-**Frm**, i.e. ϕ_A is a monomorphism in L-**Loc**, for each L-locale A. Further, Proposition 4.6 helps us prove the following corollary characterization of $\mathfrak{l}L$-spatiality (see 3.5.1):

Corollary. Let $A \in |L\text{-}\mathbf{Loc}|$. The following are equivalent:

1. A is $\mathfrak{l}L$-spatial (3.5.1).
2. ϕ_A is a isomorphism in L-**Loc**.
3. Each of $\Phi_{L\mathfrak{u}\mathfrak{l}}, \Phi$ is injective.

Proof. For (1) $\Rightarrow$ (3), suppose $(X,\tau) \in |L\text{-}\mathbf{Top}|$ and $h = \left(h^{\mathfrak{u}}, h^{\mathfrak{l}}\right)^{op} : \boldsymbol{\Omega}(X,\tau) \to A$ is an isomorphism in L-**Loc**. By the universal property of the counit $\phi_A \equiv (\Phi_{L\mathfrak{u}\mathfrak{l}}, \Phi)^{op} : L\boldsymbol{\Omega}(\boldsymbol{\Sigma}(A)) \to A$ of $L\boldsymbol{\Omega} \dashv \boldsymbol{\Sigma}$ (Proposition 4.6), $\exists! g = \left(g^{\mathfrak{u}}, g^{\mathfrak{l}}\right)^{op} : \boldsymbol{\Omega}(X,\tau) \to L\boldsymbol{\Omega}(\boldsymbol{\Sigma}(A))$, $h = \phi_A \circ g$. This means in **Frm**, we have

$$h^{\mathfrak{u}} = g^{\mathfrak{u}} \circ \Phi_{L\mathfrak{u}\mathfrak{l}}, \quad h^{\mathfrak{l}} = g^{\mathfrak{l}} \circ \Phi$$

where $h^{\mathfrak{u}}, h^{\mathfrak{l}}$ are **Frm** isomorphisms. This forces $\Phi_{L\mathfrak{u}\mathfrak{l}}, \Phi$ to be injective.

Now (3) $\Rightarrow$ (2) is immediate given the comments just above the corollary, and (2) $\Rightarrow$ (1) is trivial. □

4.8. Characterization of $\mathfrak{l}L$-Sober L-Topological Spaces. Analogous to the characterizations of traditional sobriety in 1.7 and L-sobriety in 2.2, we have the following proposition:

Proposition. Let $(X,\tau) \in |L\text{-}\mathbf{Top}|$. Then the following are equivalent:

1. (X,τ) is $\mathfrak{l}L$-sober (3.5.1).
2. $\psi_{(X,\tau)}$ is an L-homeomorphism.
3. $\psi_{(X,\tau)}$ is a bijection.

Proof. We recall that the function underlying $\psi : (X,\tau) \to \left(Pt(\iota(\tau)), \Phi^{\to}_{L\mathfrak{u}\mathfrak{l}}(\tau)\right)$ is the same as that underlying $\Psi : (X,\iota(\tau)) \to \left(Pt(\iota(\tau)), \Phi^{\to}(\iota(\tau))\right)$. Using this fact along with Definition 3.5.1 and 2.2,

$$\begin{aligned} (X,\tau) \text{ is } \mathfrak{l}L\text{-sober} \quad &\Leftrightarrow \quad (X,\iota(\tau)) \text{ is traditionally sober} \\ &\Leftrightarrow \quad \Psi \text{ is a bijection} \\ &\Leftrightarrow \quad \psi \text{ is a bijection} \end{aligned}$$

yielding (2) $\Rightarrow$ (3) $\Leftrightarrow$ (1). By Proposition 4.5, ψ is L-continuous. Assume (3). To see that ψ^{-1} is L-continuous, let $u \in \tau$, $t \in L_\top$, $p \in Pt(\iota(\tau))$, and $x \in X$ such that $\psi(x) = p$. Using the appropriate definitions along with (4.1.1), we have:

$$\begin{aligned}
\left(\psi^{-1}\right)_L^{\leftarrow}(u)(p) &> t \Leftrightarrow u\left(\psi^{-1}(p)\right) > t \Leftrightarrow u(x) > t \\
&\Leftrightarrow x \in [u > t] \Leftrightarrow \chi_{[u>t]}(x) = \top \\
&\Leftrightarrow \psi(x)[u > t] = \top \\
&\Leftrightarrow \Phi_{L\mathfrak{u}\mathfrak{l}}[u > t](\psi(x)) > t \\
&\Leftrightarrow \Phi_{L\mathfrak{u}\mathfrak{l}}[u > t](p) > t
\end{aligned}$$

By Convention 2.7, $\left(\psi^{-1}\right)_L^{\leftarrow}(u) = \Phi_{L\mathfrak{u}\mathfrak{l}}[u > t] \in \Phi_{L\mathfrak{u}\mathfrak{l}}^{\rightarrow}(\tau)$, so ψ^{-1} is L-continuous. So ψ is an L-homeomorphism. □

4.9 Characterization of L-T_0, L-Topological Spaces. Analogous to the characterizations of traditional T_0 spaces in [25] and L-T_0 spaces in 2.3.1 and [54, 55, 57], we have the following proposition:

Proposition. Let $(X, \tau) \in |L\text{-}\mathbf{Top}|$. Then the following are equivalent:

1. (X, τ) is L-T_0.
2. $\psi_{(X,\tau)}$ is an L-embedding (in the sense that ψ is an L-homeomorphism onto its image equipped with the subspace L-topology in the sense of [68] (cf. Subsection 5.2 of [60])).
3. $\psi_{(X,\tau)}$ is an injection.

Proof. Given Proposition 3.5.2(1,2), the proof becomes exactly analogous to that of Proposition 4.8 and is left to the reader. □

4.10 Categorical Equivalence Between $\mathfrak{l}L$-Spatiality and $\mathfrak{l}L$-Sobriety. Analogous to the categorical equivalences given in 1.7 and 2.2, we have the following proposition:

Proposition. The adjunction $L\mathbf{\Omega} \dashv \mathbf{\Sigma}$ restricts to an equivalence of the categories $\mathfrak{l}L$-**Spat**L-**Loc** and $\mathfrak{l}L$-**SobTop**, which are the full subcategories of L-**Loc** and L-**Top**, respectively, of $\mathfrak{l}L$-spatial L-locales and $\mathfrak{l}L$-sober topological spaces.

Proof. This is an immediate consequence of Corollary 4.7 and Proposition 4.8, once we have $L\mathbf{\Omega}$ produces $\mathfrak{l}L$-spatial L-locales—which is immediate from Definition 3.5(1)—and $\mathbf{\Sigma}$ produces $\mathfrak{l}L$-sober spaces. This latter point merits comment. Let A be an L-locale and consider

$$\mathbf{\Sigma}(A) = \left(Pt\left(A^{\mathfrak{l}}\right).\Phi_{L\mathfrak{u}\mathfrak{l}}^{\rightarrow}\left(A^{\mathfrak{u}}\right)\right)$$

This space will be $\mathbb{L}L$-sober if its topological modification

$$\left(Pt\left(A^{\mathfrak{l}}\right), \iota\left(\Phi_{\overline{L}\mathfrak{u}\mathfrak{l}}^{\rightarrow}\left(A^{\mathfrak{u}}\right)\right)\right)$$

is traditionally sober. By Lemma 4.5(1) above, this topological modification may be rewritten

$$\left(Pt\left(A^{\mathfrak{l}}\right), \Phi^{\rightarrow}\left(A^{\mathfrak{l}}\right)\right)$$

This space is precisely the traditional point-space $PT\left(A^{\mathfrak{l}}\right)$, and it is well-known (third bullet of 1.7) that such spaces are always sober. □

References

[1] J. Adámek, H. Herrlich, G. E. Strecker, Abstract And Concrete Categories: The Joy Of Cats, Wiley Interscience Pure and Applied Mathematics, John Wiley & Sons (Brisbane/Chicester/New York/Singapore/Toronto), 1990.

[2] M. H. Burton, J. Gutiérrez Garcia, *Extensions of uniform space notions*, in [13], 581–606.

[3] C. L. Chang, *Fuzzy topology*, J. Math. Anal. Appl. **24**(1968), 182–190.

[4] C. H. Dowker, D. Papert-Strauß, *Paracompact frames and closed maps*, Symposia Math. **16**(1975), 93–116.

[5] T. E. Gantner, R. C. Steinlage, R. H. Warren, *Compactness in fuzzy topological spaces*, J. Math. Anal. Appl. **62**(1978), 547–562.

[6] J. A. Goguen, *The fuzzy Tychonoff Theorem*, J. Math. Anal. Appl. **43**(1973), 734–742.

[7] J. Gutiérrez García, M. A. de Prada Vicente, A. P. Šostak, *A unified approach to the concept of a fuzzy L-uniform space*, Chapter 3 in this Volume.

[8] U. Höhle, *Probabilistic uniformization of fuzzy uniformities*, Fuzzy Sets and Systems **1**(1978), 311–332.

[9] U. Höhle, *Probabilistic metrization of fuzzy topologies*, Fuzzy Sets and Systems **8**(1982), 63–69.

[10] U. Höhle, *Fuzzy topologies and topological space objects in a topos*, Fuzzy Sets and Systems **19** (1986), 299–304.

[11] U. Höhle, *Locales and L-topologies*, Mathematik-Arbeitspapiere **48**(1997), 233-250, Universität Bremen.

[12] U. Höhle, Many Valued Topology And Its Applications, Kluwer Academic Publishers (Boston/Dordrecht/London), 2001.

[13] U. Höhle, S. E. Rodabaugh, eds, Mathematics Of Fuzzy Sets: Logic, Topology, And Measure Theory, The Handbooks of Fuzzy Sets Series, Volume **3**(1999), Kluwer Academic Publishers (Boston/Dordrecht/London).

[14] U. Höhle, S. E. Rodabaugh, *Weakening the requirement that L be a complete chain*, Appendix to Chapter 6 in this Volume.

[15] U. Höhle, A. Šostak, *Axiomatic foundations of fixed-basis fuzzy topology*, Chapter 3 in [13], 123–272.

[16] R.-E. Hoffmann, *Irreducible filters and sober spaces*, Manuscripta Math. **22**(1977), 365-380.

[17] R.-E. Hoffmann, *Soberification of partially ordered sets*, Semigroup Forum **17**(1979), 123-138.

[18] R.-E. Hoffmann, *On the soberification remainder $sX - X$*, Pacific J. Math. **83**(1979), 145–156.

[19] B. Hutton, *Normality in fuzzy topological spaces*, J. Math. Anal. Appl. **50**(1975), 74–79.

[20] B. Hutton, *Uniformities on fuzzy topological spaces*, J. Math. Anal. Appl. **58**(1977), 559–571.

[21] B. Hutton, *Products of fuzzy topological spaces*, Topology Appl. **11**(1980), 59–67.

[22] B. Hutton, I. Reilly, *Separation axioms in fuzzy topological spaces*, Fuzzy Sets and Systems **3**(1980), 93–104.

[23] J. Isbell, *Atomless parts of spaces*, Scand. Math. **31**(1972), 5–32.

[24] P.T. Johnstone, *Tychonoff's theorem without the axiom of choice,* Fund. Math. **113**(1981), 31-35.

[25] P. T. Johnstone, Stone Spaces, Cambridge University Press (Cambridge), 1982.

[26] A. J. Klein, *α-Closure in fuzzy topology*, Rocky Mount. J. Math. **11**(1981), 553–560.

[27] A. J. Klein, *Generating fuzzy topologies with semi-closure operators*, Fuzzy Sets and Systems **9**(1983), 267–274.

[28] A. J. Klein, *Generalizing the L-fuzzy unit interval*, Fuzzy Sets and Systems **12**(1984), 271–279.

[29] W. Kotzé, *Lattice morphisms, sobriety, and Urysohn lemmas*, Chapter 10 in [63], pp. 257–274.

[30] W, Kotzé, ed., Special Issue, Quaestiones Mathematicae **20**(**3**)(1997).

[31] W. Kotzé, *Uniform spaces*, Chapter 8 in [13].

[32] W. Kotzé, *Sobriety and semi-sobriety of L-topological spaces*, Quaestiones Mathematicae **24**(2001), 549–554.

[33] W. Kotzé, *Lifting of sobriety concepts with particular references to* (L, M)*-topological spaces*, Chapter 16 in this Volume.

[34] T. Kubiak, *The topological modification of the L-fuzzy unit interval*, Chapter 11 in [63], 275–305.

[35] T. Kubiak, A. P. Šostak, *Lower set-valued fuzzy topologies*, in [35], 423-429.

[36] Y.-M. Liu, M.-K. Luo, *Fuzzy Stone-Čech type compactification*, Proc. Polish Symposium on Interval and Fuzzy Mathematics (1986), 117–137.

[37] R. Lowen, *Fuzzy topological spaces and fuzzy compactness*, J. Math. Anal. Appl. **56**(1976), 621–633.

[38] R. Lowen, *A comparison of different compactness notions in fuzzy topological spaces*, J. Math. Anal. Appl. **64**(1978), 446–454.

[39] R. Lowen, R. Lowen, *Fuzzy uniform spaces*, J. Math. Anal. Appl. **82**(1981), 370–385.

[40] E. G. Manes, Algebraic Theories, Springer-Verlag (Berlin/Heidelberg/New York), 1976.

[41] S. Mac Lane, Categories For The Working Mathematician, Springer-Verlag (Berlin/Heidelberg/New York), 1974.

[42] A. Pultr, S. E. Rodabaugh, *Examples for different sobrieties in fixed-basis topology*, Chapter 17 in this Volume.

[43] A. Pultr, A. Tozzi, *Separation facts and frame representation of some topological facts*, Applied Categorical Structures **2**(1994), 107–118.

[44] A. Pultr, A. Tozzi, *A note on reconstruction of spaces and maps from lattice data*, Quaestiones Mathematicae **24**(2001), 55–63.

[45] S. E. Rodabaugh, *The Hausdorff separation axiom for fuzzy topological spaces*, Topology Appl. **11**(1980), 319-34.

[46] S. E. Rodabaugh, *Connectivity and the fuzzy unit interval*, Rocky Mount. J. Math. **12**(1982), 113–21.

[47] S. E. Rodabaugh, *A categorical accommodation of various notions of fuzzy topology*, Fuzzy Sets and Systems **9**(1983), 241–265.

[48] S. E. Rodabaugh, *A point set lattice-theoretic framework* $\mathbb{T}$ *which contains LOC as a subcategory of singleton spaces and in which there are general classes of Stone representation and compactification theorems*, first draft February 1986 / second draft April 1987, Youngstown State University Central Printing Office (Youngstown, Ohio).

[49] S. E. Rodabaugh, *Realizations of events, points of a locale, and Stone representation theorems*, Communications: IFSA Mathematics Chapter **1**(1987), 28–33.

[50] S. E. Rodabaugh, *A theory of fuzzy uniformities with applications to the fuzzy real lines*, J. Math. Anal. Appl. **129**(1988), 37–70.

[51] S. E. Rodabaugh, *Categorical aspects of realizations*, Proceedings of the Tenth International Seminar on Fuzzy Set Theory **10**(1988), Johannes Kepler Universitätsdirektion, Linz (Austria), editors S. Weber and E. P. Klement.

[52] S. E. Rodabaugh, *Lowen, para-Lowen, and* α*-level functors and fuzzy topologies on the crisp real line*, J. Math. Anal. Appl. **131**(1988), 157–169.

[53] S. E. Rodabaugh, *Functor categories in poslat topology*, Abstracts of the Fifth International Congress on Topology (17–21 September 1990, Lecce-Otranto, Italy), University of Lecce (Lecce, Italy).

[54] S. E. Rodabaugh, *Point-set lattice-theoretic topology*, Fuzzy Sets and Systems **40**(1991), 297–345.

[55] S. E. Rodabaugh, *Categorical frameworks for Stone representation theorems*, Chapter 7 in [63], 178-231.

[56] S. E. Rodabaugh, *Necessity of Chang-Goguen topologies*, Rendiconti Circolo Matematico Palermo Suppl., Serie II **29**(1992), 299-314.

[57] S. E. Rodabaugh, *Applications of localic separation axioms, compactness axioms, representations, and compactifications to poslat topological spaces*, Fuzzy Sets and Systems **73**(1995), 55–87.

[58] S. E. Rodabaugh, *Powerset operator based foundation for point-set lattice-theoretic (poslat) fuzzy set theories and topologies*, in [30], 463–530.

[59] S. E. Rodabaugh, *Powerset operator foundations for poslat fuzzy set theories and topologies*, Chapter 2 in [13], 91–116.

[60] S. E. Rodabaugh, *Categorical foundations of variable-basis fuzzy topology*, Chapter 4 in [13], 273–388.

[61] S. E. Rodabaugh, *Separation axioms: representation theorems, compactness, and compactifications*, Chapter 7 in [13], 481–552.

[62] S. E. Rodabaugh, *Axiomatic foundations for uniform operator quasi-uniformities*, Chapter 7 in this Volume.

[63] S. E. Rodabaugh, E. P. Klement, U. Höhle, eds, Application Of Category To Fuzzy Sets, Theory and Decision Library—Series B: Mathematical and Statistical Methods, Volume **14**, Kluwer Academic Publishers (Boston/Dordrecht/London), 1992.

[64] E. S. Santos, *Topology versus fuzzy topology*, preprint, Youngstown State University, 1977.

[65] F. Schulté, *The study of nested topologies through fuzzy topology*, Master's Dissertation, Youngstown State University, March 1984.

[66] A. P. Šostak, *On a fuzzy topological structure*, Rendiconti Circolo Matematico Palermo (Suppl: Serie II) **11**(1985), 89–103.

[67] S. Vickers, Topology Via Logic, Cambridge Tracts in Theor. Comp. Sci., Number 5, Cambridge University Press (Cambridge), 1985.

[68] R. H. Warren, *Neighborhoods, bases, and continuity in fuzzy topological spaces*, Rocky Mtn. J. Math. **8**(1978), 459–470.

[69] P. Wuyts, R. Lowen, *On local and global measures of separation in fuzzy topological spaces*, Fuzzy Sets and Systems **19**(1986), 51–80.

[70] L. A. Zadeh, *Fuzzy sets*, Information and Control **8**(1965), 338–353.

APPENDIX TO CHAPTER 6

Weakening The Requirement That L Be A Complete Chain [1]

U. HÖHLE AND S. E. RODABAUGH

Chapter 6, beginning with Convention 2.7, assumes (unless stated otherwise) that the base lattice L is a complete chain. As stated in the third from last paragraph of the introduction to Chapter 6, Prof. Höhle had suggested to Profs. Pultr and Rodabaugh that a complete chain is really only needed for its meet-irreducibles, and that for spatial L we also have meet-irreducibles which suffice for the constructions of Chapter 6.

This appendix makes this suggestion explicit and outlines how L a complete chain can be replaced by L a spatial frame ("spatial" being taken in the traditional or **2**-spatiality sense—see Sections 1.3, 1.7, 2.2 of Chapter 6).

The notation and section and item tags of Chapter 6 are used (unless obviously otherwise).

The following observation is constantly used and makes precise the sense in which a complete chain may be replaced by its meet-irreducibles, enabling us to use whichever approach is convenient (so long as we observe the dual order).

A.1 Observation. Let L be a complete chain. Then the following hold:

1. $L_\top$ is dually order-isomorphic to $Pt\,(L)$ (the traditional spectrum of L—Section 2.2).
2. $Pt\,(L)$ is closed under all non-empty joins.
3. $Pt\,(L)$ is order-isomorphic to $Prm{\downarrow}(L)$, the poset of prime principal ideals of L.
4. $Prm{\downarrow}(L)$ is dually order-isomorphic with the meet-irreducibles of L.

[1]This appendix stems from ideas of Prof. Höhle communicated to the authors of Chapter 6 as this volume was being prepared for publication.

S.E. Rodabaugh and E.P. Klement (eds.), Topological and Algebraic Structures in Fuzzy Sets, 189–197.
© 2003 Kluwer Academic Publishers.

Proof. Given $t < \top$, put $p_t : L \to \{0,1\}$ by

$$p_t(\alpha) = \begin{cases} 1, & \alpha > t \\ 0, & \alpha \le t \end{cases}$$

It can be shown that p_t preserves arbitrary joins and finite meets and hence is in $Pt(L)$, so the correspondence $t \mapsto p_t$ is well-defined; it is also injective. For surjectivity, let $p \in Pt(L)$, put $t = \bigvee p^{\leftarrow}\{0\}$, and note

$$\downarrow(t) = \ker(p), \quad p = \chi_{L-\downarrow(t)} = p_t$$

These comments, along with the details of II.1.3 of [8], furnish the proofs of statements (1,3,4) as to bijection. As for the bijections of statements (1,4) being dual order isomorphisms, it is straightforward to check that the above correspondence changes binary meets to binary joins ($p_{t\wedge s} = p_t \vee p_s$)—this guarantees that both the correspondence and its inverse are antitone mappings. Now (2) follows immediately since a dual isomorphism converts all existing meets to joins and $L_\top$ is closed under all non-empty meets. □

In the sequel, it is important for the reader to keep in mind the duality of the isomorphism of A.1(1) above. In particular, since $L_\top$ is closed under all non-empty meets, we have the remarkable consequence that $Pt(L)$ is closed under all non-empty joins (which is usually not true for a spectrum).

For the convenience of the reader, we repeat Definition 3.3 of L-frame from Chapter 6.

A.2 Definition[2] (***L*-Frames**). An ***L*-frame** is a system of frame morphisms

$$A \equiv \left(\varphi_t^A : A^{\mathrm{u}} \to A^{\mathrm{l}}\right)_{t\in L_\top}$$

satisfying these axioms:

(F0) For each non-empty $S \subset L_\top$, $\varphi^A_{\bigwedge S} = \bigvee_{s\in S} \varphi_s^A$ (cf. 3.3.1).

(F1) $\left(\varphi_t^A\right)_t$ is c.e.e. in **Frm**, i.e. $A^{\mathrm{l}} = \langle \bigcup_t \left(\varphi_t^A\right)^{\to}[A^{\mathrm{u}}]\rangle \equiv \langle \bigcup_t \varphi_t^A[A^{\mathrm{u}}]\rangle$ (cf. 1.4).

(F2) $\left(\varphi_t^A\right)_t$ is c.m. in **Frm**.

We now give an alternative definition of an L-frame.

[2]In the terminology of [1], (F1) is the condition of **epi-extremal sink** and (F2) is the condition of **mono-source**.

A.3 Definition. An L-**frame** is a system of frame morphisms

$$A \equiv \left(\varphi_p^A : A^{\mathfrak{u}} \to A^{\mathfrak{l}}\right)_{p \in Pt(L)}$$

satisfying these axioms:

(F0*) $\forall a \in A^{\mathfrak{u}}, \forall q \in Pt\left(A^{\mathfrak{l}}\right), \exists \alpha \in L,$

$$q\left(\varphi_{(\,)}^A(a)\right) = \Phi_{\mathbf{2}}(\alpha)$$

where $q\left(\varphi_{(\,)}^A(a)\right) : Pt(L) \to \mathbf{2}$ is the map given by $p \mapsto q\left(\varphi_p^A(a)\right)$ and $\Phi_{\mathbf{2}} : L \to \mathbf{2}^{Pt(L)}$ is the crisp (classical) first comparison map from the third paragraph of Section 2.2 given by $\Phi_{\mathbf{2}}(\alpha)(p) = p(\alpha)$.

(F2) As above.

(F3) As above.

It is also convenient to restate (F0*) thusly: $\forall a \in A^{\mathfrak{u}}, \forall q \in Pt\left(A^{\mathfrak{l}}\right), \exists \alpha \in L, \forall p \in Pt(L),$

$$q\left(\varphi_p^A(a)\right) = p(\alpha)$$

A.4 Theorem. Let L be a complete chain, let $A \equiv \left(\varphi_t^A : A^{\mathfrak{u}} \to A^{\mathfrak{l}}\right)_{t \in L_\top}$ be a system of frame morphisms, and let $A^* \equiv \left(\varphi_{p_t}^{A^*} : A^{\mathfrak{u}} \to A^{\mathfrak{l}}\right)_{p_t \in Pt(L)}$ be the system of frame morphisms—where $t \mapsto p_t$ is the order-reversing isomorphism of Observation A.1(1)—determined by the equations $\varphi_{p_t}^{A^*}(a) = \varphi_t^A(a)$ for each $t \in L_\top$ and each $a \in A^{\mathfrak{u}}$. Then the following hold:

1. A is an L-frame in the sense of Definition A.2 $\Rightarrow$ A^* is an L-frame in the sense of Definition A.3.

2. If the frame $A^{\mathfrak{l}}$ is spatial, then the converse holds—A^* is an L-frame in the sense of Definition A.3 $\Rightarrow$ A is an L-frame in the sense of Definition A.2.

Proof. We assume (F1) and (F2), the only issue then being the relationship between (F0) and (F0*). For (1), let $a \in A^{\mathfrak{u}}$ and $q \in Pt\left(A^{\mathfrak{l}}\right)$. Put

$$\alpha_{aq} = \bigvee \left\{\alpha \in L : \forall t \in L_\top, p_t(\alpha) = 1 \Rightarrow q\left(\varphi_{p_t}^A(a)\right) = 1\right\} \qquad (4.1)$$

It follows from Remark A.5 below that α_{aq} is a uniquely determined element of L.

Claim: $\forall\, s \in L_\top,\, p_s(\alpha_{aq}) = 1 \Leftrightarrow q\left(\varphi_{p_s}^{A^*}(a)\right) = 1$. Let $s \in L_\top$. By definition of A^*, it suffices to verify

$$p_s(\alpha_{aq}) = 1 \Leftrightarrow q\left(\varphi_s^A(a)\right) = 1$$

Now $\left(\varphi_t^A : A^{\text{u}} \to A^{\text{l}}\right)_{t \in L_\top}$ satisfies (F0), we have the mapping

$$q \circ \varphi_{(\,)}^A : L_\top \to Pt\,(A^{\text{u}})$$

and, since q is a frame morphism, this mapping converts non-empty meets to joins. It now follows from Lemma 2.8(2) that

$$q\left(\varphi_s^A(a)\right) = 1 \;\Leftrightarrow\; s < \bigvee \left\{w \in L_\top : \forall\, r < w \text{ in } L_\top,\, q\left(\varphi_s^A(a)\right) = 1\right\}$$

Since $w > r$ means $p_r(w) = 1$, then the join of this display is precisely α_{aq}. Thus

$$q\left(\varphi_s^A(a)\right) = 1 \Leftrightarrow \alpha_{aq} > s \Leftrightarrow p_s(\alpha_{aq}) = 1$$

Now for (2), suppose A^* satisfies (F0*). We have from Observation A.1(2) that $Pt\,(L)$ is closed under all non-empty joins. It also follows that $\forall\, \alpha \in L,\, \Phi_{\mathbf{2}}(\alpha) : Pt\,(L) \to \mathbf{2}$ preserves all joins existing in $Pt\,(L)$. Fix $a \in A^{\text{u}}$. Then from (F0*), we have that $\forall\, q \in Pt\left(A^{\text{l}}\right),\, \exists\, \alpha \in L$,

$$q\left(\varphi_{(\,)}^{A^*}(a)\right) = \Phi_{\mathbf{2}}(\alpha)$$

on $Pt\,(L)$. Thus $\forall\, q \in Pt\left(A^{\text{l}}\right),\, q\left(\varphi_{(\,)}^{A^*}(a)\right)$ preserves all joins existing in $Pt\,(L)$.
Claim: $\varphi_{(\,)}^{A^*}(a)$ preserves all joins existing in $Pt\,(L)$. Let $\emptyset \neq \{p_\gamma\}_\gamma \subset Pt\,(L)$, note $\bigvee_\gamma p_\gamma \in Pt\,(L)$ by Observation A.1(2), and suppose in A^{l} that

$$\varphi_{\bigvee_\gamma p_\gamma}^{A^*}(a) \neq \bigvee_\gamma \varphi_{p_\gamma}^{A^*}(a)$$

Since A^{l} is spatial, $\exists\, q \in Pt\left(A^{\text{l}}\right)$,

$$q\left(\varphi_{\bigvee_\gamma p_\gamma}^{A^*}(a)\right) \neq q\left(\bigvee_\gamma \varphi_{p_\gamma}^{A^*}(a)\right)$$

This is a contradiction to $q\left(\varphi_{(\,)}^{A^*}(a)\right)$ preserving all joins existing in $Pt\,(L)$.
Claim: $\varphi_{(\,)}^{A}(a)$ converts all meets existing in $L_\top$ to joins in A^{l}. This is a consequence of the previous claim and the dual isomorphism of Observation A.1(2).

Claim: $\left(\varphi_t^A : A^{\mathfrak{u}} \to A^{\mathfrak{l}}\right)_{t\in L_\top}$ satisfies (F0). This is immediate from the previous claim, letting a be free in $A^{\mathfrak{u}}$. This concludes the proof of (2) and the theorem. □

A.5 Remark.

1. If we consider a complete lattice L, consider the condition
$$q\left(\varphi_{()}^A(a)\right) = \mathbf{\Phi_2}(\alpha) \tag{5.1}$$
of (F0*), and let $a \in A^{\mathfrak{u}}$, $q \in Pt\left(A^{\mathfrak{l}}\right)$, then $\alpha \in L$ need not be uniquely determined from this condition, in which case (F0*) uniquely determines a subset of L of such α. However, if L is spatial, then (5.1)—and hence (F0*)—uniquely determines $\alpha \in L$.

2. If we consider a complete lattice L and rewrite condition (4.1) above as
$$\alpha_{aq} = \bigvee\left\{\alpha \in L : \mathbf{\Phi_2}(\alpha) = \underline{1} \Rightarrow q\left(\varphi_p^A(a)\right) = 1\right\} \tag{5.2}$$
then $\alpha_{aq} \in L$ is uniquely determined from $a \in A^{\mathfrak{u}}$ and $q \in Pt\left(A^{\mathfrak{l}}\right)$ (where $\mathbf{\Phi_2} : L \to \mathbf{2}^{Pt(L)}$ is as above and $\underline{1}$ is the constant with value 1).

3. If L-**Frm*** is defined analogously to how L-**Frm** is defined in Definition 3.3, then L-**Frm** embeds into L-**Frm***; and this embedding is an isomorphism onto that full subcategory of L-**Frm*** in which the second components of objects are spatial.

Proof. For (1), (5.1) uniquely determines α if $\mathbf{\Phi_2}$ is injective; but $\mathbf{\Phi_2}$ is injective iff L is spatial (1.7 of Chapter 6). For (2), given the completeness of L, the issue is whether the predicate "$\mathbf{\Phi_2}(\alpha) = \underline{1} \Rightarrow q\left(\varphi_p^A(a)\right) = 1$" is well-defined; and this is clear from rewriting the predicate in the form $[p \in Pt(L)] \Rightarrow \left[p(\alpha) = 1 \Rightarrow q\left(\varphi_p^A(a)\right) = 1\right]$. Statement (3) is obvious. □

A.6 Construction. The construction of L-topological spaces from L-frames in the sense of Definition A.2 is detailed in Chapter 6, along with the attendant categorical machinery and the role of ι-sobriety. We now outline how to construct L-topological spaces from L-frames in the sense of Definition A.3. The main tool is to define the "first comparison map", an important step in the constructions of Sections 1.7, 2.2, and 4.1. The first comparison map below is sufficiently similar to that of 4.1 to merit a similar notation.

1. Given an L-frame in the sense of Definition A.3, put $\Phi^*_{L\mathfrak{u}\mathfrak{l}} : A^{\mathfrak{u}} \to L^{Pt\left(A^{\mathfrak{l}}\right)}$ by $\Phi^*_{L\mathfrak{u}\mathfrak{l}}(a) : Pt\left(A^{\mathfrak{l}}\right) \to L$, which in turn is defined by
$$\Phi^*_{L\mathfrak{u}\mathfrak{l}}(a)(q) = \bigvee\left\{\alpha \in L : q\left(\varphi_{()}^A(a)\right) = \mathbf{\Phi_2}(\alpha)\right\}$$
$$\left(= \bigvee\left\{\alpha \in L : \forall p \in Pt(L),\, q\left(\varphi_p^A(a)\right) = p(\alpha)\right\}\right)$$

2. **Claim**. If L is spatial, then $\Phi^*_{L\mathfrak{u}\mathfrak{l}} : A^{\mathfrak{u}} \to L^{Pt(A^{\mathfrak{l}})}$ preserves arbitrary joins and finite meets, in which case $\Phi^{*\rightarrow}_{L\mathfrak{u}\mathfrak{l}}(A^{\mathfrak{u}})$ is an L-topology on $Pt(A^{\mathfrak{l}})$ and $\left(Pt(A^{\mathfrak{l}}), \Phi^{*\rightarrow}_{L\mathfrak{u}\mathfrak{l}}(A^{\mathfrak{u}})\right)$ is an L-topological space.

Proof. We recall from Sections 1.7 and 2.2 that L is spatial—i.e. **2**-spatial—iff $\Phi_{\mathbf{2}} : L \to \mathbf{2}^{Pt(L)}$ is injective.

Preservation of arbitrary joins. To show that $\Phi^*_{L\mathfrak{u}\mathfrak{l}}$ preserves arbitrary joins, let $\{a_\gamma\}_\gamma \subset A^{\mathfrak{u}}$. The condition

$$\Phi^*_{L\mathfrak{u}\mathfrak{l}}\left(\bigvee_\gamma a_\gamma\right) = \bigvee_\gamma \Phi^*_{L\mathfrak{u}\mathfrak{l}}(a_\gamma)$$

is equivalent—using the fact that each $q \circ \varphi^A_p$ is a frame morphism (for all q and p)—to the condition

$$\bigvee\left\{\alpha \in L : \bigvee_\gamma q\left(\varphi^A_{(\,)}(a_\gamma)\right) = \Phi_{\mathbf{2}}(\alpha)\right\} \tag{6.1}$$
$$= \bigvee_\gamma \left[\bigvee\left\{\beta^\gamma \in L : q\left(\varphi^A_{(\,)}(a_\gamma)\right) = \Phi_{\mathbf{2}}(\beta^\gamma)\right\}\right]$$

Since $\Phi_{\mathbf{2}}$ is a frame morphism,

$$\Phi_{\mathbf{2}}\,(\text{left-side of 6.1}) = \bigvee_\gamma q\left(\varphi^A_{(\,)}(a_\gamma)\right) = \Phi_{\mathbf{2}}\,(\text{right-side of 6.1})$$

Now the injectivity of $\Phi_{\mathbf{2}}$ forces equality in (6.1), which concludes the proof of preservation of joins by $\Phi^*_{L\mathfrak{u}\mathfrak{l}}$.

Preservation of finite meets. It is well known that preservation of finite meets is equivalent to the preservation of binary meets plus the preservation of the empty meet (or $\top$).

Binary case. Let $a, b \in A^{\mathfrak{u}}$ and assume $\Phi^*_{L\mathfrak{u}\mathfrak{l}}(a \wedge b) \neq \Phi^*_{L\mathfrak{u}\mathfrak{l}}(a) \wedge \Phi^*_{L\mathfrak{u}\mathfrak{l}}(b)$. Hence $\exists q \in Pt(A^{\mathfrak{l}})$ such that

$$\Phi^*_{L\mathfrak{u}\mathfrak{l}}(a \wedge b)(q) \neq \Phi^*_{L\mathfrak{u}\mathfrak{l}}(a)(q) \wedge \Phi^*_{L\mathfrak{u}\mathfrak{l}}(b)(q)$$

But $\Phi_{\mathbf{2}}$ being injective, isotone, meet-preserving implies

$$\Phi_{\mathbf{2}}\left[\Phi^*_{L\mathfrak{u}\mathfrak{l}}(a \wedge b)(q)\right] \neq \Phi_{\mathbf{2}}\left[\Phi^*_{L\mathfrak{u}\mathfrak{l}}(a)(q)\right] \wedge \Phi_{\mathbf{2}}\left[\Phi^*_{L\mathfrak{u}\mathfrak{l}}(b)(q)\right]$$

It now follows from the definition of $\Phi^*_{L\mathfrak{u}\mathfrak{l}}$ that

$$q\left(\varphi^A_{(\,)}(a \wedge b)\right) \neq q\left(\varphi^A_{(\,)}(a)\right) \wedge q\left(\varphi^A_{(\,)}(b)\right)$$

a contradiction to the fact that each $q \circ \varphi_p^A$ is a frame morphism.

Empty case. To compute $\Phi_{Lu\mathfrak{l}}^*(\top)$, let $q \in Pt\left(A^{\mathfrak{l}}\right)$. Then

$$\begin{aligned}
\Phi_{Lu\mathfrak{l}}^*(\top)(q) &= \bigvee\left\{\alpha \in L : q\left(\varphi_{()}^A(\top)\right) = \Phi_2(\alpha)\right\} \\
&= \bigvee\left\{\alpha \in L : \Phi_2(\alpha) = \underline{\top}\right\} \\
&= \top
\end{aligned}$$

where we have used that Φ_2 and each $q \circ \varphi_p^A$ are frame morphisms. It follows that $\Phi_{Lu\mathfrak{l}}^*(\top) = \underline{\top}$. □

3. **Claim**. If L is a complete chain, then $\Phi_{Lu\mathfrak{l}}^*$ coincides with the $\Phi_{Lu\mathfrak{l}}$ of Section 4.1, in which case the space $\left(Pt\left(A^{\mathfrak{l}}\right), \Phi_{Lu\mathfrak{l}}^{*\rightarrow}\left(A^{\mathfrak{u}}\right)\right)$ coincides with the L-spectrum $\left(Pt\left(A^{\mathfrak{l}}\right), \Phi_{Lu\mathfrak{l}}^{\rightarrow}\left(A^{\mathfrak{u}}\right)\right)$ of Section 4.1.

   Proof. The needed arguments are in the proof of Theorem A.4(1) above. □

A.7 Remarks.

1. Because of the above claim, we can speak of $\left(Pt\left(A^{\mathfrak{l}}\right), \Phi_{Lu\mathfrak{l}}^{*\rightarrow}\left(A^{\mathfrak{u}}\right)\right)$ as the **L-spectrum** of A. Thus we have a notion of L-spectrum for L-frames in the sense of Definition A.3 if L is spatial.

2. If $A^{\mathfrak{l}}$ is spatial, then $\Phi_{Lu\mathfrak{l}}^*$ is injective; in which case $A^{\mathfrak{u}}$ is isomorphic via (the restriction of $\Phi_{Lu\mathfrak{l}}^*$) to the L-topology $\Phi_{Lu\mathfrak{l}}^{*\rightarrow}\left(A^{\mathfrak{u}}\right)$. To see the first claim, let $a \neq b$ in $A^{\mathfrak{u}}$. Then A is c.m. implies $\exists p \in Pt(L)$, $\varphi_p^A(a) \neq \varphi_p^A(b)$. Now the spatiality of $A^{\mathfrak{l}}$ implies $\exists q \in Pt\left(A^{\mathfrak{l}}\right)$,

$$q\left(\varphi_p^A(a)\right) \neq q\left(\varphi_p^A(b)\right)$$

   Since Φ_2 applied to $\bigvee\left\{\alpha \in L : q\left(\varphi_{()}^A(a)\right) = \Phi_2(\alpha)\right\}$ yields $q\left(\varphi_{()}^A(a)\right)$, since Φ_2 applied to $\bigvee\left\{\alpha \in L : q\left(\varphi_{()}^A(b)\right) = \Phi_2(\alpha)\right\}$ yields $q\left(\varphi_{()}^A(b)\right)$, and since $q\left(\varphi_{()}^A(a)\right) \neq q\left(\varphi_{()}^A(b)\right)$—they are not equal at p, it follows that

$$\begin{aligned}
\Phi_{Lu\mathfrak{l}}^*(a)(q) &= \bigvee\left\{\alpha \in L : q\left(\varphi_{()}^A(a)\right) = \Phi_2(\alpha)\right\} \\
&\neq \bigvee\left\{\alpha \in L : q\left(\varphi_{()}^A(b)\right) = \Phi_2(\alpha)\right\} \\
&= \Phi_{Lu\mathfrak{l}}^*(b)(q)
\end{aligned}$$

   Hence $\Phi_{Lu\mathfrak{l}}^*(a) \neq \Phi_{Lu\mathfrak{l}}^*(b)$.

3. If L is spatial, then by Proposition 2.4(2), $\Phi^{*\rightarrow}_{L\mathrm{u}\mathfrak{l}}(A^{\mathrm{u}})$ is also spatial, i.e. order-isomorphic to some traditional (sober) topology. Further, the hypergraph functor [16, 9, 14, 15, 3] constructs for the L-spectrum in this case [3] an associated traditional topological space.

4. If L is not spatial, then $\Phi^{*\rightarrow}_{L\mathrm{u}\mathfrak{l}}(A^{\mathrm{u}})$ may not be an L-topology—the spatiality of L was needed in Construction A.6(2) in proving both closure under arbitrary joins and finite meets. However, in this case, we may take $\Phi^{*\rightarrow}_{L\mathrm{u}\mathfrak{l}}(A^{\mathrm{u}})$ as a subbase for an L-topology on $Pt\,(A)$.

5. It is beyond the scope of this appendix to verify that all of Chapter 6, including its categorical machinery and treatment of functor categories, carry over to the more general setting of L being a spatial frame. We note that the Kubiak-Lowen functors ι_L are available for any complete lattice L, so that the separation axiom of ι_L-sobriety is available in the more general setting of this appendix. Finally, we note that all the examples of Chapter 17 carry over to the setting of this appendix.

References

[1] J. Adámek, H. Herrlich, G. E. Strecker, Abstract and Concrete Categories: The Joy of Cats, Wiley Interscience Pure and Applied Mathematics, John Wiley & Sons (Brisbane/Chicester/New York/Singapore/Toronto), 1990.

[2] U. Höhle, Many Valued Topology And Its Applications, Kluwer Academic Publishers (Boston/Dordrecht/London), 2001.

[3] U. Höhle, *A note on the hypergraph functor*, Fuzzy Sets and Systems **131**(2002), 353–356.

[4] U. Höhle, *Many valued topologies and Borel probability measures*, Chapter 4 in this Volume.

[5] R.-E. Hoffmann, *Irreducible filters and sober spaces*, Manuscripta Math. **22**(1977), 365-380.

[6] R.-E. Hoffmann, *Soberification of partially ordered sets*, Semigroup Forum **17**(1979), 123-138.

[7] R.-E. Hoffmann, *On the soberification remainder $sX - X$*, Pacific J. Math. **83**(1979), 145–156.

[8] P. T. Johnstone, Stone Spaces, Cambridge University Press (Cambridge), 1982.

[9] R. Lowen, *A comparison of different compactness notions in fuzzy topological spaces*, J. Math. Anal. Appl. **64**(1978), 446–454.

[10] A. Pultr, S. E. Rodabaugh, *Lattice-valued frames, functor categories, and classes of sober spaces*, Chapter 6 in this Volume.

[11] A. Pultr, S. E. Rodabaugh, *Examples for different sobrieties in fixed-basis topology*, Chapter 17 in this Volume.

[12] A. Pultr, A. Tozzi, *Separation facts and frame representation of some topological facts*, Applied Categorical Structures **2**(1994), 107–118.

[13] A. Pultr, A. Tozzi, *A note on reconstruction of spaces and maps from lattice data*, Quaestiones Mathematicae **24**(2001), 55–63.

[14] S. E. Rodabaugh, *The Hausdorff axiom for fuzzy topological spaces*, Topology Appl. **11**(1980), 319–334.

[15] S. E. Rodabaugh, *Point-set lattice-theoretic topology*, Fuzzy Sets and Systems **40**(1991), 297–345.

[16] E. S. Santos, *Topology versus fuzzy topology*, preprint, Youngstown State University (Youngstown, Ohio), 1977.

CHAPTER 7

Axiomatic Foundations For Uniform Operator Quasi-Uniformities

S. E. RODABAUGH [1]

Introduction

Traditional uniformities have both the entourage approach of [33, 1], based on powersets of the form $2^{X \times X}$, as well as the uniform covering approach of [30, 10], based on double powersets of the form $2^{(2^X)}$.

Uniformities in lattice-valued set theory include the approaches of [15, 3], underpinned by powersets of the form $\mathbb{I}^{X \times X}$ or $L^{X \times X}$, as well as the approach of [9], based on exponential powersets of the form $\left(L^X\right)^{\left(L^X\right)}$ and rich in natural and canonical examples. Clearly [15, 3] are generalizations of the entourage approach, but [9] is not a generalization of either traditional approach (see [12]). For certain L, [2] puts [15, 3, 9] into a common fixed-basis framework, extending [4]; while [34] stratifies [9], providing another link with [15, 3].

This chapter completely revamps Hutton's original approach [9] with these goals: to directly motivate it from metric spaces and topological groups; to greatly generalize it by removing key restrictions; and to significantly enrich it by adding many new canonical examples. Restated, this chapter:

(1) Motivates from metric spaces and topological groups the apparently new approach of [12] to traditional uniformities, an approach using **uniform operators,** logically equivalent to traditional approaches and based on exponential powersets of the form $\left(2^X\right)^{\left(2^X\right)}$. Hutton uniformities are seen as the generalization to lattice-valued sets of such uniform operators, justifying the syntax of Hutton's axioms. Lacking this motivation, Hutton [9] mistakenly called his operators *entourages,* whereas he should have called them *uniform operators*

[1] Dedicated to Prof. S. F. Barger upon his retirement from Youngstown State University.

S.E. Rodabaugh and E.P. Klement (eds.), Topological and Algebraic Structures in Fuzzy Sets, 199–233.
© 2003 Kluwer Academic Publishers.

or at least *uniform covering operators.* Such uniformities [quasi-uniformities] are below called **uniform-operator uniformities [uniform-operator quasi-uniformities]**, denoted **uo-uniformities [uoq-uniformities].**

(2) Develops lattice-theoretic and monoidal foundations for uoq-uniformities using general tensor products $\otimes$ [8, 26] which significantly generalize Hutton's "intersection axiom" (which requires $\otimes = \wedge$). Such tensors include all t-norms (3.5–3.6 below) on $L = \mathbb{I} \equiv [0,1]$, all cross-products of t-norms on Tihonov cubes ordered coordinate-wise, and many other examples.

(3) Removes all distributivity of the base lattice L from Hutton's axioms, another significant generalization of Hutton's original setting: Hutton's original axioms depended on Raney's Lemma [17, 18, 9], and its assumption of complete distributivity, to define the "intersection axiom" for uoq-uniformities. The foundation laid in this chapter requires no distributivity of any kind—see also [2] in this regard, although in certain natural examples, infinite distributivity of the tensor w.r.t. join or meet is sometimes available and exploited (e.g. for continuous t-norms and their cross-products).

(4) Lays a foundation for categories of fixed-basis uoq-uniformities by isolating precisely those algebraic conditions on $\otimes$—associativity, commutivity, $\top$ is identity for $\otimes$—which guarantee uoq-uniformities have complete fibres, a necessary precursor for topological categories and for natural examples of bi-handed uoq-uniformities.

(5) Enriches significantly our current inventory of natural examples of Hutton and uoq-uniformities. The traditional strength of the Hutton approach has been its examples, and both the number of classes of examples, as well as the number of examples in previously known classes, are significantly enlarged using the foundations of this chapter.

These issues—symmetry axiom and uo-uniformities, morphisms and (uoq-) uniform continuity, categorical considerations such as topological constructs and variable-basis foundations—are resolved in future work. But the foundation presented in this chapter has been developed with these issues in mind—e.g. (4) above. This chapter has the following outline:

§1. Powerset operators of binary relations
§2. Uniform operators: metric spaces and topological groups
§3. Lattice-theoretic foundations: tensors and triangular norms
§4. Hutton's original axioms for uo-uniformities
§5. Intersection operators for uoq-uniformities
§6. General axioms for uoq-uniformities
§7. Complete fibres for uoq-uniformities
§8. Canonical examples of uoq-uniformities

1 Powerset operators of binary relations

In this section, $R : X \to Y$ is a relation ($R \subset X \times Y$). This section is needed for the next section and is essentially contained in [16].

1.1 Definition (Conventions for Binary Relations). The **preimage** of R, or **pre**(R), is the set $\{x \in X : \exists y \in Y,\, y \in R(x)\}$ and the **image** of R, or **im**(R), is the set $\{y \in Y : \exists x \in X,\, y \in R(x)\}$. The relation R is **panjective** [**surjective**] iff $pre(R) = X$ [$im(R) = Y$, resp.], **well-defined** iff R is a function on $pre(R)$, and **injective** or **one-to-one** iff $\forall\, (x_1, y_1), (x_2, y_2) \in R$, $y_1 = y_2 \Rightarrow x_1 = x_2$. The inverse relation $R^{-1} : X \leftarrow Y$ of R is defined by $x \in R^{-1}(y) \Leftrightarrow y \in R(x)$.

1.2 Definition (Powerset Operators of Relations). The following **powerset operators** of R may be defined:

$$R^{\rightarrow} : 2^X \to 2^Y \quad \text{by} \quad R^{\rightarrow}(A) = \{y \in Y : \exists x \in A,\, y \in R(x)\}$$

$$R^{\leftarrow}_{-1} : 2^X \leftarrow 2^Y \quad \text{by} \quad R^{\leftarrow}_{-1}(B) = \{x \in X : \exists y \in B,\, y \in R(x)\}$$

$$R^{\leftarrow}_{\dashv} : 2^X \leftarrow 2^Y \quad \text{by} \quad R^{\leftarrow}_{\dashv}(B) = \bigcup_{R^{\rightarrow}(A) \subset B} A$$

The operator $R^{\rightarrow}$ is the **forward** or **image operator** of R, and $R^{\leftarrow}_{-1}$ and $R^{\leftarrow}_{\dashv}$ are the **backward** or **preimage operators** of R.

1.3 Proposition. The following hold:

1. The operator $R^{\rightarrow}$ preserves arbitrary unions and has the alternate forms:

$$R^{\rightarrow}(A) = \bigcup_{x \in A} R^{\rightarrow}\{x\} = \bigcup_{x \in A} R(x)$$

2. The set of all relations from X to Y is bijective with the set of all arbitrary union preserving mappings from 2^X to 2^Y via $R \mapsto R^{\rightarrow}$.

3. $R^{\leftarrow}_{-1} = \left(R^{-1}\right)^{\rightarrow}$ and hence $R^{\leftarrow}_{-1}$ preserves arbitrary unions.

4. $R^{\rightarrow} \dashv R^{\leftarrow}_{\dashv}$ and hence $R^{\leftarrow}_{\dashv}$ preserves arbitrary intersections.

1.4 Comment. The choice of preimage operator for symmetry is given in 1.9.

1.5 Lemma. $\forall\, y \in Y$, $R^{\leftarrow}_{\dashv}\{y\} = (X - pre\,(R)) \cup \{z \in pre\,(R) : R\,(z) = \{y\}\}$; $\forall\, B \in 2^Y$, $R^{\leftarrow}_{\dashv}(B) \subset (X - pre\,(R)) \cup R^{\leftarrow}_{-1}(B)$; $\forall\, B \in 2^Y$, $(X - pre\,(R)) \subset R^{\leftarrow}_{\dashv}(B)$.

1.6 Theorem (Preserving Unions and Intersections). The following are equivalent:

1. $R^{\leftarrow}_{\dashv}$ preserves arbitrary unions;
2. R is well-defined (on $pre\,(R)$);
3. $\forall\, B \in 2^Y$, $R^{\leftarrow}_{\dashv}(B) = (X - pre\,(R)) \cup R^{\leftarrow}_{-1}(B)$;
4. R^{-1} is injective (on $im\,(R)$);
5. $R^{\leftarrow}_{-1}$ preserves arbitrary intersections.

1.7 Theorem (Existence of Adjunctions). R is panjective iff $\forall\, A \in 2^X$, $A \subset R^{\leftarrow}_{-1}(R^{\rightarrow}(A))$; R is well-defined (on $pre\,(R)$) iff $\forall\, B \in 2^Y$, $\left(R^{\rightarrow}\left(R^{\leftarrow}_{-1}(B)\right)\right) \subset B$. These conditions are equivalent: R is a function on X; $R^{\rightarrow} \dashv R^{\leftarrow}_{-1}$; $R^{\leftarrow}_{-1} = R^{\leftarrow}_{\dashv}$.

1.8 Theorem (Hutton's Description of Preimage Operators). The following hold:

1. $\forall\, A \in 2^X, \forall\, B \in 2^Y$, $R^{\leftarrow}_{-1}(B) \subset X - A \Leftrightarrow R^{\rightarrow}(A) \subset Y - B$.
2. $\forall\, B \in 2^Y$, $R^{\leftarrow}_{-1}(B) = \bigcap_{R^{\rightarrow}(A) \subset Y - B}(X - A) = \bigcap_{R^{\rightarrow}(X-A) \subset Y - B} A$.
3. $\forall\, B \in 2^Y$, $R^{\leftarrow}_{-1}(B) = X - R^{\leftarrow}_{\dashv}(Y - B)$, $R^{\leftarrow}_{\dashv}(B) = X - R^{\leftarrow}_{-1}(Y - B)$.

1.9 Convention (Decision on Preimage Operator). Preservation of arbitrary unions (or joins) is fundamentally important in this chapter. Therefore, we shall generally use the notation $R^{\leftarrow}$ for $R^{\leftarrow}_{-1}$ and *not* for $R^{\leftarrow}_{\dashv}$, reflecting our decision that $R^{\leftarrow}_{-1}$ is our choice of preimage operator, a collective consequence of 1.3, 1.6, 1.7. On the other hand, $R^{\leftarrow}_{\dashv}$ helps us describe $R^{\leftarrow}_{-1}$, a consequence of 1.8.

2 Uniform operators: metric spaces and topological groups

This section generates the literal syntax of Hutton's axioms [9] from metric spaces and topological groups to create the uniform operators of [12], an alternative to entourages and uniform coverings for traditional uniformities. The argument of [22]—Hutton generalizes the uniform covering approach—is corrected to better say Hutton is the generalization of these uniform operators.

Two basic insights synergize. The first insight, a comment of Prof. Mulvey, is that the binary relations on a set are bijective with the (aribitrary) union-preserving (endo-)mappings on the powerset of that set (1.3(2)). The second insight, motivated from the opening pages of [10], is that the uniformity inherent in a metric space (X, d) [topological group $(X, +, \mathfrak{T})$] is expressed by the family $\{\{N_\varepsilon(x) : x \in X\} : \varepsilon > 0\}$ $[\{\{U + x : x \in X\} : U \in \mathfrak{T}\}]$ of "uniform" coverings. Combining these insights, we find that d and $\mathfrak{T}$ are each equivalent to their respective families of uniform coverings, each of which is bijective to a certain family of binary relations, which in turn is bijective with a basis of "uniform operators" on the powerset 2^X.

From such considerations emerges the specific syntax of the axioms of uniform operators for the traditional case and hence for the Hutton uniformity axioms, which considerations we now outline without proofs.

2.1 Remark. For any set X, the collection $\mathbf{REL}(X, X)$ of all binary relations on X is in a bijection with the collection $\mathbf{CSLAT}\left(2^X, 2^X\right)$ of all (arbitrary) union preserving maps. This is an immediate consequence of 1.3(2).

2.2 Discussion (Metric Spaces). Let (X, d) be a metric space.

(1) For $\varepsilon > 0$, put the **basic uniform** (binary) **relation** $R_\varepsilon : X \to X$ by: $y \in R_\varepsilon(x) \Leftrightarrow d(x, y) < \varepsilon$. Then we have: R_ε is panjective, and hence $\forall A \in 2^X$, $A \subset [R_\varepsilon]_{-1}^{\leftarrow}([R_\varepsilon]^{\rightarrow}(A))$ (1.7); R_ε need not be well-defined on its preimage, and hence $R_\varepsilon^{\leftarrow}$ $\left(\equiv [R_\varepsilon]_{-1}^{\leftarrow}\ (1.9)\right)$ need not preserve arbitrary intersections, $[R_\varepsilon]_{\dashv}^{\leftarrow}$ need not preserve arbitrary unions, and so $R_\varepsilon^{\leftarrow}$ need not equal $[R_\varepsilon]_{\dashv}^{\leftarrow}$(1.6(1,5), 1.7); and $R_\varepsilon = R_\varepsilon^{-1}$, and hence $R_\varepsilon^{\rightarrow} = R_\varepsilon^{\leftarrow}$ (1.3(2,3)).

(2) Let $\mathcal{N}_d \equiv \{\{N_\varepsilon(x) : x \in X\} : \varepsilon > 0\}$ be the family of all **basic uniform coverings**, where $N_\varepsilon(x)$ is the ε-neighborhood centered at x. Then $\mathcal{N}_d$ is in a bijection with the family $\{R_\varepsilon : \varepsilon > 0\}$ of basic uniform relations given in (1) above.

(3) The family $\{R_\varepsilon : \varepsilon > 0\}$ of basic uniform relations is in a bijection with the induced family $\mathcal{R}_d \equiv \{R_\varepsilon^{\rightarrow} : \varepsilon > 0\} \subset \left(2^X\right)^{\left(2^X\right)}$ of **basic uniform operators** given by restricting the bijection of 1.3(2) of the previous section. Hence $\mathcal{N}_d$ is in a bijection with $\mathcal{R}_d$.

(4) Let the family $\mathcal{U}_d \subset (2^X)^{(2^X)}$ of **uniform operators** be defined by $U \in \mathcal{U}_d \Leftrightarrow \exists R_\varepsilon^\rightarrow \in \mathcal{R}_d,\ R_\varepsilon^\rightarrow \subset U$. Then $\mathcal{U}_d$ is the **uniform covering uniformity** induced on X from the **base** $\mathcal{R}_d$, and $\mathcal{U}_d$ has the following properties:

(U1) $\mathcal{U}_d \neq \emptyset$.(*non-empty*)

(U2) $\forall U \in \mathcal{U}_d,\ \forall A \in 2^X,\ A \subset U(A)$.(*enlarging*)

(U3) $\forall U \in \mathcal{U}_d$, U preserves arbitrary unions.(*join-preserving*)

(U4) $\forall V \in (2^X)^{(2^X)}$ satisfying (U2, U3), $\forall U \in \mathcal{U}_d,\ U \subset V \Rightarrow V \in \mathcal{U}_d$.(*upper-set*)

(U5) $\forall U, V \in \mathcal{U}_d,\ U \cap V \in \mathcal{U}_d$.(*intersection*)

(U6) $\forall U \in \mathcal{U}_d,\ \exists V \in \mathcal{U}_d,\ V \circ V \subset U$.(*triangle-inequality*)

(U7) $\forall U \in \mathcal{U}_d,\ U^{-1} \in \mathcal{U}_d$, where given $u \in \mathbf{REL}(X, X)$ by (1) such that $U = u^\rightarrow,\ U^{-1} \equiv u^\leftarrow$ (*symmetry*).

(5) $\forall U \in \mathcal{U}_d, \forall B \in 2^X$, 1.8(2,3) implies that

$$U^{-1}(B) = \bigcap_{U(A) \subset X-B} (X - A) = \bigcap_{U(X-A) \subset X-B} A = X - U^*(Y - B)$$

where $U \dashv U^*$ for $u \in \mathbf{REL}(X, X)$ with $U = u^\rightarrow$, $U^* = u_\dashv^\leftarrow$ as in 1.2.

(6) Given $\mathcal{U}_d$ of (5), put $\hat{d} : 2^X \times 2^X \to [0, \infty)$ by $\hat{d}(A, B) = \bigwedge_{B \subset R_\varepsilon^\rightarrow(A)} \varepsilon$. Then d is recaptured from $\mathcal{U}_d$ as follows: $\forall x, y \in X,\ d(x, y) = \hat{d}(\{x\}, \{y\})$.

2.3 Discussion (Topological Groups). Let $(X, +, \mathfrak{T})$ be a topological group.

(1) There is a local base $\mathcal{B}$ at 0 of symmetric open sets. For $B \in \mathcal{B}$, put the **basic uniform** (binary) **relation** $R_B : X \to X$ by: $y \in R_B(x) \Leftrightarrow y \in B + x$. Then 3.2(1)(1–3) hold with R_ε replaced by R_B.

(2) Let $\mathcal{N}_\mathcal{B}$ be the family of all **basic uniform coverings**, i.e. $\mathcal{N}_\mathcal{B} \equiv \{\{B + x : x \in X\} : B \in \mathcal{B}\}$. Then $\mathcal{N}_\mathcal{B}$ is in a bijection with the family $\{R_B : B \in \mathcal{B}\}$ of basic uniform relations given in (1) above.

(3) The family of basic uniform relations is in a bijection with the induced family $\mathcal{R}_\mathcal{B} \equiv \{R_B^\rightarrow : B \in \mathcal{B}\} \subset (2^X)^{(2^X)}$ of **basic uniform operators** given by restricting the bijection of 1.3(2) of the previous section. Hence $\mathcal{N}_\mathcal{B}$ is in a bijection with $\mathcal{R}_\mathcal{B}$.

(4) Let the family $\mathcal{U}_\mathcal{B} \subset (2^X)^{(2^X)}$ of **uniform operators** be defined by $U \in \mathcal{U}_\mathcal{B} \Leftrightarrow \exists R_B^\rightarrow \in \mathcal{R}_\mathcal{B},\ R_B^\rightarrow \subset U$. Then $\mathcal{U}_\mathcal{B}$ is the **uniform covering uniformity** induced on X from the **base** $\mathcal{R}_\mathcal{B}$, and $\mathcal{U}_\mathcal{B}$ has the properties (U1–U7) of 2.2(4) above with $\mathcal{U}_\mathcal{B}$ replacing $\mathcal{U}_d$.

(5) The statement of 2.2(5) above holds with $\mathcal{U}_\mathcal{B}$ replacing $\mathcal{U}_d$.

(6) Given $\mathcal{U}_\mathcal{B}$ of (5), put

$$\mathfrak{T}(\mathcal{U}_\mathcal{B}) = \left\{O \in 2^X : O = \bigcup \{A \in 2^X : \exists U \in \mathcal{U}_\mathcal{B}, U(A) \subset O\}\right\}$$

Then $\mathfrak{T}(\mathcal{U}_\mathcal{B}) = \mathfrak{T}$, i.e. $\mathfrak{T}$ is recaptured from $\mathcal{U}_\mathcal{B}$.

2.4 Definition (Uniform Operator [Quasi-] Uniformities). Given a set X, $\mathcal{U}$ is a **uniform operator quasi-uniformity** [**uniform operator uniformity**], abbreviated **uoq-uniformity** [**uo-uniformity**], if $\mathcal{U} \subset (2^X)^{(2^X)}$ and $\mathcal{U}$ satisfies (U1–U6) [(U1–U7)] of 2.2(4) and 2.3(4) above with $\mathcal{U}$ replacing $\mathcal{U}_d$ and $\mathcal{U}_\mathcal{B}$; and $\mathfrak{T}(\mathcal{U})$ denotes the topology

$$\left\{O \in 2^X : O = \bigcup \{A \in 2^X : \exists U \in \mathcal{U}_\mathcal{B}, U(A) \subset O\}\right\}$$

generated by $\mathcal{U}$. Each **uniform operator** $U \in \mathcal{U}$ **uniformly generates** a **uniform cover** $\{U(A) : A \in 2^X\}$ of **uniform neighborhoods** of subsets (including singletons); cf. [22].

From 2.4, one obtains uniform covers in the usual sense from uniform operators by restricting the uniform covers of 2.4 to singletons. In metric spaces, on the other hand, $A \subset R_\varepsilon^{\rightarrow}(A) \equiv N_\varepsilon(A)$, the latter being the ε-neighborhood of subset A, a widely exploited notion in geometric topology (where one needs ε-neighborhoods of starlike continua); cf. [19, 20].

3 Lattice-theoretic foundations: tensors and triangular norms

3.1 Definition (Complete Quasi-Monoidal Lattices). CQML, the category of **complete quasi-monoidal lattices** [26], comprises the following data, where composition and identities are taken from **SET**:

1. **Objects**: Complete lattices L equipped with a binary operation or **tensor product** $\otimes : L \times L \to L$ such that $\otimes$ is isotone in both arguments and $\top \otimes \top = \top$, where $\top$ is the upper bound of the lattice L.

2. **Morphisms**: All **SET** morphisms, between the above objects, which preserve $\otimes$ and $\top$ and arbitrary $\bigvee$.

3.2 Notation. An object of **CQML** may be denoted L, $(L, \otimes)$, $(L, \leq, \otimes)$, according to convenience or emphasis; and given a concrete category $\mathbf{X}$ and subcategories $\mathbf{A}, \mathbf{B}$, then $\mathbf{A} \cap \mathbf{B}$ denotes the largest subcategory of $\mathbf{X}$ which is a subcategory of both $\mathbf{A}$ and $\mathbf{B}$, if such subcategory exists.

3.3 Definition (Categories Related to CQML).

(1) $\mathbf{CQML}_{\le}$ [$\mathbf{CQML}_{\top}$, $\mathbf{CQML}_{\bot}$] is the full subcategory of **CQML** for which $\otimes \le \wedge$ [$\top$ is the identity for $\otimes$, $\bot$ is the zero of $\otimes$, resp.]. Subscripts may be combined, and other categories may be modified in the same way by these subscripts. Note the "$\le$" condition implies the "$\bot$" condition.

(2) The categoiry **QUML** of **quasi-uniform monoidal lattices** is the full subcategory of **CQML** for which $\otimes$ is associative, commutative, and $\top$ is the identity.

(3) The category **QUANT** [8] [**COQUANT**] of **quantales** [**co-quantales**] is the full subcategory of **CQML** for which $\otimes$ distributes across arbitrary $\bigvee$ [$\bigwedge$]; and **BIQUANT** = **QUANT** $\cap$ **COQUANT**.

(4) The category **DQML** of **deMorgan quasi-monoidal lattices** is the subcategory of **CQML** in which each object is additionally equipped with an order-reversing involution (or **polarity**) and each morphism additionally preserves the involution. No distributivity is assumed (3.4 below).

(5) The category **ORTHODQML** of **orthocomplemented deMorgan quasi-monoidal lattices** is the full subcategory of **DQML** in which the polarity $'$ additionally satisfies $\forall\, a,\ a \vee a' = \top,\ a \wedge a' = \bot$. No distributivity is assumed (3.4 below).

(6) **SFRM (semiframes)** [**FRM (frames)**, **DMRG (deMorgan algebras)**, **ORTHODMRG (orthocomplemented deMorgan algebras)**] is isomorphic to that isomorphism closed subcategory of **CQML** [**QUANT**, **DQML**, **ORTHODQML**, resp.] in which $\otimes = \wedge$. The category **HUT** of Hutton algebras is the full subcategory of **DMRG** in which each object is completely distributive.

3.4 Examples (Justification of Non-Distributive deMorgan Algebras). The non-symmetric five-point diamond, a fundamental example of a non-distributive lattice, is a deMorgan algebra in the sense of **DMRG**, but the polarity is not an orthocomplementation. The symmetric six-point diamond is a non-distributive, orthocomplemented deMorgan algebra; and the closed linear subspaces of Hilbert space (dim ≥ 2) form a non-distributive, orthocomplemented, deMorgan algebra with the ordering by inclusion, the meet by intersection, and the join by the closed linear hull (communicated by Prof. Höhle).

3.5 Examples (Quasi-Uniform and deMorgan Monoidal Lattices).

(1) Let $L = \mathbb{I} \equiv [0,1]$ and let $\otimes$ be any triangular norm (t-norm) on $\mathbb{I}$. Then $(L, \le, \otimes) \in \left|(\mathbf{QUML} \cap \mathbf{DQML})_{\le}\right|$. As studied extensively in [11], t-norms include many important families: basic, Schweizer-Sklar, Hamacher, Frank, Yager, Dombi, Sugeno-Weber, Aczél-Alsina, Mayer-Torrens, Dubois-Prade, etc; see also [32]. The basic t-norms $T_{\wedge}$ (minimum), $T_{Ł}$ (Łukasiewicz), T_{π} (prod-

uct), and T_D (drastic product), especially needed for this chapter, are defined as follows:

$$\begin{aligned} T_\wedge(a,b) &= a \wedge b, \quad T_{Ł}(a,b) = (a+b-1) \vee 0, \quad T_\pi(a,b) = ab, \\ T_D(a,b) &= \begin{cases} 0, & (a,b) \in [0,1)^2 \\ a \wedge b, & (a,b) \notin [0,1)^2 \end{cases} \end{aligned}$$

(2) $\left|(\mathbf{QUML} \cap \mathbf{DQML})_{\leq}\right|$ is closed under direct products with the product tensor defined coordinate-wise. This generates many non-t-norm examples of $(\mathbf{QUML} \cap \mathbf{DQML})_{\leq}$ objects; e.g., $(\mathbb{I} \times \mathbb{I}, T_D \times T_{Ł}) \in \left|(\mathbf{QUML} \cap \mathbf{DQML})_{\leq}\right|$.

3.6 Examples (Quasi-Uniform Quantales, Co-Quantales, Bi-Quantales)

1. **Lemma.** Each of $T_\wedge, T_{Ł}, T_\pi$ is continuous on $\mathbb{I}^2$; and each continuous t-norm on $\mathbb{I}$ is an ordinal sum of $T_\wedge$, $T_{Ł}$, T_π ([11], pp. 81ff).

   Proof. The proof comprises these results in [11]: Theorem 2.18, Corollary 5.7, Theorem 5.11. □

2. **Lemma.** Let $f : \mathbb{I} \to \mathbb{I}$ be isotone and continuous. Then f preserves arbitrary $\bigvee$ and $\bigwedge$.

   Proof. In each case, isotonicity gives one inequality and continuity the other inequality. □

3. **Corollary.** Each continuous t-norm T on $\mathbb{I}$ exhibits infinite distributivity over both arbitrary $\bigvee$ and arbitrary $\bigwedge$. Hence for each continuous t-norm T, including $T = T_\wedge$, $T_{Ł}$, T_π,

$$(\mathbb{I}, T) \in \left|(\mathbf{QUML} \cap \mathbf{DQML} \cap \mathbf{BIQUANT})_{\leq}\right|$$

   Proof. Fixing $a \in \mathbb{I}$, $T(a, \ ) : \mathbb{I} \to \mathbb{I}$ is isotone and continuous; so apply the previous lemma. □

4. **Fact.** $\left|\mathbf{QUANT}_{\leq}\right|$ $[\left|\mathbf{COQUANT}_{\leq}\right|, \left|\mathbf{BIQUANT}_{\leq}\right|]$ is closed under direct products. Hence $\left|(\mathbf{QUML} \cap \mathbf{DQML} \cap \mathbf{BIQUANT})_{\leq}\right|$ is closed under direct products.

5. **Corollary.** $(\mathbb{I} \times \mathbb{I} \times \mathbb{I}, T_\wedge \times T_{Ł} \times T_\pi)$ is a non-t-norm example of $\left|(\mathbf{QUML} \cap \mathbf{DQML} \cap \mathbf{BIQUANT})_{\leq}\right|$.

4 Hutton's original axioms for uo-uniformities

4.1 Hutton-Raney Lemma. Let X be a set and let L be completely distributive. Then $\forall\, a \in L^X,\ \exists\, \mathcal{C}(a) \subset L^X$, the following hold: (1) $\bigvee \mathcal{C}(a) = a$; and (2) $\mathcal{A} \subset L^X$ and $\bigvee \mathcal{A} = a \Rightarrow \forall\, b \in \mathcal{C}(a),\ \exists\, c \in \mathcal{A},\ b \leq c$.

4.2 Delta Operator Lemma. Let X be a set, let L be completely distributive, and let $U, V \in \left(L^X\right)^{\left(L^X\right)}$ satisfy these properties: (U2) $\forall\, a \in L^X$, $a \leq U(a)$, $a \leq V(a)$; and (U3) U, V preserve arbitrary $\bigvee$. Further, let $U \,\Delta\, V \in \left(L^X\right)^{\left(L^X\right)}$ by $(U \,\Delta\, V)(a) = \bigvee \{(U \wedge V)(b) : b \in \mathcal{C}(a)\}$. Then $U \,\Delta\, V$ has the following properties: $U \,\Delta\, V$ satisfies (U2), (U3), and

$$U \,\Delta\, V = \bigvee \left\{ W \in \left(L^X\right)^{\left(L^X\right)} : W \text{ satisfies (U2,U3) and } W \leq U \wedge V \right\}$$

where "$\wedge$" is taken in $\left(L^X\right)^{\left(L^X\right)}$.

4.3 Hutton Axioms for UO[Q]-Unifomities. Let X be a set, let $L \in |\mathbf{HUT}|$ (3.3(6) above), and let $\mathcal{U}$ be a subset of $\left(L^X\right)^{\left(L^X\right)}$ satisfying (U1–U6) of 2.2(4) and 2.4 above with L replacing 2, L^X replacing 2^X, $\leq$ replacing $\subset$:, $\bigvee$ replacing $\bigcup$, and Δ replacing $\cap$. Then $\mathcal{U}$ is an (L-)**uniform operator q(uasi)-uniformity** or (L-)**uoq-uniformity** on X and $(X, \mathcal{U})$ is a (L-)**uniform operator q(uasi)-uniform space** or (L-)**uoq-uniform space.** For $U \in \left(L^X\right)^{\left(L^X\right)}$, define $U^{-1} \in \left(L^X\right)^{\left(L^X\right)}$ by $U^{-1}(a) = \bigwedge \{b : U(b') \leq a'\}$, where $'$ is the polarity on L. If $\mathcal{U}$ further satisfies (U7) of 2.2(4) with this definition of U^{-1}, then $\mathcal{U}$ is a (L-)**uniform operator uniformity** or (L-)**uo-uniformity** on X and $(X, \mathcal{U})$ is a (L-)**uniform operator uniform space** or (L-)**uo-uniform space.** For $\mathcal{U}$ an (L-)**uoq-uniformity** on X, $\tau(\mathcal{U})$ denotes the L-topology

$$\left\{ u \in L^X : u = \bigvee \left\{ a \in L^X : \exists\, U \in \mathcal{U},\ U(a) \leq u \right\} \right\}$$

on X generated by $\mathcal{U}$. Each **uniform operator** $U \in \mathcal{U}$ **uniformly generates** a **uniform cover** $\left\{ U(A) : A \in 2^X \right\}$ of **uniform neighborhoods** of L-subsets (including fuzzy singletons) (cf. [22]).

Because of 1.8, 2.2(5), the symmetry axiom (U7) of Hutton is the generalization of the symmetry axiom of 2.2(4), 2.4 for traditional uniform operators. The uniform neighborhoods of 4.3 and their neighborhood theory in [22] are subsumed in [5]. In contrast to Sections 1,2, this section and Section 6 are not based on binary relations and the powerset monads of [16].

4.4 Examples. From [28] and its references, we have the following examples. $\mathbb{R}(L)$ and $\mathbb{I}(L)$ are uo-uniform spaces (for L a Hutton algebra); there are uoq-uniformities generating the left and right hand L-topologies on $\mathbb{R}(L)$ and $\mathbb{I}(L)$; and fuzzy addition $\oplus : \mathbb{R}(L) \times \mathbb{R}(L) \to \mathbb{R}(L)$ is L-uniformly continuous for L a deMorgan chain. Also, $\mathbb{R}$ and $\mathbb{I}$ with the co-fuzzy dual topologies are uo-uniform spaces denoted $\mathbb{R}_L$ and $\mathbb{I}_L$ in the terminology of Section 8; there are uoq-uniformities generating the left and right hand L-topologies on $\mathbb{R}_L$ and $\mathbb{I}_L$; and ordinary addtion $+ : \mathbb{R}_L \times \mathbb{R}_L \to \mathbb{R}_L$ is L-uniformly continuous. Further, Hutton uniformities construct middle and high order separation axioms, including a metrizability intimately connected to covering uniformities and the **Erceg metric** defined by

$$d(a,b) = \bigwedge \{r \in \mathbb{R} : b \le D_r(a)\}$$

where $\{D_r\}_{r \in \mathbb{R}^+}$ is a "basis" of symmetric elements of the covering uniformity satisfying $D_r \circ D_s \le D_{r+s}$. (It is assumed in metrizability that a space is L-T_0 in the sense that $\forall x \ne y$, $\exists u$ open, $u(x) \ne u(y)$—see [27] and its references). Finally, $\mathbb{R}(L), \mathbb{I}(L), \mathbb{R}_L, \mathbb{I}_L$ are all metric spaces with Erceg metrics extending the usual Euclidean metric *à la* 2.2(6).

5 Intersection operators for uoq-uniformities

In Section 6 below we significantly generalize and extend the Hutton axioms (U1–U6) of 4.3 for uoq-uniformities by weakening Hutton's requirements on the underlying lattice L. Two of the Hutton's most important requirements are the following: (1) complete distributivity of L; and (2) choice of binary meet for intersection in L.

We note the complete distributivity requirement is mandated by Hutton for the delta operator Δ needed in his intersection axiom ((U5) in 4.3 above) due to his dependence on Raney's Lemma. Thus, the two issues of complete distributivity and choice of binary meet are linked in Hutton's approach.

This section lays the foundation for a two-fold generalization of Hutton's axioms by extending Hutton's Δ operator to a general "box" operator $\boxtimes$, corresponding to a tensor $\otimes$ on L, which requires no distributivity whatsoever, but which reduces to Hutton's operator if $\otimes = \wedge$ and L is completely distributive. Thus the main goal of this section is to define the $\boxtimes$ operators, develop their properties, and show for $(L, \le, \otimes) \in |\mathbf{QUML}|$, $\boxtimes$ is associative, commutative, and has constant map $\underline{\top}$ as identity.

Throughout the sequel, X is any set, $(L, \le, \otimes) \in |\mathbf{CQML}|$ with additional assumptions as needed, and the same symbols are used of relations or operations on L as of their liftings to both L^X and $\left(L^X\right)^{\left(L^X\right)}$.

We begin with the restriction of the exponential powerset $\left(L^X\right)^{\left(L^X\right)}$ (using the same labeling of axioms as in 4.3) implicit in Hutton's axioms and which is essential for understanding the $\boxtimes$ operators in our generalizations of Hutton.

5.1 Definition (Restrictions $\mathcal{W}$ and $\mathcal{W}_\otimes$ of $\left(L^X\right)^{\left(L^X\right)}$). $\mathcal{W}$ is that subset of $\left(L^X\right)^{\left(L^X\right)}$ whose members W satisfy (U2,U3) of 4.3. And $\mathcal{W}_\otimes$ is that subset of $\mathcal{W}$ whose members W satisfy the addtional axiom:

(W) W is a *semi-monoidal morphism*, or simply *semi-morphism*, i.e.

$$\forall\, a, b \in L^X,\ W(a \otimes b) \leq W(a) \otimes W(b)$$

5.2 Proposition (Comparison of $\mathcal{W}$ with $\mathcal{W}_\otimes$). Generally, $\mathcal{W}_\otimes \subset \mathcal{W}$; and if $\otimes = \wedge$, then $\mathcal{W}_\otimes = \mathcal{W}$, i.e. $\mathcal{W} = \mathcal{W}_\wedge$.

Proof. (U3) implies each W in $\mathcal{W}$ is isotone; so for $a, b \in L^X$, $W(a \wedge b) \leq W(a), W(b)$. It follows that $W(a \wedge b) \leq W(a) \wedge W(b)$. □

5.3 Examples (Operators in $\mathcal{W}_\otimes$). The (W) axiom is needed to establish the associativity of the $\boxtimes$ operators in this section and the existence of complete fibres in Section 7. It is automatically satisfied (5.2) in the Hutton axioms of Section 4 (where $\otimes = \wedge$). Section 8 establishes a rich inventory for general tensors $\otimes$ of canonical uniform operators (*à fortiori* members of $\mathcal{W}_\otimes$ by Section 6). For now we note the identity id and the constant map $\underline{\top}$ are in $\mathcal{W}_\otimes$. For non-trivial members of $\mathcal{W}_\otimes$, if $g : L \to L$ is a **generator**, i.e. a map satisfying (U2,U3,W) of 4.3 with L^X replaced by L, then $W_g : L^X \to L^X$, defined by $W_g(a) = g \circ a$, is in $\mathcal{W}_\otimes$. Such generators include the following: fix $\alpha \in L$ and put $g_\alpha(\beta) = \alpha \vee \beta$ (where L satisfies $\alpha \vee (\beta \otimes \gamma) \leq (\alpha \otimes \beta) \vee (\alpha \otimes \gamma)$); for $L = \mathbb{I}$, $\otimes = T_\pi$ $[\otimes = T_{\text{Ł}}]$ (3.5.1), choose $g_n(x) = x^{1/n}$ (for $n \in \mathbb{N}$) $[g(\alpha) \equiv S(\alpha) = \frac{\alpha+1}{2}$ of Example 1.2.3 in [8]]; for $L = \mathbb{I} \times \mathbb{I} \times \mathbb{I}$, $\otimes = T_\wedge \times T_\pi \times T_{\text{Ł}}$, choose $g_\alpha \times g_n \times S$; and generally, any direct product of generators is a generator for any collection of **CQML** objects. It is an open question to characterize members of $\mathcal{W}_\otimes$ having generators (but each such member of $\mathcal{W}_\otimes$ preserves all constant maps, i.e. $W_g(\underline{\alpha}) = \underline{g(\alpha)}$).

5.4 Proposition (Basic Properties of $\mathcal{W}_\otimes$).

1. Each $\mathcal{W}_\otimes$ is a complete lattice with relative order induced from $\left(L^X\right)^{\left(L^X\right)}$.
2. $\forall\, W \in \mathcal{W}_\otimes$, $W(\underline{\top}) = \underline{\top}$.
3. $\mathcal{W}_\otimes$ is closed under $\circ$.
4. $\circ : \mathcal{W}_\otimes \times \mathcal{W}_\otimes \to \mathcal{W}_\otimes$ is isotone in both arguments (with product order on $\mathcal{W}_\otimes \times \mathcal{W}_\otimes$).

Proof. *Ad (1).* Let $\{W_\gamma\}_\gamma \subset \mathcal{W}_\otimes$, $a, b \in L^X$, $\{a_\beta\}_\beta \subset L^X$. Note: $a \le \bigvee_\gamma a \le \bigvee_\gamma W_\gamma(a) = \left(\bigvee_\gamma W_\gamma\right)(a)$, so $\bigvee_\gamma W_\gamma$ has (U2); $\left(\bigvee_\gamma W_\gamma\right)\left(\bigvee_\beta a_\beta\right) = \left(\bigvee_\gamma W_\gamma\left(\bigvee_\beta a_\beta\right)\right) = \bigvee_\gamma\left(\bigvee_\beta W_\gamma(a_\beta)\right) = \bigvee_\beta\left(\bigvee_\gamma W_\gamma(a_\gamma)\right) = \bigvee_\beta\left(\left(\bigvee_\gamma W\right)(a_\gamma)\right)$, so (U3) holds; and $\left(\bigvee_\gamma W_\gamma\right)(a \otimes b) = \bigvee_\gamma W_\gamma(a \otimes b) \le \bigvee_\gamma (W_\gamma(a) \otimes W_\gamma(b)) \le \left[\bigvee_\gamma W_\gamma(a)\right] \otimes \left[\bigvee_\gamma W_\gamma(b)\right] = \left(\bigvee_\gamma W_\gamma\right)(a) \otimes \left(\bigvee_\gamma W_\gamma\right)(b)$, where we use isotonicity of $\otimes$ in each argument, so (W) holds. Hence $\mathcal{W}_\otimes$ is a complete join semilattice and hence a complete lattice.

Ad (2). This is a consequence of (U2): $\underline{\top} \le W(\underline{\top}) \le \underline{\top}$.

Ad (3). Let $W_1, W_2 \in \mathcal{W}_\otimes$ and $a, b \in L^X$. Then W_2 preserves arbitrary joins implies W_2 is order-preserving. Since W_1, W_2 both satisfy (U2), $a \le W_2(a) \le W_2(W_1(a)) = (W_2 \circ W_1)(a)$. So $W_2 \circ W_1$ satisfies (U2). It is straightforward to see that W_1, W_2 preserving arbitrary joins implies $W_2 \circ W_1$ preserves arbitrary joins and so satisfies (U3). Since W_2 is order-preserving and W_1, W_2 both satisfy (W), it follows $W_2(W_1(a \otimes b)) \le W_2(W_1(a) \otimes W_1(b)) \le W_2(W_1(a)) \otimes W_2(W_1(b))$, so that $W_2 \circ W_1 \in \mathcal{W}_\otimes$.

Ad (4). Let $W_1 \le \widehat{W_1}$, $W_2 \le \widehat{W_2}$. Since W_2 preserves order, it follows that $W_2(W_1(a)) \le W_2\left(\widehat{W_1}(a)\right) \le \widehat{W_2}\left(\widehat{W_1}(a)\right)$. So $W_2 \circ W_1 \le \widehat{W_2} \circ \widehat{W_1}$. □

5.5 Definition (Delta and Box Operators). Let $U, V \in \mathcal{W}_\otimes$. Then define $U \triangle V$, $U \boxtimes V$ as follows: $U \triangle V = \bigvee_{W \in \mathcal{W},\, W \le U \wedge V} W$, and

$$U \boxtimes V = \bigvee \left\{ \begin{array}{c} W \in \mathcal{W}_\otimes : \forall a, b \in L^X, \\ W(a \otimes b) \le U(a) \otimes V(b) \end{array} \right\}$$

We call the former the **delta operator** and the latter the **box operator**. Further, if we drop reference to $W \in \mathcal{W}_\otimes$ (or $\mathcal{W}$) and the universal quantifier on a, b in the condition $W(a \otimes b) \le U(a) \otimes V(b)$, then these definitions shorten as follows: $U \triangle V = \bigvee_{W \le U \wedge V} W$, $U \boxtimes V = \bigvee_{W(a \otimes b) \le U(a) \otimes V(b)} W$.

5.6 Comment. The condition defining $\triangle$ is function-theoretic. The condition defining $\boxtimes$ can also be stated function-theoretically as $W \circ \otimes \le \otimes \circ (U \times V)$, thus yielding $U \boxtimes V = \bigvee_{W \circ \otimes \le \otimes \circ (U \times V)} W$.

5.7 Acknowledgements. Our delta operator stems from Section 4, but *sans* complete distributivity; and if complete distributivity is assumed, then the two operators coincide by 4.2. The box operator stems from a suggestion of Prof. U. Höhle, who originally proposed it to the author for all operators in $\left(L^X\right)^{\left(L^X\right)}$ satisfying (U3).

5.8 Proposition (Basic Properties of Δ and $\boxtimes$).

1. $\mathcal{W}_{\otimes}$ $[\mathcal{W}]$ is closed under $\boxtimes$ $[\Delta$, respectively$]$.

2. Let $(L, \otimes) \in |\mathbf{CQML}_{\top}|$ be the identity for $\otimes$. Then the following comparisons hold:

   a. $[W \circ \otimes \leq \otimes \circ (U \times V)] \Rightarrow [W \leq U \wedge V]$.

   b. $\boxtimes \leq \Delta$.

   c. If $\otimes = \wedge$, then $\boxtimes \geq \Delta$, in which case $\boxtimes = \Delta$.

3. $\boxtimes$ is isotone in both variables.

Proof. (1) follows from Proposition 5.4 and (3) is straightforward. As for (2), let $U, V \in \mathcal{W}_{\otimes}$ and consider these two conditions: (A) $W \circ \otimes \leq \otimes \circ (U \times V)$; (B) $W \leq U \wedge V$.

Ad (2)(a,b). It suffices to prove (A) $\Rightarrow$ (B). We observe $\forall\, a \in L^X$ that

$$W(a) = W(a \otimes \underline{\top}) \leq U(a) \otimes V(\underline{\top}) = U(a) \otimes \underline{\top} = U(a)$$

So $W \leq U$. Similarly, $W \leq V$. Hence $W \leq U \wedge V$.

Ad (2)(c). It suffices to prove (B) $\Rightarrow$ (A). We observe $\forall\, a, b \in L^X$ that the isotonicity of U, V implies $W(a \wedge b) \leq U(a \wedge b) \wedge V(a \wedge b) \leq U(a) \wedge V(b)$. $\square$

5.9 Comment. The delta operator defined above subsumes Hutton's delta operator (4.2, 5.7) and the box operator subsumes the delta operator when $\otimes = \wedge$ (5.2). Therefore, we now focus on the box operator.

5.10 Lemma (Fundamental Algebraic Properties of Box Operators). Let $(L, \leq, \otimes) \in |\mathbf{QUML}|$ and $X \in |\mathbf{SET}|$. Then $(\mathcal{W}_{\otimes}, \leq, \boxtimes) \in |\mathbf{QUML}|$, i.e. the following hold:

1. $\boxtimes$ has the constant map $\underline{\top}$ as identity.

2. $\boxtimes$ is commutative.

3. The following conditions are equivalent:

   a. $W \circ [\otimes \circ (\otimes \times id)] \leq \otimes \circ [(U \circ \otimes) \times V]$;

   b. $W \circ \otimes \leq \otimes \circ (U \times V)$;

   c. $W \circ [\otimes \circ (id \times \otimes)] \leq \otimes \circ [U \times (V \circ \otimes)]$.

4. $\boxtimes$ is associative.

Proof. *Ad (1).* Trivial.

Ad (2). Let $\otimes$ be commutative. Then clearly the two bi-implication

$$[W \circ \otimes \leq \otimes \circ (U \times V)] \Leftrightarrow [W \circ \otimes \leq \otimes \circ (V \times U)]$$

holds; and from this the commutivity of the box operator follows.

Ad (3). (b) $\Leftrightarrow$ (c) is analogous to (a) $\Leftrightarrow$ (b), so we only discuss the latter. Assuming that all variables are universally quantified, we are to prove the following bi-implication:

$$W\left((a \otimes b) \otimes c\right) \leq U\left(a \otimes b\right) \otimes V\left(c\right) \;\Leftrightarrow\; W\left(a \otimes b\right) \leq U\left(a\right) \otimes V\left(b\right)$$

Now ($\Leftarrow$) is immediate. And for ($\Rightarrow$), specify $b = \top$ and use that $\top$ is the identity for $\otimes$.

Ad(4). Let $U, V, Y \in \mathcal{W}_{\otimes}$. We are to show that

$$(U \boxtimes V) \boxtimes Y = U \boxtimes (V \boxtimes Y) \tag{A0}$$

Our strategy is to show that the left and right sides are each equal to the same operator in $\mathcal{W}_{\otimes}$.

Step I of Proof. We begin by working with $(U \boxtimes V) \boxtimes Y$. Using the definition of $\boxtimes$ and the Lemma 5.14(1) above, we have

$$\begin{aligned}(U \boxtimes V) \boxtimes Y &= \bigvee_{W \circ \otimes \leq \otimes \circ ((U \boxtimes V) \times Y)} W &\text{(A1)}\\ &= \bigvee_{W \circ \otimes \leq \otimes \circ \left(\left(\bigvee_{W^* \circ \otimes \leq \otimes \circ (U \times V)} W^*\right) \times Y\right)} W \\ &= \bigvee_{W \circ [\otimes \circ (\otimes \times id)] \leq \otimes \circ \left[\left(\left(\bigvee_{W^* \circ \otimes \leq \otimes \circ (U \times V)} W^*\right) \circ \otimes\right) \times Y\right]} W\end{aligned}$$

We now claim

$$\begin{aligned}&\bigvee_{W \circ [\otimes \circ (\otimes \times id)] \leq \otimes \circ \left[\left(\left(\bigvee_{W^* \circ \otimes \leq \otimes \circ (U \times V)} W^*\right) \circ \otimes\right) \times Y\right]} W &\text{(A2)}\\ = &\bigvee_{W \circ [\otimes \circ (\otimes \times id)] \leq \otimes \circ [(\otimes \circ (U \times V)) \times Y]} W\end{aligned}$$

and hence claim

$$(U \boxtimes V) \boxtimes Y = \bigvee_{W \circ [\otimes \circ (\otimes \times id)] \leq \otimes \circ [(\otimes \circ (U \times V)) \times Y]} W \tag{A3}$$

To verify (A2), it suffices to prove the following bi-implication of conditions in which the fuzzy subset variables a, b, c, e, f are to be understood as universally quantified:

$$\left[W\left((a \otimes b) \otimes c\right) \leq \left[\bigvee_{W^*(e\otimes f) \leq U(e)\otimes V(f)} W^*\left(a \otimes b\right)\right] \otimes Y\left(c\right)\right] \quad \text{(A4)}$$
$$\iff \quad \left[W\left((a \otimes b) \otimes c\right) \leq \left[U\left(a\right) \otimes V\left(b\right)\right] \otimes Y\left(c\right)\right]$$

Ad ($\Rightarrow$) *of (A4).* Assume that W satisfies the left-hand side of (A3). Now the defining condition of the join of that left-hand side, with e, f instantiated as a, b, respectively, yields

$$\bigvee_{W^*(e\otimes f) \leq U(e)\otimes V(f)} W^*\left(a \otimes b\right) \leq U\left(a\right) \otimes V\left(b\right)$$

Applying the isotonicity of $\otimes$, we have

$$\left[\bigvee_{W^*(e\otimes f) \leq U(e)\otimes V(f)} W^*\left(a \otimes b\right)\right] \otimes Y\left(c\right) \leq \left[U\left(a\right) \otimes V\left(b\right)\right] \otimes Y\left(c\right)$$

and so W satisfies the right-hand side of (A3).

Ad ($\Leftarrow$) *of (A4).* Assume that W satisfies the right-hand side of (A4). Then from previous Lemmas and W being a semi-morphism from its membership in $\mathcal{W}_\otimes$ (Axiom (W)), we have the following facts:

1. $\forall\, a, b,\ W\left(a \otimes b\right) \leq U\left(a\right) \otimes V\left(b\right);$
2. $\forall\, c,\ W\left(c\right) \leq Y\left(c\right);$
3. $\forall a, b, c,\ W\left((a \otimes b) \otimes c\right) \leq W\left(a \otimes b\right) \otimes W\left(c\right).$

We therefore have for this W that

$$\forall\, a, b, c,\ W\left((a \otimes b) \otimes c\right) \leq W\left(a \otimes b\right) \otimes W\left(c\right) \leq W\left(a \otimes b\right) \otimes Y\left(c\right)$$

and that

$$\forall\, e, f,\ W\left(e \otimes f\right) \leq U\left(e\right) \otimes V\left(f\right)$$

It follows that W is a W^* of the defining condition of the join of the left-hand side of (A3), and hence that

$$\begin{aligned} W\left((a \otimes b) \otimes c\right) \quad &\leq \bigvee_{W^*(e\otimes f) \leq U(e)\otimes V(f)} \left[W^*\left(a \otimes b\right) \otimes Y\left(c\right)\right] \\ &\leq \left[\bigvee_{W^*(e\otimes f) \leq U(e)\otimes V(f)} W^*\left(a \otimes b\right)\right] \otimes Y\left(c\right) \end{aligned}$$

where the latter inequality follows from the isotonicity of $\otimes$ (and so infinite distributivity of $\otimes$ over $\bigvee$ is not needed).

This concludes the proof of (A4) and hence of (A2,A3).

Step II. We now invoke the associativity of $\otimes$ to claim that

$$\bigvee_{W \circ [\otimes \circ (\otimes \times id)] \leq \otimes \circ [(\otimes \circ (U \times V)) \times Y]} W = \bigvee_{W \circ [\otimes \circ (\otimes \times id)] \leq \otimes \circ [U \times (\otimes \circ (V \times Y))]} W \tag{A5}$$

Since $(U(a) \otimes V(b)) \otimes Y(c) = U(a) \otimes (V(b) \otimes Y(c))$, (A5) follows.

Step III. The purpose of this step is to work with $U \boxtimes (V \boxtimes Y)$ and prove the counterpart to (A3), namely the claim that

$$U \boxtimes (V \boxtimes Y) = \bigvee_{W \circ [\otimes \circ (\otimes \times id)] \leq \otimes \circ [U \times (\otimes \circ (V \times Y))]} W \tag{A6}$$

But the proof is completely analogous to the proof of (A3), and is left to the reader to check.

Step IV. To summarize from the previous steps, (A3,A5,A6) allow us to say that

$$\begin{aligned} (U \boxtimes V) \boxtimes Y &= \bigvee_{W \circ [\otimes \circ (\otimes \times id)] \leq \otimes \circ [(\otimes \circ (U \times V)) \times Y]} W \\ &= \bigvee_{W \circ [\otimes \circ (\otimes \times id)] \leq \otimes \circ [U \times (\otimes \circ (V \times Y))]} W \\ &= U \boxtimes (V \boxtimes Y) \end{aligned}$$

Thus (A0) is verified and the proof of the associativity of $\boxtimes$, and this lemma, is complete. □

5.11 Examples of Box Operators. The conjunction of 3.5, 3.6, 5.3.4, 5.10 guarantees existence of many non-trivial examples of box operators $\boxtimes$ for which $(\mathcal{W}_\otimes, \leq, \boxtimes) \in |\mathbf{QUML}|$, including many examples in which $\otimes$ is not $\wedge$.

5.12 Open Question:Alternative Generalized Delta Operators $\triangle_\otimes$. As seen above, $\boxtimes$ is a generalization of Δ—if $\otimes = \wedge$, then $\boxtimes = \Delta$. Another generalization of Δ is possible. Assume $(L, \leq, \otimes) \in |\mathbf{QUML}|$ and define $\triangle_\otimes$ for $U, V \in \mathcal{W}_\otimes$ by $U \triangle_\otimes V = \bigvee_{W \leq U \otimes V} W$. It follows from Lemma 5.10 that if $\otimes = \wedge$, then $(\mathcal{W}_\otimes, \leq, \triangle_\otimes) \in |\mathbf{QUML}|$ for each set X. We conjecture "yes" to the following open question concerning the converse: let $(L, \leq, \otimes) \in |\mathbf{QUML}|$ and assume that $(\mathcal{W}_\otimes, \leq, \triangle_\otimes) \in |\mathbf{QUML}|$ for each set X; must $\otimes = \wedge$?

6 General axioms for uoq-uniformities

This section extends the Hutton axioms for uoq-uniformities to a setting in which the underlying lattice-theoretic base is a complete quasi-monoidal lattice with tensor product. Sections 1,2,4 above motivate the syntax of the following axioms; Section 5 above shows that these axioms are well-defined for lattice-theoretic bases from **CQML**; Section 7 below uses Section 5 to show that these axioms are sufficient to guarantee complete fibres for lattice-theoretic bases from **QUML**; and Section 8 below uses Sections 3 and 7 to give an extensive list of important examples. The reader should recall from Section 5 the restriction $\mathcal{W}_{\otimes}$ (of the exponential powerset $\left(L^X\right)^{\left(L^X\right)}$) and the box operator $\boxtimes$.

6.1 Axioms for UOQ-Unifomities. Let X be a set, let $L \in |\mathbf{CQML}|$, and let $\mathcal{U}$ be a subset of $\mathcal{W}_{\otimes}$ satisfying (U1,U4,U5,U6) of 4.3. Then $\mathcal{U}$ is an (L-)**uniform operator q(uasi)-uniformity** or (L-)**uoq-uniformity** on X and $(X, \mathcal{U})$ is a (L-)**uniform operator q(uasi)-uniform space** or (L-)**uoq-uniform space**. Members U of $\mathcal{U}$ are called **uniform operators**. The the induced L-topology on X is

$$\tau(\mathcal{U}) = \left\langle\left\langle \left\{ u \in L^X : u = \bigvee \left\{ a \in L^X : \exists U \in \mathcal{U}, \, U(a) \le u \right\} \right\} \right\rangle\right\rangle$$

where "$\langle\langle \quad \rangle\rangle$" indicates the L-topology generated from the subbasis listed inside "$\langle\langle \quad \rangle\rangle$" (3.2.2 of [26]).

Note that (U2,U3) of the Hutton axioms, plus the semi-morphism property (W), have been subsumed in the assumption that $\mathcal{U} \subset \mathcal{W}_{\otimes}$; and this accounts for our numbering the axioms of 6.1 as we have. This means in the sequel we can appeal, for example, to (U2) in regard to a uniform operator without confusion.

6.2 Proposition (Reconciliation with Hutton). If L is a frame, then

$$\tau(\mathcal{U}) = \left\{ u \in L^X : u = \bigvee \left\{ a \in L^X : \exists U \in \mathcal{U}, \, U(a) \le u \right\} \right\}$$

Proof. This is a corollary of the paragraph following 3.2.2 of [26]. □

7 Complete fibres for uoq-uniformities

Throughout this section, X is a set, and $(L, \le, \otimes) \in |\mathbf{QUML}|$ unless indicated otherwise. This section proves the collection of all L-uoq-uniformities on X—as defined in Section 6—form a complete lattice with the inclusion order, i.e. L-uoq-uniform spaces yield complete fibres. Complete fibres are closely related to topological constructs and topological categories, topics deferred to later work; and complete fibres are applied in Section 7 to construct natural examples.

We proceed as follows: first, given $\mathcal{U}, \mathcal{V}$ are L-uoq-uniformities on X, we define the notions of subbase and base for L-uoq-uniformities and show $\mathcal{U} \cup \mathcal{V}$ is a subbase for an L-uoq-uniformity on X having base $\mathcal{U} \boxtimes \mathcal{V}$; second, this binary procedure is generalized to any family $\{\mathcal{U}_\gamma : \gamma \in \Gamma\}$ of L-uoq-uniformities on X; and this general construction is used to show the set of all L-uoq-uniformities on X, when ordered by inclusion, is a complete lattice.

We note two things for this section: on one hand, no distributivity of any kind is needed, and in this respect L-uoq-uniform spaces behave as topological spaces [8, 26]; and on the other hand, the algebraic conditions on $\otimes$—associativity, commutivity, and $\top$ as identity—distinguish the uoq-uniform case from the topological case. Also note 3.5–3.6 give many such tensor products.

7.1 Definition (Subbase of UOQ-Uniformity). We say $\mathcal{S} \subset \mathcal{W}_\otimes$ is a **subbase** for a uoq-uniformity if $\mathcal{S}$ satisfies (U1,U6) of Section 6. Restated, $\mathcal{S} \subset \left(L^X\right)^{\left(L^X\right)}$ and $\mathcal{S}$ satisfies (U1–U3,W,U6).

7.2 Lemma (Subbase from UOQ-Uniformities). Let $\mathcal{U}, \mathcal{V}$ be q-uniformities on X. Then $\mathcal{U} \cup \mathcal{V}$ satisfies (U1–U3,W,U6) and is a subbasis for a uoq-uniformity on X.

7.3 Construction: L-UOQ-Uniformity on X Generated from $\mathcal{U} \cup \mathcal{V}$. We proceed in several steps.

Step I: Construction of "base" $\mathcal{U} \boxtimes \mathcal{V}$.

7.3.0 Definition. Put $\mathcal{U} \boxtimes \mathcal{V} \equiv \langle \mathcal{U} \cup \mathcal{V} \rangle = \{U \boxtimes V : U, V \in \mathcal{U} \cup \mathcal{V}\}$.

This notation is justified by the following lemma.

7.3.1 Lemma. $\mathcal{U} \boxtimes \mathcal{V} = \{U \boxtimes V : U \in \mathcal{U},\ V \in \mathcal{V}\}$.

Proof. The inclusion from right to left is trivial. For the inclusion from left to right, the only detail is that if $U \in \mathcal{U}$, $V \in \mathcal{V}$, then we want $U, V \in \mathcal{U} \boxtimes \mathcal{V}$. But by (U1,U4), $\underline{\top}$ is in each of $\mathcal{U}$ and $\mathcal{V}$. Now by Lemma 5.10 above, we have that $U = U \boxtimes \underline{\top}$, $V = \underline{\top} \boxtimes V$, which yield the desired conclusion. □

Step II: $\mathcal{U} \boxtimes \mathcal{V}$ is closed under $\boxtimes$ and satisfies (U5). To show that $\mathcal{U} \boxtimes \mathcal{V}$ is closed under $\boxtimes$ requires several cases, most of which are immediate or use part or all of the argument for the following, hardest case: let $U_1, U_2 \in \mathcal{U}$, $V_1, V_2 \in \mathcal{V}$; and then show

$$(U_1 \boxtimes U_2) \boxtimes (V_1 \boxtimes V_2) = (U_1 \boxtimes V_1) \boxtimes (U_2 \boxtimes V_2)$$

This follows immediately from Lemma 5.10. □

Step III: $\mathcal{U} \boxtimes \mathcal{V}$ satisfies (U6). Showing that $\mathcal{U} \boxtimes \mathcal{V}$ satisfies (U6) requires several cases, most of which are immediate or use part or all of the argument for thefollowing, hardest case: let $U \in \mathcal{U}$, $V \in \mathcal{V}$; and then show $\exists\, W \in \mathcal{U} \boxtimes \mathcal{V}$,

$$W \circ W \leq U \boxtimes V \tag{7.3.III.1}$$

Since $\mathcal{U}, \mathcal{V}$ are uoq-uniformities, $\exists U^* \in \mathcal{U}$, $V^* \in \mathcal{V}$ such that $U^* \circ U^* \leq U$, $V^* \circ V^* \leq V$. Set $W = U^* \boxtimes V^*$. Clearly $W \in \mathcal{U} \boxtimes \mathcal{V}$. We now show that 7.3.III.1 holds for this W. As a first step, we verify

$$(W \circ W) \circ \otimes \leq \otimes \circ ((U^* \circ U^*) \times (V^* \circ V^*)) \tag{7.3.III.2}$$

Let $a, b \in L^X$. Then repeatedly using axiom (W) and the fact that $\boxtimes$ is isotone (Lemma 5.8(3)), we have

$$\begin{aligned} W(W(a \otimes b)) &= (U^* \boxtimes V^*)((U^* \boxtimes V^*)(a \otimes b)) \\ &\leq (U^* \boxtimes V^*)((U^*(a) \otimes V^*(b))) \\ &= U^*(U^*(a)) \otimes V^*(V^*(a)) \end{aligned}$$

It now follows from the definition of $\boxtimes$ and its isotonicity that

$$\begin{aligned} W \circ W &\leq \bigvee_{Z \circ \otimes \leq \otimes \circ ((U^* \circ U^*) \times (V^* \circ V^*))} Z \\ &= (U^* \circ U^*) \boxtimes (V^* \circ V^*) \\ &\leq U \boxtimes V \end{aligned}$$

verifying 7.3.III.1. This completes the proof of Step III. □

Step IV: $\mathcal{U} \boxtimes \mathcal{V}$ generates an L-uoq-uniformity on X.

7.3.2 Definition (Base for UOQ-Uniformity). We say $\mathcal{B} \subset \mathcal{W}_\otimes$ is a **base** for an L-uoq-uniformity if $\mathcal{B}$ satisfies (U1,U5,U6) of Section 6. Restated, $\mathcal{B} \subset (L^X)^{(L^X)}$ and $\mathcal{B}$ satisfies (U1–U3,W,U5,U6).

7.3.3 Lemma. Let $(L, \leq, \otimes) \in |\mathbf{CQML}_\top|$ and $\mathcal{B}$ be a base for an uoq-uniformity on X. Then

$$\langle \mathcal{B} \rangle \equiv \{W \in \mathcal{W}_\otimes : \exists\, B \in \mathcal{B},\, B \leq W\}$$

is an L-uoq-uniformity on X, called the **uoq-uniformity generated from $\mathcal{B}$.**

Proof. Clearly, (U1-U4,W) are satisfied. It can be shown that (U5) follows from the assumption that $\mathcal{B}$ satisfies (U5) and the fact that $\boxtimes$ is isotone in both arguments (5.8(3)). And (U6) follows since $\mathcal{B} \subset \langle \mathcal{B} \rangle$, $\mathcal{B}$ satisfies (U6), and $\circ : \mathcal{W}_\otimes \times \mathcal{W}_\otimes \to \mathcal{W}_\otimes$ is isotone in both arguments (5.4(4)). $\square$

7.3.4 Notation. If a basis $\mathcal{B}$ is **generated** from a subbase $\mathcal{S}$, i.e. $\mathcal{B} = \{U \boxtimes V : U, V \in \mathcal{S}\}$, then we may write both $\langle \mathcal{B} \rangle$ and $\langle\langle \mathcal{S} \rangle\rangle$ for the L-uoq-uniformity generated from $\mathcal{B}$ and also speak of it as the L-uoq-uniformity generated from $\mathcal{S}$ (cf. [22]).

7.3.5 Lemma. $\mathcal{U} \boxtimes \mathcal{V}$ is a base for an L-uoq-uniformity on X. Hence $\langle \mathcal{U} \boxtimes \mathcal{V} \rangle \equiv \langle\langle \mathcal{U} \cup \mathcal{V} \rangle\rangle$ is an L-uoq-uniformity on X.

Proof. Since $\top$ is idempotent and $\underline{\top}$ is in each of $\mathcal{U}$ and $\mathcal{V}$ (by (U1,U4) applied to $\mathcal{U}, \mathcal{V}$), it follows $\underline{\top} \boxtimes \underline{\top} = \underline{\top}$ and $\underline{\top} \in \mathcal{U} \boxtimes \mathcal{V}$, so $\mathcal{U} \boxtimes \mathcal{V}$ satisfies (U1). Now since each $U \boxtimes V \in \mathcal{W}_\otimes$, it follows from Lemma 7.3.1 that $\mathcal{U} \boxtimes \mathcal{V} \subset \mathcal{W}_\otimes$ and hence $\mathcal{U} \boxtimes \mathcal{V}$ satisfies (U2,U3,W). From Steps II–III, $\mathcal{U} \boxtimes \mathcal{V}$ satisfies (U5,U6). So $\mathcal{U} \boxtimes \mathcal{V}$ is a base. Now apply Lemma 7.3.3. $\square$

7.4. General Case of Construction of L-UOQ-Uniformity Generated from $\{\mathcal{U}_\gamma\}_{\gamma\in\Gamma}$.

7.4.0 Definition. Let $\{\mathcal{U}_\gamma\}_{\gamma\in\Gamma}$ be a collection of L-uoq-uniformities on X. Then we form $\boxtimes_{\gamma\in\Gamma} \mathcal{U}_\gamma$ as follows: $\boxtimes_{\gamma\in\Gamma} \mathcal{U}_\gamma = \left\{U \boxtimes V : U, V \in \bigcup_{\gamma\in\Gamma} \mathcal{U}_\gamma\right\}$. In keeping with Construction 7.3, we also write $\boxtimes_{\gamma\in\Gamma} \mathcal{U}_\gamma = \left\langle \bigcup_{\gamma\in\Gamma} \mathcal{U}_\gamma \right\rangle$.

Arguments analogous to those in 7.3 establish the following results:

7.4.1 Lemma. Let $\{\mathcal{U}_\gamma\}_{\gamma\in\Gamma}$ be a collection of L-uoq-uniformities on X. The following hold:

1. $\bigcup_{\gamma\in\Gamma} \mathcal{U}_\gamma$ satisfies (U1–U4,W,U6), so is a subbase for an L-uoq-uniformity on X.

2. $\boxtimes_{\gamma\in\Gamma} \mathcal{U}_\gamma \equiv \left\langle \bigcup_{\gamma\in\Gamma} \mathcal{U}_\gamma \right\rangle$ satisfies (U1–U3,W,U5,U6) and so is a base for an L-uoq-uniformity on X.

3. $\langle \boxtimes_{\gamma\in\Gamma} \mathcal{U}_\gamma \rangle \equiv \left\langle\left\langle \bigcup_{\gamma\in\Gamma} \mathcal{U}_\gamma \right\rangle\right\rangle$ is an L-uoq-uniformity on X.

7.4.2 Theorem (UOQ-Uniformities Form Complete Fibres). Let X be a set and $(L, \otimes) \in |\mathbf{QUML}|$. Then the family of all L-uoq-uniformities on X, when ordered by inclusion, form a complete lattice.

Proof. We need only show that this family is closed under arbitrary joins—duality then furnishes closure under arbitrary meets. For a subfamily $\{\mathcal{U}_\gamma\}_{\gamma\in\Gamma}$, we have from Lemma 7.4.1 that $\mathcal{U} \equiv \left\langle\left\langle \bigcup_{\gamma\in\Gamma} \mathcal{U}_\gamma \right\rangle\right\rangle$ is an L-uoq-uniformity on X. Using (U1,U4) and the fact $\underline{\top}$ is the identity for $\boxtimes$ (Lemma 5.10), it follows $\bigcup_{\gamma\in\Gamma} \mathcal{U}_\gamma \subset \boxtimes_{\gamma\in\Gamma} \mathcal{U}_\gamma$. Further, $\boxtimes_{\gamma\in\Gamma} \mathcal{U}_\gamma$ being a base (7.4.1) implies $\boxtimes_{\gamma\in\Gamma} \mathcal{U}_\gamma \subset \mathcal{U}$. Thus $\bigcup_{\gamma\in\Gamma} \mathcal{U}_\gamma \subset \mathcal{U}$, which implies $\mathcal{U}$ is an upper bound for $\{\mathcal{U}_\gamma\}_{\gamma\in\Gamma}$ w.r.t. inclusion. Now suppose that $\mathcal{V}$ is any upper bound of $\{\mathcal{U}_\gamma\}_{\gamma\in\Gamma}$ w.r.t. inclusion. Then $\bigcup_{\gamma\in\Gamma} \mathcal{U}_\gamma \subset \mathcal{V}$. The (U5) axiom now forces $\boxtimes_{\gamma\in\Gamma} \mathcal{U}_\gamma \subset \mathcal{V}$, and the (U6) axiom then forces $\mathcal{U} = \langle \boxtimes_{\gamma\in\Gamma} \mathcal{U}_\gamma \rangle \subset \mathcal{V}$. So we have $\mathcal{U}$ is the least upper bound (or join) of the subfamily $\{\mathcal{U}_\gamma\}_{\gamma\in\Gamma}$. □

8 Canonical examples of uoq-uniformities

The following important classes of examples are catalogued in this section: right [left] uoq-uniformities on the L-fuzzy real line $\mathbb{R}(L)$ and L-fuzzy unit interval $\mathbb{I}(L)$ generating the standard right [left] L-topology, for $L \in |\mathbf{HUT}|$ ($\otimes = \wedge$); right [left] uoq-uniformities on the L-real line $\mathbb{R}_L$ and L-unit interval $\mathbb{I}_L$ [23, 28] generating the standard right [left] weakly stratified L-topology on $\mathbb{R}$ and $\mathbb{I}$, for $L \in |\mathbf{COQUANT}_\top|$ (cf. 3.6); and right [left] uoq-uniformities on the alternative L-fuzzy real line $\mathbb{R}^*(L)$ and the alternative L-fuzzy unit interval $\mathbb{I}^*(L)$ generating the right [left, resp.] L-topology on these spaces for $L \in |\mathbf{CQML}_{\le,\top}|$ (see 3.5–3.6), where $\mathbb{R}^*(L)$ and $\mathbb{I}^*(L)$ are the L-2 soberifications of $\mathbb{R}$ and $\mathbb{I}$, respectively, using the L-2 soberification functors.

Furthermore, Theorem 7.4.2 guarantees the **bi-handed** uoq-uniformity, i.e. the smallest uoq-uniformity containing the right and left uoq-uniformities in each of these spaces, as follows: for each $L \in |\mathbf{HUT}|$, there is a bi-handed uoq-uniformity containing the right and left uoq-uniformities on $\mathbb{R}(L)$ and $\mathbb{I}(L)$; $\forall\, L \in |\mathbf{COQUANT} \cap \mathbf{QUML}|$, there is a bi-handed uoq-uniformity containing the right and left uoq-uniformities on $\mathbb{R}_L$ and $\mathbb{I}_L$; and $\forall\, L \in |\mathbf{QUML}_{\le}|$, there is a bi-handed uoq-uniformity containing the right and left uoq-uniformities on $\mathbb{R}^*(L)$ and $\mathbb{I}^*(L)$.

These classes of examples justify the generalization of the Hutton axioms given in this chapter: no required distributivity, the (W) axiom, and allowance of $\otimes$ [$\mathcal{W}_\otimes$, $\boxtimes$] other than $\wedge$ [$\mathcal{W}_\wedge$, Δ, resp.].

We summarize the construction of each type of space, develop the uoq-uniformities for that space, and then justify important technical details needed for the operators of a particular uoq-uniformity. We appeal throughout to the fact that L-topology has complete fibres for each $L \in |\mathbf{CQML}|$ [8, 26] and that L-uoq-uniformities have complete fibres for each $L \in |\mathbf{QUML}|$ (7.4.2 above).

8.1 Basic set-up of $\mathbb{R}(L)$, $\mathbb{I}(L)$, $\mathbb{R}_L$, $\mathbb{I}_L$, $\mathbb{R}^*(L)$, $\mathbb{I}^*(L)$

Literature for this subsection includes: [13, 14, 6, 22, 23, 24]; Section 7 of [26]; Subsections 2.15–2.16 of [27]; [28]; and their bibliographies. The well-known definition of the L-**fuzzy real line** $\mathbb{R}(L)$, found in these references, is here taken assuming $L \in |\mathbf{DQML}|$ with tensor $\otimes$ and antitone involution $'$. The canonical **right** [**left**] L-topology $\{\mathfrak{R}_t : t \in \mathbb{R}\}$ $[\{\mathcal{L}_t : t \in \mathbb{R}\}]$ on $\mathbb{R}(L)$ is denoted $\tau^r(L)$ $[\tau^l(L)]$ and the canonical **symmetric** L-topology $\langle\langle \tau^r(L) \cup \tau^l(L) \rangle\rangle$ is denoted $\tau(L)$.

The definition of the canonical right [left, symmetric] L-topology on $\mathbb{R}$ is found in [23, 24, 28]. Letting $[\lambda] \in \mathbb{R}(L)$, put $\mathfrak{R}_{[\lambda]} : \mathbb{R} \to L$ $[\mathfrak{L}_{[\lambda]} : \mathbb{R} \to L]$ by $\mathfrak{R}_{[\lambda]}(t) = \lambda(t+)$ $[\mathfrak{L}_{[\lambda]}(t) = \lambda'(t-)]$. The right [left, symmetric] L-topology τ_L^r $[\tau_L^l, \tau_L]$ is $\{\mathfrak{R}_{[\lambda]} : [\lambda] \in \mathbb{R}(L)\}$ $[\{\mathfrak{L}_{[\lambda]} : [\lambda] \in \mathbb{R}(L)\}, \langle\langle \tau_L^r \cup \tau_L^l \rangle\rangle]$. In this context, we speak of the space $\mathbb{R}_L$, called the L-**real line.** This simplifies the terminology of [23, 24, 28].

The L-**fuzzy unit interval** $\mathbb{I}(L)$ and the L-**unit interval** $\mathbb{I}_L$ are the L-subspaces of $\mathbb{R}(L)$ and $\mathbb{R}_L$ resp. (equipped in context with one of the available topologies), based on the subsets $\mathbb{I}(L)$ and $\mathbb{I} \equiv [0, 1]$, resp. (cf. Subsection 5.2 of [26]).

The polarity $'$ on L can be removed in the definitions of τ_L^r, τ_L^l, τ_L. Letting $L \in |\mathbf{CQML}|$, we only require antitonicity [isotonicity] of the maps $\lambda : \mathbb{R} \to L$, but use the same equivalence relation as in $\mathbb{R}(L)$. We obtain $\mathbb{E}_{\searrow}(L)$ $[\mathbb{E}_{\nearrow}(L)]$, models for the L-**fuzzy extended real line.** In the case of $\mathbb{E}_{\searrow}(L)$, we define (as above) $\mathfrak{R}_{[\rho]}^{\searrow} : \mathbb{R} \to L$ by $\mathfrak{R}_{[\rho]}^{\searrow}(t) = \rho(t+)$—and if L has polarity $'$ we also define $\mathfrak{L}_{[\rho]}^{\searrow} : \mathbb{R} \to L$ by $\mathfrak{L}_{[\rho]}(t) = \rho'(t-)$; and in the case of $\mathbb{E}_{\nearrow}(L)$, we define $\mathfrak{L}_{[\rho]}^{\nearrow} : \mathbb{R} \to L$ by $\mathfrak{L}_{[\rho]}(t) = \rho(t-)$—and if L has polarity $'$ we also define $\mathfrak{R}_{[\rho]}^{\nearrow} : \mathbb{R} \to L$ by $\mathfrak{R}_{[\rho]}^{\nearrow}(t) = \rho'(t+)$. By a suitable modification of the proof of Theorem 3.1.2 of [23] the following can be proved:

$$\begin{aligned} \tau_L^r &= \{\mathfrak{R}_{[\lambda]} : [\lambda] \in \mathbb{R}(L)\} = \left\{\mathfrak{R}_{[\rho]}^{\searrow} : [\rho] \in \mathbb{E}_{\searrow}(L)\right\}, \\ \tau_L^l &= \{\mathfrak{L}_{[\lambda]} : [\lambda] \in \mathbb{R}(L)\} = \left\{\mathfrak{L}_{[\rho]}^{\nearrow} : [\rho] \in \mathbb{E}_{\nearrow}(L)\right\} \end{aligned}$$

Sinceτ_L^r, τ_L^l, τ_L can now be obtained in the deMorgan case without the polarity, we then define τ_L^r, τ_L^l, τ_L in the general **CQML** case by the right hand sides of the above displays.

Finally, $\mathbb{R}^*(L)$ and $\mathbb{I}^*(L)$ are the L-2 soberifications of the ordinary topological spaces $\mathbb{R}$ and $\mathbb{I}$ given by the L-2 soberification functors. To detail these functors, the reader is referred to: Section 7 of [26]; Subsections 2.15–2.16 of [27]; and [6]. Note [27] for L a frame (with $\otimes = \wedge$), $\mathbb{R}^*(L)$ and $\mathbb{I}^*(L)$ are respectively L-homeomorphic to $\mathbb{R}(L)$ and $\mathbb{I}(L)$ iff L is a Boolean algebra; generalizations of this result are given in [6]. To define $\mathbb{R}^*(L)$ and $\mathbb{I}^*(L)$, let $\mathfrak{I}$

be the traditional topology on $\mathbb{R}$, note that the tensor on $\Im$ is the binary $\cap$, put $Lpt(\Im) = \mathbf{CQML}(\Im, L)$ (so that p converts $\bigcup, \cap, \mathbb{R}$ of $\Im$ to $\bigvee, \otimes, \top$ of L), define $\Phi_L : \Im \to L^{Lpt(\Im)}$ by $\Phi_L(G)(p) = p(G)$, and set $\tau^*(L) = \Phi_L^{\to}(\Im)$. Then $\mathbb{R}^*(L)$ is the L-topological space $(Lpt(\Im), \tau^*(L))$. Because of the functoriality of the L-2 soberification functor, $\mathbb{I}^*(L)$ may be equivalently defined either by repeating this construction using the traditional topology on $\mathbb{I}$ or by specifying the appropriate subspace of $\mathbb{R}^*(L)$ (cf. Subsection 5.2 of [26]).

8.2 UOQ-Uniformities for $\mathbb{R}(L)$ and $\mathbb{I}(L)$

Let $L \in |\mathbf{HUT}|$, $\otimes = \wedge$, $a \in L^{\mathbb{R}(L)}$, define $B_\varepsilon, C_\varepsilon : L^{\mathbb{R}(L)} \to L^{\mathbb{R}(L)}$ by

$$B_\varepsilon(a) = \mathfrak{R}_{t-\varepsilon}, \quad t = \bigvee \{s \in \mathbb{R} : a \le (\mathfrak{L}_s)'\}$$

$$C_\varepsilon(a) = \mathfrak{L}_{t+\varepsilon}, \quad t = \bigwedge \{s \in \mathbb{R} : a \le (\mathfrak{R}_s)'\}$$

and set

$$\mathcal{B}(L) \equiv \left\{B_\varepsilon : L^{\mathbb{R}(L)} \to L^{\mathbb{R}(L)} \mid \varepsilon \in \mathbb{R},\, \varepsilon > 0\right\}$$

$$\mathcal{C}(L) \equiv \left\{C_\varepsilon : L^{\mathbb{R}(L)} \to L^{\mathbb{R}(L)} \mid \varepsilon \in \mathbb{R},\, \varepsilon > 0\right\}$$

Then $\mathcal{B}(L)$ [$\mathcal{C}(L)$] is an L-uoq-uniformity on $\mathbb{R}(L)$ inducing via 6.1 the right [left] topology $\tau^r(L)$ [$\tau^l(L)$] on $\mathbb{R}(L)$ and hence is designated the **right [left]** L-uoq-uniformity on $\mathbb{R}(L)$. The bi-handed L-uoq-uniformity $\mathcal{U}(L)$, the smallest L-uoq-uniformity containing both $\mathcal{B}(L)$ and $\mathcal{C}(L)$, also satisfies (U7) of 2.2(4) (and hence is an "L-uo-uniformity") which induces via 6.1 the symmetric topology $\tau(L)$ on $\mathbb{R}(L)$.

The definitions of $\mathcal{B}(L)$, $\mathcal{C}(L)$, $\mathcal{U}(L)$ come from [9] and the proofs are found in [21]. Appropriate modifications yield right [left] L-uoq-uniformities and bi-handed L-uo-uniformities on $\mathbb{I}(L)$ yielding the appropriate right [left] and symmetric (subspace) L-topologies.

8.3 UOQ-Uniformities for $\mathbb{R}_L$ and $\mathbb{I}_L$

8.3.1. Definition (Restricted Fuzzy Addition). Given $r \in \mathbb{R}$, we define the "crisp" number in $\mathbb{R}(L)$ to have representative $\lambda_r(t) = \begin{cases} \top, & t < r \\ \bot, & t > r \end{cases}$. Then given $[\lambda] \in \mathbb{E}_\searrow(L)$, we define $[\lambda] \oplus [\lambda_r]$ to be $[\lambda \oplus \lambda_r]$, where $(\lambda \oplus \lambda_r)(t) = \lambda(t-r)$. This is a special case of general L-fuzzy addition of all fuzzy numbers given in reference [30] of [23], except L need not be a chain; cf. 3.4.3 of [23]. This special case is needed for defining uoq-uniformities on $\mathbb{R}$ and $\mathbb{I}$.

The notation in the sequel generally blurs the distinction between a fuzzy number and its representatives. We also note that each such fuzzy number has right [left]-continuous members.

8.3.2 Construction of Right UOQ-Uniformity $\mathcal{Q}_L^r$. Let $L \in |\mathbf{CQML}|$, $\otimes$ be the tensor on L, $\varepsilon > 0$ (in $\mathbb{R}$), and $a \in L^{\mathbb{R}}$. Put

$$\lambda^a = \bigwedge \{\mu \in \mathbb{E}_{\searrow}(L) : a \le \mathfrak{R}_\mu\}$$

$$B_{\lambda_\varepsilon} : L^{\mathbb{R}} \to L^{\mathbb{R}} \text{ by } B_{\lambda_\varepsilon}(a) = \mathfrak{R}_{\lambda^a \oplus \lambda_\varepsilon}$$

$$\mathcal{Q}_L^r = \{U \in \mathcal{W}_\otimes : \exists \varepsilon > 0,\ B_{\lambda_\varepsilon} \le U\}$$

We show $\mathcal{Q}_L^r$ is an L-uoq-uniformity, with base $\mathcal{B}_L \equiv \{B_{\lambda_\varepsilon} : L^{\mathbb{R}} \to L^{\mathbb{R}} \mid \varepsilon > 0\}$, generating the right L-topology τ_L^r on $\mathbb{R}_L$. Some details are in [23].

8.3.2.1 Discussion (Reconciliation of 8.3.2 with [23]). The following sequence of definitions for $\mathcal{Q}_L^r$—requiring $L \in |\mathbf{DQML}|$, $\otimes = \wedge$—are given in [23]:

$$\lambda^a = \bigwedge \{\mu \in \mathbb{E}_{\searrow}(L) : a \le \mathfrak{L}'_\mu\}$$

$$B_{\lambda_\varepsilon}(a) = \mathfrak{R}_{\lambda^a \oplus \lambda_\varepsilon}$$

$$\mathcal{Q}_L^r = \{U \in \mathcal{W}_\otimes : \exists \varepsilon > 0,\ B_{\lambda_\varepsilon} \le U\}$$

The main syntactic discrepancy is in the definition of λ^a. The following proposition reconciles these two definitions by showing that the definition of [23] is a special case of that given in 8.3.2:

8.3.2.2 Proposition. Let $L \in |\mathbf{DQML}|$. Then $\forall\, a \in L^{\mathbb{R}}$,

$$\bigwedge \{\mu \in \mathbb{E}_{\searrow}(L) : a \le \mathfrak{L}'_\mu\} = \bigwedge \{\nu \in \mathbb{E}_{\searrow}(L) : a \le \mathfrak{R}_\nu\}$$

Proof. Let M, N be the following sets of (extended) L-fuzzy numbers:

$$M = \{\mu \in \mathbb{E}_{\searrow}(L) : a \le \mathfrak{L}'_\mu\}, \quad N = \{\nu \in \mathbb{E}_{\searrow}(L) : a \le \mathfrak{R}_\nu\}$$

and let $a \in L^{\mathbb{R}}$ and $t \in \mathbb{R}$ be given. Then the inequality $a(t) \le \nu(t+) \le v(t-)$ shows that $N \subset M$ and so $\bigwedge M \le \bigwedge N$. For the reverse inequality, suppose the inequality $a(t) \le \mu(t-)$ is given. Now suppose WLOG that μ is a left-continuous member of the class $[\mu]$ so that $a(t) \le \mu(t)$. Define a sequence $\{\mu_n : n \in \mathbb{N}\}$ of fuzzy number (representatives) by $\forall\, s \in \mathbb{R}$, $\mu_n(s) = \mu\left(s - \frac{1}{n}\right)$. Then each μ_n is a left-continuous member of its class and $\mu_n \searrow \mu$ pointwise in this sense: $\mu_1 \ge \mu_2 \ge \ldots$, $\bigwedge_{n \in \mathbb{N}} \mu_n = \mu$. Now $\forall\, n \in \mathbb{N}$, we have $a(t) \le \mu(t) \le \mu_n(t+)$.

Letting t go free, we have that each $\mu_n \in N$. It follows $\bigwedge N \leq \mu_n$ and hence that $\bigwedge N \leq \bigwedge \mu_n = \mu$. We conclude that $\bigwedge N \leq \bigwedge M$. □

8.3.3 Lemma (Properties of $\mathcal{B}_L$). Let $(L, \leq, \otimes) \in |\mathbf{CQML}|$. Then the following hold:

1. $\mathcal{B}_L$ satisfies the conditions (U1–U3,U6).
2. If $(L, \leq, \otimes) \in |\mathbf{COQUANT}_\top|$, then these statements hold:
   (a) $\mathcal{B}_L$ satisfies (W) and hence $\mathcal{B}_L \subset \mathcal{W}_\otimes$;
   (b) $\mathcal{B}_L$ satisfies (U5).
   (c) $\mathcal{B}_L$ is a base for an L-uoq-uiniformity on $\mathbb{R}_L$.

Proof. *Ad (1).* The proofs of (U1–U3,U6) are in 3.4.6–3.4.8 of [23] with this modification: in those proofs, the deMorgan polarity is assumed and used in the definition of $\mathcal{B}_L$ and $\mathcal{Q}_L^r$; but using 8.3.3 without the deMorgan polarity, the same proofs are simpler and still work.

Ad(2.1). Let $a, b \in L^\mathbb{R}$ and $\varepsilon > 0$. Our goal is to show $B_{\lambda_\varepsilon}(a \otimes b) \leq B_{\lambda_\varepsilon}(a) \otimes B_{\lambda_\varepsilon}(b)$. Consider the following sets of fuzzy numbers:

$$\begin{aligned} M &= \{\mu \in \mathbb{E}_\searrow(L) : a \leq \mathfrak{R}_\mu\}, \\ N &= \{\nu \in \mathbb{E}_\searrow(L) : b \leq \mathfrak{R}_\nu\}, \\ S &= \{\sigma \in \mathbb{E}_\searrow(L) : a \otimes b \leq \mathfrak{R}_\sigma\} \end{aligned}$$

And put $\lambda^a = \bigwedge_{\mu \in M} \mu$, $\lambda^b = \bigwedge_{\nu \in N} \nu$, $\lambda^{a \otimes b} = \bigwedge_{\sigma \in S} \sigma$. Now let $\mu \in M$, $\nu \in N$, and $t \in \mathbb{R}$ be given, and assume WLOG that λ, ν are right-continuous. Then $\otimes$ is isotone in both variables implies $a(t) \otimes b(t) \leq \lambda(t) \otimes \nu(t)$. This implies that $\mu \otimes \nu \in S$. Using the infinite distributivity of $\otimes$ over $\bigwedge$, it follows

$$\begin{aligned} \lambda^{a \otimes b}(t) &\leq \left(\bigwedge_{\mu \in M, \nu \in N} \mu \otimes \nu\right)(t) = \bigwedge_{\mu \in M, \nu \in N} (\mu(t) \otimes \nu(t)) \\ &= \left(\bigwedge_{\mu \in M} \mu(t)\right) \otimes \left(\bigwedge_{\nu \in N} \nu(t)\right) = \lambda^a(t) \otimes \lambda^b(t) \end{aligned}$$

And so $\lambda^{a \otimes b} \leq \lambda^a \otimes \lambda^b$. To finish the proof, we let $t \in \mathbb{R}$ and note (using right-continuous representatives)

$$\begin{aligned} (B_{\lambda_\varepsilon}(a \otimes b))(t) &= \mathfrak{R}_{\lambda^{a \otimes b} \oplus \lambda_\varepsilon}(t) = \lambda^{a \otimes b}(t - \varepsilon) \leq \lambda^a(t - \varepsilon) \otimes \lambda^b(t - \varepsilon) \\ &= \mathfrak{R}_{\lambda^a \oplus \lambda_\varepsilon}(t) \otimes \mathfrak{R}_{\lambda^b \oplus \lambda_\varepsilon}(t) = (B_{\lambda_\varepsilon}(a))(t) \otimes (B_{\lambda_\varepsilon}(b))(t) \end{aligned}$$

from which W for B_{λ_ε} now follows.

Ad(2.2–2.3). It is our first goal to show that $\forall \varepsilon, \delta > 0$, $B_{\lambda_{\varepsilon \wedge \delta}} \leq B_{\lambda_\varepsilon} \boxtimes B_{\lambda_\delta}$. Toward that end, let $a, b \in L^{\mathbb{R}}$. Then invoking (W) for members of $\mathcal{B}_L$ from 8.3.3(2.1), the isotonicity of $\otimes$ in both variables, and the antitonicity of fuzzy numbers, we have

$$B_{\lambda_{\varepsilon \wedge \delta}}(a \otimes b) \leq B_{\lambda_{\varepsilon \wedge \delta}}(a) \otimes B_{\lambda_{\varepsilon \wedge \delta}}(b) \leq B_{\lambda_\varepsilon}(a) \otimes B_{\lambda_\delta}(b)$$

Since $B_{\lambda_{\varepsilon \wedge \delta}} \in \mathcal{W}_\otimes$ (8.3.3(2.1) just above), it follows

$$B_{\lambda_{\varepsilon \wedge \delta}} \in \left\{W \in \mathcal{W}_\otimes : \forall a, b \in L^{\mathbb{R}},\ W(a \otimes b) \leq B_{\lambda_\varepsilon}(a) \otimes B_{\lambda_\delta}(b)\right\}$$

Therefore,

$$\begin{aligned} B_{\lambda_{\varepsilon \wedge \delta}} &\leq \bigvee \left\{W \in \mathcal{W}_\otimes : \forall a, b \in L^{\mathbb{R}},\ W(a \otimes b) \leq B_{\lambda_\varepsilon}(a) \otimes B_{\lambda_\delta}(b)\right\} \\ &= B_{\lambda_\varepsilon} \boxtimes B_{\lambda_\delta} \end{aligned}$$

On the other hand, applying 5.8(2)(b) above and assuming WLOG right-continuous representatives, we have

$$\begin{aligned} (B_{\lambda_\varepsilon} \boxtimes B_{\lambda_\delta})(a) &\leq (B_{\lambda_\varepsilon} \triangle B_{\lambda_\delta})(a) \leq B_{\lambda_\varepsilon}(a) \wedge B_{\lambda_\delta}(a) \\ &= \lambda^a(t - \varepsilon) \wedge \lambda^a(t - \delta) = \lambda^a(t - (\varepsilon \wedge \delta)) \\ &= B_{\lambda_{\varepsilon \wedge \delta}}(a) \end{aligned}$$

This completes the proof of 8.3.3(2.2). Clearly, 7.3.2 is now satisfied, which confirms 8.3.3(2.3). □

8.3.4 Corollary (Properties of $\mathcal{Q}_L^r$). Let $L \in |\mathbf{COQUANT}_\top|$. Then $\mathcal{Q}_L^r$ satisfies (U1–U3,W,U4–U6) and hence is an L-uoq-uniformity on $\mathbb{R}$ with base $\mathcal{B}_L$.

Proof. Conjoin 8.3.3 with 7.3.3. □

8.3.5 Corollary (Coincidence of $\triangle$ and $\boxtimes$). Let $L \in |\mathbf{COQUANT}_\top|$. Then $\boxtimes = \triangle$ on $\mathcal{B}^r(L)$.

Proof. For $\varepsilon, \delta > 0$ and $B_\varepsilon, B_\delta \in \mathcal{B}^r(L)$, the proof of 8.3.2 for (U5) reveals that

$$(B_\varepsilon \boxtimes B_\delta) \leq B_\varepsilon \triangle B_\delta \leq B_\varepsilon \wedge B_\delta \leq B_\varepsilon \boxtimes B_\delta \quad \square$$

8.3.6 Theorem (Right UOQ-Uniformities and Right L-Topologies). Let $L \in |\mathbf{COQUANT}_\top|$. Then

$$\tau(\mathcal{Q}_L^r) = \tau_L^r = \{\mathfrak{R}_\rho : \rho \in \mathbb{E}_\searrow(L)\}$$

Proof. We modify the proof of 3.4.10(1) of [23]. For "$\subset$", let u be in the subbasis of $\tau(\mathcal{Q}_L^r)$. It follows $u \leq \bigvee_{\exists U \in \mathcal{Q}_L^r,\, U(a) \leq u} a \leq \bigvee_{\exists B_{\lambda_\varepsilon} \in \mathcal{B}_L,\, B_{\lambda_\varepsilon}(a) \leq u} B_{\lambda_\varepsilon}(a) \leq$

u. So u is a join of $\mathfrak{R}_\rho$'s since each $B_{\lambda_\varepsilon}(a) \in \{\mathfrak{R}_\rho : \rho \in \mathbb{E}_\searrow(L)\}$.. Now for "$\supset$", let $\rho \in \mathbb{E}_\searrow(L)$ be given and consider $\mathfrak{R}_\rho$. Then it follows $\rho \oplus \lambda_{(-\varepsilon)} = \bigwedge\{\mu : \mathfrak{R}_{\rho\oplus\lambda_{(-\varepsilon)}} \le \mathfrak{R}_\rho\}$, hence $B_{\lambda_\varepsilon}\left(\mathfrak{R}_{\rho\oplus\lambda_{(-\varepsilon)}}\right) = \mathfrak{R}_\rho$ and hence

$$\begin{aligned}
\mathfrak{R}_\rho &= \bigvee\{\mathfrak{R}_{\rho\oplus\lambda_{(-\varepsilon)}} : \varepsilon > 0\} \\
&= \bigvee\{\mathfrak{R}_{\rho\oplus\lambda_{(-\varepsilon)}} : \varepsilon > 0 \text{ and } B_{\lambda_\varepsilon}\left(\mathfrak{R}_{\rho\oplus\lambda_{(-\varepsilon)}}\right) = \mathfrak{R}_\rho\} \\
&\le \bigvee\{a \in L^X : \exists\varepsilon > 0,\, B_{\lambda_\varepsilon}(a) \le \mathfrak{R}_\rho\} \\
&\le \bigvee\{B_{\lambda_\varepsilon}(a) : \forall a \in L^X,\, \exists\varepsilon > 0,\, B_{\lambda_\varepsilon}(a) \le \mathfrak{R}_\rho\} \\
&\le \mathfrak{R}_\rho
\end{aligned}$$

The equality forced at the fourth line of this display completes the proof of "$\supset$" and the theorem. □

We now outline the construction of left uoq-uniformities on $\mathbb{R}_L$ generating the left L-topologies. This subsection concludes with the construction of the bi-handed L-uoq-uniformity containing the right and left L-uoq-uniformities on $\mathbb{R}_L$. Finally, all these constructions can be appropriately modified to give analogous uoq-uniformities on $\mathbb{I}_L$.

8.3.8 Construction of Left UOQ-Uniformity $\mathcal{Q}_L^l$. Let $L \in |\mathbf{CQML}|$, $\otimes$ be the tensor on L, $\varepsilon > 0$, $a \in L^{\mathbb{R}}$, define $\rho_\varepsilon(t) = \begin{cases} \bot, & t < r \\ \top, & t > r \end{cases}$, and put

$$\rho^a = \bigwedge\{\rho \in \mathbb{E}_\nearrow(L) : a \le \mathfrak{L}_\rho^\nearrow\}$$

$$C_{\rho_\varepsilon} : L^{\mathbb{R}} \to L^{\mathbb{R}} \text{ by } C_{\rho_\varepsilon}(a) = \mathfrak{L}_{\rho^a\oplus\rho_{(-\varepsilon)}}$$

$$\mathcal{Q}_L^l = \{U \in \mathcal{W}_\otimes : \exists\varepsilon > 0,\, C_{\rho_\varepsilon} \le U\}$$

By proofs analogous to the right-hand case (which are omitted), $\mathcal{Q}_L^l$ is an L-uoq-uniformity on $\mathbb{R}_L$, having base $\mathcal{C}_L \equiv \{C_{\rho_\varepsilon} : L^{\mathbb{R}} \to L^{\mathbb{R}} \mid \varepsilon > 0\}$, which generates the left L-topology τ_L^l. The reconciliation with [23] is a little more delicate than in the right-hand case. The following sequence of definitions for $\mathcal{Q}_L^l$—requiring $L \in |\mathbf{DQML}|$, $\otimes = \wedge$—are given in [23]:

$$\lambda^a = \bigvee\{\mu \in \mathbb{E}_\searrow(L) : a \le \mathfrak{R}'_\mu\}$$

$$C_{\lambda_\varepsilon}(a) = \mathfrak{L}_{\lambda^a\oplus\lambda_{(-\varepsilon)}}$$

$$\mathcal{Q}_L^r = \{U \in \mathcal{W}_\otimes : \exists\varepsilon > 0,\, C_{\lambda_\varepsilon} \le U\}$$

As in the right-hand case above, the main syntactic discrepancy is in the definition of λ^a *vis-a-vis* ρ^a. The following proposition reconciles these two definitions by showing that the definition of [23] is a special case of that given just above.

8.3.8.1 Proposition. Let $L \in |\mathbf{DQML}|$ and $a \in L^{\mathbb{R}}$, and set $\nu^a = \bigvee \{\nu \in \mathbb{E}_{\searrow}(L) : a \leq \mathfrak{L}_{\nu}^{\searrow}\}$. Then $\lambda^a = \nu^a = \rho^a$ and $\mathfrak{L}_{\lambda^a \oplus \lambda_{(-\varepsilon)}} = \mathfrak{L}_{\nu^a \oplus \lambda_{(-\varepsilon)}} = \mathfrak{L}_{\rho^a \oplus \rho_{(-\varepsilon)}}$.

Proof. The proof of $\lambda^a = \nu^a$ is analogous to that of 8.3.7.1 above; the proof of $\nu^a = \rho^a$ rests on the deMorgan laws; and the proof of $\mathfrak{L}_{\lambda^a \oplus \lambda_{(-\varepsilon)}} = \mathfrak{L}_{\nu^a \oplus \lambda_{(-\varepsilon)}} = \mathfrak{L}_{\rho^a \oplus \rho_{(-\varepsilon)}}$ is computation. $\square$

8.3.9 Theorem (Left UOQ-Uniformities and Left L-Topologies). Let $L \in |\mathbf{COQUANT}_{\top}|$. Then

$$\tau\left(\mathcal{Q}_L^l\right) = \tau_L^l = \{\mathfrak{L}_{\rho} : \rho \in \mathbb{E}_{\nearrow}(L)\} = \tau_L^l$$

8.3.10 Summary for $\mathbb{R}_L$ and $\mathbb{I}_L$. For each $L \in |\mathbf{COQUANT}_{\top}|$, there are right and left L-uoq-uniformities $\mathcal{B}_L$ and $\mathcal{C}_L$ on $\mathbb{R}_L$ inducing via 6.1 the right and left topologies $\tau_L^r(L)$ and $\tau_L^l(L)$ on $\mathbb{R}_L$. Hence by 7.4.2, there is for each $L \in |\mathbf{COQUANT} \cap \mathbf{QUML}|$ a bi-handed L-uoq-uniformity on $\mathbb{R}_L$ which is the smallest uoq-uniformity containing $\mathcal{B}_L$ and $\mathcal{C}_L$. These three uoq-uniformities may be appropriately modified for $\mathbb{I}_L$ with analogous results. For both $\mathbb{R}_L$ and $\mathbb{I}_L$, the symmetric L-topology $\tau_L(L)$ is the smallest L-topology containing both $\tau_L^r(L)$ and $\tau_L^l(L)$.

8.4 UOQ-Uniformities for $\mathbb{R}^*(L)$ and $\mathbb{I}^*(L)$

Recall $\mathbb{R}^*(L)$ is the L-topological space $(Lpt(\Im), \tau^*(L))$, where $\Im$ is the traditional topology on $\mathbb{R}$, $Lpt(\Im) = \mathbf{CQML}(\Im, L)$, $\Phi_L : \Im \to L^{Lpt(\Im)}$ by $\Phi_L(G)(p) = p(G)$, and $\tau^*(L) = \Phi_L^{\rightarrow}(\Im)$. The right and left L-uoq-uniformities on $\mathbb{R}^*(L)$ only require $L \in |\mathbf{CQML}_{\leq,\top}|$, while the bi-handed L-uoq-uniformities on $\mathbb{R}^*(L)$ require $L \in |\mathbf{QUML}_{\leq}|$.

The L-2 soberification functor $LPT \circ \Omega$ and the relationship between 2-spatiality and L-spatiality (cf. Section 7 of [26], Subsections 2.15–2.16 of [27], reference [42] of [27]) force the tensor $\otimes$ to coincide with the binary meet $\wedge$ on $\tau^*(L)$ for *all* $(L, \leq, \otimes) \in |\mathbf{CQML}_{\perp}|$ (8.4.0). This leads to the coincidence of $\triangle$ and $\boxtimes$ on basic L-uoq-uniform operators constructed below on $L^{\mathbb{R}^*(L)}$ for $(L, \leq, \otimes) \in |\mathbf{CQML}_{\leq,\top}|$ (8.4.5), a result analogous to 8.3.5 of the previous subsection.

8.4.0 Proposition (Coincidence of $\wedge$ and $\otimes$). Let $(L, \leq, \otimes) \in |\mathbf{CQML}_{\perp}|$ and $\mathcal{W}_{\otimes}$ be considered with $X = L^{\mathbb{R}^*(L)}$. Then $\otimes = \wedge$ on $\tau^*(L)$ (where $\otimes$ and $\wedge$ are lifted to $L^{\mathbb{R}^*(L)}$ and then restricted to $\tau^*(L)$).

Proof. We first show $\Im$ is L-spatial. (Slightly reworked, this proof shows that 2-spatiality implies L-spatiality for all $(L, \leq, \otimes) \in |\mathbf{CQML}_{\perp}|$, a generalization of the references cited above which require $\otimes = \wedge$.) Let $U, V \in \Im$, $U \neq V$. Then WLOG $\exists x \in \mathbb{R}$, $x \in U - V$. Put $p : \Im \to L$ by $p(G) = \chi_G(x)$, where χ_G is the characteristic of G mapping into L. It follows $p(U) \neq p(V)$. Now clearly p preserves $\bigcup$ to $\bigvee$ and $\mathbb{R}$ to $\top$. The preservation by p of $\cap$ to $\otimes$ follows from $\bot$ being the zero of $\otimes$. Hence $p \in Lpt(\Im)$. Since $\Phi_L(U)(p) = p(U) \neq p(V) = \Phi_L(V)(p)$, Φ_L is injective, i.e. $\Im$ is L-spatial. It follows Φ_L is an order isomorphism onto its image $\tau^*(L) \equiv \Phi_L^{\to}(\Im)$. Therefore Φ_L preserves meets. But Φ_L also preserves tensors, i.e. converts the tensor $\cap$ of $\Im$ to the (lifted) tensor $\otimes$ of $L^{\mathbb{R}^*(L)}$—this is a consequence of Φ_L being defined by evaluation and the L-points of $Lpt(\Im)$ preserving tensors. Letting $A, B \in \Im$, we now have

$$\Phi_L(A) \otimes \Phi_L(B) = \Phi_L(A \cap B) = \Phi_L(A) \wedge \Phi_L(B) \quad \square$$

We now introduce "subbasic" notation for $\mathbb{R}^*(L)$ and $\mathbb{I}^*(L)$ which is identical to that for $\mathbb{R}(L)$ and $\mathbb{I}(L)$—we depend on the context to keep us straight: let $t \in \mathbb{R}^e \equiv [-\infty, \infty]$ and define $\mathfrak{L}_t, \mathfrak{R}_t : \mathbb{R}^*(L) \to L$ by

$$\mathfrak{L}_t = \Phi_L(-\infty, t), \quad \mathfrak{R}_t = \Phi_L(t, \infty)$$

8.4.1 Construction of Right UOQ-Uniformities on $\mathbb{R}^*(L)$. Let $a \in L^{\mathbb{R}^*(L)}$ and $\varepsilon > 0$, put

$$t_a = \bigvee_{a \le \mathfrak{R}_s} s, \quad B_\varepsilon : L^{\mathbb{R}^*(L)} \to L^{\mathbb{R}^*(L)} \text{ by } B_\varepsilon(a) = \mathfrak{R}_{t_a - \varepsilon}$$

$$\mathcal{B}^*(L) = \{B_\varepsilon : \varepsilon > 0\}, \quad \mathcal{Q}^{*r}(L) = \{U \in \mathcal{W}_\otimes : \exists \varepsilon > 0, B_\varepsilon \le U\}$$

We now show that $\mathcal{Q}^{*r}(L)$ is an L-uoq-uniformity, with base $\mathcal{B}^*(L)$, generating the **right L-topology** $\tau^{*r}(L) \equiv \{\mathfrak{R}_t : t \in \mathbb{R}^e\}$ on $\mathbb{R}^*(L)$.

8.4.2 Lemma (Properties of $\mathcal{B}^*(L)$). Let $(L \le, \otimes) \in |\mathbf{CQML}_{\le,\top}|$. Then $\mathcal{B}^*(L)$ satisfies (U1–U3,W,U5,U6) and is a base for an L-uoq-uniformity on $\mathbb{R}^*(L)$.

Proof. (U1) is obvious. As for (U2), let $\varepsilon > 0$, $a \in L^{\mathbb{R}^*(L)}$. Now $t_a = \bigvee_{a \le \mathfrak{R}_s} s$ implies $\exists s_\varepsilon$, $s_\varepsilon > t_a - \varepsilon$. Since $a \le \mathcal{R}_{s_\varepsilon}$, we have $a \le \Phi_L(s_\varepsilon, \infty)$. Now $(s_\varepsilon, \infty) \subset (t_{a-\varepsilon}, \infty)$ and Φ_L is isotone, so $a \le \Phi_L(s_\varepsilon, \infty) \le \Phi_L(t_{a-\varepsilon}, \infty)$, i.e. $a \le \mathcal{R}_{t_a - \varepsilon} = B_{\varepsilon(a)}$.

For (U3), let $\varepsilon > 0$, $\{a_\gamma\}_\gamma \subset L^{\mathbb{R}^*(L)}$, put $t_{\bigvee_\gamma a_\gamma} = \bigvee_{\bigvee_\gamma a_\gamma \le \mathcal{R}_s} s$, and put $\forall \gamma$, $t_{a\gamma} = \bigvee_{a_\gamma \le \mathcal{R}_{s(a_\gamma)}} s(a_\gamma)$. We wish to show $B_\varepsilon\left(\bigvee_\gamma a_\gamma\right) = \bigvee_\gamma B_\varepsilon(a_\gamma)$, i.e. we want $\mathcal{R}_{t_{\bigvee_\gamma a_\gamma} - \varepsilon} = \bigvee_\gamma \mathcal{R}_{t_{a_\gamma} - \varepsilon}$. Given that $\mathcal{R}_{t_{\bigvee_\gamma a_\gamma} - \varepsilon} = \Phi_L\left(t_{\bigvee_\gamma a_\gamma} - \varepsilon, \infty\right)$

and

$$\bigvee_{\gamma} \mathcal{R}_{t_{a_\gamma}-\varepsilon} = \bigvee_{\gamma} \Phi_L\left(t_{a_\gamma}-\varepsilon,\infty\right) = \Phi_L\left[\bigcup_{\gamma}\left(t_{a_\gamma}-\varepsilon,\infty\right)\right] = \Phi_L\left(\bigwedge_{\gamma} t_{a_\gamma}-\varepsilon,\infty\right)$$

then it suffices to prove $t_{\bigvee_\gamma a_\gamma} = \bigwedge_\gamma t_{a_\gamma}$. It is easy to show the map $T : L^{\mathbb{R}^*(L)} \to L$—defined by $T(a) = t_a \equiv \bigvee_{a\le\Re_s} s$—is antitone, it follows that $T\left(\bigvee_\gamma a_\gamma\right) \le \bigwedge_\gamma T(a_\gamma)$, i.e. $t_{\bigvee_\gamma a_\gamma} \le \bigwedge_\gamma t_{a_\gamma}$. To show that $t_{\bigvee_\gamma a_\gamma} \ge \bigwedge_\gamma t_{a_\gamma}$, for each γ put $S_\gamma = \left\{s_\gamma : a_\gamma \le \mathcal{R}_{s_\gamma}\right\}$. Invoking the complete distributivity of $\mathbb{R}^e$ as a complete chain, we have that

$$\bigwedge_{\gamma}\left[\bigvee_{s_\gamma\in S_\gamma} s_\gamma\right] = \bigvee_{f\in\prod_\gamma S_\gamma}\left[\bigwedge_{\gamma} f(\gamma)\right]$$

Now $\bigvee_\gamma a_\gamma \le \bigvee_\gamma \mathcal{R}_{f(\gamma)} = \mathcal{R}_{\bigwedge_\gamma f(\gamma)}$, where the equality follows since

$$\bigvee_{\gamma} \Phi_L\left(f(\gamma),\infty\right) = \Phi_L\left[\bigcup_{\gamma}\left(f(\gamma),\infty\right)\right] = \Phi_L\left(\bigwedge_{\gamma} f(\gamma),\infty\right)$$

Thus $\bigwedge_\gamma f(\gamma) \in \left\{s : \bigvee_\gamma a_\gamma \le \mathcal{R}_s\right\}$. It follows $t_{\bigvee_\gamma a_\gamma} \ge \bigwedge_\gamma f(\gamma)$ and hence

$$t_{\bigvee_\gamma a_\gamma} \ge \bigvee_{f\in\prod_\gamma S_\gamma}\left[\bigwedge_{\gamma} f(\gamma)\right] = \bigwedge_{\gamma}\left[\bigvee_{s_\gamma\in S_\gamma} s_\gamma\right] = \bigwedge_{\gamma} t_{a_\gamma}$$

For the (W) axiom, let $\varepsilon > 0$ and $a, b \in L^{\mathbb{R}^*(L)}$. Using the assumption that $\otimes \le \wedge$, the fact that B_ε is isotone (a consequence of (U3) proved above), and the fact that $\otimes = \wedge$ on $\Phi_L^{\to}(\Im)$ (8.4.0 above), we have:

$$B_\varepsilon(a\otimes b) \le B_\varepsilon(a\wedge b) \le B_\varepsilon(a)\wedge B_\varepsilon(b) = B_\varepsilon(a)\otimes B_\varepsilon(b)$$

where the middle inequality follows from isotonicity of B_ε and proof of 5.2.

For (U5), let $\varepsilon, \delta > 0$ and $a, b \in L^{\mathbb{R}^*(L)}$. It suffices to show $B_{\varepsilon\wedge\delta} = B_\varepsilon \boxtimes B_\delta$. Note

$$B_{\varepsilon\wedge\delta}(a\otimes b) \le B_{\varepsilon\wedge\delta}(a)\otimes B_{\varepsilon\wedge\delta}(b) \le B_\varepsilon(a)\otimes B_\delta(b)$$

where the first inequality follows from $B_{\varepsilon\wedge\delta}$ satisfying the (W) axiom from above and the second inequality can be proved from the isotonicity of Φ_L. Since $B_{\varepsilon\wedge\delta} \in \mathcal{W}_\otimes$ (a consequence of (U2,U3,W) above), then

$$B_{\varepsilon\wedge\delta} \in \left\{W \in \mathcal{W}_\otimes : \forall\, c, d \in L^{\mathbb{R}^*(L)},\ W(c\otimes d) \le B_\varepsilon(c)\otimes B_\delta(d)\right\}$$

which implies $B_{\varepsilon\wedge\delta} \le \bigvee_{W(c\otimes d)\le B_\varepsilon(c)\otimes B_\delta(d)} W = B_\varepsilon \boxtimes B_\delta$. For the reverse inequality, we invoke 5.8(2)(b) and 8.4.0 to obtain these steps:

$$\begin{aligned}
(B_\varepsilon \boxtimes B_\delta)(a) &\le (B_\varepsilon \bigtriangleup B_\delta)(a) \le B_\varepsilon(a) \wedge B_\delta(a) \\
&= \Phi_L(t_a - \varepsilon, \infty) \wedge \Phi_L(t_a - \delta, \infty) \\
&= \Phi_L[(t_a - \varepsilon, \infty) \cap (t_a - \delta, \infty)] \\
&= \Phi_L((t_a - \varepsilon) \vee (t_a - \delta), \infty) \\
&= \Phi_L(t_a - (\varepsilon \wedge \delta), \infty) \\
&= B_{\varepsilon\wedge\delta}(a)
\end{aligned}$$

For (U6), let $\varepsilon, \delta > 0$ and $a \in L^{\mathbb{R}^*(L)}$. It suffices to show $B_\varepsilon \circ B_\delta = B_{\varepsilon+\delta}$. It is helpful to first prove:

Claim. Let $\varepsilon > 0$ and $a \in L^{\mathbb{R}^*(L)}$. Then $t_{\mathfrak{R}_{t_a-\varepsilon}} = t_a - \varepsilon$.

Proof of Claim. First , we calculate:

$$t_{\mathfrak{R}_{t_a-\varepsilon}} = t_{\Phi_L(t_a-\varepsilon,\infty)} = \bigvee_{\Phi_L(t_a-\varepsilon,\infty)\le\Phi_L(s,\infty)} s$$

Now invoking that Φ_L is an order-isomorphism onto its image (see the first part of the proof of 8.4.0), we have that

$$\begin{aligned}
\{s \in \mathbb{R}^e : \Phi_L(t_a - \varepsilon, \infty) \le \Phi_L(s, \infty)\} &= \{s \in \mathbb{R}^e : (t_a - \varepsilon, \infty) \subset (s, \infty)\} \\
&= \{s \in \mathbb{R}^e : s \le t_a - \varepsilon\}
\end{aligned}$$

It follows $\bigvee_{\Phi_L(t_a-\varepsilon,\infty)\le\Phi_L(s,\infty)} s = t_a - \varepsilon$, completing the proof of the claim.

The proof of (U6) is completed by these steps:

$$B_\varepsilon(B_\delta(a)) = B_\varepsilon(\mathfrak{R}_{t_a-\delta}) = \mathfrak{R}_{t_{\mathfrak{R}_{t_a-\delta}}-\varepsilon} = \mathfrak{R}_{t_a-\delta-\varepsilon} = \mathfrak{R}_{t_a-(\varepsilon+\delta)} = B_{\varepsilon+\delta}(a) \quad \square$$

8.4.3 Corollary (Properties of $\mathcal{Q}^{*r}(L)$). Let $L \in |\mathbf{CQML}_{\le,\mathsf{T}}|$. Then $\mathcal{Q}^{*r}(L)$ satisfies (U1–U3,W,U4–U6) and hence is an L-uoq-uniformity on $\mathbb{R}^*(L)$ with base $\mathcal{B}^*(L)$.

Proof. Conjoin 8.4.3 with 7.3.3. □

8.4.4 Corollary (Coincidence of $\bigtriangleup$ and $\boxtimes$). Let $L \in |\mathbf{CQML}_{\le,\mathsf{T}}|$. Then $\boxtimes = \bigtriangleup$ on $\mathcal{B}^*(L)$.

Proof. Analogous to 8.3.5. □

8.4.5 Theorem (Right UOQ-Uniformities and Right L-Topologies). Let $L \in |\mathbf{CQML}_{\leq,\top}|$. Then

$$\tau\left(\mathcal{Q}^{*r}(L)\right) = \tau^{*r}(L) \equiv \{\mathfrak{R}_t : t \in \mathbb{R}^e\}$$

Proof. The precise details are analogous to those of the proof 8.3.6. Roughly speaking, we use 6.1 to write every open set in $\tau\left(\mathcal{Q}^{*r}(L)\right)$ as a join of right open sets which is then in $\Phi_L^{\rightarrow}(\Im)^r$; and the reverse inclusion follows using the trick $\mathcal{R}_t = \mathcal{R}_{(t+\varepsilon)-\varepsilon}$ and observing $\mathcal{R}_{t+\varepsilon} \leq \mathcal{R}_t$. □

8.4.6 Construction of Left UOQ-Uniformities on $\mathbb{R}^*(L)$. Let $a \in L^{\mathbb{R}^*(L)}$ and $\varepsilon > 0$, put

$$s_a = \bigwedge_{a \leq \mathfrak{L}_s} s, \quad C_\varepsilon : L^{\mathbb{R}^*(L)} \to L^{\mathbb{R}^*(L)} \text{ by } C_\varepsilon(a) = \mathfrak{L}_{t_a+\varepsilon}$$

$$\mathcal{C}^*(L) = \{C_\varepsilon : \varepsilon > 0\}, \quad \mathcal{Q}^{*l}(L) = \{U \in \mathcal{W}_\otimes : \exists \varepsilon > 0, C_\varepsilon \leq U\}$$

By proofs analogous to the right-hand case (which are omitted), $\mathcal{Q}^{*l}(L)$ is an L-uoq-uniformity, with base $\mathcal{C}^*(L)$, which generates the **left L-topology** $\tau^{*l}(L) \equiv \{\mathfrak{L}_t : t \in \mathbb{R}^e\}$ on $\mathbb{R}^*(L)$. We therefore state:

8.4.7 Theorem (Left UOQ-Uniformities and Left L-Topologies). Let $L \in |\mathbf{CQML}_{\leq,\top}|$. Then

$$\tau\left(\mathcal{Q}^{*l}(L)\right) = \tau^{*l}(L) \equiv \{\mathfrak{L}_t : t \in \mathbb{R}^e\}$$

8.4.8 Summary for $\mathbb{R}^*(L)$ and $\mathbb{I}^*(L)$. For each $L \in |\mathbf{CQML}_{\leq,\top}|$, there are right and left L-uoq-uniformities $\mathcal{B}^*(L)$ and $\mathcal{C}^*(L)$ on $\mathbb{R}^*(L)$ inducing via 6.1 the right and left topologies $\tau^{*r}(L)$ and $\tau^{*l}(L)$ on $\mathbb{R}^*(L)$. Hence by 7.4.2, there is for each $L \in |\mathbf{QUML}_{\leq}|$ a bi-handed L-uoq-uniformity on $\mathbb{R}^*(L)$ which is the smallest uoq-uniformity containing $\mathcal{B}^*(L)$ and $\mathcal{C}^*(L)$. These three L-uoq-uniformities may be appropriately modified for $\mathbb{I}^*(L)$ with analogous results. For both $\mathbb{R}^*(L)$ and $\mathbb{I}^*(L)$, it can be shown that the symmetric L-topology $\tau^*(L)$ is the smallest L-topology containing both $\tau^{*r}(L)$ and $\tau^{*l}(L)$.

References

[1] N. Bourbaki, Topologie Générale, Hermann Press (Paris), 1940, 1965; Springer-Verlag (Berlin/Heidelberg/New York), 1980.

[2] J. Gutiérrez García, M. A. de Prada Vicente, A. P. Šostak, *A unified approach to the concept of fuzzy L-uniform space*, Chapter 3 in this Volume.

[3] U. Höhle, *Probabilistic uniformization of fuzzy uniformities*, Fuzzy Sets and Systems **1**(1978), 311–332.

[4] U. Höhle, *Probabilistic metrization of fuzzy topologies*, Fuzzy Sets and Systems **8**(1982), 63–69.

[5] U. Höhle, *Characterization of L-topologies by L-valued neighborhoods*, Chapter 5 in [7], 389–432.

[6] U. Höhle, Many Valued Topology And Its Applications, Kluwer Academic Publishers (Dordrecht/Boston/London), 2001.

[7] U. Höhle, S. E. Rodabaugh, eds, Mathematics Of Fuzzy Sets: Logic, Topology, And Measure Theory, The Handbooks of Fuzzy Sets Series, Volume **3**(1999), Kluwer Academic Publishers (Dordrecht/Boston/London).

[8] U. Höhle, A. Šostak, *Axiomatic foundations of fixed-basis fuzzy topology*, Chapter 3 in [7], 123–272.

[9] B. Hutton, *Uniformities on fuzzy topological spaces*, J. Math. Anal. Appl. **58**(1977), 559–571.

[10] J. R. Isbell, Uniform Spaces, Math. Surveys **12**(1964), American Mathematical Society (Providence, R.I.).

[11] E. P. Klement, R. Mesiar, E. Pap, Triangular Norms, Trends in Logic, Studia Logica Library, Volume **8**(2000), Kluwer Academic Publishers (Dordrecht/Boston/London).

[12] W. Kotzé, *Uniform spaces*, Chapter 8 in [7], 553–580.

[13] T. Kubiak, *The topological modification of the L-fuzzy unit interval*, Chapter 11 in [29], 275–305.

[14] T. Kubiak, *Fuzzy reals: Brouwer Fixed Point Theorem, open questions*, Chapter 5 in this Volume.

[15] R. Lowen, *Fuzzy uniform spaces*, J. Math. Anal. Appl. **82**(1981), 370–385.

[16] E. G. Manes, Algebraic Theories, Springer-Verlag (Berlin/Heidelberg/New York), 1976.

[17] G. N. Raney, *Completely distributive complete lattices*, Proc. Amer. Math. Soc. **3**(1952), 677–680.

[18] G. N. Raney, *A subdirect-union representation for completely distributive complete lattices*, Proc. Amer. Math. Soc. **4**(1953), 518–522.

[19] S. E. Rodabaugh, *U.s.c. decompositions of convex metric spaces*, Fundamenta Mathematicae **105**(1980), 213–227.

[20] S. E. Rodabaugh, *Locally starlike decompositions of separable metric spaces*, Fundamenta Mathematicae **112**(1981), 61–68.

[21] S. E. Rodabaugh, *Separation axioms and the L-fuzzy real lines*, Fuzzy Sets and Systems **11**(1983), 163–183.

[22] S. E. Rodabaugh, *A theory of fuzzy uniformities with applications to the fuzzy real lines*, J. Math. Anal. Appl. **129**(1988), 37–70.

[23] S. E. Rodabaugh, *Dynamic topologies and their applications to crisp topologies, fuzzification of crisp topologies, and fuzzy topologies on the crisp real line*, J. Math. Anal. Appl. **131**(1988), 25–66.

[24] S. E. Rodabaugh, *Point-set lattice-theoretic topology*, Fuzzy Sets and Systems **40**(1991), 297–345.

[25] S. E. Rodabaugh, *Powerset operator foundations for poslat fuzzy set theories and topologies*, Chapter 2 in [7], 91–116.

[26] S. E. Rodabaugh, *Categorical foundations of variable-basis fuzzy topology*, Chapter 4 in [7], 273–388.

[27] S. E. Rodabaugh, *Separation axioms: representation theorems, compactness, and compactifications*, Chapter 7 in [7], 481–552.

[28] S. E. Rodabaugh, *Fuzzy real lines and dual real lines as poslat topological, uniform, and metric ordered semirings with unity*, Chapter 10 in [7], 607–632.

[29] S. E. Rodabaugh, E. P. Klement, U. Höhle, eds, Application Of Category To Fuzzy Sets, Theory and Decision Library—Series B: Mathematical and Statistical Methods, Volume **14**, Kluwer Academic Publishers (Boston/Dordrecht/London), 1992.

[30] J. W. Tukey, Convergence And Uniformity In Topology, Ann. Math. Studies, No. **1**, Princeton University Press (1940).

[31] P. Viceník, *Additive generators of non-continuous triangular norms*, Chapter 18 in this Volume.

[32] C. Walker, E. Walker, *Groups, t-norms, and families of De Morgan systems*, Chapter 19 in this Volume.

[33] A. Weil, Sur les Espaces à Structure Uniforme et sur la Topologie Générale, Act. Sci. et Ind. **551**, Hermann Press (Paris), 1937.

[34] D. Zhang, *Stratified Hutton uniform spaces*, Fuzzy Sets and Systems (to appear).

CHAPTER 8

Fully Fuzzy Topology

L. N. STOUT

Introduction

Fuzzy topology has used a definition which has an L-topological space consisting of a crisp set of fuzzy subsets of a crisp set. The notion of a topological space object in SET(L) provides a generalization of this concept to that of a fuzzy set of fuzzy subsets of a fuzzy set. This chapter extends the introduction given in [2] and provides connections with more mainstream fuzzy topology.

1 Categorical setting

In several previous papers ([8, 2, 9]) I have discussed the logic of unbalanced subobjects of fuzzy sets, showing how a form of internal second order logic can be developed which can be used, among other things, to do topology internally. My object in this chapter is to take the simple case of fuzzy sets with values in the unit interval equipped with a left continuous t-norm $*$ (as in [4]) and show concretely what these constructions give. Specifically, let $*$ be a semigroup operation on [0,1] such that

$$\begin{aligned} a * b &= b * a \\ a * 1 &= a \\ a * 0 &= 0 \\ a * \sup_{y \in Y} y &= \sup_{y \in Y}(a * y) \end{aligned}$$

S.E. Rodabaugh and E.P. Klement (eds.), Topological and Algebraic Structures in Fuzzy Sets, 235–253.
© 2003 *Kluwer Academic Publishers.*

Hereafter we will refer to this structure (the lattice [0,1] equipped with a t-norm) as L. No harm will be done if the reader thinks of the examples $a * b = \max(a + b - 1, 0)$ or $a * b = ab$. Any such structure will have two implications:

$$a \Rightarrow b = \bigvee\{x \mid \min(a, x) \leq b\}$$

(the right adjoint to the minimum operation) and

$$a \rightarrow b = \bigvee\{x \mid a * x \leq b\}$$

(the right adjoint to the $*$ operation).

Example:

Let's look at $\mathbf{L} = \{0, \frac{1}{4}, \frac{1}{2}, \frac{3}{4}, 1\}$ with $a * b = \max(a + b - 1, 0)$. Here

$$a \Rightarrow b = \begin{cases} 1 & \text{if } a \leq b \\ b & \text{otherwise} \end{cases}$$

and

$$a \rightarrow b = \min(1, 1 - a + b)$$

We get two negations by

$$\neg a = a \Rightarrow 0 = \begin{cases} 1 & \text{if } a = 0 \\ 0 & \text{otherwise} \end{cases}$$

and

$$\sim a = a \rightarrow 0 = 1 - a$$

Notice that $\neg\neg h = 1$ if $h > 0$ and $\neg\neg 0 = 0$, so double negation using this negation can be used to pick out supports of fuzzy subsets. The Łukasiewicz negation has $\sim\sim h = h$. We will follow this example throughout this chapter. It is a system rather similar to the Likert scale used in opinion research (strongly disagree = 0, disagree = $\frac{1}{4}$, neutral = $\frac{1}{2}$, agree = $\frac{3}{4}$, strongly agree = 1), so we will call it the Likert lattice. ♢

Definition 1.1 *The category* SET(L) *has*

- *objects: pairs (A, α) where A is a set and $\alpha : A \rightarrow \mathrm{L}$ is any map*
- *morphisms from (A, α) to (B, β): functions $f : A \rightarrow B$ such that $\alpha(a) \leq \beta(f(a))$.*

This category has lots of structure (see [9, 2] for details) which makes it possible to interpret second order statements in a consistent, intrinsic way.

Theorem 1.2 SET(L) *has the following structures:*

Products given by	$(A,\alpha)\times(B,\beta)=(A\times B,(a,b)\mapsto \min(\alpha(a),\beta(b)))$
Tensor products given by	$(A,\alpha)\otimes(B,\beta)=(A\times B,(a,b)\mapsto \alpha(a)*\beta(b))$
Exponentials given by	$(B,\beta)^{(A,\alpha)}=(\{f:A\to B\},\xi)$ $\xi: f\mapsto \bigwedge_{a\in A}(\alpha(a)\Rightarrow\beta(f(a)))$
Hom given by	$(B,\beta)\uparrow(A,\alpha)=(\{f:A\to B\},\zeta)$ $\zeta: f\mapsto \bigwedge_{a\in A}(\alpha(a)\to\beta(f(a)))$
unbalanced monos of the form	$(A,\alpha')\ggg\!\!\longrightarrow(A,\alpha)$
where	$\alpha'(a)\leq\alpha(a)$ *for all* $a\in A$
weakly represented by	$\begin{array}{rccc}\chi_{\alpha'}: & (A,\alpha) & \to & (\mathrm{L},1)\\ & a & \mapsto & \alpha'(a)\end{array}$
lattice of unbalanced subobjects	$\mathcal{U}(A,\alpha)$
with a second operation	$*$
unbalanced powerobject	$P(A,\alpha)=(\mathrm{L},1)^{(A,\alpha)}=(\mathrm{L},1)\uparrow(A,\alpha)$
or equivalently	$(\{f:A\to\mathrm{L}\mid f(a)\leq\alpha(a)\},1)$
membership subobject given by	$\epsilon_{(A,\alpha)}\ggg\!\!\longrightarrow P(A,\alpha)\otimes(A,\alpha)$ $\epsilon_{(A,\alpha)}=(\{(f,a)\},(f,a)\mapsto f(a))$
quantifiers given as functors	$\exists_f:\mathcal{U}(A,\alpha)\to\mathcal{U}(B,\beta)$ $\forall_f:\mathcal{U}(A,\alpha)\to\mathcal{U}(B,\beta)$
left and right adjoint to	$f^{-1}:\mathcal{U}(B,\beta)\to\mathcal{U}(A,\alpha)$
where	$f:(A,\alpha)\to(B,\beta)$
union operator	$\bigoplus: P(P(A,\alpha))\to P(A,\alpha)$
representing	$\exists_{\pi_2}((\epsilon_{P(A,\alpha)}\otimes(A,\alpha))*(P(A,\alpha)\otimes\epsilon_{(A,\alpha)}))$

Most of this is fairly straightforward. For clarity in what follows we include more concrete descriptions: the theorem gives prescriptions for how to calculate products, tensor products, and both exponentials. The unbalanced powerobject $P(A,\alpha)$ is the crisp set of all unbalanced subobjects of (A,α). The functors f^{-1}, $\exists_f$, and $\forall_f$ have the following effects:

$$f^{-1}(B,\beta')=(A,\beta'\circ f)$$

$$\exists_f(A,\alpha')=(B,b\mapsto \bigvee_{\{a\mid f(a)=b\}}\alpha'(a))$$

$$\forall_f(A,\alpha')=(B,b\mapsto \bigwedge_{\{a\mid f(a)=b\}}\alpha'(a))$$

Note that these also give functions between $P(A,\alpha)$ and $P(B,\beta)$, indicated later by $\underline{f^{-1}}$, $\underline{\exists_f}$, and $\underline{\forall_f}$.

The union operator is a bit more complicated. The way we have defined $P(A,\alpha)$ always gives a crisp set, so $P(P(A,\alpha))$ is the (crisp) set of fuzzy subsets of the (crisp) set of unbalanced subobjects of (A,α). To say what the union operator gives, it will suffice to say what unbalanced subobject it assigns to a given fuzzy subset of $P(A,\alpha)$. The description given in the theorem is embodied in the formula

$$\bigoplus(\nu)(a) = \bigvee_{\alpha'\leq\alpha} \nu(\alpha') * \alpha'(a)$$

If ν is the characteristic function of a crisp subset, this gives us the supremum of its members. If ν is a fuzzy subset, we use $*$ to weight the contribution of a member of ν according to its degree of membership.

The existence of all of this structure makes it possible to express most mathematical notions internally: references to powersets are replaced by the internal weak power object; lattices of subsets are replaced with lattices of unbalanced (fuzzy) subobjects. Things become a little more complicated because of the existence of two (potentially) different products—hence an additional logical connective $*$.

2 Topological space objects

2.1 Definition and examples

There are several different ways to approach the definition of a topological space: on can characterize the properties of open subsets, closed subsets, interior operators, closure operators, or neighborhood systems. We know from the example of topos topology ([7, 6]) that description using closed sets and closure operators can give different results than description using open sets and interior operators, at least when using intuitionistic connectives. In that work (and in subsequent work since) I chose to use the description in terms of open subsets as being closest to classical topology. Hence the definition:

Definition 2.1 *A **topological space object** in* SET(L) *is an object* (A,α) *equipped with an unbalanced subobject* $T = (PA,\tau) \overset{L}{\rightarrowtail} P(A,\alpha)$ *such that*

1. *The unbalanced subobjects* $(A,0)$ *and* (A,α) *both have* τ *with value* $\top$. *This can be expressed by asking that the names of these subobjects factor through* T:

$$\begin{array}{ccc} 1 & \longrightarrow & T \\ \downarrow & & \downarrow \\ 1 & \overset{\ulcorner(A,0)\urcorner}{\longrightarrow} & P(A,\alpha) \end{array}$$

and

$$\begin{array}{ccc} 1 & \longrightarrow & T \\ \downarrow & & \downarrow \\ 1 & \stackrel{\ulcorner(A,\alpha)\urcorner}{\longrightarrow} & P(A,\alpha) \end{array}$$

2. *T is closed under pairwise $*$: this says there is a lifting*

$$\begin{array}{ccc} T \times T & \stackrel{*}{\longrightarrow} & T \\ \downarrow & & \downarrow \\ P(A,\alpha) \times P(A,\alpha)) & \stackrel{*}{\longrightarrow} & P(A,\alpha) \end{array}$$

This tells us that

$$\tau((A,\alpha')) * \tau((A,\alpha'')) \leq \tau((A,\alpha' * \alpha''))$$

3. *T is closed under arbitrary internal unions: this says there is a lifting*

$$\begin{array}{ccc} PT & \stackrel{\bigoplus}{\longrightarrow} & T \\ \exists_\iota \downarrow & & \downarrow \\ P(P(A,\alpha)) & \stackrel{\bigoplus}{\longrightarrow} & P(A,\alpha) \end{array}$$

Definition 2.2 *A morphism $f : (A,\alpha) \to (B,\beta)$ is continuous with respect to the topologies T_A and T_B if there is a lifting*

$$\begin{array}{ccc} T_B & \longrightarrow & T_A \\ \downarrow & & \downarrow \\ PB & \stackrel{f^{-1}}{\longrightarrow} & PA \end{array}$$

The category FFTOP(L) *is the category of fully fuzzy topological space objects in* SET(L) *with continuous morphisms.*

These axioms have some unusual consequences, so let us look closely at some examples, using the "Likert" lattice.

Example:

An indiscrete space: Suppose that (A,α) is a two element set $\{a,b\}$ with $\alpha(a) = 1$ and $\alpha(b) = \frac{1}{2}$. The indiscrete topology on (A,α) should be the smallest topology possible. Let us describe the values of τ which are forced on us. First we get

$$\tau(A,\alpha) = 1 \text{ and } \tau(A,0) = 1$$

Closure under $*$ tells us that

$$\tau\begin{pmatrix} a & b \\ 1 & 0 \end{pmatrix} \geq 1$$

since this subobject is $(A, \alpha) * (A, \alpha)$. Further application of the rule on closure under $*$ gives no further information.

Applying the union axiom is more tricky. Suppose that (PA, ν) is an unbalanced subobject of (PA, τ). Applying $\bigoplus$ gives the unbalanced subobject

$$\bigoplus(\nu)(x) = \bigvee_{t \in PA} (\nu(t) * t(x))$$

Consequences of this depend on how we choose ν. If ν is the crisp characteristic function of a subset S of the full members of T (those t with $\tau(t) = 1$), then closure under $\bigoplus$ requires that the supremum of the fuzzy sets in S be fully open.

Now suppose that we know that $\tau(\alpha') = 1$; then given $h \in$ L we can define a $\nu_{\alpha',h}$ by taking $\nu_{\alpha',h}(\alpha') = h$ and $\nu_{\alpha',h}(\alpha'') = 0$ for all other α''. Taking the union of this gives

$$\begin{aligned} \bigoplus(\nu_{\alpha',h})(x) &= \bigvee_{t \in PA} (\nu_{\alpha',h}(t) * t(x)) \\ &= h * \alpha'(x) \end{aligned}$$

Thus we must also have τ take all of

$$\begin{pmatrix} a & b \\ \frac{3}{4} & \frac{1}{4} \end{pmatrix}, \begin{pmatrix} a & b \\ \frac{1}{2} & 0 \end{pmatrix}, \begin{pmatrix} a & b \\ \frac{1}{4} & 0 \end{pmatrix}$$

to 1.

If we had any unbalanced subobjects with $0 < \tau(\alpha') < 1$, the construction in the previous paragraph would require that, given $h \leq \tau(\alpha')$, the subobject $\bigoplus \nu_{\alpha',h}(x) = \alpha'(x) * h$ has membership in T at least as great as that of α'. For the indiscrete topology we need not have any subobjects with partial membership, so this does not give anything new in this example.

We can summarize this example in a table giving the values assigned to all of the unbalanced subobjects of (A, α). Such subobjects are given by giving a membership value for a in the range $0 \leq \alpha'(a) \leq 1$ and a membership value for b with $0 \leq \alpha'(b) \leq .5$. This gives $5 \times 3 = 15$ subobjects we must give values of τ for; these can be easily represented in a table listing the membership degrees for a in increasing order downward and those for b in increasing order from left to right. The truncation condition tells us that as you move along the diagonal up and to the left, the values must not decrease. So the conditions on known full members and their truncations give the 1's in the following table.

	$\alpha'(b) =$	0	.25	.5
	0	1		
	.25	1		
$\alpha'(a) =$	.5	1		
	.75		1	
	1			1

Now a consequence of closure under $*$ is shrinkage of fuzzy members of opens:

$$\tau(A, \alpha' * \alpha') \geq \tau(A, \alpha') * \tau(A, \alpha')$$

So if

$$\tau\left(\begin{pmatrix} a & b \\ 1 & .75 \end{pmatrix}\right) = 1$$

then we also get

$$\tau\left(\begin{pmatrix} a & b \\ 1 & .75 \end{pmatrix} \otimes \begin{pmatrix} a & b \\ 1 & .75 \end{pmatrix}\right) = \tau\left(\begin{pmatrix} a & b \\ 1 & .5 \end{pmatrix}\right) = 1$$

and

$$\tau\left(\begin{pmatrix} a & b \\ 1 & .25 \end{pmatrix}\right) = 1$$

and

$$\tau\left(\begin{pmatrix} a & b \\ 1 & 0 \end{pmatrix}\right) = 1$$

Using this and making the remaining entries 0 gives the indiscrete fully fuzzy topology.

	$\alpha'(b) =$	0	.25	.5
	0	1	0	0
	.25	1	0	0
$\alpha'(a) =$	.5	1	0	0
	.75	1	1	0
	1	1	1	1

$\diamond$

For this particular $*$ fully fuzzy topologies on a fuzzy set will always have an open fuzzy subset with value 1 at the elements with crisp membership and 0 elsewhere. This is because this $*$ has a finite power of each non-1 value giving 0 (it is nilpotent).

Example:

A topology with fuzzy membership:

$\alpha'(a) =$ \ $\alpha'(b) =$	0	.25	.5
0	1	.25	0
.25	1	.5	.25
.5	1	.5	.5
.75	1	1	.5
1	1	1	1

Here we have taken the indiscrete topology and increased the level of membership in the upper right subject to constraints imposed as follows:

- truncation by h of a fuzzy subset with degree of openness t must be open to a tleast degree t: this made the choices in the interior of the tabe constrained to be greater than or equal to the entry on the right edge diagonally down from the interior entry.
- Unions of fuzzy subsets must be open to the minimal degree of the openness of the terms in the union: this limited the choices in the last column to a nondecreasing sequence of values. $\diamondsuit$

Example:

Given three different topologies $T_1 \subseteq T_2 \subseteq T_3$ on a set A we can produce a fully fuzzy topological space structure such that

$$\{ \text{ crisp subsets } A' \subseteq A \mid \tau(\chi_{A'}) \geq .75\} = T_1$$
$$\{ \text{ crisp subsets } A' \subseteq A \mid \tau(\chi_{A'}) \geq .5\} = T_2$$
$$\{ \text{ crisp subsets } A' \subseteq A \mid \tau(\chi_{A'}) \geq .25\} = T_3$$

We start by specifying the value of $\tau(\chi_{A''})$ for all possible crisp subsets. It is clear that χ_A and $\chi_\emptyset$ must give 1. The remaining elements of T_1 then get assigned .75, then the remaining elements of T_2 get assigned .5, then the remaining elements of T_3 get assigned .25. Any remaining crisp subsets of A are sent to 0.

This tells us how to assign values to the crisp subsets, so all that remains is the values for fuzzy subsets. Next we handle the truncations: a fuzzy set $\chi_{A'} * h$ is given the same truth value as $\chi_{A'}$. Now any fuzzy subset of A can be represented as the supremum of fuzzy sets of the form $\chi_{A'} * h$ (perhaps in many different ways). We assign a fuzzy set the largest degree of membership which occurs as the minimum of $\tau(\chi_{A'})$ for such a representation.

The resulting fuzzy subset of $P(A,1)$ will be closed under $*$ and internal union.

For example, on the set $\{a,b\}$ let T_1 be the indiscrete topology, T_2 be the Sierpinski space with $\{a\}$ open, and T_3 be the discrete topology. The table for the fully fuzzy topology giving this stack of topologies is

	0	.25	.5	.75	1
0	1	.25	.25	.25	.25
.25	.5	1	.25	.25	.25
.5	.5	.5	1	.25	.25
.75	.5	.5	.5	1	.25
1	.5	.5	.5	.5	1

◇

2.2 Relations with topological spaces and fuzzy topology

The last example of the previous section illustrates one of the ways that ordinary topological spaces show up in fully fuzzy topological spaces. The approach used is different from the α-cuts approach of fuzzy topology in that instead of asking for fuzzy sets which are (crisp) members of the topology and considering the sets of elements with membership at least α we look at crisp subsets with membership at least α in the fully fuzzy topology.

Theorem 2.3 *If F gives an order reversing function from* L *to the lattice of topologies on A such that if $h = \bigwedge\{t \mid A' \in \mathrm{L}(t)\}$ then $A' \in F(h)$ and $F(\bot)$ is the indiscrete topology, then there is a fully fuzzy topology $P((A,1),\tau)$ such that $T_h = \{A' \subseteq A \mid \tau(\chi_{A'}) \geq h\} = F(h)$.*

Proof. Let $\tau(\chi_{A'}) = \bigwedge\{h \mid A' \in F(h)\}$. This has $T_h = F(h)$. It remains to define τ for fuzzy subsets and then show that the result is a fully fuzzy topology. Now suppose that (A,α) is a fuzzy subset A. We can think of (A,α) as the internal union of the subobject of PA which is non-zero only on subobjects of the form

$$\alpha'_h(a) = \begin{cases} h & \text{if } \alpha(a) \geq h \\ \bot & \text{otherwise} \end{cases}$$

Such a fuzzy set is the h truncation of a crisp subset of A, the h cut of (A,α), and thus can be assigned the same degree of membership in our topology as that crisp subset. Applying the union axiom then tells us the smallest degree of membership that (A,α') must be assigned. What results will be a fully fuzzy topology. □

Theorem 2.4 *If $* = \min$, then each T_h is a topology.*

Proof. It is clear that A and $\emptyset$ will be in each T_h, since both must get membership $\top$ in (PA, τ). If A_1 and A_2 are both in T_h, then $\chi_{A_1 \cap A_2} = \chi_{A_1} * \chi_{A_2}$, so it must have membership at least as large as those of A_1 and A_2. If $F = \{A_i \mid i \in I\}$ is a family of subobjects all of which are in T_h, then if we define

$$\nu(t) = \begin{cases} h & \text{if } t \in F \\ 0 & \text{otherwise} \end{cases}$$

then

$$\bigoplus(\nu)(a) = \bigvee_{t \in PA} \nu(t) * t(a) = \begin{cases} h & \text{if } \exists_t a \in t \in F \\ 0 & \text{otherwise} \end{cases}$$

Thus the union of F is in T_h. □

Example:

> For L $= \{0, .5, 1\}$ with $*$ given by the Łukasiewicz t-norm there is a fully fuzzy topology on $\{a, b, c\}$ for which $T_{.5}$ is not closed under intersection. We assign a value of 1 to the fuzzy sets $\begin{pmatrix} a & b & c \\ 1 & 1 & 1 \end{pmatrix}$, $\begin{pmatrix} a & b & c \\ .5 & .5 & .5 \end{pmatrix}$, and $\begin{pmatrix} a & b & c \\ 0 & 0 & 0 \end{pmatrix}$; a value of .5 to fuzzy subsets with two elements having membership 1 and the remaining element having either 0 or .5; and a value of 0 to the rest. This is closed under $\bigoplus$ and $*$. It is in fact a fully fuzzy topology, but $T_{.5}$ is not closed under intersection. ◇

The next theorem shows part of the connection with interval valued fuzzy topological spaces.

Theorem 2.5 *A fully fuzzy topological space structure on a crisp set* $(A, \top)$ *with only crisp members is a stratified* $[0, 1]$*-topology in the sense of Lowen [3] if* $* = \min$.

Proof. The requirements that $(A, 1)$ and $(A, 0)$ be open are common to both definitions. The requirement of closure under $*(= \min)$ gives closure under pairwise minima. Any family of opens can be represented as a crisp subobject of PT, the union of which gives the supremum. Truncations of the open $(A, 1)$ give the constants condition. □

If you do not use min for $*$, then an $[0, 1]$-topological space need not give a subobject closed under $*$. If $[0, 1]$-topological space is not fully stratified (the Lowen constants condition), then the resulting subobject of $P(A)$ will not be closed under internal unions.

2.3 Initial structures

It is clear that there is a forgetful functor from the category of fully fuzzy topological spaces and continuous maps to the category SET(L). To see that this functor behaves like the forgetful functor from TOP to SETS, recall the definition of a topological functor (from [1], p.333) that a concrete category $U : \mathcal{T} \to \text{SET(L)}$ is called topological if every U-structured source $A \to U(A_i)$ has a unique U-initial lift.

Theorem 2.6 *The category of fully fuzzy topological spaces is topological over* SET(L)*. That is, given any U-structured source $f_i : (A, \alpha) \to U((B_i, \beta_i), \tau_{B_i})$ there is a unique smallest structure of a fully fuzzy topology on (A, α) making all of the f_i continuous.*

Proof. What we need is a construction analogous to that of a topology from a subbase. Given a subobject of PA we wish to find the smallest fully fuzzy topology containing it. Basically what must be done is to close under $*$ and then close under $\bigoplus$.

We start by obtaining the fuzzy subobject of PA which contains all of the fuzzy subsets which must be in the topology in order for all of the maps to be continuous:

$$\sigma(\alpha') = \bigvee_{\{\beta'_i \mid \alpha' = f_i^{-1}\beta'_i\}} \tau_i(\beta'_i)$$

In general this will not be a fully fuzzy topology. To make it one we first make (A, α) and $(A, 0)$ have membership $\top$. Next increase $\sigma(\alpha')$ to

$$\sigma_1(\alpha') = \bigvee_{\{\alpha_1 \ldots \alpha_n \mid \alpha_1 * \ldots * \alpha_n = \alpha'\}} (\bigwedge_{k=1}^{n} \sigma(\alpha_k))$$

This gives a fuzzy subset closed under pairwise $*$.

Now define the fully fuzzy topology to be the image under $\bigoplus$ of $P(\sigma_1)$. Because $*$ distributes over $\bigoplus$ (which follows from the associativity and commutativity of $*$ and the fact that it distributes over $\bigvee$) the resulting object will still be closed under $*$. □

Having a topological functor is equivalent to every U-structured sink having a unique U-final lift. Limits and colimits can be found by taking the limit or colimit in SET(L) and giving it the appropriate final or initial structure. The forgetful functor preserves and reflects epi-sinks and mono-sources, hence both epis and monos.

Topological functors have fibers which are complete lattices, hence have both largest and smallest structures available. We will call the largest structure (that is, the final structure for the sink consisting of all maps with underlying set map the identity) the indiscrete structure. The smallest is called the discrete structure. Notice that the inclusions here are not subobject inclusions, but rather the ordering which results from the identity being continuous.

Product topologies are obtained from the initial structure as in ordinary topology. Quotients arise from the final topology.

3 Interior operators

Given a topology in terms of open sets it is fairly easy to see how to get an interior operator: just take the union of the family of all open subsets of an unbalanced subobject of (A, α) as the interior of that subobject. We need some definitions and some basic properties of the downsegment map to make interior work.

3.1 Extentional order, down segments, and singletons

Definition 3.1 *The* ***extensional order*** *on $P(A, \alpha)$ is given by the unbalanced subobject*

$$(\{(\alpha', \alpha'')\}, (\alpha', \alpha'') \mapsto \bigwedge_{a \in A} (\alpha'(a) \to \alpha''(a))) \rightarrowtail P(A, \alpha) \times P(A, \alpha)$$

Example:

The extensional order on $\begin{pmatrix} a & b \\ 1 & .5 \end{pmatrix}$ takes a pair $((A, \alpha'), (A, \alpha''))$ to 1 if both $\alpha'(a) \leq \alpha''(a)$ and $\alpha'(b) \leq \alpha''(b)$. The interseting cases are where this is not true and fuzzy values are given. Examples are:

$$\begin{aligned} \left(\begin{pmatrix} a & b \\ .5 & .25 \end{pmatrix}, \begin{pmatrix} a & b \\ .25 & .5 \end{pmatrix}\right) &\mapsto \min(.5 \to .25, .25 \to .5) \\ &= .75 \\ \left(\begin{pmatrix} a & b \\ .5 & .5 \end{pmatrix}, \begin{pmatrix} a & b \\ 1 & 0 \end{pmatrix}\right) &\mapsto \min(.5 \to 1, .5 \to 0) \\ &= .5 \\ \left(\begin{pmatrix} a & b \\ 1 & .25 \end{pmatrix}, \begin{pmatrix} a & b \\ 0 & 0 \end{pmatrix}\right) &\mapsto \min(1 \to 0, .25 \to 0) \\ &= 0 \end{aligned}$$

Definition 3.2 *The smallest characteristic map of the extensional order has exponential adjoint*

$$\downarrow seg : P(A,\alpha) \to P^2(A,\alpha)$$

If we restrict the first factor to a fully fuzzy topology τ *we get the open down segment map*

$$\downarrow seg^\circ : P(A,\alpha) \to P(\tau)$$

Definition 3.3 *The extensional equality on* $P(A,\alpha)$ *is given by*

$$(\{(\alpha',\alpha'')\},(\alpha',\alpha'') \mapsto \bigwedge_{a\in A}(\alpha'(a) \leftrightarrow \alpha''(a))) \rightarrowtail P(A,\alpha) \times P(A,\alpha)$$

Example:

On $(A,\alpha) = \begin{pmatrix} a & b \\ 1 & .5 \end{pmatrix}$ this gives the degree of equality of $\begin{pmatrix} a & b \\ .5 & .25 \end{pmatrix}$ and $\begin{pmatrix} a & b \\ .75 & .5 \end{pmatrix}$ to be

$$\min(.5 \leftrightarrow .75, .25 \leftrightarrow .5) = .75. \quad \diamondsuit$$

Note that the crisp elements of $\downarrow$ seg(α') are the unbalanced subobjects of (A,α'); other subobjects α'' of (A,α) are assigned membership to the extent that the pair (α'',α') is in the extensional order relation. This membership uses the Łukasiewcz implication.

Many definitions in ordinary topology are given in terms of points of the space. In fuzzy topology one often replaces these with definitions given in terms of fuzzy points. In topos topology it is more common to work with singletons. These three notions can be made distinct and related in the fully fuzzy category in the following fashion:

Definition 3.4 *A fuzzy point of a set A is a function* $p_a^h : A \to \mathrm{L}$ *such that* $p_a^h(a) = h$ *and* $p_a^h(a') = \bot$ *whenever* $a' \neq a$. *A fuzzy point* p_a^h *belongs to a fuzzy set* (A,α) *precisely when* $\alpha(a) \geq h$.

Note that a fuzzy point p_a^h belongs to a fuzzy set (A,α) precisely if, when the fuzzy point is thought of as a fuzzy subset of A, it is an unbalanced subobject of (A,α). The fuzzy points of a fuzzy set completely determine that fuzzy set.

Proposition 3.5 *The fuzzy set* (A,α) *is the direct limit of the system of fuzzy points belonging to it thought of as unbalanced subobjects.*

Proof. For each $a \in A$ there is a largest h such that p_a^h is in (A, α), namely $\alpha(a)$. The direct limit of the points is the smallest fuzzy set structure on A such that all of the points map into it, hence it must have degree of membership at least as big as α. Since each point does in fact map into (A, α) this is sufficient to give the proposition. □

Definition 3.6 *A partial section σ of a fuzzy set (A, α) with support h is a morphism of* SET(L) *from $(\{*\}, h)$ to (A, α). We say that the value of the section is a if $\sigma(*) = a$.*

Proposition 3.7 *There is a one to one correspondence given by the image factorization between the fuzzy points belonging to a fuzzy set (A, α) and the partial sections of that fuzzy set.*

If we combine this with the result that each fuzzy set is the direct limit of its points, we can see that the category SET(L) is generated by the unbalanced subobjects of the terminal.

Two more notions related to singletons are also relevant.

Definition 3.8 *The singleton map $\{\}_A : A \to PA$ is obtained as the exponential adjoint of the least characteristic map of the unbalanced subobject*

$$\left(A \times A, (a, b) \mapsto \begin{cases} \alpha(a) & \text{if } a = b \\ \bot & \text{otherwise} \end{cases} \right) \rightarrowtail (A, \alpha) \times (A, \alpha)$$

The subobject characterized here is the unbalanced image of the diagonal map. Its unbalanced image of the singleton map is the crisp subset of all fuzzy points of maximal support.

Definition 3.9 *The extensional singleton map $\{\}_{ext} : PA \to P^2(A)$ is obtained as the exponential adjoint of the smallest characteristic map of the extensional equality.*

This gives a fuzzy singleton instead of a crisp one.

Either singleton map gives $\{\alpha'\} \leq \downarrow \mathrm{seg}(\alpha')$ and either singleton has $\bigoplus(\{\alpha'\}) = \alpha'$. Combining these results with the fact that $\bigoplus$ preserves order we get:

Proposition 3.10 *Both of the following compositions give the identity:*

$$\tau \longrightarrow P(A, \alpha) \stackrel{\downarrow seg^\circ}{\longrightarrow} P(\tau) \stackrel{\bigoplus}{\longrightarrow} \tau$$

and

$$P(A, \alpha) \stackrel{\downarrow seg}{\longrightarrow} P^2(A, \alpha) \stackrel{\bigoplus}{\longrightarrow} P(A, \alpha)$$

3.2 Definition and properties of the interior operator

We are now in a position to define the interior operator:

Definition 3.11 *The* ***interior operator*** $()^{\circ}_{\tau}$ *is the map*

$$P(A,\alpha) \xrightarrow{\downarrow seg^{\circ}} P(\tau) \xrightarrow{\bigoplus} \tau$$

This assigns each unbalanced subobject of (A, α) a new unbalanced subobject which is the union of the open unbalanced subobjects contained in it.

Theorem 3.12 *The interior operator* $()^{\circ}_{\tau}$ *satisfies the following:*

1. *It fixes* τ *in the sense that the following diagram commutes:*

$$\begin{array}{ccc} \tau & \xrightarrow{\mathrm{id}_{\tau}} & \tau \\ \downarrow & & \downarrow \\ P(A,\alpha) & \xrightarrow{()^{\circ}_{\tau}} & \tau \end{array}$$

2. *it is idempotent:* $((A,\alpha')^{\circ}_{\tau})^{\circ}_{\tau} = (A,\alpha')^{\circ}_{\tau}$ *for all* α'
3. *it preserves order: if* $(A,\alpha') \leq (A,\alpha'')$, *then* $(A,\alpha')^{\circ}_{\tau} \leq (A,\alpha'')^{\circ}_{\tau}$
4. $(A,\alpha')^{\circ}_{\tau} \gg\!\!\longrightarrow (A,\alpha')$
5. $(A,\alpha' * \alpha")^{\circ}_{\tau} \geq (A,\alpha')^{\circ}_{\tau} * (A,\alpha")^{\circ}_{\tau}$

Proof. Proposition 3.10 gives (1). Idempotency follows from the fact that

$$\{(A,\alpha')^{\circ}_{\tau}\} \leq \downarrow \mathrm{seg}^{\circ}(A,\alpha')$$

and

$$\tau \xrightarrow{\{\}} P(\tau) \xrightarrow{\bigoplus} \tau = \tau \xrightarrow{\mathrm{id}_{\tau}} \tau$$

The preservation of order by $\bigoplus$ and the fact that open downsegments are smaller than downsegments give (3) and (4). The last property follows from the preservation of order by $\bigoplus$ and the fact that fully fuzzy topologies are closed under $*$. □

4 Neighborhoods and convergence

4.1 Neighborhoods

The interior operator gives us the means to make the following definition of neighborhood, which says that a fuzzy subset (A, α') is a neighborhood of a to the extent that a is in the interior of (A, α').

Definition 4.1 *The neighborhood map $\nu_\mathcal{T} : (A, \alpha) \to P^2(A, \alpha)$ is the exponential adjoint of the smallest classifying map for the unbalanced subobject*

$$\begin{array}{cccc} nbd & \twoheadrightarrow & P(A,\alpha)\times(A,\alpha) & \longrightarrow & (\mathrm{L},\top) \\ & & (\alpha', a) & \longmapsto & (\alpha')^{\circ}_{\mathcal{T}}(a) \\ \hline\hline & & (A,\alpha) & \xrightarrow{\nu} & P^2(A,\alpha) \end{array}$$

It is fundamental for a theory of convergence that the neighborhoods of a point form a filter. Since classically elements of filters are not allowed to be empty, this means we need to see how to deal with the partial membership of fuzzy sets.

4.2 Non-emptiness conditions

In order to talk about quasi-neighborhoods, convergence, and adherence we need to talk first about non-emptiness conditions, since different notions give rise to different convergence theories.

Definition 4.2 *The fuzzy set (A, α) has global support if $\alpha(a) \neq 0$ for all $a \in A$.*

Definition 4.3 *The degree of non-emptiness of a fuzzy set (A, α) is*

$$\mathfrak{n}(\alpha) = \bigvee_{a \in A} \alpha(a)$$

Definition 4.4 *The fuzzy set (A, α) is weakly non-empty if $\alpha(a) \neq 0$ for some $a \in A$. This can be generalized to the notion of strong h-nonemptiness where (A, α) is strong h-nonempty if $\alpha(a) > h$ for some $a \in A$. We will use $N_h(A, \alpha) \twoheadrightarrow P(A, \alpha)$ for the crisp subobject of strong h-nonempty subobjects of (a, α).*

Definition 4.5 *We can also define weak h-nonemptiness by requiring $\alpha(a) \geq h$ If $h = 1$, this gives strong non-emptiness (that is, (A, α) has crisp members). Any of these notions can be used to define a crisp subsct $P_h(A, \alpha) \twoheadrightarrow P(A, \alpha)$ consisting of those unbalanced subobjects of (A, α) which are weakly h-nonempty.*

Notions of quasi-coincidence are important in some of the convergence theory:

Definition 4.6 *A fuzzy point p_a^h is quasi-coincident with a fuzzy set (A, α) if $h * \alpha(a) > 0$.*

Note that the usual definition of quasi-coincident (as in [5]) is equivalent to the one just given: asking that $h + \alpha(a) > 1$ is equivalent to asking that $h * \alpha(a) > 0$—one focuses on the Łukasiewicz union of the point with the fuzzy subset having strong membership and the other focuses on the Łukasiewicz intersection being weakly nonempty.

Definition 4.7 *A fuzzy set (A, α) is quasi-non-empty if $\alpha * \alpha$ is not always 0. This is the same as saying that there is a fuzzy point belonging to (A, α) which is also quasi-coincident with (A, α).*

Definition 4.8 *The degree of quasi-non-emptiness of a fuzzy set (A, α) is*

$$\mathfrak{n}_q(\alpha) = \bigvee_{a \in A} \alpha(a) * \alpha(a)$$

Note that for the Łukasiewcz $*$, quasi-non-emptiness is equivalent to $> .5$-non-emptiness.

With the notion of quasi-coincidence in hand we can define quasi-neighborhoods by saying that a fuzzy set (A, α') is a quasineighborhood of s fuzzy point p_a^h if p_a^h is quasicoincident with the interior of (A, α').

Definition 4.9 *A pair (α', a) is in **qnbd** to the extent $\alpha(a) * (\alpha')^{\circ}_{\mathcal{T}}(a)$. By taking the exponential adjoint of the smallest characteristic map of **qnbhd** this gives rise to the quasi-neighborhood map*

$$\nu_q : (A, \alpha) \to P^2(A, \alpha)$$

4.3 Filters

Definition 4.10 *A **filter** on (A, α) is a function $\phi : P(A, \alpha) \to \mathrm{L}$ such that*

1. *ϕ preserves order: if $\alpha' \geq \alpha$", then $\phi(\alpha') \geq \phi(\alpha$").*
2. *ϕ respects $*$: $\phi(\alpha' * \alpha") \geq \phi(\alpha') * \phi(\alpha")$*
3. *$\phi(\alpha') \leq \mathfrak{n}(\alpha')$ this says that a fuzzy set can only be in a filter to its degree of non-emptiness. This makes our filters proper.*

We can internalize this to give an object of fuzzy filters.

Example:

The neighborhood filter of a point $a \in A$ is

$$N_a^{\tau}(\alpha') = (\alpha')^{\circ}_{\tau}(a) \quad \diamondsuit$$

Example:

The quasi-neighborhood filter of a point $a \in A$ is

$$QN_a^{\tau}(\alpha') = \alpha(a) * (\alpha')^{\circ}_{\tau}(a)$$

Notice that in general $QN_a(\alpha') \leq N_a(\alpha')$. $\diamondsuit$

Example:

The principal filter ϕ_a has $\phi_a(\alpha') = \alpha'(a)$. $\diamondsuit$

Example:

For any unbalanced subobject (A, α') the function

$$\upsilon_{\alpha'}(\alpha'') = \bigvee_{a \in A} (\alpha''(a) \wedge (\alpha'(a) \to \alpha''(a)))$$

gives the filter of unbalanced subobjects containing (A, α'). This generalizes the principle filter, which can be thought of as the filter of unbalanced subobjects containing the singleton given by the fuzzy point $p_a^{\alpha(a)}$. $\diamondsuit$

In ordinary topology a point is a limit point of a filter if all of its neighborhoods are in the filter. Since we need to allow for fuzzy subsets which are neighborhoods only to some degree, we express this with an implication. Since we do not want elements to be limit points to a greater extent than they are elements of the space we need to restrict with the $\alpha(a)$ term in the following definition:

Definition 4.11 *A filter ϕ on (A, α) with fully fuzzy topology τ has limit fuzzy set*

$$\lim_{\tau}(\phi) \rightarrowtail (A, \alpha)$$

taking

$$a \mapsto \alpha(a) \wedge \bigwedge_{\alpha'} N_a(\alpha') \to \phi(\alpha'))$$

A related definition gives quasi-convergence in terms of quasi-neighborhoods:

Definition 4.12 *A filter ϕ on (A, α) with fully fuzzy topology τ has quasi-limit fuzzy set*

$$qlim_{\tau}(\phi) \gg\!\!\longrightarrow (A, \alpha)$$

taking

$$a \mapsto \alpha(a) \wedge \bigwedge_{\alpha'} (QN_a(\alpha') \to \phi(\alpha'))$$

References

[1] J. Adámek, H. Herrlich, G. E. Strecker, Abstract and Concrete Categories: The Joy of Cats, Wiley Interscience Pure and Applied Mathematics, John Wiley & Sons (Brisbane/Chicester/New York/Singapore/Toronto), 1990.

[2] U. Höhle, L. N. Stout, *Foundations of fuzzy sets*, Fuzzy Sets and Systems **40**(2)(1991), 257–296.

[3] R. Lowen, *Fuzzy topological spaces and fuzzy compactness*, Journal of Mathematical Analysis and its Applications **56**(3)(1976), 621–633.

[4] B. Schweitzer, A. Sklar, Probabilistic Metric Spaces, North Holland (Amsterdam), 1983.

[5] A. P. Šostak, *Two decades of fuzzy topology: basic ideas, notions, and results*, Russian Math. Surveys **44**(6)(1989), 125–186.

[6] L. N. Stout, *Topology in a topos II: $\mathcal{E}$-completeness and $\mathcal{E}$ -cocompleteness*, Manuscripta Mathematica, **17**(1975), 1–14.

[7] L. N. Stout, *Quels sont les espaces topologiques dans les topos?* Annales des sciences mathématiques de Québec **II**(1)(1978), 123–141.

[8] L. N. Stout, *Fuzzy logic from a second closed structure on a quasitopos,* in Proceedings of the Third IFSA World Congress (1989), 458–461, IFSA (August 1989, Seattle).

[9] L. N. Stout, *The logic of unbalanced subobjects in a category with two closed structures*, Chapter 3 in: S. E. Rodabaugh, E. P. Klement, U. Höhle, eds, Application Of Category To Fuzzy Sets, Theory and Decision Library—Series B: Mathematical and Statistical Methods, Volume **14**, Kluwer Academic Publishers (Boston/Dordrecht/London), 1992, pp. 73–105.

Part II:

Algebraic Structures In Fuzzy Sets

CHAPTER 9

Fuzzy Logics Arising From Strict De Morgan Systems

M. GEHRKE, C. WALKER, AND E. WALKER

Introduction

In the application of fuzzy logic to expert systems, fuzzy control, and the like, it is not a single logic that is used, but a plethora of distinct logics. The choice used in a specific application is often ad hoc—decided on the basis of empirical factors or familiarity.

There is a technical commonality to these logics in that they all arise in the same manner through the specification of an algebra of truth values—that is, a set of truth values equipped with algebraic operations corresponding to each of the logical connectives in question. In our paper [9] we described in detail the technical set-up common to these logics and, using tools from universal algebra, we showed how one can use algebra to answer many of the central questions about the logics thus obtained. This chapter is a continuation of the work started there.

Using our results, it is our hope that the users of fuzzy logics will have the option of making their choices of fuzzy logic based on properties of the logics and that they will have tools available to use these logics more efficiently.

In this chapter, after reviewing some of the results and definitions from [9], we apply our methods to the study of fuzzy logics arising from truth value algebras on the unit interval that are strict De Morgan systems. This class of logics includes the ones in which the fuzzy conjunction is given by the usual multiplication of real numbers, a common choice in applications. In [9] we mainly considered applications to the original fuzzy logic introduced by Zadeh and interval-valued fuzzy logic. For these logics the conjunction is idempotent and we showed that in the idempotent setting, these two logics and classical

S.E. Rodabaugh and E.P. Klement (eds.), Topological and Algebraic Structures in Fuzzy Sets, 257–276.
© 2003 Kluwer Academic Publishers.

Boolean logic are the only possibilities. As we will see here things are much more unwieldy once we drop idempotence.

The main question we will address is when two choices of truth value algebras within the class of strict De Morgan systems yield the same logic or even comparable logics. Surprisingly, two strict De Morgan systems yield comparable logics if and only if the two algebras are isomorphic. This leaves us with uncountably many logics of this type. We will give a countable set of equational properties that separate these logics and we will explore questions of common extensions and other logical properties.

1 Propositional logic

Propositional logic is the study of the properties of logical connectives. The first step in studying propositional connectives is to choose which ones one wants to study, and which ones will be needed as basic connectives. In our study for example, we will not consider implication. One can either consider our setting as the reduced logic not involving implication, or, as is often the case in fuzzy logic applications, we are assuming a choice of implication given by a formula over our other connectives, such as material implication. In a De Morgan system, which is the kind of truth value algebra usually considered in fuzzy logic, conjunction (denoted by $\circ$), disjunction ($\diamond$), negation (η), and the constants 0 and 1 for the universally false and true statements are taken as basic. However, because we will have De Morgan's laws

$$\eta(x \circ y) = \eta(x) \diamond \eta(y)$$
$$\eta(x \diamond y) = \eta(x) \circ \eta(y)$$

and since the negation is an involution

$$\eta(\eta(x)) = x$$

it follows that the disjunction is definable by a formula over the other two connectives:

$$x \diamond y = \eta(\eta(x) \circ \eta(y))$$

Consequently we can leave disjunction out as a basic connective. On the other hand, in a De Morgan system, the conjunction and negation are linked to the conjunction and disjunction given by minimum (denoted by $\wedge$) and maximum ($\vee$), respectively, which were used in Zadeh's original fuzzy logic by the t-norm and order reversal properties among others, e.g.,

$$x \circ (y \wedge z) = (x \circ y) \wedge (x \circ z)$$
$$\eta(x \wedge y) = \eta(x) \vee \eta(y)$$

For this reason we include these as basic connectives. That is, we will take as the basic connectives $\vee, \wedge, 0, 1, \circ$, and η of arity 2,2,0,0,2,1, respectively. Once the connectives have been chosen, we construct the set $\mathcal{F}$ of formulas over a set X of propositional variables. It is usually described inductively as the smallest subset of the set of all strings in the variables, connective symbols, and parentheses satisfying:

1. If x is a propositional variable, then x is a well-formed formula;
2. If α and β are well-formed formulas, then so are $(\alpha \wedge \beta)$, $(\alpha \vee \beta)$, 0, 1, $(\alpha \circ \beta)$, $\eta(\alpha)$.

The set $\mathcal{F}$ together with the connectives $\vee, \wedge, 0, 1, \circ$, and η is an algebra of type $(2, 2, 0, 0, 2, 1)$.

Logical meaning is imparted by giving a notion of logical equivalence. For all the logics we deal with here this is achieved by specifying a truth value algebra of the form $\mathbb{A} = ([0,1], \wedge, \vee, 0, 1, \circ, \eta)$ and defining the relation of logical equivalence with respect to the truth value algebra $\mathbb{A}$ by

$$\alpha \sim_{\mathbb{A}} \beta \text{ iff } h(\alpha) = h(\beta) \text{ for all homomorphisms } h : \mathcal{F} \to \mathbb{A}$$

A map $h : \mathcal{F} \to \mathbb{A}$ being a homomorphism corresponds to saying that h is completely determined by the truth valuation $h\restriction X : X \to A$ on the propositional variables by the inductive definition $h((\alpha \wedge \beta)) = h(\alpha) \wedge h(\beta)$, $h((\alpha \vee \beta)) = h(\alpha) \vee h(\beta)$, $h(0) = 0$, $h(1) = 1$, $h((\alpha \circ \beta)) = h(\alpha) \circ h(\beta)$, and $h(\eta(\alpha)) = \eta(h(\alpha))$. In algebraic terms what this means is that formulas α and β are logically equivalent with respect to the truth value algebra $\mathbb{A}$ if and only if the equation $\alpha = \beta$ holds in the algebra $\mathbb{A}$—that is,

$$\alpha \sim_{\mathbb{A}} \beta \text{ iff } \alpha = \beta \text{ is in } Eq(\mathbb{A})$$

The logic then is the algebra $L_{\mathbb{A}}$ of the same type as $\mathbb{A}$ obtained by considering two formulae in $\mathcal{F}$ equal whenever they are logically equivalent, that is,

$$L_{\mathbb{A}} = \mathcal{F}/Eq(\mathbb{A})$$

It is a central theorem in universal algebra that this algebra $L_{\mathbb{A}}$ is the free algebra on $|X|$ generators in the variety generated by the algebra $\mathbb{A}$.

The variety generated by the algebra $\mathbb{A}$ may be obtained as the class of all algebras satisfying all the equations satisfied in $\mathbb{A}$, that is, $\mathcal{V}(\mathbb{A}) = Mod(Eq(\mathbb{A}))$. But more importantly for our purposes, Birkhoff's theorem tells us that $\mathcal{V}(\mathbb{A})$ may be obtained by taking all homomorphic images of all subalgebras of all Cartesian products of $\mathbb{A}$, that is, $\mathcal{V}(\mathbb{A}) = HSP(\mathbb{A})$. The upshot of this is that the logic $L_{\mathbb{A}}$ is completely determined by the class of algebras $\mathcal{V}(\mathbb{A}) = HSP(\mathbb{A})$ and that many of the questions of logical interest one can ask about $L_{\mathbb{A}}$ may be answered by studying the algebraic properties of this class $\mathcal{V}(\mathbb{A}) = HSP(\mathbb{A})$. In the next section we illustrate these ideas by a brief overview of the results obtained in [9].

2 De Morgan logics—idempotent case

In this section we only deal with one conjunction, rather than two, so our type is just $\wedge, \vee, \neg, 0, 1$ with arity 2, 2, 1, 0, 0. Classical propositional logic is obtained when we choose $\mathbb{A}$ to be the two element Boolean algebra

$$\mathbf{2} = (\{0,1\}, \wedge, \vee, \neg, 0, 1)^1$$

In this case

$$L_\mathbf{2} = \mathcal{F}/Eq(\mathbf{2}) = \mathcal{F}_\mathbb{B}(X)$$

is the free Boolean algebra freely generated by the set X of variables. And we get the exact same propositional logic no matter which Boolean algebra we choose as the set of truth values since all (non-trivial) Boolean algebras have the same equational theory, or equivalently, they all generate the same variety.

The propositional logic underlying what people loosely call fuzzy logic, or Lee-Chang fuzzy logic [5], is the logic obtained by taking $\mathbb{I} = ([0,1], \wedge, \vee, 0, 1, \neg)$ to be the algebra of truth values. Here $\mathbb{I}$ is the unit interval of real numbers with the lattice operations induced by the natural order, that is $x \wedge y = \min\{x, y\}$ and $x \vee y = \max\{x, y\}$, and $\neg x = 1 - x$ as the negation. To understand which kind of logic this choice of truth values gives rise to we need to understand which kind of algebra we are dealing with.

Definition 2.1 *A De Morgan algebra is an algebra* $\mathbb{L} = (L, \vee, \wedge, 0, 1)$ *of the type* $(2, 2, 0, 0, 1)$ *satisfying*

1. $(L, \vee, \wedge, 0, 1)$ *is a distributive lattice with* $0, 1$.
2. $\neg(x \vee y) = \neg x \wedge \neg y$ *and* $\neg(x \wedge y) = \neg x \vee \neg y$ *are identities.*
3. $\neg(\neg x) = x$ *is an identity.*

We denote the equational class of De Morgan algebras by $\mathcal{M}$. A unary operation on a distributive lattice satisfying 2. and 3. is called a **negation**.

Example 2.1 *Let* **3** *denote the three element chain* $\{0, u, 1\}$ *with its unique involution. When we choose* A *to be the De Morgan algebra* **3**, *then we get what is known as Kleene three-valued logic.*

Example 2.2 *The algebra* $\mathbb{I} = ([0,1], \vee, \wedge, \neg, 0, 1)$ *as described above forms a De Morgan algebra. Using* $\mathbb{I}$ *as the algebra of truth values we get the propositional logic known as (classical) fuzzy logic.*

[1] As the reader may have noticed, we are using the same label for the operation in a specific algebra as for the corresponding function symbol, assuming that it will be clear from the context which we are talking about.

Example 2.3 *Given any De Morgan algebra* $\mathbb{L}$*, let* $\mathbb{L}^{[2]} = (L^{[2]}, \vee, \wedge, \neg, 0, 1)$ *where* $L^{[2]} = \{(x, y) : x, y \in L \text{ and } x \leq y\}$*,* $\vee$ *and* $\wedge$ *are defined coordinate-wise,* $\neg(x, y) = (\neg y, \neg x)$*,* $0_{L^{[2]}} = (0_L, 0_L)$ *and* $1_{L^{[2]}} = (1_L, 1_L)$*. Then* $\mathbb{L}^{[2]}$ *is again a De Morgan algebra. Since the pairs* (x, y) *satisfy* $x \leq y$*, they can be thought of as subintervals of* L*. Indeed the De Morgan algebra* $\mathbb{I}^{[2]}$ *is the one used as the algebra of truth values in the propositional logic known as interval fuzzy logic or practical fuzzy logic* [9, 16].

The nature of each of the logics above is completely determined by which variety of De Morgan algebras is generated by the truth value algebra $\mathbb{A}$ in question. Therefore it is of interest to know the subvarieties, that is, the equational subclasses of the class of De Morgan algebras. This has long since been worked out and it turns out that there are very few subvarieties.

We denote the variety of De Morgan algebras by $\mathcal{M}$; the trivial subvariety of $\mathcal{M}$, consisting of all one-element algebras, by $\mathcal{M}_{-1}$; the subvariety of $\mathcal{M}$ generated by **2**, **3**, and $\diamond$, by $\mathcal{M}_0$, $\mathcal{M}_1$, and $\mathcal{M}_2$, respectively, where

$$\diamond \;=\; \begin{array}{ccccc} & & 1 & & \\ & / & & \backslash & \\ v & & & & w \\ & \backslash & & / & \\ & & 0 & & \end{array}$$

with $\neg v = v$, and $\neg w = w$.

Theorem 2.1 [13] *The only subvarieties of* $\mathcal{M}$ *are*

$$\mathcal{M}_{-1} \subsetneq \mathcal{M}_0 \subsetneq \mathcal{M}_1 \subsetneq \mathcal{M}_2 = \mathcal{M}$$

Moreover, for $\mathbb{L} \in \mathcal{M}$*,* $\mathbb{L} \in \mathcal{M}_0$ *if and only if* $x \wedge \neg x = 0$ *is an identity in* $\mathbb{L}$ *(i.e.,* M_0 *is the class of Boolean algebras), and* $\mathbb{L} \in \mathcal{M}_1$ *if and only if*

$$[(x \vee \neg x) \wedge (y \wedge \neg y)] \vee [(y \vee \neg y) \wedge (x \wedge \neg x)] = (y \wedge \neg y) \vee (x \wedge \neg x)$$

is an identity in $\mathbb{L}$*.*

The latter equality is equivalent to the inequality

$$x \wedge \neg x \leq y \vee \neg y$$

[13] and consequently to the equality

$$(x \wedge \neg x) \wedge (y \vee \neg y) = (x \wedge \neg x)$$

A De Morgan algebra in which the inequality $x \wedge \neg x \leq y \vee \neg y$ holds is called a **normal** i**-lattice** [13] or a **Kleene algebra** [2, 14].

Before giving the logic consequences of this algebraic result, we make precise a few concepts:

Definition 2.2 *Two logics $L_{\mathbb{A}}$ and $L_{\mathbb{B}}$ are* **equal** *if they are equal as quotients of the set of well-formed formulas $\mathcal{F}/Eq(\mathbb{A}) = \mathcal{F}/Eq(\mathbb{B})$, which is the same as saying that $Eq(\mathbb{A}) = Eq(\mathbb{B})$. The logic $L_{\mathbb{A}}$ is* **weaker** *than the logic $L_{\mathbb{B}}$ provided $Eq(A) \subseteq Eq(B)$, which is the same as saying that $\mathcal{V}(A) \supseteq \mathcal{V}(B)$. Finally we say that $L_{\mathbb{A}}$ and $L_{\mathbb{B}}$ are* **comparable** *provided $L_{\mathbb{A}}$ is weaker than $L_{\mathbb{B}}$ or vice versa.*

It is clear from the above theorem that the logics given by De Morgan algebras are all comparable, with Boolean logic being the strongest and De Morgan logic the weakest.

Corollary 2.2 *There are only three distinct non-trivial De Morgan logics, classical logic, Kleene logic and De Morgan logic, and these form a chain.*

It follows that the other logics mentioned above, namely, Chang-Lee fuzzy logic and interval-valued fuzzy logic must each be equal to one of classical logic, Kleene logic and De Morgan logic.

Theorem 2.3 *The De Morgan algebras* $\mathbf{3}$ *and* $\mathbb{I}$ *both generate the subvariety $\mathcal{M}_1$ of $\mathcal{M}$. That is, $L_{\mathbf{3}} = L_{\mathbb{I}}$ and classical fuzzy logic is the same as Kleene three-valued logic.*

Proof. We know that $\mathbf{3}$ generates $\mathcal{M}_1$ by the definition of $\mathcal{M}_1$. Also, since $\mathbb{I}$ satisfies the inequality $x \wedge \neg x \leq y \vee \neg y$ we have $\mathcal{V}ar(\mathbb{I}) \subseteq \mathcal{M}_1$. On the other hand, $\mathbb{I}$ is not Boolean so $\mathcal{M}_0$ is a proper subvariety of $\mathcal{V}ar(\mathbb{I})$. It follows that $\mathcal{V}ar(\mathbb{I}) = \mathcal{M}_1$. □

Theorem 2.4 *The De Morgan algebras $\diamond$ and $\mathbb{I}^{[2]}$ both generate the equational class $\mathcal{M}$ of all De Morgan algebras. That is, $L_{\diamond} = L_{\mathbb{I}^{[2]}}$ and interval-valued fuzzy logic is the same as what is sometimes known as diamond logic.*

Proof. We know that $\diamond$ generates $\mathcal{M}$. And it is easy to see that $\diamond$ can be realized as a subalgebra of $\mathbf{3}^{[2]}$ which can be realized as a subalgebra of $\mathbb{I}^{[2]}$. Thus $\mathcal{M} = \mathcal{V}ar(\diamond) \subseteq \mathcal{V}ar(\mathbf{3}^{[2]}) \subseteq \mathcal{V}ar(\mathbb{I}^{[2]})$. On the other hand, $\mathbb{I}^{[2]}$ is a De Morgan algebra, so $\mathcal{V}ar(\mathbb{I}^{[2]}) \subseteq \mathcal{M}$. It follows that $\mathcal{V}ar(\mathbf{3}^{[2]}) = \mathcal{V}ar(\mathbb{I}^{[2]}) = \mathcal{M}$. □

Apart from determining when two truth value algebras yield the same logic, and how many distinct logics are potentially available within a certain equational class, the algebraic theory can also help us determine other properties of logical interest for these logics. For example, notice that if $\mathbb{A}$ is any finite algebra, then there is a finite process for checking whether or not $\alpha \sim_{\mathbb{A}} \beta$. The variables occurring in α together with the ones occurring in β form a finite subset Y of X. It follows that $\alpha \sim_{\mathbb{A}} \beta$ if and only if $\tilde{t}(\alpha) = \tilde{t}(\beta)$ for each $t \in A^Y$, and A^Y is finite since both A and Y are finite. Thus the logical equivalence of the propositional logic given by an algebra $\mathbb{A}$ is finitely checkable if the variety generated by $\mathbb{A}$ is also generated by a finite algebra. Thus we get:

Corollary 2.5 *The logical equivalence relation $\sim_{\mathbb{I}}$ of Lee-Chang fuzzy logic is decidable via three-valued truth tables.*

Corollary 2.6 *The logical equivalence relation $\sim_{\mathbb{I}^{[2]}}$ of interval-valued fuzzy logic is decidable via four-valued truth tables.*

For more details on this and further results and applications in this direction, including complete descriptions of normal forms for each of these logics, see our paper [8].

Another logical application of algebraic information is to determining syntactic calculi describing these logics. Birkhoff's Theorem on the completeness of equational logic gives a scheme for obtaining a syntactic deductive system which will derive all equational consequences of a given set of equations, see, for example, [3, chapter II, section 14]. Thus we may obtain syntactic counterparts to our logics by specifying equational bases for the corresponding varieties of algebras. In each of the idempotent cases our varieties have nice finite equational bases, so we get syntactic counterparts for these logics:

Corollary 2.7 *All logical equivalences for Chang-Lee fuzzy logic can be derived from the set of logical equivalences corresponding to saying that $\mathbb{I}$ is a Kleene algebra, that is the set*

$$\begin{array}{cc} x \wedge y \sim_{\mathbb{I}} y \wedge x & x \vee y \sim_{\mathbb{I}} y \vee x \\ x \wedge 1 \sim_{\mathbb{I}} x & x \vee 0 \sim_{\mathbb{I}} x \\ x \wedge x \sim_{\mathbb{I}} x & x \vee x \sim_{\mathbb{I}} x \end{array}$$

$$\begin{array}{c} x \wedge (y \wedge z) \sim_{\mathbb{I}} (x \wedge y) \wedge z \\ x \vee (y \vee z) \sim_{\mathbb{I}} (x \vee y) \vee z \\ x \wedge (y \vee z) \sim_{\mathbb{I}} (x \wedge y) \vee (x \wedge z) \\ x \vee (y \wedge z) \sim_{\mathbb{I}} (x \vee y) \wedge (x \vee z) \\ \neg(x \wedge y) \sim_{\mathbb{I}} \neg x \vee \neg y \\ \neg(x \vee y) \sim_{\mathbb{I}} \neg x \wedge \neg y \\ \neg\neg x \sim_{\mathbb{I}} x \\ (x \wedge \neg x) \wedge (y \vee \neg y) \sim_{\mathbb{I}} (x \wedge \neg x) \end{array}$$

Corollary 2.8 *All logical equivalences for interval-valued fuzzy logic can be derived from the set of logical equivalences corresponding to saying that $\mathbb{I}^{[2]}$ is a De Morgan algebra, that is the set*

$$\begin{array}{cc} x \wedge y \sim_{\mathbb{I}} y \wedge x & x \vee y \sim_{\mathbb{I}} y \vee x \\ x \wedge 1 \sim_{\mathbb{I}} x & x \vee 0 \sim_{\mathbb{I}} x \\ x \wedge x \sim_{\mathbb{I}} x & x \vee x \sim_{\mathbb{I}} x \end{array}$$

$$x \wedge (y \wedge z) \sim_{\mathbb{I}} (x \wedge y) \wedge z$$
$$x \vee (y \vee z) \sim_{\mathbb{I}} (x \vee y) \vee z$$
$$x \wedge (y \vee z) \sim_{\mathbb{I}} (x \wedge y) \vee (x \wedge z)$$
$$x \vee (y \wedge z) \sim_{\mathbb{I}} (x \vee y) \wedge (x \vee z)$$
$$\neg(x \wedge y) \sim_{\mathbb{I}} \neg x \vee \neg y$$
$$\neg(x \vee y) \sim_{\mathbb{I}} \neg x \wedge \neg y$$
$$\neg\neg x \sim_{\mathbb{I}} x$$

In conclusion, when looking for an appropriate propositional logic structure for a particular situation, one may study the universal algebraic properties of the corresponding algebras of truth values. If it has already been decided that the algebra of truth values is to be in a certain class for example, then there will be as many choices of distinct logics as there are varieties generated by elements of that class. A common choice of truth value structure in fuzzy logic is a De Morgan system (see [10]). In [10, 11] we determined which De Morgan systems on $\mathbb{I}$ and on $\mathbb{I}^{[2]}$ are isomorphic. Of course isomorphic algebras generate the same variety, but non-isomorphic algebras can generate the same variety also, and hence determine the same logic. Any two Boolean algebras generate the same variety, for example, and hence the same propositional logic. As we shall see here though, any two non-isomorphic strict De Morgan systems give rise to non-comparable logics. Consequently, there are uncountably many distinct such logics. This is in sharp contrast to our result in [9] that the conjunctive reduct $L_{(\mathbb{I},\circ)}$ is independent of the choice of t-norm, strict or nilpotent.

3 De Morgan logics—strict case

We now consider different negations, and from this point on, the notation $\mathbb{I}$ will be used to denote the unit interval as a bounded lattice—not including a negation. That is, $\mathbb{I} = ([0,1], \vee, \wedge, 0, 1)$.

Definition 3.1 *A* ***t-norm*** $\circ$ *on* $\mathbb{I}$ *is a binary operation* $\circ : [0,1] \times [0,1] \longrightarrow [0,1]$ *that is associative, commutative, order preserving in each coordinate, and for which the right end point* 1 *is an identity. A t-norm* $\circ$ *is* ***convex*** *if whenever* $x \circ y \leq c \leq x_1 \circ y_1$*, then there is an* r *between* x *and* x_1 *and an* s *between* y *and* y_1 *such that* $c = r \circ s$*. A t-norm* $\circ$ *is* ***Archimedean*** *if it is convex and for each* $a, b \in (0,1)$*, there is a positive integer* n *such that* $a^n = \overbrace{a \circ a \circ \cdots \circ a}^{n \text{ times}} < b$*. An Archimedean t-norm is* ***strict*** *if in addition we have for all* $x \in (0,1)$*,* $x \circ x < x$*. A* ***strict De Morgan system*** *is an algebra* $\mathbb{A} = (\mathbb{I}, \circ, \eta)$ *where* $\mathbb{I}$ *is the unit interval as a bounded lattice,* $\circ$ *is a strict t-norm, and* η *is a negation on* $\mathbb{I}$.

In our paper [10, Theorem 29] we showed exactly when two De Morgan systems are isomorphic. We state the following corollary of that characterization here as we will use it in proving the main result of this chapter.

Theorem 3.1 *Each strict De Morgan system is isomorphic to a unique De Morgan system $\mathbb{I}_\beta = (\mathbb{I}, \cdot, \beta)$ where the t-norm $\cdot$ is the usual multiplication on the unit interval, and β is a negation with fixed-point* $1/2$.

We are now ready to state and prove our main result.

Theorem 3.2 *Two strict De Morgan systems give rise to comparable logics if and only if the systems are isomorphic.*

Proof. Since it is clear that isomorphic algebras give the same logic, we may assume we are dealing with algebras of the form $\mathbb{I}_\beta = (\mathbb{I}, \cdot, \beta)$ and $\mathbb{I}_\gamma = (\mathbb{I}, \cdot, \gamma)$ where $\cdot$ is the usual multiplication on the unit interval, and β and γ are negations with fixed-points at $1/2$. We need to show that if $\mathcal{V}(\mathbb{I}_\beta) \subseteq \mathcal{V}(\mathbb{I}_\gamma)$—that is, $\mathbb{I}_\beta \in \mathcal{V}(\mathbb{I}_\gamma)$, then $\beta = \gamma$. In the following lemmas, we show that if $\mathbb{I}_\beta \in \mathcal{V}(\mathbb{I}_\gamma)$, then β and γ agree on a dense set. Then, since both functions preserve arbitrary meets and joins (or alternatively because they are continuous), it follows that $\beta = \gamma$. □

Lemma 3.3 *If $\mathbb{I}_\beta \in \mathcal{V}(\mathbb{I}_\gamma)$, then $\beta(1/2^n) = \gamma(1/2^n)$ for each $n \in \mathbb{N}$.*

Proof. We denote the operation symbols for the t-norm and the negation by $\cdot$ and η, respectively (Using $\cdot$ should not cause any confusion as it will be interpreted as actual multiplication in each of our algebras). For $n \in \mathbb{N}$ we write x^n for $\overbrace{x \cdot x \cdot \cdots \cdot x}^{n \text{ times}}$.

Now consider the terms

$$s_{n,l}(x) = (\eta((x \vee \eta(x))^n))^\ell$$

and

$$t_n(x) = (x \vee \eta(x))^n$$

The equation $s_{n,\ell}(x) \le t_k(y)$ (that is, $s_{n,\ell}(x) \vee t_k(y) = t_k(y)$) holds in an algebra $\mathbb{I}_\beta$ if and only if for all $a, b \in \mathbb{I}$ we have

$$(\beta((a \vee \beta(a))^n))^\ell \le (b \vee \beta(b))^k$$

But this is equivalent to the statement:

$$\max_{a \in I} (\beta((a \vee \beta(a))^n))^\ell \le \min_{b \in I} (b \vee \beta(b))^k$$

It is clear that both of these extrema are attained at the fixed point of β, which we are assuming to be $1/2$. Thus we get, for all $n, \ell, k \in \mathbb{N}$,

$$s_{n,\ell}(x) \le t_k(y) \text{ holds in } \mathbb{I}_\beta$$

if and only if

$$(\beta((1/2)^n))^\ell \leq (1/2)^k$$

if and only if

$$\beta((1/2)^n) \leq (1/2)^{\frac{k}{\ell}}$$

Similarly we have for $n, \ell, k \in \mathbb{N}$

$$s^\partial_{n,\ell}(x) \geq t^\partial_k(y) \text{ holds in } \mathbb{I}_\beta$$

if and only if

$$\beta((1/2)^n) \geq (1/2)^{\frac{k}{\ell}}$$

where

$$s^\partial_{n,\ell}(x) = (\eta((x \wedge \eta(x))^n))^\ell$$

is the term dual to $s_{n,\ell}$, and

$$t^\partial_k(x) = (x \wedge \eta(x))^n$$

is the term dual to $t_k(x)$. Now, since we are assuming that $\mathbb{I}_\beta \in \mathcal{V}(\mathbb{I}_\gamma)$, we have that $\mathbb{I}_\beta$ satisfies every equation satisfied by $\mathbb{I}_\gamma$. In particular for all $n, \ell, k \in \mathbb{N}$

$$s_{n,\ell}(x) \leq t_k(y) \text{ holds in } \mathbb{I}_\gamma \text{ implies it holds in } \mathbb{I}_\beta$$

and

$$s^\partial_{n,\ell}(x) \geq t^\partial_k(y) \text{ holds in } \mathbb{I}_\gamma \text{ implies it holds in } \mathbb{I}_\beta$$

so for all $n, \ell, k \in \mathbb{N}$ we have

$$\gamma((1/2)^n) \leq (1/2)^{\frac{k}{\ell}} \text{ if and only if } \beta((1/2)^n) \leq (1/2)^{\frac{k}{\ell}}$$

and

$$\gamma((1/2)^n) \geq (1/2)^{\frac{k}{\ell}} \text{ if and only if } \beta((1/2)^n) \geq (1/2)^{\frac{k}{\ell}}$$

Since the set $\{(1/2)^{\frac{k}{\ell}} : \ell, k \in \mathbb{N}\}$ is dense in the unit interval, it follows that for each $n \in \mathbb{N}$, $\gamma((1/2)^n) = \beta((1/2)^n)$. □

Now for a negation β on $\mathbb{I}$ with fixed point $1/2$, define

$$D(\beta) = \{(\beta((1/2)^n))^m : n, m \in \mathbb{N}\}$$

Lemma 3.4 *Let β be any negation on $\mathbb{I}$. Then the set $D(\beta)$ is dense in I.*

Proof. The sequence $\{(1/2)^n\}_{n\in\mathbb{N}}$ is a decreasing sequence whose meet is 0, and since β is an anti-automorphism of $\mathbb{I}$, it follows that $\{\beta((1/2)^n)\}_{n\in\mathbb{N}}$ is an increasing sequence whose join is 1. Thus if we take any $\varepsilon > 0$, there is $n_0 \in \mathbb{N}$ such that $1 - \eta((1/2)^{n_0}) < \varepsilon$. But $1 - b < \varepsilon$ implies $b^m - b^{m+1} = b^m(1-b) < 1 - b < \varepsilon$, so it follows that the sequence$\{(\eta((1/2)^{n_0}))^m\}_{m\in\mathbb{N}}$ gets within ε of each element of I. □

Lemma 3.5 *If $\mathbb{I}_\beta \in \mathcal{V}(\mathbb{I}_\gamma)$, then β and γ agree on $D(\gamma)$.*

Proof. Suppose $\mathbb{I}_\beta \in \mathcal{V}(\mathbb{I}_\gamma)$ and let $d \in D(\gamma)$. Then there are $m, n \in \mathbb{N}$ so that $d = (\gamma(1/2^n))^m$ and by Lemma 3.3 we have $\beta(1/2^n) = \gamma(1/2^n)$ so it follows that $(\beta(1/2^n))^m = (\gamma(1/2^n))^m$. Now, similarly to our approach in Lemma 3.3 we consider, for $k, l \in \mathbb{N}$, the equations $\left(\eta\left((\eta((x \wedge \eta(x))^n))^m\right)\right)^\ell \leq (y \vee \eta(y))^k$ and $\left(\eta\left((\eta((x \vee \eta(x))^n))^m\right)\right)^\ell \geq (y \wedge \eta(y))^k$. We obtain

$$\left(\eta\left((\eta((x \wedge \eta(x))^n))^m\right)\right)^\ell \leq (y \vee \eta(y))^k \text{ holds in } \mathbb{I}_\beta$$

if and only if

$$\beta((\beta(1/2^n))^m) \leq (1/2)^{\frac{k}{\ell}}$$

and

$$\left(\eta\left((\eta((x \vee \eta(x))^n))^m\right)\right)^\ell \geq (y \wedge \eta(y))^k \text{ holds in } \mathbb{I}_\beta$$

if and only if

$$\beta((\beta(1/2^n))^m) \geq (1/2)^{\frac{k}{\ell}}$$

Naming the terms $\left(\eta\left((\eta((x \wedge \eta(x))^n))^m\right)\right)^\ell = u_{n,m,\ell}(x)$ and $(x \vee \neg x)^k = t_k(x)$, we get for each $k, \ell \in \mathbb{N}$

$$\begin{aligned}
\gamma(d) \leq (1/2)^{\frac{k}{\ell}} \quad & \Leftrightarrow \gamma((\gamma(1/2^n))^m) \leq (1/2)^{\frac{k}{\ell}} \\
& \Leftrightarrow u_{n,m,\ell}(x) \leq t_k(y) \text{ holds in } \mathbb{I}_\gamma \\
& \Rightarrow u_{n,m,\ell}(x) \leq t_k(y) \text{ holds in } \mathbb{I}_\beta \\
& \Leftrightarrow \beta((\beta(1/2^n))^m) \leq (1/2)^{\frac{k}{\ell}} \\
& \Leftrightarrow \beta((\gamma(1/2^n))^m) \leq (1/2)^{\frac{k}{\ell}} \\
& \Leftrightarrow \beta(d) \leq (1/2)^{\frac{k}{\ell}}
\end{aligned}$$

Similarly we get for all $k, \ell \in \mathbb{N}$ that

$$\gamma(d) \geq (1/2)^{\frac{k}{\ell}} \text{ implies } \beta(d) \geq (1/2)^{\frac{k}{\ell}}$$

Now since $\left\{(1/2)^{\frac{k}{\ell}}\right\}_{k,\ell\in\mathbb{N}}$ is dense in I, it follows that $\gamma(d) = \beta(d)$ and we have showed that γ and β agree on $D(\gamma)$ as desired. □

The two families of equations have played different roles in the proofs, so we have considered them separately. However, as you may have observed by now, the first family is a special case of the second family. Namely, taking $n = 1$ and applying the De Morgan laws gives

$$(\eta((\eta((x \wedge \eta(x))^1))^m))^l = (\eta((x \vee \eta(x))^m))^l$$

so only the one family of equations is required to separate the logics. Also, by Theorem 3.1, it follows that there are as many isomorphism classes of strict De Morgan systems as there are negations β on $\mathbb{I}$ which fix 1/2. Of course there are uncountably many of these. We conclude:

Corollary 3.6 *The fuzzy logics arising from strict De Morgan systems form an uncountable set of pair-wise incomparable logics. Furthermore, these logics are separated by the following family of equations:*

$$\{(\eta((\eta((x \wedge \eta(x))^n))^m))^l \leq (y \vee \eta(y))^k : k, l, m, n \in \mathbb{N}\}$$

4 Properties of strict De Morgan logics

The fact that there are so many of these logics means that uncountably many won't have finite equational bases and such nice logical features as we saw in the idempotent case (simply due to cardinality considerations). A priori, there is no reason that some of them can't be finitely based (have finite equational bases). The variety of algebras generated by the nilpotent De Morgan system obtained by taking the Lukasiewicz t-norm and the negation α where $\alpha(x) = 1 - x$ is the well-known variety of MV-algebras. It is finitely based. However, it is our conjecture that none of the varieties generated by strict De Morgan systems are finitely based, but we leave this question for future work. Another question is that of finite generation: Does there exist a finite algebra generating a given variety corresponding to a strict De Morgan logic? If this is the case, recall that we can use truth tables to determine logical equivalence in the corresponding logic. In the remainder of this chapter we start an exploration of the subdirectly irreducible elements in the varieties generated by strict De Morgan algebras, allowing us to answer the question of finite generation in the negative.

Given a strict De Morgan system $\mathbb{I}_\beta$ we are interested in knowing whether or not there is a finite algebra $\mathbb{A}$ such that $L_{\mathbb{I}_\beta} = L_{\mathbb{A}}$, or in algebraic terms, we want $\mathcal{V}(\mathbb{I}_\beta) = \mathcal{V}(\mathbb{A})$. Here again we have a powerful theory from universal algebra to help us answer this question. Another celebrated result of Birkhoff's is that any variety is generated by its so called subdirectly irreducible members so we can restrict our attention to these - or to a subclass known to contain these, and Jónsson's Lemma supplies us with such a class: It tells us that if we are dealing with a congruence-distributive variety (which all of these are), then the subdirectly irreducible algebras lie within the class of all homomorphic images of subalgebras of ultraproducts of any generating class for the variety, that is, in our situation:

$$Subdir.Irred.(\mathcal{V}(\mathbb{I}_\beta)) \subseteq HSP_\mu(\mathbb{I}_\beta)$$

Given this result, we don't necessarily need to worry about exactly what subdirectly irreducible algebras are, we can just look at all algebras in $HSP_\mu(\mathbb{I}_\beta)$. So first of all, what are the elements of $P_\mu(\mathbb{I}_\beta)$? These are ultraproducts of algebras from the class $\{\mathbb{I}_\beta\}$. But since there is only one algebra, we are really dealing with ultrapowers of $\mathbb{I}_\beta$. These are either $\mathbb{I}_\beta$ itself (if the ultrafilter used is principal) or what can be thought of as a non-standard extension $\mathbb{I}_\beta^*$ of $\mathbb{I}_\beta$ (if the ultrafilter used is free). This situation allows us to use the methods of non-standard analysis in the study of the varieties generated by members of $P_\mu(\mathbb{I}_\beta)$ and more generally of $HSP_\mu(\mathbb{I}_\beta)$. For general references on Non-standard Analysis, see for example [17] or [12]. For a brief introduction to these methods as applied to lattice theory in particular, see [7]. Before jumping in to the non-standard extensions, let's consider the trivial ultrapower of $\mathbb{I}_\beta$, namely $\mathbb{I}_\beta$ itself. What are the members of $HS(\mathbb{I}_\beta)$?

Example 4.1 *Notice that for each $\mathbb{I}_\beta = ([0,1], \vee, \wedge, 0, 1, \cdot, \beta)$, the subset $\{0,1\}$ gives a subalgebra $\mathbf{2} = (\{0,1\}, \vee, \wedge, 0, 1, \wedge, \neg)$. Notice that on the set $\{0,1\}$ multiplication has the exact same action as* min *or $\wedge$, and any negation β has the same action as the negation α given by $1-x$, and on $\{0,1\}$ any of these are simply the usual Boolean negation, $\neg$. That is, apart from having the conjunction repeated twice in its type (which plays no role to speak of), this is just the two-element Boolean algebra. So $\mathcal{V}(\mathbf{2}) \subseteq \mathcal{V}(\mathbb{I}_\beta)$ which says that Boolean logic is a common strengthening of all of these strict De Morgan logics, or in turn that the strict De Morgan logics are all generalizations of Boolean logic. This of course should come as no surprise.*

However, subalgebras of $\mathbb{I}_\beta$'s yield no other logics.

Proposition 4.1 *If $\mathbb{A} \in S(\mathbb{I}_\beta)$ with $\mathbb{A} \neq \mathbf{2}$, then $\mathcal{V}(\mathbb{A}) = \mathcal{V}(\mathbb{I}_\beta)$.*

Proof. Let $\mathbb{A} \in S(\mathbb{I}_\beta)$ with $\mathbb{A} \neq \mathbf{2}$. The first thing to observe is that then the set A is dense in $[0,1]$. The argument for this is identical to the one given in Lemma 3.3: Let $x \in A$ with $0 \neq x \neq 1$. Then $\{x^n\}_{n \in \mathbb{N}}$ is a sequence that converges to 0, and thus $\{\beta(x^n)\}_{n \in \mathbb{N}}$ is a sequence that converges to 1. Finally, this implies that the set $\{(\beta(x^n))^m\}_{n,m \in \mathbb{N}}$, which is a subset of A, is dense in

$[0,1]$. Now let $s = t$ be an equation which holds in the subalgebra $\mathbb{A}$. That means that the two term functions $s^{\mathbb{A}}, t^{\mathbb{A}} : A^n \to A$ (where n is the total number of variables involved in the two terms) are identical, that is, for all $a_1, \ldots, a_n \in A$ we have $s^{\mathbb{A}}(a_1, \ldots, a_n) = t^{\mathbb{A}}(a_1, \ldots, a_n)$. Furthermore, $\mathbb{A}$ being a subalgebra of $\mathbb{I}_\beta$ exactly means that the term functions on $\mathbb{A}$ are just the term functions on $\mathbb{I}_\beta$ restricted to A. But now since $s^{\mathbb{I}_\beta}$ and $t^{\mathbb{I}_\beta}$ are continuous functions, being the composition of continuous functions (since the basic operations on $\mathbb{I}_\beta$ are all continuous), it follows that they must be identical as soon as they agree on a dense set. That is, $s = t \in Eq(\mathbb{A})$ implies $s = t \in Eq(\mathbb{I}_\beta)$. But since $\mathbb{A}$ is a subalgebra of $\mathbb{I}_\beta$, the other inclusion $Eq(\mathbb{I}_\beta) \subseteq Eq(\mathbb{A})$ comes for free. That is, $Eq(\mathbb{A}) = Eq(\mathbb{I}_\beta)$ and thus $\mathcal{V}(\mathbb{A}) = Mod(Eq(\mathbb{A})) = Mod(Eq(\mathbb{I}_\beta)) = \mathcal{V}(\mathbb{I}_\beta)$ as desired. □

Example 4.2 *For each* $\mathbb{I}_\beta = ([0,1], \vee, \wedge, 0, 1, \cdot, \beta)$*, the algebra*

$$\mathbf{3} = (\{0, u, 1\}, \vee, \wedge, 0, 1, \wedge, \neg)$$

obtained by repeating the lattice meet twice for the three-element Kleene algebra is a homomorphic image using the map

$$\begin{aligned} h : [0,1] &\to \mathbf{3} \\ 0 &\mapsto 0 \\ x &\mapsto u, \textit{ for } x \in (0,1) \\ 1 &\mapsto 1 \end{aligned}$$

So $\mathcal{V}((\mathbb{I}, \alpha)) = \mathcal{V}(\mathbf{3}) \subseteq \mathcal{V}(\mathbb{I}_\beta)$ *which says that Chang-Lee Fuzzy Logic, or equivalently, Kleene logic, is a common strengthening of all of these strict De Morgan logics, or in turn that the strict De Morgan logics are all generalizations of Kleene logic. This also should come as no surprise given that these logics were conceived as weakenings of Zadeh's classical fuzzy logic.*

In the interest of studying the homomorphic images of an algebra, we use the universal algebraic concept of a congruence. Given a homomorphic image $h : \mathbb{A} \to \mathbb{B}$ of $\mathbb{A}$, we can recover $\mathbb{B}$, up to isomorphism, from the equivalence relation $\ker(h) = \{(a,b) \in A \times A : h(a) = h(b)\}$ on A induced by the map h. Namely, $\mathbb{B}$ is isomorphic to the algebra of equivalence classes of $\ker(h)$ with operations defined on the classes by taking the corresponding operation in $\mathbb{A}$ applied to any choice of representatives of the equivalence classes in question. Furthermore, an equivalence relation θ on $\mathbb{A}$ corresponds to a homomorphic image of $\mathbb{A}$ if and only if whenever $f : A^n \to A$ is a basic operation of $\mathbb{A}$ and $a_1, \ldots, a_n, b_1, \ldots, b_n \in A$ with $a_i \theta b_i$ for $i = 1, \ldots, n$ we have $f(a_1, \ldots, a_n) \theta f(b_1, \ldots, b_n)$. Such equivalence relations on $\mathbb{A}$ are called **congruence relations** of $\mathbb{A}$. The set of all congruences of an algebra $\mathbb{A}$, $Con(\mathbb{A})$, ordered by inclusion, forms a complete lattice. In particular we have a smallest congruence on $\mathbb{A}$ and that is of course the diagonal $\Delta_A = \{(a,a) : a \in A\}$ which corresponds to the homomorphic image $id_A : \mathbb{A} \to \mathbb{A}$. We also have a largest congruence and that is the all-relation on A: $\nabla_A = \{(a,b) : a,b \in A\} = A \times A$. It corresponds to the trivial homomorphic image of $\mathbb{A}$, the one element algebra of the same type as $\mathbb{A}$. The non-trivial homomorphic images of $\mathbb{A}$ correspond to whatever congruences lie between these two extremes.

Since all of our algebras are among other things lattices, the congruence classes corresponding to any congruence are convex in the underlying order. That is, if two elements a and b with $a \leq b$ are related by a congruence θ, then $[a,b] \times [a,b] = \{(c,d) : a \leq c \leq b, a \leq d \leq b\}$ is contained in θ. This will be quite useful to us here.

Proposition 4.2 *Let $\mathbb{A} \neq \mathbf{2}$ be a subalgebra of the strict De Morgan systems $\mathbb{I}_\beta$. Then $\mathbb{A}$ has only one non-trivial congruence and that is the congruence $(A \cap (0,1))^2 \cup \{(0,0),(1,1)\}$ corresponding to the homomorphic image $\mathbf{3}$.*

Proof. Let $\mathbb{A} \neq \mathbf{2}$ be a subalgebra of the strict De Morgan systems $\mathbb{I}_\beta$ and let θ be a congruence with $\Delta_A \neq \theta$. Then there are $a,b \in A$ with $a \neq b$ but with $a\theta b$. We may assume without loss of generality that $a < b$. In addition, since $a \neq 1/2$ or $b \neq 1/2$, using $\beta(a)\theta\beta(b)$ we may assume $1/2 \leq a < b$. Now pick $c \in A$ with $1 - c < \frac{b-a}{4}$. Then there is $n \in \mathbb{N}$ with $c^n, c^{n+1} \in (a,b)$. That is, $c^n \theta c^{n+1}$. Multiplying repeatedly by c on both sides we get $c^n\theta c^{n+1}, c^{n+1}\theta c^{n+2}, c^{n+2}\theta c^{n+3}, \ldots,$ and by transitivity we have $c^n\theta c^{n+k}$ for all $k \geq 0$; and in particular, $c^n\theta\left(c^n\right)^i$ for all $i \in \mathbb{N}$. That is, if we take $d = c^n$, then $1/2 < d$ and $d\theta d^i$ for all $i \in \mathbb{N}$. Now since $\left\{d^i\right\}_{i\in\mathbb{N}}$ converges to 0 and since the θ-equivalence classes are convex, it follows that $(A \cap (0,d))^2 \subseteq \theta$. Applying the negation β we conclude that $(A \cap (\beta(d),1))^2 \subseteq \theta$ as well. Now since $\frac{1}{2} < d$, it follows that $\beta(d) < d$ and thus $(A \cap (0,1))^2 \subseteq \theta$. It is clear that if either $(A \cap [0,1))^2 \subseteq \theta$ or $(A \cap (0,1])^2 \subseteq \theta$, then $\theta = \nabla_A$ since using β we have $0\theta\frac{1}{2}$ if and only if $1\theta\frac{1}{2}$. That is, we have shown that $Con(\mathbb{A}) = \{\Delta_A, (A \cap (0,1))^2 \cup \{(0,0),(1,1)\}, \nabla_A\}$ as desired. □

Now we proceed to consider homomorphic images of subalgebras of non-standard extensions $\mathbb{I}_\beta^*$ of $\mathbb{I}_\beta$. The first thing to notice is that every element of $\mathbb{I}_\beta^*$ is near-standard, that is, for each x in $[0,1]^*$ there is a (necessarily unique) x^o in the standard interval $[0,1]$ with $x \approx x^o$, i.e., x infinitely close to x^o. This follows from the general theory of non-standard analysis for example because the unit interval is compact. For each x in $[0,1]^*$ we call x^o the 'standard part' of x. This standard part map will be quite useful to us because in our setting it is a homomorphism:

Proposition 4.3 *The standard-part map*

$$\begin{aligned} st : \mathbb{I}^*_\beta &\to \mathbb{I}_\beta \\ x &\mapsto x^o \end{aligned}$$

is a homomorphism.

Proof. To see this, notice that the basic operations $\vee$, $\wedge$, $\cdot$, and β are all continuous, and thus uniformly continuous on $[0,1]$. In non-standard terms we have that a function $f : X \to Y$ among topological spaces is uniformly continuous if and only if

$$\text{For all } x, y \in X^* \quad x \approx y \text{ implies } f^*(x) \approx f^*(y)$$

Applying this here we get for $x, y \in [0,1]^*$, $x \wedge y \approx x^o \wedge y^o$, $x \vee y \approx x^o \vee y^o$, $x \cdot y \approx x^o \cdot y^o$, and $\beta(x) \approx \beta(x^o)$ since $x \approx x^o$ and $y \approx y^o$. That is, $(x \wedge y)^o = x^o \wedge y^o$, $(x \vee y)^o = x^o \vee y^o$, $(x \cdot y)^o = x^o \cdot y^o$, and $(\beta(x))^o = \beta(x^o)$. Finally it is clear that $0^o = 0$ and $1^o = 1$ so h is a homomorphism of De Morgan systems. □

Corollary 4.4 *Let $\mathbb{A}$ be a subalgebra of a non-standard extension $\mathbb{I}^*_\beta$ of $\mathbb{I}_\beta$ with $st(\mathbb{A}) \neq \mathbf{2}$. Then $\mathcal{V}(\mathbb{A}) = \mathcal{V}(\mathbb{I}_\beta)$ and thus $L_\mathbb{A} = L_{\mathbb{I}_\beta}$.*

Proof. Since $st(\mathbb{A})$ is a subalgebra of $\mathbb{I}_\beta$ and $st(\mathbb{A}) \neq \mathbf{2}$ it follows by Proposition 4.1 that $\mathcal{V}(st(\mathbb{A})) = \mathcal{V}(\mathbb{I}_\beta)$. But since $st(\mathbb{A})$ is in $\mathcal{V}(\mathbb{A})$ and $\mathbb{A}$ is in $\mathcal{V}(\mathbb{I}_\beta)$ it follows that $\mathcal{V}(st(\mathbb{A})) \subseteq \mathcal{V}(\mathbb{A}) \subseteq \mathcal{V}(\mathbb{I}_\beta)$. Putting these two facts together we have $\mathcal{V}(\mathbb{A}) = \mathcal{V}(\mathbb{I}_\beta)$ as desired. □

Example 4.3 *For each De Morgan system $\mathbb{I}_\beta$, define $S_\beta = st^{-1}(2) = \{x \in [0,1]^* : x \approx 0 \text{ or } x \approx 1\}$. Then $\mathbb{S}_\beta$ is a subalgebra of $\mathbb{I}^*_\beta$ since for $x, y \in [0,1]^*$ we have $x \approx 0$ or $y \approx 0$ implies $x \cdot y \approx 0$ and $x \wedge y \approx 0$, $x \approx 1$ or $y \approx 1$ implies $x \vee y \approx 1$, $x \approx 1$ and $y \approx 1$ implies $x \cdot y \approx 1$ and $x \wedge y \approx 1$, and $x \approx 0$ and $y \approx 0$ implies $x \vee y \approx 0$. Clearly $\mathbb{S}_\beta$ is the largest subalgebra of $\mathbb{I}^*_\beta$ so that $st(\mathbb{A}) = \mathbf{2}$. This is, so to speak, the non-standard version of $\mathbf{2}$.*

We also have a non-standard version of **3** in this sense. We use the commonly accepted notation from non-standard analysis $x \ll y$ to denote $x < y$ and x not infinitely close to y. Also $((0,1))$ denotes the set $\{x \in [0,1]^* : 0 \ll x \ll 1\}$.

Example 4.4 *For each De Morgan system $\mathbb{I}_\beta$, define $\theta_\beta = ((0,1))^2 \cup \{(x,x) : x \in S_\beta = st^{-1}(2)\}$. Then it is straightforward to check that θ_β is a congruence of $\mathbb{I}^*_\beta$. Denote the corresponding homomorphic image by $\mathbb{H}_\beta$. Notice that $\mathbb{H}_\beta$ satisfies the equation $x \wedge \eta(x) \leq (y \vee \eta(y))^2$ (in fact, it satisfies $x \wedge \eta(x) \leq (y \vee \eta(y))^n$ for each $n \in \mathbb{N}$) but does not satisfy $x^2 = x$. So we have*

$$\mathcal{V}(\mathbf{1}) \subsetneq \mathcal{V}(\mathbf{2}) \subsetneq \mathcal{V}(\mathbf{3}) \subsetneq \mathcal{V}(\mathbb{H}_\beta) \subsetneq \mathcal{V}(\mathbb{I}_\beta)$$

Of course we don't know yet that the two last inclusions are covering pairs, meaning there are no other subvarieties in between.

Remark 4.1 *Given two De Morgan systems $\mathbb{I}_\beta$ and $\mathbb{I}_\gamma$, notice that if $\beta(x) = \gamma(x)$ for all $x \approx 0$, then $\beta(x) = \gamma(x)$ for all $x \in st^{-1}(2) = \{x \in [0,1]^* : x \approx 0$ or $x \approx 1\}$ and thus $\mathbb{H}_\beta = \mathbb{H}_\gamma$. Also, by the transfer principle the condition $\beta(x) = \gamma(x)$ for all $x \approx 0$ is equivalent to*

$$\text{there exists } \varepsilon > 0 \text{ such that for all } x < \varepsilon, \quad \beta(x) = \gamma(x)$$

That is β and γ agree on some neighborhood of 0.

So if β and γ agree on some neighborhood of 0, then $\mathcal{V}(\mathbb{H}_\beta) = \mathcal{V}(\mathbb{H}_\gamma) \subseteq \mathcal{V}(\mathbb{I}_\beta) \cap \mathcal{V}(\mathbb{I}_\gamma)$ and thus we do sometimes at least have non-trivial common extensions of two strict De Morgan logics.

Proposition 4.5 *Let $\mathbb{A}$ be a subalgebra of a non-standard extension $\mathbb{I}^*_\beta$ of $\mathbb{I}_\beta$, and let θ be a congruence on $\mathbb{A}$ with $\theta \nsubseteq \ker(st)$ and $\mathbb{A}/\theta \neq \mathbf{2}$. Then*

$$(A \cap ((0,1)))^2 \subseteq \theta,$$

that is, $\theta_\beta \subseteq \theta$ and thus $\mathcal{V}(\mathbb{A}/\theta) \subseteq \mathcal{V}(\mathbb{H}_\beta)$.

Proof. The proof of this result is completely parallel to the proof for $\mathbb{I}_\beta$ in Proposition 4.2. Since $\theta \nsubseteq \ker(st)$, there are $x, y \in A$ with $x \ll y$ but $x\theta y$. Using β, we may also assume that $1/2 \ll y$. Also since $st(A)$ is dense in $[0,1]$, there is $x' \in A$ with $x \leq x' \ll y$ and $1/2 \leq x'$. So, as in Proposition 4.2, we may assume that $1/2 \leq x \ll y$. Now, using again the denseness of $st(A)$ in $[0,1]$, pick $z \in A$ with $1 - z^o < \frac{x^o - y^o}{4}$. Then there is $n \in \mathbb{N}$ with $(z^o)^n, (z^o)^{n+1} \in (x^o, y^o)$ and thus $(z^o)^n, (z^o)^{n+1} \in ((x, y))$. By continuity of multiplication, it follows that $z^n \approx (z^o)^n$ and $z^{n+1} \approx (z^o)^{n+1}$ so that $z^n > 1/2$ and $z^n \theta (z^n)^i$ for all $i \in \mathbb{N}$. As in the standard argument, the convex hull of the sequence $\left\{(z^n)^i\right\}_{i \in \mathbb{N}}$ is $((0, z^n]$ and it follows that $(A \cap ((0, z^n])^2 \subseteq \theta$. Using β we get $(A \cap [\beta(z^n), 1)))^2 \subseteq \theta$ and since $\beta(z^n) \leq z^n$ the result follows. □

Corollary 4.6 *For any De Morgan system $\mathbb{I}_\beta$, $\mathcal{V}(\mathbb{H}_\beta)$ is the largest proper subvariety of $\mathcal{V}(\mathbb{I}_\beta)$ and thus all proper extensions of $L_{\mathbb{I}_\beta}$ satisfy $x \wedge \eta(x) \leq (y \vee \eta(y))^2$.*

Proof. Let $\mathcal{V}$ be a proper subvariety of $\mathcal{V}(\mathbb{I}_\beta)$. Then $\mathcal{V}$ is generated by the subdirectly irreducible algebras in $\mathcal{V}$. Let $\mathbb{A}$ be such a subdirectly irreducible element of $\mathcal{V}$. Then $\mathbb{A} = \mathbb{B}/\theta$ for some subalgebra $\mathbb{B}$ of a non-standard extension $\mathbb{I}^*_\beta$ of $\mathbb{I}_\beta$. If $\mathbb{A} = \mathbb{B}/\theta \cong \mathbf{2}$, then clearly $\mathbb{A}$ is in $\mathcal{V}(\mathbb{H}_\beta)$. If $\mathbb{A} = \mathbb{B}/\theta \ncong \mathbf{2}$ but $\theta \subseteq \ker(st)$, then st factors through the quotient given by θ, whence $st(\mathbb{B})$

which is a subalgebra different from **2** is in any variety containing $\mathbb{A}$. But $\mathcal{V}(st(\mathbb{B})) = \mathcal{V}(\mathbb{I}_\beta)$, and thus $\mathbb{I}_\beta$ is in $\mathcal{V}$, a contradiction of $\mathcal{V}$ being proper. Thus we must have $\theta \not\subseteq \ker(st)$ and $\mathbb{A} = \mathbb{B}/\theta \not\cong \mathbf{2}$ and so by the above result $\theta_\beta \subseteq \theta$ and $\mathcal{V}(\mathbb{A}) \subseteq \mathcal{V}(\mathbb{H}_\beta)$. The corollary follows since $\mathcal{V} \subseteq \mathcal{V}(\mathbb{H}_\beta)$. □

Corollary 4.7 *For any De Morgan system $\mathbb{I}_\beta$, $\mathcal{V}(\mathbb{I}_\beta)$ is not finitely generated, in fact, it is not generated by its finite members.*

Proof. If $\mathcal{V}(\mathbb{I}_\beta)$ were generated by its finite members, then it would be generated by finite subdirectly irreducibles $\mathbb{A}$ in $HSP_\mu(\mathbb{I}_\beta)$. But if $\mathbb{A} = \mathbb{B}/\theta$ for some subalgebra $\mathbb{B}$ of a non-standard extension $\mathbb{I}^*_\beta$ of $\mathbb{I}_\beta$ and $\mathbb{A}$ is finite, then either $\mathbb{A} \cong \mathbf{2}$ or $\theta \not\subseteq \ker(st)$, so that $\mathcal{V}(\mathbb{A}) \subseteq \mathcal{V}(\mathbb{H}_\beta)$. Thus the finite subdirectly irreducible algebras in $\mathcal{V}(\mathbb{I}_\beta)$ do not generate $\mathcal{V}(\mathbb{I}_\beta)$. □

5 Further questions

As can be seen from the results in the previous section, all indications are pointing in the direction of these logics not acting much like logics. There are no truth tables for checking logical equivalence for any of them. They are all distinct but the least extension for any of them already bunches them together, at least in batches corresponding to $\mathbb{H}_\beta = \mathbb{H}_\gamma$, that is, into batches for which β and γ agree on some neighborhood of 0. Of great importance to judging these structures as logics is knowing whether or not there are finite equational bases for these systems. This is definitely an important open question. Of interest is also a full understanding of the subvariety structure of these varieties. For example, we have seen that they are not finitely generated and that they have the finitely generated varieties of Boolean algebras and Kleene algebras in common. One question is whether there are any other finitely generated subvarieties. Looking in the other direction, one may ask for common generalizations. In particular it would be interesting to know what the variety generated by all strict De Morgan systems is.

These are just a few of the questions still open for varieties of strict De Morgan systems. In addition, it would be very interesting to study these same questions in the nilpotent case. The equations used here to separate the varieties generated by strict systems are certainly not enough to separate the nilpotent ones, but maybe some similar methods can be used. Also, we already know of at least one variety generated by a nilpotent De Morgan system which is finitely based, namely the one generated by the system given by the Lukasiewicz t-norm and the standard negation α. This system generates the variety of MV-algebras [4]. So the nilpotent case looks quite different.

Finally, we have included in our study the lattice operations on the unit interval in our type. This is of course not strictly necessary. One may want to study these logical questions in the setting of the ordered algebraic structures $\mathbb{I}_\beta = (I, \leq, 0, 1, \circ, \beta)$.

References

[1] R. Balbes, P. Dwinger, Distributive Lattices, University of Missouri Press (Columbia, Missouri), 1974.

[2] D. Brignole, A. Monteiro, *Characterization des algèbres de Nelson par des égalités*, Notas de Lógica Matemática, Instituto de Matemática Universidad del sur Bahía Blanca **20**(1964).

[3] S. Burris, H. D. Sankappanavar, A Course In Universal Algebra, Springer Verlag (Berlin/Heidelberg/New York), 1980.

[4] C. C. Chang, *Algebraic analysis of many-valued logics*, Trans. Amer. Math. Soc. **88**(1958), 467–490.

[5] C. L. Chang, Lee, R. C. T., *Some properties of fuzzy logic*, Information and Control **19**(1971), 417–431.

[6] C. Elkan, *The paradoxical success of fuzzy logic*, Proceedings of AAAI-93, American Association for Artificial Intelligence 1993 Conference, 698–703.

[7] M. Gehrke, *Robinson lattices and their spectra*, Algebra Universalis **32**(1994), 204–223.

[8] M. Gehrke, C. Walker, E. Walker, *Some comments on fuzzy normal forms*, Fuzzy Sets and Systems (to appear).

[9] M. Gehrke, C. Walker, E. Walker, *A mathematical setting for fuzzy logics*, International Journal of Uncertainty, Fuzziness and Knowledge-Based Systems **5**(1997), 223–238.

[10] M. Gehrke, C. Walker, E. Walker, *De Morgan systems on the unit interval*, International Journal of Intelligent Systems **11**(1996), 733–750.

[11] M. Gehrke, C. Walker, E. Walker, *Some comments on interval-valued fuzzy sets*, International Journal of Intelligent Systems **11**(1996), 751–759.

[12] R. Goldblatt, Lectures On The Hyperreals: An Introduction To Nonstandard Analysis, Graduate Texts in Mathematics **188**, Springer-Verlag (Berlin/Heidelberg/New York), 1998.

[13] J. A. Kalman, *Lattices with involution*, Trans. Amer. Math. Soc. **87**(1958), 485–491.

[14] S. C. Kleene, Introduction To Metamathematics, North-Holland Elsevier (Amsterdam/New York), 1952.

[15] D. E. Muller, *Treatment of transition signals in electronic switching circuits by algebraic methods*, IRE Transactions on Electronic Computers, **8**(1959).

[16] H. T. Nguyen, O. M. Kosheleva, V. Kreinovich, *Is the success of fuzzy logic really paradoxical? or What is the actual logic behind expert systems?*, International Journal of Intelligent Systems **11**(1996), 295–326.

[17] A. Robinson, Non-Standard Analysis, reprint of the second (1974) edition, Princeton Landmarks in Mathematics, Princeton University Press (Princeton, New Jersey), 1996.

[18] L. A. Zadeh, *Fuzzy sets*, Information and Control **8**(1965), 338-353.

CHAPTER 10

Structure Of Girard Monoids On $[0,1]$

S. JENEI [1]

Introduction

The structure of commutative, residuated, zero closed, lattice-ordered, integral monoids—in other words, Girard monoids—is investigated, when the underlying universe is the unit interval $[0,1]$. On $[0,1]$, the notion of a Girard monoid coincides with the notion of a left-continuous triangular norm with strong induced negation. Thus, this chapter investigates the structure of left-continuous triangular norms with strong induced negations. Based on an exhaustive geometrical description we discuss how to construct and how to decompose such triangular norms. Further, theorems are established on their continuity and integrality.

The general case. A *commutative lattice-ordered semigroup* (or *commutative ℓ-semigroup*) is a lattice L with a commutative semigroup multiplication $\circledast$ which satisfies the following conditions:

- the isotonicity condition

$$a \le b \text{ implies } a \circledast x \le b \circledast x$$

  holds for all $a, b, x \in L$

- products are distributive across binary joins, i.e.

$$x \circledast (y \vee z) = x \circledast y \vee x \circledast z$$

  for all $x, y, z \in L$

[1]Supported by the National Scientific Research Fund Hungary (OTKA F/032782) and by the Higher Education Research and Development Programme Hungary (FKFP 0051/2000).

S.E. Rodabaugh and E.P. Klement (eds.), Topological and Algebraic Structures in Fuzzy Sets, 277–308.
© 2003 *Kluwer Academic Publishers.*

An ℓ-semigroup is called an *ℓ-monoid* if there is an element in L such that it acts as a (left and right) neutral element of $\circledast$. An ℓ-monoid is called *integral* if the neutral element of the multiplication is the greatest element of L. For any commutative ℓ-semigroup $(L, \circledast)$, the operation

$$\rightarrow_{\circledast} : L \times L \rightarrow L$$

is called the *residual pair* of $\circledast$ if the following condition

$$x \circledast y \leq z \Leftrightarrow x \rightarrow_{\circledast} z \geq y$$

for all $x, y, z \in L$) is satisfied. Equivalently,

$$x \rightarrow_{\circledast} y$$

is the largest $z \in L$ such that

$$x \circledast z \leq y$$

If such a residual pair exists, then L is called *residuated.* For any $c \in L$, where $(L, \circledast)$ is a commutative, residuated ℓ-semigroup,

$$(x \rightarrow_{\circledast} c) \rightarrow_{\circledast} c \geq x$$

holds for all $x \in L$; and an element $x \in L$ is called *c-closed* if

$$(x \rightarrow_{\circledast} c) \rightarrow_{\circledast} c = x$$

We call a bounded, residuated ℓ-semigroup *zero closed*—where 0 is the least element if all the elements of L are 0-closed. A commutative, residuated, zero closed, integral ℓ-monoid is called a *Girard* monoid. Some works use the word "classical residuated lattice" for what we call Girard monoid; we refer to [29]. In any Girard monoid we can define residual complements by

$$x' = x \rightarrow_{\circledast} 0$$

and the dual semigroup operator $\diamond$ with the de Morgan rule:

$$x \diamond y = (x' \circledast y')'$$

Zero closed property of L yields that the residual complement operation $'$ is an involution (that is, an order reversing bijection) of L. Girard monoids form the algebraic counterpart of Girard's commutative affine linear logic [11] and are generalizations of MV-algebras [2], structures which play a central role in many-valued logic [33, 34, 3]. In addition, Girard monoids arise naturally in the field of measure-free conditioning (see, e.g., [39], [14]). When the underlying lattice L is the unit interval $[0, 1]$ of the real numbers, the notion of a Girard monoid coincides with the notion of a *left-continuous triangular norm with strong induced negation.* Due to the wide spectrum of fields that use triangular norms we will use the traditional $T(x,y)$ notation instead of $x \circledast y$.

Motivation for the $[0,1]$ **case.** T-norms are two-place functions that were first introduced by B. Schweizer and A. Sklar [36] in 1958 (following ideas of K. Menger [31] from 1942) in order to formulate properly the triangle inequality in probabilistic metric spaces. Since then, t-norms have been applied in various other mathematical disciplines including game theory, the theory of non-additive measures and integrals, the theory of measure-free conditioning, fuzzy set theory, fuzzy logic, fuzzy control, preference modeling and decision analysis, and artificial intelligence: being fully ordered semigroups they belong to classical algebra. A t-norm is defined on the unit square $[0,1]^2$ so it can be considered as solution of the associativity equation; this is the link to functional equations [1]. In probabilistic (statistical) metric spaces [36] they model the 'triangle inequality' of a probabilistic metric space, where distance of two objects is described by a probability distribution instead of a real number. In fuzzy sets theory, together with their duals—the triangular conorms—they are extensively used to model the intersection and union of fuzzy subsets, respectively. In fuzzy logic (which is a many valued propositional logic with a continuum of truth values modeled by the unit interval) t-norms and t-conorms model the (semantic) interpretation of the logical conjunction and disjunction, respectively. In the field of decision making, fuzzy preference modeling [6, 9] uses t-norms and t-conorms as well. T-norms are applied in control, in the theory of non-additive measures and integrals [35] and so on.

The condition of left-continuity is a frequently cited property and plays a central role in all the fields that use t-norms. The role of left-continuous t-norms with strong induced negations is even more relevant:

- First of all, the crisp (ordinary) intersection admits special continuity and discontinuity properties. To show this, let A be a proper subset of X (the universe of discourse). Then

$$A \cap B \begin{cases} = \emptyset, & B \subset A^c \\ \neq \emptyset, & B \not\subset A^c \end{cases}$$

  where A^c is the complement of A in X. Therefore the crisp intersection is left-continuous but not right-continuous operator at $B = A^c$ in the inclusion order topology. This fact suggests that a fuzzy intersection should be left-continuous.

- A recent result in the field of fuzzy preference modeling [38] draws the following conclusion: a nice mathematical model for fuzzy preference structures, where the preferences of individuals between two alternatives are described by a real number from $[0, 1]$ rather than from $\{0, 1\}$, requires the usage of a left-continuous t-norm with strong induced negation.
- Assuming the left-continuity condition we are entering the field of *residuated* semigroups in algebra.
- In probabilistic metric spaces left-continuity of the probability distributions requires only left-continuity of the t-norm (when a t-norm is used).
- In the context of fuzzy logic, left-continuity makes possible the evaluation of the interpretation of the implication by the residuum of the t-norm. In addition, the logical disjunction has to be *dual* to the logical conjunction. That is,
$$\varphi \vee \psi = \neg(\neg\varphi \wedge \neg\psi)$$
holds for all well-formed formulas φ and ψ. This is the unique approach that allows very nice and deep logical results to be proven [28]. From this fact one may conclude that only the Łukasiewicz logic is reasonable, since in this case the negation which is given by
$$\neg\varphi = \varphi \rightarrow 0$$
is strong, which means
$$\neg\neg\varphi = \varphi$$
for all well-formed formulas φ. The particular importance of this strongness condition comes from the fact that if we define disjunction to be dual to conjunction we get a reasonable connective if and only if the negation is interpreted by a strong negation. The translation of all the arguments above to the semantic stage yields that for a nice and reasonable fuzzy logic we have to require that the conjunction be a left-continuous t-norm which has strong induced negation. Otherwise the negation has to be chosen independently from $. \rightarrow 0$, the induced negation. This is the 'secret' of the Łukasiewicz logic: the Łukasiewicz t-norm is left-continuous and its induced negation is strong. Moreover, as an additional considerable argument we recall that the interpretation of the implication admits the contrapositive symmetry property with respect to the negation: that is,
$$I_T(x, y) = I_T(N(y), N(x))$$
for all $x, y \in [0, 1]$ if and only if the interpretation of the conjunction is a left-continuous t-norm T with strong induced negation N (see [19]). Therefore, the class of left-continuous t-norms with strong induced negations should play an important role in the field of non-classical logic.

In spite of their significance, the knowledge about left-continuous t-norms is rather poor at present; until recently there were no results in the literature where left-continuous t-norms stood as the focus of interest. Moreover, until recently there were no known examples left-continuous t-norms, except for the standard class of continuous t-norms.

Continuous t-norms have become well understood as ordinal sums of continuous Archimedean t-norms (see the famous and widely cited paper of Ling [30]) and have been used in several applications. The poor knowledge about left-continuous t-norms on one hand and the good understanding of continuous t-norms on the other hand result in the use of continuous t-norms when left-continuity would be sufficient in theory. This very much restricts the freedom of choice when the proper operation has to be found in the mathematical setting in question. In other words, this makes modeling in probabilistic metric spaces, in game theory, in the theory of non-additive measures and integrals, in the theory of measure-free conditioning, in fuzzy set theory, in fuzzy logic, in fuzzy control, in preference modeling and decision analysis, and in artificial intelligence much less flexible.

This chapter is organized as follows. After recalling the necessary definitions in the rest of this section, we introduce the rotation invariance and the quasi-inverse properties in Section 2. We investigate their geometrical meaning and their correspondence with left-continuous t-norms with strong induced negations. All the other results of this chapter are motivated by the observations of this section. Then, in Section 3 two theorems are established which are almost straight consequences of the previous results. In Section 4 a new method is introduced which—from any left-continuous t-norm which has no zero divisors—produces a left-continuous (but not continuous) t-norm with strong induced negation. The method is called rotation. Then in Section 5 a second method is introduced which produces as well left-continuous (but not continuous) t-norms with strong induced negations. The method is called rotation-annihilation. An infinite number of new families of left-continuous t-norms with strong induced negations can be generated with these constructions and this provides a tremendously wide spectrum of choice for the proper operation in the above-mentioned fields; this fairly enlarges the set of the three families which were until recently the only known examples. On the other hand, the introduced constructions bring us closer to the understanding of the structure of these operations and this may have consequences to the better understanding of e.g. the corresponding logics too (see [25, 12]). In Section 6 we define (in)decomposability and present the decomposition theorems. We illustrate the results with 3-dimensional plots.

Results of the present chapter have been presented in international conferences (first in [24]) and have been generalized in the setting of partially-ordered semigroups in [20]. A further method for constructing left-continuous t-norms is in [26].

1 Preliminaries

In the forthcoming definitions, the unit interval $[0,1]$ plays a central role. T-norms, implication functions, negations and all further definitions concerning them are defined on $[0,1]$. These notions and the corresponding definitions can be defined on any interval $[a,b] \subset \mathbb{R}$ and what we get is the linear transformation of the same notions on $[0,1]$. Since linear transformations are order-isomorphisms from the semigroup theoretic sense, it is not surprising that a certain operation admits a certain property if and only if the linear transformation of the operation admits the same property rewritten to $[a,b]$.

To repeat the essential definitions, a *triangular norm* (t-norm for short) is a *binary operation* T (that is, a function $T : [0,1]^2 \to [0,1]$) such that for all $x, y, z \in [0,1]$ the following four axioms (T1)-(T4) are satisfied:

(T1)	Symmetry	$T(x,y) = T(y,x)$
(T2)	Associativity	$T(x, T(y,z)) = T(T(x,y), z)$
(T3)	Monotonicity	$T(x,y) \leq T(x,z)$ whenever $y \leq z$
(T4)	Boundary condition	$T(x,1) = x$
(T4′)	Boundary condition	$T(0,x) = 0$
(T4″)	Range condition	$T(x,y) \leq \min(x,y)$.

It is immediate to see that (T3) and (T4) imply (T4') and that (T1), (T3) and (T4) imply (T4"). Now, we introduce a new class of two-place functions. This class will play a key-role from Section 5.

Definition 1. A *triangular subnorm* (t-subnorm for short) is a function $T : [0,1]^2 \to [0,1]$ such that for all $x, y, z \in [0,1]$ axioms (T1), (T2), (T3) and (T4") are satisfied.

Any t-norm is a t-subnorm. We say that a t-subnorm T has *zero divisors* if there are $x, y \in \,]0,1]$ such that $T(x,y) = 0$. A t-subnorm is said to be *continuous*, resp. *left-continuous*, if it is continuous, resp. left-continuous, as a two-place function. A continuous t-norm T is called Archimedean if $T(x,x) < x$ for all $x \in \,]0,1[$. As we have already mentioned one can define t-subnorms on any $[a,b] \subset \mathbb{R}$ and gets the notion of a t-subnorm on $[a,b]$. Then *for any t-(sub)norm T, the function $T_{[a,b]}$: $[a,b] \times [a,b] \to [a,b]$ defined by*

$$T_{[a,b]}(x,y) = a + (b-a) \cdot T\left(\frac{x-a}{b-a}, \frac{y-a}{b-a}\right)$$

is a t-(sub)norm on $[a,b]$. If $T_{[a,b]}$ is a t-(sub)norm on $[a,b]$, then the function

$$T : [0,1] \times [0,1] \to [0,1]$$

defined by

$$T(x,y) = \frac{T_{[a,b]}(a + x(b-a), a + y(b-a)) - a}{b-a}$$

is a t-(sub)norm. Call $T_{[a,b]}$ the *linear transformation* of T into $[a,b]$. Similarly, call T the linear transformation of $T_{[a,b]}$ into $[0,1]$.

A *negation* [37] N is a non-increasing function on $[0,1]$ with boundary conditions $N(0) = 1$ and $N(1) = 0$. A negation is called *strong* if N is an involution; that is, if in addition $N(N(x)) = x$ holds for all $x \in [0,1]$. A negation is strong if and only if its graph is invariant with respect to the reflection at the median (given by $y = x$). A strong negation is automatically a strictly decreasing, continuous function and hence it has exactly one fixed point.

Let $T : [0,1]^2 \to [0,1]$ be a function satisfying (T1) and (T3). The implication $I_T : [0,1]^2 \to [0,1]$ generated by T is given by

$$I_T(x,y) = \sup\{t \in [0,1] \mid T(x,t) \leq y\}$$

If T is left-continuous, then I_T is called the *residual implication* generated by T. An equivalent formulation of left-continuity is the so-called "adjoint pair condition": for all $x, y, z \in [0,1]$, we have

(T5) *Adjoint pair condition.* $T(x,y) \leq z \Leftrightarrow I_T(x,z) \geq y$.

For a left-continuous $T : [0,1]^2 \to [0,1]$ which admits (T1) and (T3) define

$$N_T(x) = I_T(x,0) \tag{1}$$

for $x \in [0,1]$. If N_T is a negation (this holds, e.g. if T in addition satisfies (T4)), then N_T is called the induced negation of T. We say that a t-norm T has strong induced negation if N_T is a strong negation, that is, if for all $x \in [0,1]$ it satisfies the involution condition:

(T6) *Involution.* $N_T(N_T(x)) = x$.

In Figure 1 we present the three-dimensional plots of the product t-norm given by $T(x,y) = x \cdot y$ and an ordinal sum with one Łukasiewicz summand on $[0, 0.4]$ (see Section 1.1 and (3)).

One can see easily the induced negation of them on the plane which is spanned by the axes X and Y (see the bold lines in Figure 1). It follows the border between the regions where T is zero and where T is positive except for the 'vertical parts'. The border belongs to the zero region because of left continuity. None of the negations in the figures is strong. T-norms with strong induced negation will be shown later, e.g. in Figure 2.

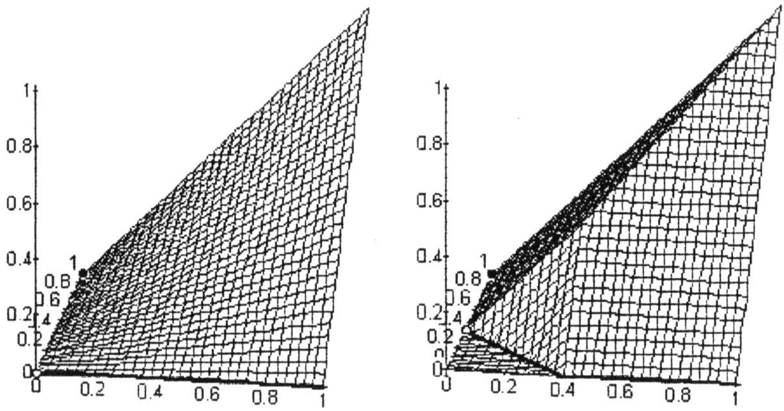

Figure 1: How the induced negation can be seen from the graph of the t-norm

Definition 2. Let N be a strong negation and t be its unique fixed point. Let $d \in]t, 1]$. Then $N_d : [0,1] \to [0,1]$ defined by

$$N_d(x) = \frac{N(\, x \cdot (d - N(d)) + N(d)) - N(d)}{d - N(d)}$$

is a strong negation. Call N_d the *zoomed d-negation* of N.

Definition 3. $T : [0,1]^2 \to [0,1]$ is a *left-continuous t-norm with strong induced negation* if it satisfies (T1)-(T6).

1.1 Structure of continuous t-norms

The following method of constructing a new t-norm from a family of given t-norms is based on the results of Climescu [5] and Clifford [4] concerning ordinal sums of semigroups (see also [30]). Here, we state the form of the theorem which is applied to t-norms (see [16]for an up-to-date discussion).

Theorem 1. *Suppose that* $\{[a_i, b_i]\}_{i\in K}$ $(a_i < b_i)$ *is a countable family of non-overlapping, closed subintervals of* $[0,1]$, *denoted by* $\mathcal{I}$. *With each* $[a_i, b_i] \in \mathcal{I}$ *associate a t-norm* T_i. *Let* $T_{[a_i,b_i]}$ *be the linear transformation of* T_i *into* $[a_i, b_i]$. *Let* T *be a function defined on* $[0,1]^2$ *by*

$$T(x,y) = \begin{cases} T_{[a_i,b_i]}(x,y) & \text{if } (x,y) \in [a_i,b_i]^2 \\ \min(x,y) & \text{otherwise} \end{cases} \tag{2}$$

Then T *is a t-norm.*

In this case T is called the *ordinal sum* of $\{([a_i,b_i],T_i)\}_{i\in\mathcal{I}}$ and each T_i is called a *summand.* We present below the theorem which describes the structure of continuous t-norms. The original result of Mostert and Shields [32] corresponded to I-semigroups on $[0,1]$.

Theorem 2. *Suppose T is a continuous t-norm. Then either T is continuous Archimedean; or $T = \min$; or there exists a family $\{([a_i,b_i],T_i)\}_{i\in\mathcal{I}}$ with continuous Archimedean t-norms T_i such that T is the ordinal sum of this family.*

Continuous Archimedean t-norms (as ordered semigroups) are isomorphic to either the product t-norm $T(x,y) = x\cdot y$ (those called *strict* t-norms) or to the Łukasiewicz t-norm (see (3)) (those called *nilpotent* t-norms).

1.2 Known examples of left-continuous t-norms with strong induced negations

Only three families are known. By "family" we understand a t-norm T together with its f-transformations which are t-norms as well and are defined by $T_f(x,y) = f^{-1}(T(f(x),f(y)))$, where f is any increasing bijection of $[0,1]$. That is, a given family is a set of t-norms which are order-isomorphic in the semigroup theoretic sense. As a consequence, we can choose an arbitrary member from a family and we can call it the representative member of the family. This member is usually chosen so that its defining formula is the simplest possible. According to this terminology and by the last sentence of Section 1.1, all the strict t-norms (and all the nilpotent t-norms) form a single family.

The first family is the family of nilpotent t-norms, and their representative is the Łukasiewicz t-norm given by

$$T(x,y) = \max(x+y-1,0). \tag{3}$$

Another one is the family of nilpotent minimum (introduced in [8]), its representative is given by

$$T(x,y) = \begin{cases} 0 & \text{if } y \le 1-x \\ \min(x,y) & \text{otherwise} \end{cases} \tag{4}$$

and its three-dimensional plot can be seen in Figure 7. The third family was introduced in [19], a representative is given by (5), and its plot can also be seen in Figure 7.

$$T(x,y) = \begin{cases} 0 & \text{if } x \le 1-y \\ \frac{1}{3}+x+y-1 & \text{if } \frac{1}{3} \le x,y \le \frac{2}{3} \text{ and } x > 1-y \\ \min(x,y) & \text{otherwise} \end{cases} \tag{5}$$

2 Geometry of left-continuous triangular norms with strong induced negations

The purpose of this section is to make the first step toward understanding the structure of left-continuous t-norms with strong induced negations. For two-placed functions on $[0,1]$ two properties are introduced: the rotation invariance property and the quasi-inverse property. It is proved that these properties are characteristic for the class left-continuous t-norms with strong induced negations. The two properties turn out to be equivalent on the class of symmetric, non-decreasing two-place functions on $[0,1]$; that is, such a function admits the rotation invariance property if and only if it admits the quasi-inverse property. These properties have equivalent geometrical counterparts which are investigated, explained in detail, and for which examples are given. These geometrical counterparts can be represented in 3 and 2 dimensions. The missing proofs can be found in [18].

2.1 Rotation invariance and quasi-inverse properties

First, we define two properties:

Definition 4. Let $T : [0,1]^2 \to [0,1]$ be a function satisfying (T3), and let N be a strong negation. We say that T admits the *rotation invariance property* with respect to N or is *rotation invariant* with respect to N if for all $x, y, z \in [0,1]$, we have

$$T(x,y) \leq z \Leftrightarrow T(y, N(z)) \leq N(x)$$

In addition, suppose that T satisfies (T1) and (T5). We say that T admits the *quasi-inverse property* with respect to N if for all $x, y, z \in [0,1]$. we have

$$I_T(x,y) = z \Leftrightarrow T(x, N(y)) = N(z)$$

Theorem 3. *Any rotation invariant monotone binary operation is left-continuous. That is, any function $T : [0,1]^2 \to [0,1]$ which satisfies (T3) and which is rotation invariant with respect to a strong negation N is left-continuous.*

Proposition 1 given below claims that the only possible choice for the negation in Definition 4 is the induced negation N_T if T is a left-continuous t-norm. In the light of this proposition it is sufficient to say (without mentioning explicitly the negation) that a t-norm is rotation invariant (then it is left-continuous by Theorem 3 and thus the induced negation of it exists) or a left-continuous t-norm admits the quasi-inverse property.

The importance of the two properties above is due to the fact that, as we will see in Theorem 4, any left-continuous t-norm with strong induced negation has the rotation invariance property and the quasi-inverse property. Moreover, a left-continuous t-norm which admits one of the above defined properties turns out to be a left-continuous t-norms with strong induced negation. Hence, these properties are characteristic for the class of left-continuous t-norms with strong induced negations (Corollary 1).

Proposition 1. *Let T be a t-norm and N be a strong negation. If T admits the rotation invariance property with respect to N, or T admits the quasi-inverse property with respect to N then $N = N_T$.*

Theorem 4. *Let $T : [0,1]^2 \to [0,1]$ be a left-continuous function which satisfies (T1), (T2) and (T3) and suppose that N_T (given by (1)) is a strong negation. Then for all $x, y, z \in [0,1]$ the following assertions hold:*

i. $I_T(x,y) = N_T(T(x, N_T(y)))$, *($I_T = I_S$)*[2]

ii. $I_T(x,y) = z \Leftrightarrow T(x, N_T(y)) = N_T(z)$, *(quasi-inverse)*

iii. $I_T(x,y) = I_T(N_T(y), N_T(x))$, *(contrapositive symmetry of I_T)*

iv. $T(x,y) \le z \Leftrightarrow T(y, N_T(z)) \le N_T(x)$. *(rotation invariance)*

In particular, if T is a left-continuous t-norm with strong induced negation N_T, then (i–iv) *hold.*

Corollary 1. *Let N be a strong negation and T be a t-norm. The following three assertions are equivalent:*

i. *T is rotation invariant (with respect to N),*

ii. *T is left-continuous and admits the quasi-inverse property (with respect to N),*

iii. *T is left-continuous and its induced negation equals N.*

Moreover, the equivalence between (i) *and* (ii) *holds for any $T : [0,1]^2 \to [0,1]$ which satisfies (T1) and (T3).*

Proof. First notice that T is left-continuous in each case (see Theorem 3 for case (i)). Proposition 1 and Theorem 4 prove the first statement. The proof of the second statement then follows from a result in [7], a second proof is given in [18]. □

---

[2] It expresses that the R-implication and the S-implication coincide.

2.2 Geometrical interpretation of the rotation invariance and the quasi-inverse property

The given names and the geometrical interpretation of the rotation invariance property and quasi-inverse property are explained now:

Rotation invariance property: Let N be a strong negation. First observe that the transformation

$$\sigma : [0,1] \times [0,1] \times [0,1] \to [0,1] \times [0,1] \times [0,1]$$

given by

$$(x, y, z) \mapsto (y, N(z), N(x))$$

is of order 3. Therefore it can be considered as a rotation. In order to make this observation more visual, let $N(x) = 1 - x$ (the standard negation). Then easy calculation verifies that σ is indeed a rotation of $[0,1]^3$ with angle $\frac{2\pi}{3}$ around the axis which is based on the points $(0,0,1)$ and $(1,1,0)$. By virtue of the above, the formula

$$T(x, y) \leq z \Leftrightarrow T(y, N(z)) \leq N(x)$$

means exactly that the part of the space $[0,1]^3$ which is *above the graph of* T remains invariant under σ. Equivalently, we have

$$T(x, y) > z \Leftrightarrow T(y, N(z)) > N(x)$$

for all $x, y, z \in [0,1]$, and this means exactly that the part of the space $[0,1]^3$ which is *strictly below the graph of* T remains invariant under σ.

Self quasi-inverse property: Now, we recall the definition of quasi-inverses of decreasing real functions [36] in the form which is restricted to functions of type $[0,1] \to [0,1]$. Let $f : [0,1] \to [0,1]$ be a decreasing function, and let $f^* : [0,1] \to [0,1]$ be a function fulfilling the following conditions:

i. If y is in the range of f, then $f^*(y)$ is in $f^{-1}(\{y\})$.

ii. If y in not in the range of f, then (by declaring $\sup \emptyset = 0$)

$$f^*(y) = \sup\{t \in [0,1] \mid f(t) \geq y\}$$

Call f^* a *quasi-inverse* of f. Generally, such an f^* is not unique. If f is a decreasing bijection of $[0,1]$, then f^* is unique and coincides with f^{-1}, the usual inverse of f.

It can be proved that for any left-continuous t-norm T, the negation of any partial mapping $T(x,\cdot)$ of T is quasi-inverse of itself if and only if T admits the quasi-inverse property. This fact motivates the name of this property in Definition 4.

Proposition 2. *Let T be left-continuous t-norm. Then the (induced) negation of the x-partial mapping defined by $f_x : [0,1] \to [0,1]$, $y \mapsto N_T(T(x,y))$ is quasi-inverse of itself for all $x \in [0,1]$ if and only if T admits the quasi-inverse property.*

In the frame of real functions the quasi-inverse has a geometrical interpretation: there is a simple geometrical way to construct the graph of the quasi-inverse f^* from the graph of f:

i. Draw vertical line segments at the discontinuities of f.

ii. Reflect the obtained graph at the first median, i.e., the graph of the identity function.

iii. Remove all vertical line segments from the reflected graph except for one point (this has to be done in such a way that $f^*(y) \in f^{-1}(\{y\})$) is satisfied).

Now, let $N(x) = 1 - x$. Then the geometrical interpretation of the negation is the reflection of the graph at the line given by $y = \frac{1}{2}$. In this case, the graph of any partial mapping $T(x,\cdot)$ has the following geometrical property: first extend the discontinuities of $T(x,\cdot)$ with vertical line segments; then the obtained graph is invariant under the reflection at the second median (given by $y = 1 - x$).

In Figure 2 below, we demonstrate the geometrical meaning of the rotation invariance and quasi-inverse properties. The first row presents the three-dimensional plots of the t-norms given by (3), (4) and (5), respectively; and since their induced negations equal $1 - x$ one can recognize easily the geometrical meaning of the rotation invariance property. The second row presents plots of the partial mappings $T\left(x, \frac{1}{2}\right)$ of the t-norms given by (3), (4) and (5), respectively; and again, since their induced negations equal $1 - x$ one can recognize easily the geometrical meaning of the quasi-inverse property.

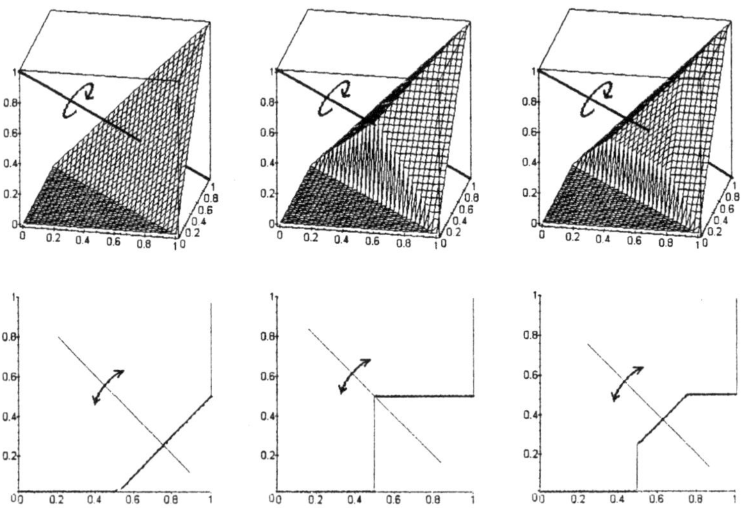

Figure 2: Geometrical interpretation of the rotation invariance property (first row) and the quasi-inverse property (second row)

3 Continuity of left-continuous triangular norms with strong induced negations and their boundary condition

Based on the results of the previous section, we present here two theorems. First, we assert that continuity of a left-continuous t-norm with strong induced negation is equivalent with the strict increasing of it on the domain where its value is positive. Second, the connection between the concept of strongness of the induced negation and the boundary condition is discovered: we show that axioms of a left-continuous t-norm with strong induced negation are not independent, namely, the boundary condition follows from the other axioms. The details are in [17].

3.1 Continuity

Remark 1. From the geometrical interpretation of the quasi-inverse property one observes that the graph of a partial mapping has jump discontinuity ("vertical segment") if and only if it has constant part ("horizontal segment") (see

Figure 2). This leads to the following conjecture: continuity is equivalent with strict monotonicity on the domain where the value of T is bigger than 0. We turn now to the discussion of this topic.

Definition 5. A t-norm T admits the *cancellation law on the positive domain* if $T(x, y) = T(x, z) > 0$ implies $y = z$. Equivalently, a t-norm T is *strict on the positive domain* if $T(x, y) > 0$ implies $T(x, y^*) > T(x, y)$ whenever $y^* > y$.

Theorem 5. *Suppose T is a left-continuous t-norm with strong induced negation. Then T is strict on the positive domain if and only if T is continuous.*

The 'if' part of Theorem 5 can be derived easily from deep theorems which have quite long proofs: a continuous t-norm with strong induced negation is necessarily *Archimedean* and hence *nilpotent* by Theorem 2. Then strict increasing on the positive domain can be derived from the well-known representation theorem of continuous Archimedean t-norms (see, e.g., [30]). Therefore the important message of Theorem 5 is the other direction: if we want to find new left-continuous (but not continuous) t-norms with strong induced negations, then those t-norms can not be strict on the positive domain. Although the 'if' part is a consequence of deep theorems, the idea behind the proof of 'only if' of Theorem 5 is quite simple (see Remark 1) and its proof is quite short. This proof is based on a single fact which has a one line proof; namely, that any left-continuous t-norm with strong induced negation admits the quasi-inverse property (see Theorem 4).

3.2 The boundary condition

The geometrical interpretation of the rotation invariance property motivates the observation of the (otherwise surprising) connection between the concept of boundary condition and the concept of strongness of the induced negation:

Lemma 1. *Let $T : [0, 1]^2 \to [0, 1]$ be a function which satisfies (T1) and (T3), and which is rotation invariant with respect to a strong negation N. Then the following statements hold true:*

i. *$T(x, y) \leq \min(x, y)$ for all $x, y \in [0, 1]$ if and only if $T(x, N(x)) = 0$ for all $x \in [0, 1]$.*

ii. *(T4) holds if and only if N equals N_T, the induced negation of T. (In this case the induced negation of T is obviously strong.)*

Proof. $T(x, y) \leq \min(x, y)$ for all $x, y \in [0, 1]$ if and only if for all $x, y \in [0, 1]$ we have $T(y, x) \leq x$. This holds if and only if $T(1, x) \leq x$ for all $x \in [0, 1]$

by (T3). This is equivalent with $T(x, N(x)) \leq 0$ by the rotation invariance property and it is equivalent with $T(x, N(x)) = 0$ for all $x \in [0,1]$. This proves (i). For the proof of (ii), observe first that T is left-continuous by Theorem 3. Therefore N is the induced negation if and only if

$$T(x,y) = 0 \Leftrightarrow y \leq N(x)$$

holds. $T(x,y) = 0$ is equivalent with $T(x,y) \leq 0$ which is equivalent with $T(y,1) \leq N(x)$ by the rotation invariance property. The boundary condition is equivalent with $T(y,1) = y$; hence we have $T(x,y) = 0 \Leftrightarrow y \leq N(x)$ if and only if (T4) holds. □

Theorem 6. *Let $T : [0,1]^2 \to [0,1]$ be a function satisfying (T1), (T2), (T3), (T5) and (T6). Then T satisfies (T4).*

Proof. Theorem 4 shows that T admits the rotation invariance property with respect to N_T. Then Lemma 1(ii) ends the proof. □

We briefly recall the generalization of this theorem to the algebraic setting [20]: *a commutative, residuated, zero closed po-semigroup is necessarily an integral monoid. A commutative, residuated, zero closed ℓ-semigroup is necessarily a Girard monoid.*

4 Rotation construction

A new algebraic construction, called rotation, is introduced in this section which, from any left-continuous t-norm which has no zero divisors, produces a left-continuous (but not continuous) t-norm with strong induced negation. The name is motivated by a geometrical feature of the resulted t-norm; namely, its graph is produced via the 'rotation' of the graph of some linear transformation of the starting t-norm. For the details, we refer to [21].

Theorem 7. *Let N be a strong negation, t its unique fixed point, and T be a left-continuous t-norm. Let T_1 be the linear transformation of T into $[t,1]$, $I^+ =]t,1]$, and $I^- = [0,t]$. Define $T_{\mathbf{J}}$ and $I_{T_{\mathbf{J}}}$ (of types $[0,1] \times [0,1] \to [0,1]$) by*

$$T_{\mathbf{J}}(x,y) = \begin{cases} T_1(x,y) & \text{if } x,y \in I^+ \\ N\left(I_{T_1}(x, N(y))\right) & \text{if } x \in I^+,\ y \in I^- \\ N\left(I_{T_1}(y, N(x))\right) & \text{if } x \in I^-,\ y \in I^+ \\ 0 & \text{if } x,y \in I^- \end{cases} \qquad ((6a))$$

$$I_{T_J}(x,y) = \begin{cases} I_{T_1}(x,y) & \text{if } x,y \in I^+ \\ N(T_1(x, N(y))) & \text{if } x \in I^+,\ y \in I^- \\ 1 & \text{if } x \in I^-,\ y \in I^+ \\ I_{T_1}(N(y), N(x)) & \text{if } x,y \in I^- \end{cases} \qquad ((6b))$$

T_J is a left-continuous t-norm if and only if either

C1. *T has no zero divisors, or*

C2. *there exists $c \in]0,1]$ such that for any zero divisor x of T we have $I_T(x,\ 0) = c$.*

In this case, the induced negation of T is N and the residual implication generated by T_J is I_{T_J}.

Proof. For the sufficiency of C1, see [21]. The rest of the proof is in [15]. □

Definition 6. Call T_J given by (6a) the *N-rotation* of T. If $N(x) = 1 - x$ (the standard negation), then call T_J simply the *rotation* of T.

Remark 2.

i. For the justification of the name 'N-rotation' observe that in formulae (6) the definition on the domains $\mathbf{I}^+ \times \mathbf{I}^-$ and $\mathbf{I}^- \times \mathbf{I}^+$ is just the extension of the values on $\mathbf{I}^+ \times \mathbf{I}^+$ which induces quasi-inverse property of T_J with respect to N. Then the definition on the domain $\mathbf{I}^- \times \mathbf{I}^-$ is again the extension of the values on $\mathbf{I}^+ \times \mathbf{I}^-$ and $\mathbf{I}^- \times \mathbf{I}^+$ which induces the quasi-inverse property of T_J with respect to N. So T_J admits the quasi-inverse property with respect to N, hence Corollary 1 ensures that the graph of T_J is 'rotation invariant' in the sense which is described at the beginning of Section 2.2.

ii. It is immediate to see that (t,t) (t is the fixed point of N) is always a discontinuity point of T_J and that it is the unique discontinuity point of T_J if and only if T_1 is a strict t-norm.

iii. From the proof of Theorem 7 the following fact is easily derived: if T is a left-continuous t-subnorm without zero divisors or T is the t-subnorm defined by $T(x,y) \equiv 0$ in Theorem 7, then T_J given by (6a) is a left-continuous t-subnorm which is rotation invariant with respect to N and its residual implication is given by (6b). This observation will be useful in the examples of Section 5.

Examples: When we evaluate T_1 only with continuous t-norms (that is, with ordinal sums of continuous Archimedean summands) without zero divisors we can obtain by rotation an infinite number of new families of left-continuous but not continuous t-norms with strong induced negations. Moreover T_1 can be evaluated with ordinal sums of *left*-continuous (or/and continuous) t-norms. The left-continuous t-norms may be generated by a previous rotation. Alternatively, T_1 can be evaluated with ordinal sums of (left-)continuous t-*sub*norms. The interested reader can find details of this topic in [16]. The only thing we have to pay attention is that either the ordinal sums don't have zero divisors, or the zero values of the graph of T_1 has to be in a square (compare with the left-hand side of Figure 3 and with condition C2 in Theorem 7). Iteration of these ideas leads to quite 'exotic and beautiful' t-norms. In Figure 3 the rotation of the minimum t-norm (center) and the rotation of the product t-norm (right) can be seen. We have highlighted the rotated units. Observe that the nilpotent minimum t-norm (given by (4)) is just the rotation of the minimum t-norm. In Figure (4a) a t-norm and its rotation is seen. The two figures of Figure (4b) present the rotations of certain ordinal sums: first, the ordinal sum has two summands—the Łukasiewicz t-norm and the product t-norm; and on the rightmost figure the summand is the rotation of the product.

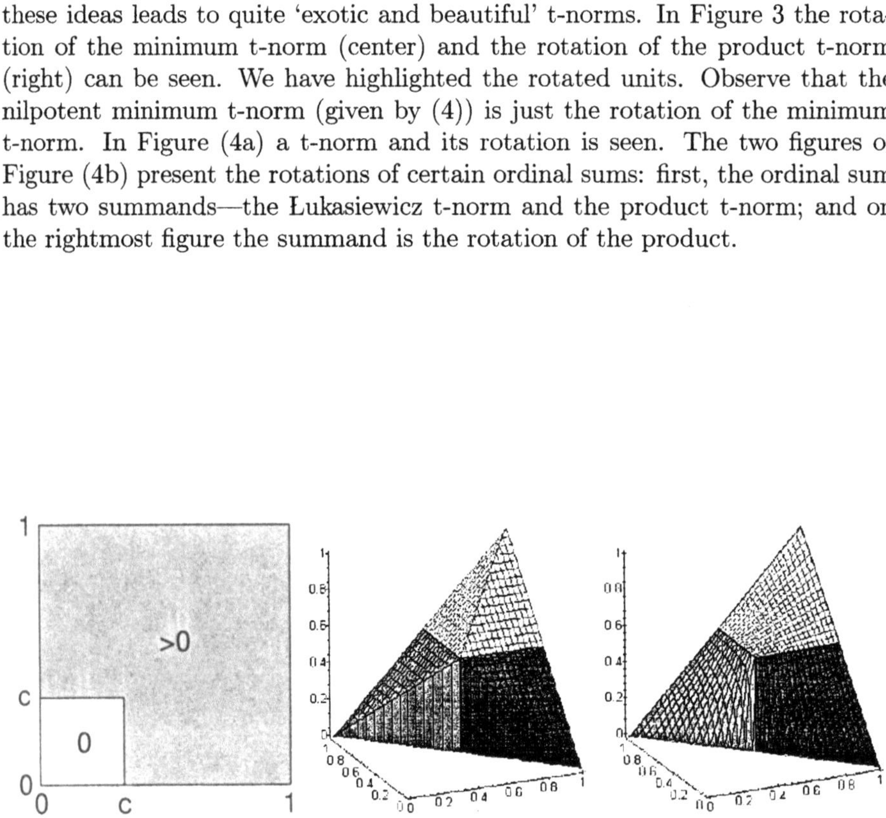

Figure 3: Zero values of t-norms suitable for rotation construction (left), rotation of minimum t-norm (center), and rotation of product t-norm (right)

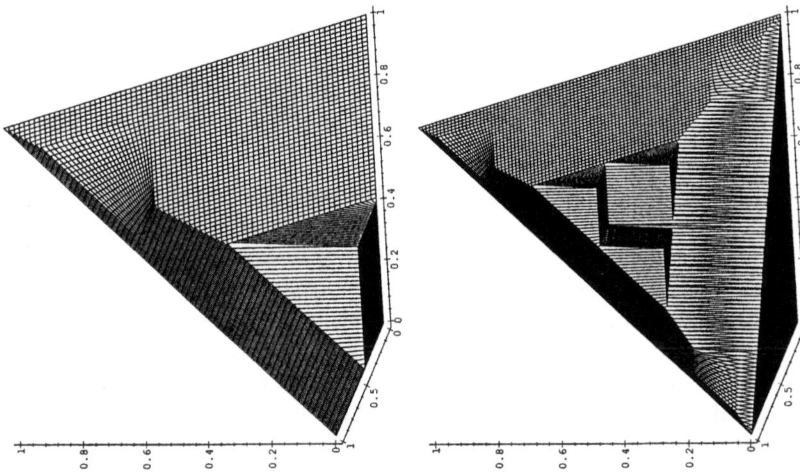

Figure 4a: T-norm with zero divisors and its rotation

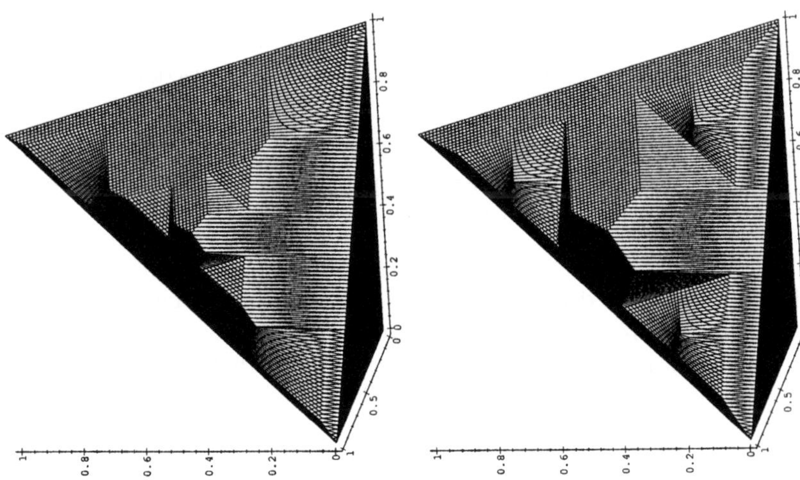

Figure 4b: T-norms generated by combination of rotation and ordinal sums

5 Rotation-annihilation construction

In this section a second construction, called rotation-annihilation, is introduced. Let T_1 be a left-continuous t-norm. If T_1 has no zero divisors, then let T_2 be a left-continuous rotation invariant t-subnorm. If T_1 has zero divisors then let T_2 be a left-continuous rotation invariant t-norm. From each such pair (T_1, T_2) the rotation-annihilation construction produces a left-continuous t-norm with strong induced negation.

The *rotation* part of the name is motivated by a geometrical feature of the resulted t-norm, namely, a part of the graph of the resulted t-norm is produced via the 'rotation' of the graph of some linear transformation of a left-continuous t-norm T_1. Annihilation is a certain transformation: Some values of an associative two-place function are replaced by the least element of the underlying universe. The question is whether the obtained operation inherits the associativity of the starting operation. Some results corresponding to this problem have already been published in [19] and led to the discovery of the family, which is defined in (5). We slightly modify the problem of annihilation here and prove that preservation of associativity is equivalent with the rotation invariance property of the starting function. The *annihilation* part of the name is due to the fact that a part of the resulted t-norm is derived as the annihilation of some linear transformation of a rotation invariant t-subnorm T_2. Briefly, we rotate T_1, annihilate T_2, and the resulted t-norm is built with the help of them.

Definition 7. Let N be a strong negation, t its unique fixed point, $d \in]t, 1[$, and N_d be the zoomed d-negation of N. Let T_1 be a left-continuous t-norm.

i. If T_1 has no zero divisors, then let T_2 be a left-continuous t-subnorm which admits the rotation invariance property with respect to N_d. Further, let $\mathbf{I}^- = [0, N(d)[$, $\mathbf{I}^0 = [N(d), d]$, and $\mathbf{I}^+ =]d, 1]$.

ii. If T_1 has zero divisors, then let T_2 be a left-continuous t-norm which admits the rotation invariance property with respect to N_d (that is, by Corollary 1, a left-continuous t-norm with strong induced negation equal with N_d). Further, let $\mathbf{I}^- = [0, N(d)]$, $\mathbf{I}^0 =]N(d), d[$, and $\mathbf{I}^+ = [d, 1]$.

Let T_3 be the linear transformation of T_1 into $[d, 1]$, T_4 be the linear transformation of T_2 into $[N(d), d]$, and T_5 be the annihilation of T_4 given by

$$T_5(x, y) = \begin{cases} 0 & \text{if } x, y \in [N(d), d] \text{ and } x \leq N(y) \\ T_4(x, y) & \text{if } x, y \in [N(d), d] \text{ and } x > N(y) \end{cases}$$

Define $T_{\mathbf{J}} : [0,1] \times [0,1] \to [0,1]$ and $I_{T_{\mathbf{J}}} : [0,1] \times [0,1] \to [0,1]$ by

$$T_{\mathbf{J}}(x,y) = \begin{cases} T_3(x,y) & \text{if } x,y \in \mathbf{I}^+ \\ N\left(I_{T_3}\left(x, N\left(y\right)\right)\right) & \text{if } x \in \mathbf{I}^+,\ y \in \mathbf{I}^- \\ N\left(I_{T_3}\left(y, N\left(x\right)\right)\right) & \text{if } x \in \mathbf{I}^-,\ y \in \mathbf{I}^+ \\ 0 & \text{if } x,y \in \mathbf{I}^- \\ T_5\left(x,y\right) & \text{if } x,y \in \mathbf{I}^{\mathbf{0}} \\ y & \text{if } x \in \mathbf{I}^+,\ y \in \mathbf{I}^{\mathbf{0}} \\ x & \text{if } x \in \mathbf{I}^{\mathbf{0}},\ y \in \mathbf{I}^+ \\ 0 & \text{if } x \in \mathbf{I}^-,\ y \in \mathbf{I}^{\mathbf{0}} \\ 0 & \text{if } x \in \mathbf{I}^{\mathbf{0}},\ y \in \mathbf{I}^- \end{cases} \tag{(7a)}$$

$$I_{T_{\mathbf{J}}}(x,y) = \begin{cases} I_{T_3}\left(x,y\right) & \text{if } x,y \in \mathbf{I}^+ \\ N\left(T_3(x, N\left(y\right))\right) & \text{if } x \in \mathbf{I}^+,\ y \in \mathbf{I}^- \\ 1 & \text{if } x \in \mathbf{I}^-,\ y \in \mathbf{I}^+ \\ I_{T_3}\left(N\left(y\right), N\left(x\right)\right) & \text{if } x,y \in \mathbf{I}^- \\ I_{T_4}\left(x,y\right) & \text{if } x,y \in \mathbf{I}^{\mathbf{0}},\ x > y \\ 1 & \text{if } x,y \in \mathbf{I}^{\mathbf{0}},\ x \le y \\ y & \text{if } x \in \mathbf{I}^+,\ y \in \mathbf{I}^{\mathbf{0}} \\ N\left(x\right) & \text{if } x \in \mathbf{I}^{\mathbf{0}},\ y \in \mathbf{I}^- \\ 1 & \text{if } x \in \mathbf{I}^-,\ y \in \mathbf{I}^{\mathbf{0}} \\ 1 & \text{if } x \in \mathbf{I}^{\mathbf{0}},\ y \in \mathbf{I}^+ \end{cases} \tag{(7b)}$$

Call $T_{\mathbf{J}}$ the *N-d-rotation-annihilation* of T_1 and T_2. If $N(x) = 1 - x$ (the standard negation), then call $T_{\mathbf{J}}$ simply the *d-rotation-annihilation* of T_1 and T_2.

Theorem 8. *Let N be a strong negation, t its unique fixed point, $d \in]t, 1[$, and T_1 be a left-continuous t-norm. Take T_2, depending on the zero divisors of T_1—as it is taken in Definition 7, and let $T_{\mathbf{J}}$ be the N-d-rotation-annihilation of T_1 and T_2. Then $T_{\mathbf{J}}$ is a left-continuous t-norm, its induced negation is N, and the residual implication generated by $T_{\mathbf{J}}$ is given by (7b).*

Remark 3.

i. It is immediate to see that each point in $\{(x, N(x)) \mid x \in [N(d), d]\}$ is a discontinuity point of $T_{\mathbf{J}}$ and that these are the only discontinuity points of $T_{\mathbf{J}}$ if and only if T_1 is a strict t-norm and T_2 is continuous.

ii. Similar to Remark 2/iii observe that the proof of Theorem 8 induces the following fact: if T_1 is a left-continuous t-*sub*norm in Theorem 8 (the rest is unchanged), then $T_{\mathbf{J}}$ given by (7a) is a left-continuous t-*sub*norm which is rotation invariant with respect to N and its residual implication is given by (7b). This observation will be useful in the following examples:

Examples. To show an example for a left-continuous rotation invariant t-subnorm (which is not a t-norm), consider the following operation:

$$T(x,y) = \max(x+y-1-\varepsilon, 0), \qquad ((8))$$

where ε is in $[0,1]$. Observe that the case $\varepsilon = 0$ is just the Łukasiewicz t-norm, while $\varepsilon = 1$ defines the drastic t-subnorm $T : [0,1]\times[0,1] \to [0,1], \ T(x,y) \equiv 0$ (which is not the so called drastic t-norm!).

For the illustration of the rotation-annihilation construction we present two examples. On the left-hand side of Figure 5 the 0.7-rotation-annihilation of T_1 and T_2 is presented, where T_1 is an ordinal sum defined by a Łukasiewicz t-norm and a product t-norm and T_2 is the rotation of the product. On the right-hand side of Figure 5 the 0.7-rotation-annihilation of T_1 and T_2 is presented, where T_1 is the same as in the previous example and T_2 is the t-subnorm given by (Łukasiewicz) with $\varepsilon = \frac{1}{2}$.

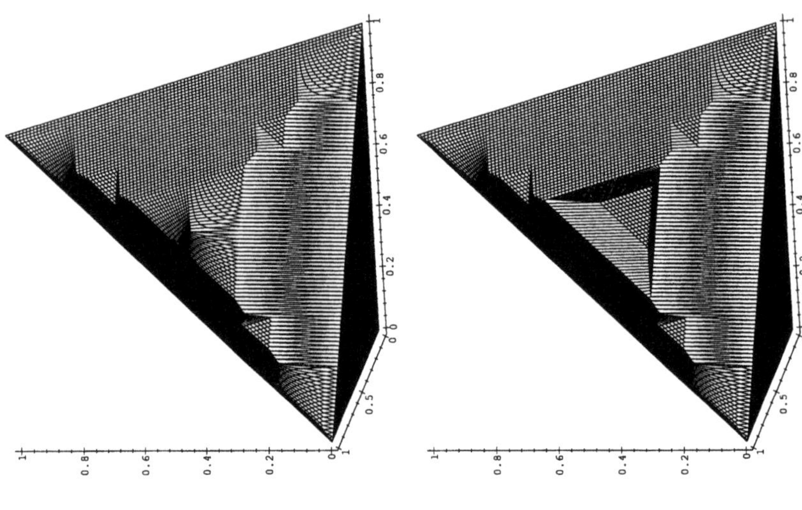

Figure 5: T-norms generated by rotation-annihilation construction

The main conclusion of the previous sections is the following: if we have certain operations T_1 and T_2, then we can produce by rotation-annihilation $T_{\mathbf{J}}$ which is a left-continuous t-norm with strong induced negation. In the rest of this section we illustrate the power of the rotation-annihilation construction by emphasizing the wide spectrum of the appropriate T_1's and T_2's.

When we evaluate T_1 only with continuous t-norms (that is, with ordinal sums of continuous Archimedean t-norms) we can obtain by rotation-annihilation an infinite number of new families of left-continuous but not continuous t-norms with strong induced negations. Moreover T_1 can be evaluated with ordinal sums of *left*-continuous (or/and continuous) t-norms. The left-continuous t-norms may be generated by a previous rotation or rotation-annihilation. Then it can be used as a summand in an ordinal sum again and so on. Alternatively, T_1 can be evaluated with ordinal sums of (left-)continuous t-*sub*norms (see [16]). Another way to construct an appropriate T_1 is to rotate a t-norm which was produced by one of the previously mentioned way.

In this way any t-norm which is the result of subsequent applications of rotation, rotation-annihilation and ordinal sums (again, the summands may be t-subnorms as well, with some restrictions) can play the role of T_1.

The role of T_2 in Theorem 8 can be played by e.g. any left-continuous t-norm with strong induced negation (see Corollary 1). In particular, by any t-norm, which is the result of a rotation or a rotation-annihilation. The rotation of any left-continuous t-subnorm (see Remark 2/iii) or a t-subnorm which is generated by the rotation-annihilation construction (when T_1 is a left-continuous t-subnorm, see Remark 3/ii can also play the role of T_2. In addition T_2 can be evaluated with the operation which is introduced in (Łukasiewicz). In the last three cases attention should be paid for the zero divisors of T_1.

Iteration of these ideas leads to quite 'exotic and beautiful' t-norms. Finally, we present such an 'exotic' example via the corresponding 3-dimensional plot in Figure 6:

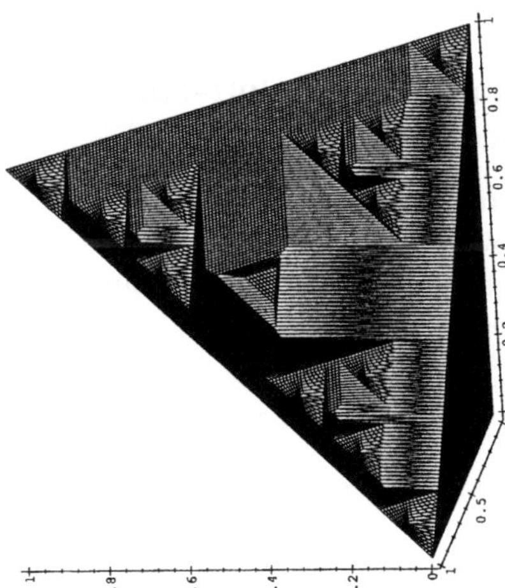

Figure 6: T-norm generated by combination of rotation, rotation-annihilation, and ordinal sums

6 Decomposition

The rest of this chapter is based on [23]. We recall that in case of continuous t-norms the description of *structure* is the following: the indecomposable class is the class of continuous Archimedean t-norms, decomposition and construction is answered by the ordinal sum theory (see Section 1.1). Since the constructions have already been presented in Sections 4 and 5, the only thing that we need to know is: what is indecomposability and how to decompose into indecomposable units. Below we define (in)decomposability and present the decomposition theorems. We show that each left-continuous t-norm with strong induced negation which is decomposable according to our definition, can be derived as the result of one of the above-mentioned constructions. This proves that our construction methods are "sharp" in the following sense: T_1 and T_2 can not be chosen from a wider class of operations. In addition, the decomposition can be defined in a unique way, namely, when the decomposition is done by the minimal decomposition point. At this decomposition one of the resulted operations turns out to be indecomposable.

Given a left-continuous t-norm with strong induced negation, as many decompositions are possible as the cardinality of the set of its decomposition points. If first we decompose and than we construct from the obtained operation(s) we get back the starting operation. If we take certain operation(s) and we construct first, we can decompose the constructed t-norm in such a way that we get back the starting operation(s). In this sense the presented constructions and decompositions are inverse operations of each other.

6.1 Indecomposability of left-continuous t-norms with strong induced negations

In this section for a left-continuous rotation invariant t-subnorm we define its decomposition points by using the concept of the skeleton function. As a special case, this definition yields the decomposition points of a left-continuous t-norm with strong induced negation.

Notation. We denote 't converges to x with $t > x$' by $t \downarrow x$. For any $H_1, H_2 \subset [0,1]$, let

$$T(H_1, H_2) = \{T(x, y) \mid x \in H_1,\ y \in H_2\}$$

and we write $H_1 < H_2$ (resp. $H_1 \leq H_2$) if $t < s$ (resp. $t \leq s$) holds for all $t \in H_1$, $s \in H_2$. For $c \in [0,1]$ usually we identify c with $\{c\}$ and write e.g. $T(H_1, H_2) = c$ instead of $T(H_1, H_2) = \{c\}$ for sake of simplicity. Similarly, for one-place functions, if A is a subset of the domain of f, then the meaning of $f(A)$ is $\{f(x) \mid x \in A\}$.

Definition 8. Let N be a strong negation and T be a left-continuous t-subnorm which is rotation invariant with respect to N. Define $\chi :]0,1[\to [0,1]$ by

$$\chi(x) = \inf T\left(x,]N\left(x\right), 1]\right)$$

or equivalently,

$$\chi(x) = \inf_{t \downarrow N(x)} T\left(x, t\right)$$

(We can converge to $N\left(x\right)$ strictly from above in $[0,1]$ and $]N\left(x\right), 1]$ is not empty since 0 is excluded from the domain of χ and hence $N\left(x\right)$ is not 1; therefore, these definitions are meaningful.) A third equivalent definition of the skeleton function is

$$N\left(\chi(x)\right) = \inf\{c \in [0,1] \mid T\left(c, x\right) = x\}$$

Call χ the *skeleton function* of T, and call the following subset of the graph of T the *skeleton* of T: $\{(x, y, T\left(x, y\right)) \mid T\left(x, y\right) = \min(x, y)\}$. The name 'skeleton function' is motivated by the last equivalent definition, since the skeleton function determines the 'skeleton', that is, the part of the graph of the t-norm where its value coincides with the minimum t-norm.

Definition 9. Let N be a strong negation, T be a left-continuous t-subnorm which is rotation invariant with respect to N, its skeleton function is χ and let t be the unique fixed point of N. Define the set of *decomposition points* of T by

$$D = \{d \in [t, 1[\mid N\left(d\right) \leq \chi([t, d])\}$$

Call T *indecomposable* if $D = \emptyset$, *decomposable* if $D \neq \emptyset$ and *totally decomposable* if $t \in D$. An equivalent definition of the set of decomposition points is:

$$D = \{d \in [t, 1[\mid T\left(y, z\right) = y, \text{ for all } y \in \{t\} \cup]N(d), d], \ z \in]d, 1]\}$$

6.2 Decompositions

Based on the results of the previous subsection we now present the decomposition theorems. The first such theorem states that with each decomposition point d which is different from the fixed point of the induced negation of T, we can decompose T into a left-continuous t-norm T_1 and a left-continuous rotation invariant t-subnorm T_2 which is, in addition, a t-norm if T_1 has zero divisors.

Theorem 9 (General decomposition). *Let T be a decomposable left-continuous t-norm with strong induced negation N. Suppose that $d \in D$ is a decomposition point of T which is different from t the fixed point of N, and let N_d be the zoomed d-negation of N. Define $T_3 : [d,1] \times [d,1] \to [0,1]$ and $T_4 : [N(d), d] \times [N(d), d] \to [0,1]$ by*

$$T_3\ (x,y) = \begin{cases} T(x,y) & \text{if } x,y > d \\ d & \text{otherwise} \end{cases} \tag{(9a)}$$

$$T_4\ (x,y) = \begin{cases} N(d) & \text{if } x \le N(y) \\ T(x,y) & \text{otherwise} \end{cases} \tag{(9b)}$$

Then T_3 is a binary operation on $[d,1]$ and T_4 is a binary operation on $[N(d), d]$. Let T_1 (resp. T_2) be the linear transformation of T_3 (resp. T_4) into $[0,1]$. Then

i. *T_1 is a left-continuous t-norm.*

ii. *If T_1 has no zero divisors, then T_2 is a left-continuous t-subnorm fulfilling the rotation invariance property with respect to N_d. If T_1 has zero divisors, then T_2 is a left-continuous t-norm fulfilling the rotation invariance property with respect to N_d (that is, a left-continuous t-norm with strong induced negation equal with N_d, by Corollary 1).*

Definition 10. Taking the notations of Theorem 9, call the pair (T_1, T_2) the *d-decomposition* of T.

By *maximal decomposition* we mean that the decomposition is done by the least decomposition point. We first state a theorem investigating the case when the fixed point of the induced negation is a decomposition point; and subsequent to this theorem we consider the case when the fixed point of the induced negation is not a decomposition point.

Theorem 10 (Total decomposition). *Let T be a totally decomposable left-continuous t-norm with strong induced negation N. Let $d = t$ and define a binary operation T_3 on $[d,1]$ by (9a). Let T_1 be the linear transformation of T_3 into $[0,1]$. Then T_1 is a left-continuous t-norm without zero divisors.*

Definition 11. Taking the notations of Theorem 10, call T_1 the *total decomposition* of T.

Lemma 2. Let T be a decomposable left-continuous t-norm with strong induced negation N and D be the set of its decomposition points. Then for any $\emptyset \neq H \subset D$ we have $\inf H \in D$.

The above lemma proves that the set of decomposition points is a closed set in the right-continuity topology. This fact makes it generally possible to define maximal decomposition when the fixed point of the induced negation is not a decomposition point. For this decomposition, roughly speaking, we take as 'big' T_1 and as 'small' T_2 as it is possible. The main gain is that T_2 is indecomposable if and only if the decomposition of T is done by the minimal element of D.

Illustrations of total and maximal decompositions. Refer to Figure 7 below.

- The total decomposition of the nilpotent minimum t-norm is seen in the left drawing of Figure 7.
- In the right drawing of Figure 7, a possible decomposition of the t-norm in the center drawing (given by (5)) is presented. The related strong negation is $1 - x$ by Lemma 1(ii). The chosen decomposition point is $\frac{2}{3}$, T_1 is the minimum t-norm, and T_2 is the Łukasiewicz t-norm. The set of decomposition points is $[\frac{2}{3}, 1[$. This decomposition is at the same time the maximal decomposition, since $\frac{2}{3}$ is the least decomposition point of this t-norm. (Observe that looking at first the right drawing and then the center one, the *construction* of the t-norm (center) is seen; namely, we start from the minimum t-norm and the Łukasiewicz t-norm and apply the rotation-annihilation construction.)

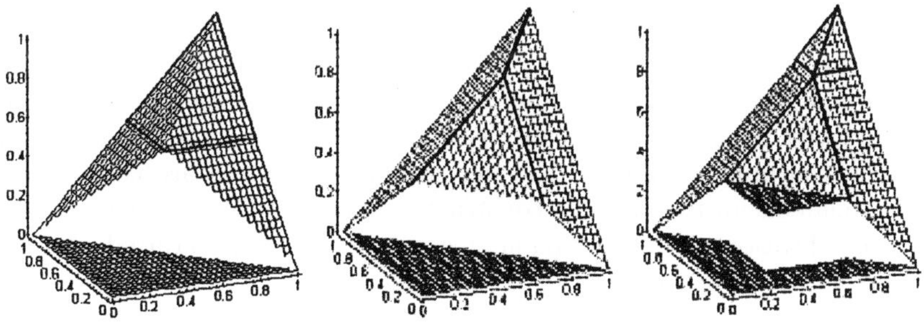

Figure 7: Total decomposition of nilpotent minimum t-norm (left), t-norm given in (5) (center), and its maximal decomposition (right)

6.3 The relation between the constructions and the decompositions

Theorem 11. *The following statements hold.*

i. *Let T be a totally decomposable left-continuous t-norm with strong induced negation N. Let T_1 be the total-decomposition of T. Then the N-rotation of T_1 equals T.*

ii. *Let T be a decomposable left-continuous t-norm with strong induced negation N. Suppose that $d \in D$ is a decomposition point of T which is different from the fixed point of N. Let (T_1, T_2) be the d-decomposition of T. Then the N-d-rotation-annihilation of T_1 and T_2 equals T.*

iii. *Let T_1 be a left-continuous t-norm without zero divisors and N be a strong negation. Then the total-decomposition of the N-rotation of T_1 equals T_1.*

iv. *Let N be a strong negation, t its unique fixed point, $d \in]t, 1]$ and T_1 be a left-continuous t-norm. Take T_2 depending on the zero divisors of T_1, as it is taken in Definition 7. Then the d-decomposition of the N-d-rotation-annihilation of T_1 and T_2 equals (T_1, T_2).*

Remark 4.

1. Statements (i) and (ii) in Theorem 11 show that the operation(s) resulting from the decomposition determine T uniquely.

2. Generally, the operation(s) arising in the decompositions is (are) not from a smaller class than the class of left-continuous t-norms with strong induced negations. (Recall that this in not the case when decomposing continuous t-norms; the decomposed operators are continuous and Archimedean there.) The constructive part of this work reveals that the classification of left-continuous t-norms with strong induced negations is as hard as the classification of left-continuous t-norms.

3. An open problem is the characterization of the class which is called indecomposable in this chapter. There may exist further decomposition methods for (a subset of) this class. Whence, the notion of indecomposability/decomposability (introduced here) may be changed.

7 Conclusion

An exhaustive discussion of the structure of Girard monoids on [0, 1] (another terminology is left-continuous t-norms with strong induced negations) is given in this chapter.

Two properties (the quasi-inverse property and the rotation invariance property with respect to a strong negation) are introduced (Section 2). Each of them has been proved to be characteristic for the class of left-continuous t-norms with strong induced negations. By understanding the geometrical content of those properties, a geometrical description of left-continuous t-norms with strong induced negations is given. This geometrical understanding turns out to be useful in proving two theorems in Section 3. The first one states that in this context continuity is equivalent with strict monotonicity on the positive domain. The second one asserts that the boundary condition of a left-continuous t-norm with strong induced negation follows from its other axioms: a monotone, symmetric, associative, left-continuous two-place function which has strong induced negation necessarily admits the boundary condition.

In other formulation, the concept of neutral element and its integrality are quite natural notions in the field of Girard monoids; there exists no "Girard semigroup" which is not an integral monoid. Then a new method, called rotation, is introduced in Section 4 (Definition 6). From any left-continuous t-norm which has no zero divisors this method produces a left-continuous (but not continuous) t-norm with strong induced negation. By introducing the class of t-subnorms, a second method, called rotation-annihilation, is defined in Section 5. From any pair of a left-continuous t-norm T_1 and a left-continuous rotation invariant t-subnorm T_2—which has to be a t-norm, when T_1 has zero divisors—this construction produces a left-continuous (but not continuous) t-norm with strong induced negation. An infinite number of new families of left-continuous t-norms with strong induced negations can be produced with these constructions, which t-norms can be starting points of many further investigations.

As some type of 'inverse' operation, for the introduced constructions the corresponding decomposition theorems are presented in Section 6. (In)decomposability is defined. It is established that each decomposable left-continuous t-norm with strong induced negation can be decomposed with one of the above mentioned decomposition theorems. That is, each such t-norm can be described as result of one of the mentioned constructions.

Throughout the chapter the results are illustrated with 3-dimensional plots.

It is very important to emphasize that the results of the algebraic investigations of this chapter may open new perspectives in all the fields mentioned in the Introduction of this chapter. Below we list several possible outcomes of this study on left-continuous t-norms (and Girard monoids):

i. The discovery of new non-classical logics with nice logical properties.

ii. The discovery of more general mathematical models for fuzzy preference structures.

iii. The discovery of conditioning operators with nice mathematical properties.

References

[1] J. Aczél, *Sur les opérations definies pour des nombres réels*, Bull. Soc. Math. France **76**(1949), 59–64.

[2] C. C. Chang, *Algebraic analysis of many valued logics*, Trans. Amer. Math. Soc. **88**(1958), 467–490.

[3] R. Cignoli, I. M. L. D'Ottaviano, D. Mundici, Algebraic Foundations Of Many-Valued Reasoning, Kluwer (Boston/Dordrecht/London), 2000.

[4] A. H. Clifford, *Naturally totally ordered commutative semigroups*, Amer. J. Math. **76**(1954), 631–646.

[5] A. C. Climescu, *Sur l'équation fonctionelle de l'associativité*, Bull. École Polytechnique Iassy textbf1(1946), 1–16.

[6] B. De Baets, J. C. Fodor, *Twenty years of fuzzy preference structures (1978-1997)*, Belg. J. Oper. Res. Statist. Comput. Sci. **37**(1997), 61–82.

[7] J. C. Fodor, *A new look at fuzzy connectives*, Fuzzy Sets and Systems **57**(1993), 141–148.

[8] J. C. Fodor, *Contrapositive symmetry of fuzzy implications*, Fuzzy Sets and Systems **69**(1995), 141–156.

[9] J. C. Fodor, M. Roubens, Fuzzy Preference Modeling And Multicriteria Decision Support, Kluwer Academic Publishers (Boston/Dordrecht/London), 1994.

[10] L. Fuchs, Partially Ordered Algebraic Systems, Pergamon Press (London/New York/Oxford/Paris), 1963.

[11] J. Y. Girard, *Linear logic*, Theor. Comp. Sci. **50**(1987), 1–102.

[12] S. Gottwald, S. Jenei, *A new axiomatization for Involutive Monoidal T-norm-based Logic*, Fuzzy Sets and Systems (to appear).

[13] U. Höhle, *Commutative, residuated l-monoids*, Chapter IV in: U. Höhle, E. P. Klement, eds, Non-Classical Logics And Their Applications To Fuzzy Subsets—A Handbook of the Mathematical Foundations of Fuzzy Set Theory, Theory and Decision Library—Series B: Mathematical and Statistical Methods, Volume **32**, Kluwer Academic Publishers (Boston/Dordrecht/London), 1995, pp. 53–106.

[14] U. Höhle, S. Weber, *On conditioning operators*, Chapter 12 in: U. Höhle, S. E. Rodabaugh, eds, Mathematics Of Fuzzy Sets: Logic, Topology, And Measure Theory, The Handbooks of Fuzzy Sets Series, Volume **3**(1999), Kluwer Academic Publishers (Dordrecht/Boston/London), pp. 653–673.

[15] S. Jenei, *A characterization theorem on the rotation construction for triangular norms*, (submitted).

[16] S. Jenei, *A note on the ordinal sum theorem and its consequence for the construction of triangular norms*, Fuzzy Sets and Systems (to appear).

[17] S. Jenei, *Continuity of left-continuous triangular norms with strong induced negations and their boundary condition*, Fuzzy Sets and Systems, **124**(2001), 35-41.

[18] S. Jenei, *Geometry of left-continuous t-norms with strong induced negations*, Belg. J. Oper. Res. Statist. Comput. Sci., **38**(1998), 5–16.

[19] S. Jenei, *New family of triangular norms via contrapositive symmetrization of residuated implications*, Fuzzy Sets and Systems, **110**(2000), 157–174.

[20] S. Jenei, *On the structure of rotation-invariant semigroups*, (submitted).

[21] S. Jenei, *Structure of left-continuous t-norms with strong induced negations. (I) Rotation construction*, Journal of Applied Non-Classical Logics, **10**(2000), 83–92.

[22] S. Jenei, *Structure of left-continuous t-norms with strong induced negations. (II) Rotation-annihilation construction*, Journal of Applied Non-Classical Logics (to appear).

[23] S. Jenei, *Structure of left-continuous t-norms with strong induced negations. (III) Construction and decomposition*, Fuzzy Sets and Systems (to appear).

[24] S. Jenei, *The structure of Girard monoids on* $[0,1]$, in: E. P. Klement, S. E. Rodabaugh, eds, Proc. 20th Linz Seminar on Fuzzy Set Theory (Linz, Austria, February 1999), 21–33.

[25] S. Jenei, F. Montagna, *A proof of standard completeness of Esteva and Godo's monoidal logic MTL*, Studia Logica (to appear).

[26] S. Jenei, F. Montagna, *A general method for constructing left-continuous t-norms*, (submitted).

[27] E. P. Klement, R. Mesiar, E. Pap, Triangular Norms, Kluwer Academic Publishers (Boston/Dordrecht/London), 2000.

[28] E. P. Klement, M. Navara, *A survey on different triangular norm-based fuzzy logics*, Fuzzy Sets and Systems **101**(1999), 241–251.

[29] T. Kowalsky, H. Ono, *Variety of residuated lattices is generated by its finite simple members*, Reports on Mathematical Logic **34**(2000), 57–75.

[30] C-H. Ling, *Representation of associative functions*, Publ. Math. Debrecen **12**(1965), 189–212.

[31] K. Menger, *Statistical metrics*, Proceedings of the National Academy of Sciences **28**(1942), 535–537.

[32] P. S. Mostert, A. L. Shields, *On the structure of semigroups on a compact manifold with boundary*, Ann. Math. **65**(1957), 117–143.

[33] D. Mundici, *Interpretation of AF C^*-algebras in Łukasiewicz sentential calculus*, J. Funct. Anal. **65**(1986), 15–63.

[34] D. Mundici, *Ulam games, Łukasiewicz logic, and AF C^*-algebras*, Fund. Inf. **18**(1993), 151–161.

[35] E. Pap, Null-Additive Set Functions, Kluwer Academic Publishers (Boston/Dordrecht/London), 1995.

[36] B. Schweizer, A. Sklar, Probabilistic Metric Spaces, North-Holland Elsevier (Amsterdam), 1983.

[37] E. Trillas, *Sobre funciones de negación en la teoría de conjuntas difusos*, Stochastica **3**(1979), 47–60.

[38] B. Van de Walle, B. De Baets, E. Kerre, *A plea for the use of Łukasiewicz triplets in the definition of fuzzy preference structures. (I) General argumentation*, Fuzzy Sets and Systems **97**(1998), 349–359.

[39] S. Weber, *Conditioning on MV-algebras and additive measures, further results*, in: D. Dubois, H. Prade, E. P. Klement, eds, Fuzzy Sets, Logics And Reasoning About Knowledge, Kluwer Academic Publishers (Boston/Dordrecht/London), 1999, pp. 175–199.

CHAPTER 11

On The Geometry Of Choice

C. J. Mulvey [1]

Introduction

The Axiom of Choice has, since its conception, formed an important part of the framework within which much of mathematics is done. Not only has its application allowed the proof of results to which it is an apparently essential pre-requisite, but its availability has shaped the way in which the mathematics which it proves has been formulated. There is no need to carry along the way the impedimenta of accumulated detail when selection of an arbitrarily chosen instance can be relied upon in every situation due to the presence of an Axiom of Choice which guarantees its retrieval from a range of possibilities. There is a received wisdom which says that this shaping of mathematics is fine and good, and that the avoidance of intricacy which this encourages leads to a minimality of expression which is characteristic of elegance in mathematics. There is also a reality, perhaps less well received, which says that all may not be quite so fine, quite so good, and that the intricacies avoided in the cause of elegance may be waiting just round the next conceptual corner to pounce on the unsuspecting. Indeed, the joys of life with the Axiom of Choice may actually limit, rather than expand, the extent to which mathematics may be developed.

This chapter is about the impact of the Axiom of Choice on one particular part of functional analysis, namely that surrounding the Hahn-Banach theorem. It describes the way in which the Hahn-Banach theorem, which classically is almost, but not quite, equivalent to the Axiom of Choice, may be reformulated in such a way that it may be proved entirely constructively, in particular without application of the Axiom of Choice, while remaining classically equivalent, in the presence of the Axiom of Choice, to its classical formulation. At the centre of this

[1] Dedicated to the memory of Japie Vermeulen.

S.E. Rodabaugh and E.P. Klement (eds.), Topological and Algebraic Structures in Fuzzy Sets, 309–336.
© 2003 *Kluwer Academic Publishers.*

reformulation is the idea that examining the ways in which the theorem fails to remain valid under the change to a context that is propositionally non-classical may itself indicate the path that must be followed towards constructivity. In this sense, the development of mathematical ideas within such contexts achieves a powerful extrinsic value, in addition to its intrinsic merits.

Although the case to be made is specifically aimed at the Hahn-Banach theorem, the ideas behind it may be applied to a wide range of mathematical problems. Indeed, an indication of the extent to which this has already been carried out will be given towards the end of the chapter. The temptation, on the basis of the evidence that already exists, is to feel that almost any mathematical result, at least of the kind that is central to the development of the subject and its applications, may with a little care be treated in this way. Moreover, that re-examining mathematics thoughtfully, within a more ascetic world in which use of the Axiom of Choice is replaced by a constructive approach, may lead to greater insight into what is actually taking place within it.

The form of the chapter is largely historical, following the development of the ideas which it examines. With the exception of the final stage of the argument, of which details have yet to be published in a joint paper with J.J.C. Vermeulen [22], the ideas have already been presented in papers by various authors [10, 20, 24, 25]. The presentation here will be kept to the level of detail needed to follow the development of the ideas, the reader being referred to the relevant original sources for a more thorough discussion. The aim instead will be to take the broader view, focusing on the techniques that have been developed and the principles that lie behind. For making those clearer to me, I have to thank principally Bill Lawvere and Japie Vermeulen, whose mathematical insights have that clarity of vision and intensity of perspective that open the eyes of others to new things. Finally, to Bernhard Banaschewski and to Joan Pelletier, as well as to Japie Vermeulen, without whom the work would never have been done and with whom it variously was done, I give warm thanks, both for mathematics and for friendship.

Sadly, as this chapter was in proof, Japie Vermeulen died in a tragic accident in South Africa: his profound and insightful contributions to this and other fields will be greatly missed.

1 The classical Hahn-Banach theorem

Before considering its constructive aspects, we shall recall the Hahn-Banach theorem in the form with which we shall be concerned, and remind ourselves of its proof. The theorem is straightforwardly stated:

Theorem 1.1 *For any bounded linear functional φ on a linear subspace A of a seminormed linear space B, there exists a norm-preserving extension to a linear functional ψ defined on the seminormed space B itself.*

The theorem is also straightforwardly proved, by applying Zorn's Lemma, which is classically equivalent to the Axiom of Choice. The proof is divided into two stages, of which the first is that which depends on applying Zorn's Lemma, and the second that which acts on the result of doing so. The collection of norm-preserving extensions

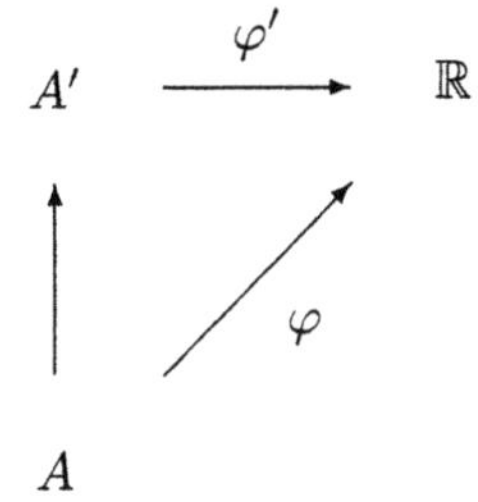

of the linear functional $\varphi : A \to \mathbb{R}$ to linear functionals $\varphi' : A' \to \mathbb{R}$ defined on subspaces A' of B containing the subspace A, partially ordered by extension, is an inductive set, in which therefore, by Zorn's Lemma, there exist maximal members. Choosing such a maximal norm-preserving extension, which we shall denote by $\varphi' : A' \to \mathbb{R}$, we assert that the linear subspace A' on which the extension is defined is actually the seminormed space B, from which the theorem follows.

To see this, we argue by contradiction, assuming that there exists an element $b \in B$ which lies outside the subspace A'. In this case, it is possible to extend the linear functional $\varphi' : A' \to \mathbb{R}$ to a linear functional $\varphi'' : A'' \to \mathbb{R}$ defined on the subspace A'' obtained by extending A' by the element $b \in B$ by assigning to $b \in B$ an arbitrarily chosen element $\beta \in \mathbb{R}$. The criterion for the extension thus defined to be norm-preserving is that $\beta \in \mathbb{R}$ be chosen satisfying the condition that

$$\sup_{a \in A} \{ \varphi(a) - \| a - b \| \} \leq \beta \leq \inf_{a \in A} \{ -\varphi(a) + \| a + b \| \}$$

It may be shown fairly easily that indeed one has that $\sup_{a \in A} \{ \varphi(a) - \| a - b \| \}$ is not greater than $\inf_{a \in A} \{ -\varphi(a) + \| a + b \| \}$, allowing the choice of such an element $\beta \in \mathbb{R}$, from which the result follows by contradiction of the maximality of the extension $\varphi' : A' \to \mathbb{R}$. Hence, there exists a norm-preserving extension to a linear functional $\psi : B \to \mathbb{R}$. Of course, we have been dealing here with the case of a seminormed linear spaces over the reals, from which the result for the complex case is easily deduced.

The objective that we have is to obtain a proof of the Hahn-Banach theorem which is constructively valid [22]. It is evident, from whatever constructive standpoint one may take, that the proof just outlined is not constructive. Most immediately, the application of the Axiom of Choice in the form of Zorn's Lemma renders the proof decidedly non-constructive. The method of proof by contradiction is also likely to be an obstacle to constructivity, although this is a difficulty which is quite often surmountable. More subtly, a dependence on the construction of suprema and infima in the set of real numbers will cause anxiety in the constructive mind, since $\mathbb{R}$ constructively is not order-complete. At another level of concern, there is the question of what exactly should be meant by the terms *seminormed linear space* and *bounded linear functional* within the constructive context, since these may classically be expressed in many syntactically different, but semantically equivalent, ways, which may no longer be semantically equivalent from a constructive viewpoint. Indeed, we may wish later to apply the same remark to the form of the *statement* of the Hahn-Banach theorem.

Before any of this, we should first specify precisely what we shall be meaning by *constructive*, and perhaps indicate why we are interested in finding a constructive form of the Hahn-Banach, or indeed any other, theorem. The answers to these questions are essentially the same: by *constructive mathematics* we shall mean mathematics that is valid in any topos having a natural number object. The motivation for deciding on this description is that it encompasses many situations that occur in practice, in terms of extending mathematics to cases involving questions of continuity in parameters, or of equivariance with respect to the action of a group or monoid. For a discussion of this particular viewpoint, the reader is referred to Mulvey and Pelletier [20] and to Banaschewski and Mulvey [4, 6]. For the moment, suffice it to say that the category of sheaves on a topological space X and the category of M-sets for any monoid M are instances of the concept of a topos. For a leisurely discussion of the interpretation of mathematics in a category of sheaves, the reader is referred to Mulvey [17]. The formalities may be found in Johnstone [13].

Pragmatically, and to a great extent formally, this means that we may develop mathematics within a context in which we may still talk of sets and mappings, while limiting ourselves to constructions that are based on the formation of cartesian products of pairs of sets, of the power set of a set, and of the subset of a set consisting of elements satisfying a formula in the usual language of sets. Additionally, we may assume the existence of a set of natural numbers [13], satisfying the Peano axioms in the usual way. Logically, we may use all the usual rules of deduction of the Predicate Calculus, with the exception of those derived from the Law of the Excluded Middle, asserting of any proposition φ that the disjunction

$$\varphi \vee \neg\varphi$$

of the proposition φ with its negative is necessarily valid. In particular, we must make use neither of the Axiom of Choice, nor of any of its variants, such as Zorn's Lemma. Indeed, even the Axiom of Countable Choice, which allows the choosing of a sequence of elements from a sequence of sets, each having at least one element, is not valid within this context. It may be remarked in passing that, within a topos the validation of the Axiom of Choice implies that of the Law of the Excluded Middle [13].

It is already evident that the proof of the Hahn-Banach theorem outlined above fails to remain valid constructively on each and every of the grounds just considered. Indeed, within the slightly different context of the constructivity considered in intuitionistic mathematics, it became clear that there was little hope of pushing the conventional proof through in anything beyond the case of a seminormed space of countable dimension. In fact, even that case depends on the acceptance within intuitionistic mathematics of the Axiom of Countable Choice, and even the Axiom of Countable Dependent Choice, in which the choosing of each element of a sequence is dependent on the choices made thus far. The reader is referred to Bishop [8] and to Bridges [9] for various valiant attempts to push through some form of the Hahn-Banach theorem in greater generality, by requiring only that the extending functional is approximately equal to that which is being extended. For an insight into the desperation experienced in trying to achieve even this kind of progress, one should read Bishop's introductory comments on Heyting's endeavours in respect of the continuum [8].

One part of the difficulty that faced the intuitionistic mathematician seems to have arisen from a philosophical position which still tried to prove mathematical results in a form as close as possible to that considered classically, with little attempt being made to consider the meaning of the theorem, rather than simply its proof. There was also little attempt made to adapt the definitions involved in the mathematics being considered to the context in which that consideration was taking place. In the case of seminormed spaces, taking the naive definition of a seminorm leads fairly immediately to difficulties, for instance in a seminormed space admitting quotient spaces only by subspaces satisfying additional, non-intuitive, conditions [9]. Explicitly, the quotient space of a seminormed linear space B by a linear subspace A will not admit a seminorm unless the subspace is located, in the sense that the infimum

$$\inf_{a \in A} \| a + b \|$$

exists for each $b \in B$. Again, this is a difficulty that arises through the set of real numbers failing to be order-complete, in this case in the sense of intuitionistic constructivity.

To avoid difficulties of this kind, and to provide the conceptual setting within which one can begin to constructivise the Hahn-Banach theorem, means adapting the definition of the concept of seminormed linear space to the constructive context. In doing this, one may also be guided by insights into what seminormed spaces in certain toposes should turn out to be. In particular, one believes that seminormed, normed, and Banach spaces in the category of sheaves over a topological space X should correspond to bundles of seminormed, normed, and

Banach spaces over X in the more conventional sense arising in parametrising these structures continuously over X. The detail of the elucidation of these concepts may be found elsewhere [11], together with a discussion of their equivalence with the corresponding concepts of bundles over the topological space X when interpreted in the category of sheaves on X. For the moment, it suffices to observe the following:

Definition 1.1 *By a* seminormed linear space A *will be meant a linear space over the field* $\mathbb{Q}$ *of rational numbers, together with a mapping*

$$N : \mathbb{Q}^+ \to \Omega^A$$

from the set of positive rationals to the power set of the set A, satisfying the following conditions:

1. $0 \in N(q)$;
2. $a \in N(q) \;\to\; pa \in N(pq)$;
3. $a \in N(q) \wedge a' \in N(q') \;\to\; a + a' \in N(q + q')$;
4. $\exists q \in \mathbb{Q}^+ \; a \in N(q)$;
5. $a \in N(q) \;\leftrightarrow\; \exists\, q' < q \;\; a \in N(q')$,

for all $a, a' \in A$ and for all $p, q, q' \in \mathbb{Q}^+$.

It should be noted that in this and other definitions, those concepts not defined *ab initio* are to be interpreted just as in classical mathematics. Thus, the set of rational numbers is constructed from the set of natural numbers in the usual way, and the subset of positive rationals is carved out from it by interpreting the conventional formula expressing positivity [17].

The fact that we have considered linear spaces over the rational numbers, rather than the reals, will simplify the constructions at a later point, while extending slightly the scope of the Hahn-Banach theorem. More importantly, the axiomatisation of the notion of a seminorm is centred around the concept of the open balls which classically a seminorm determines. For each positive rational $q \in \mathbb{Q}^+$, the subset

$$N(q)$$

of the linear space A assigned by the seminorm mapping is to be thought of as the open ball of radius q. The axioms express the usual properties of a seminorm in terms of these open balls. Of course, classically one may retrieve a seminorm in the more conventional sense, by defining

$$\|a\| = \inf \{\, q \in \mathbb{Q}^+ \mid a \in N(q) \,\}$$

for each $a \in A$. The problem constructively is that such an infimum need not necessarily exist in the set $\mathbb{R}$ of real numbers. Indeed, by now one may be beginning to have some doubts about what exactly the set $\mathbb{R}$ of real numbers is, since this involves a far more sophisticated construction than that of the set $\mathbb{Q}$ of rational numbers. This question, and that of whether the seminorm may indeed be described in terms of a seminorm function, is one to which we must later return.

2 Extending the Hahn-Banach theorem

Having defined the context, one may examine the limitations that this is known to impose upon the problem. In fact, these are severe, although they also contain the hint of a way forward. In a word, it may be shown that it is not possible to prove the Hahn-Banach theorem constructively in the form stated. Explicitly, one may determine precisely the conditions under which the Hahn-Banach theorem stated above, which we shall refer to as the *naive* Hahn-Banach theorem, is valid in two contexts to which constructive methods may be applied. The first of these is that of the category of sheaves over a topological space X, the context in which the Hahn-Banach theorem for bundles of seminormed linear spaces over X may be examined. The result, due to Banaschewski, is the following:

Theorem 2.1 *In the category of sheaves on a topological space X, the naive Hahn-Banach theorem is valid if, and only if, the topological space X is extremally disconnected.*

It may be recalled that a topological space X is said to be *extremally disconnected* provided that the closure of every open subset is again open.

On the one hand, this extinguishes for once and for all any hope that the theorem might be constructively provable, since extremally disconnected topological spaces constitute a fairly extreme class of spaces. On the positive side, it may be observed that the toposes corresponding to these topological spaces are distinguished amongst other categories of sheaves on topological spaces by being those in which De Morgan's Law, the weakening of the Law of the Excluded Middle that asserts that

$$\neg\varphi \vee \neg\neg\varphi$$

is valid for any proposition φ, is validated. Tantalisingly, this is also the condition for the set $\mathbb{R}$ of real numbers in a topos to be order-complete. That this is the case is no surprise in the light of work by Burden [10] establishing an extension of the Hahn-Banach theorem which is valid in the category of sheaves over any topological space X.

The set $\mathbb{R}$ of real numbers that the statement of the Hahn-Banach theorem refers to is that obtained from the set $\mathbb{Q}$ of rational numbers by constructing the set of Dedekind cuts on the rationals, in the sense of the following:

Definition 2.1 *By a* Dedekind cut $x \in \mathbb{R}$ *on the set* $\mathbb{Q}$ *of rational numbers is meant an ordered pair* $(\underset{\vee}{x}, \hat{x})$ *consisting of subsets of* $\mathbb{Q}$ *satisfying the following conditions:*

1. $\exists\, p \in x$ *and* $\exists\, q \in \hat{x}$;
2. $p' < p \wedge p \in \underset{\vee}{x} \rightarrow p' \in \underset{\vee}{x}$ *and* $q \in \hat{x} \wedge q < q' \rightarrow q' \in \hat{x}$;
3. $p' \in \underset{\vee}{x} \rightarrow \exists\, p > p'\; p \in \underset{\vee}{x}$ *and* $q' \in \hat{x} \rightarrow \exists\, q < q'\; q \in \hat{x}$;
4. $p \in \underset{\vee}{x} \wedge q \in \hat{x} \rightarrow p < q$;
5. $p < q \rightarrow p \in \underset{\vee}{x} \vee q \in \hat{x}$.

It may be remarked that the set $\mathbb{R}$ of *Dedekind real numbers* thereby defined may be canonically given the algebraic and order structure normally associated with the set of real numbers. Moreover, there is a canonical inclusion

$$\mathbb{Q} \rightarrow \mathbb{R}$$

of the set of rational numbers in the set of real numbers, with respect to which, in the seminorm canonically determined by this order-preserving, algebraic embedding, the seminormed space of real numbers is exactly the completion of the seminormed space $\mathbb{Q}$. However, although classically this seminormed completion of the set $\mathbb{Q}$ of rational numbers is also order-complete, constructively this is not the case. Indeed, the equivalence of these conditions in any particular topos may be shown to be equivalent to the validation there of De Morgan's Law, in the sense referred to above.

There is, however, also a constructive order-completion $^*\mathbb{R}$ of the set $\mathbb{Q}$ of rational numbers, obtained by a slight adaptation of a construction originally investigated by Staples [23], and consisting of Dedekind *cuts on the rationals, in the sense of the following:

Definition 2.2 *By a* Dedekind *cut $x \in {}^*\mathbb{R}$ *on the set* $\mathbb{Q}$ *of rational numbers is meant an ordered pair consisting of subsets of* $\mathbb{Q}$ *satisfying the conditions for a Dedekind cut, with the last condition replaced by the following:*

6. $\exists p' > p(\forall q\,(q \in \hat{x} \rightarrow p' < q)) \rightarrow p \in \underset{\vee}{x}$ and $\exists q' < q(\forall p\,(p \in \underset{\vee}{x} \rightarrow p < q') \rightarrow q \in \hat{x}$.

Once again, it may be remarked that the set $^*\mathbb{R}$ of *Dedekind *real numbers* thereby defined may be canonically endowed with algebraic and order structure, together with a canonical inclusion

$$\mathbb{Q} \rightarrow {}^*\mathbb{R}$$

of the set of rational numbers in the set of *real numbers, with respect to which, in the order canonically inherited from that of the set $\mathbb{Q}$, the partially-ordered set $^*\mathbb{R}$ of *real numbers is exactly the completion of the partially-ordered set $\mathbb{Q}$. Moreover, it may straightforwardly be shown that the canonical inclusion of $\mathbb{Q}$ naturally factors through that of $\mathbb{Q}$ in the set $\mathbb{R}$ of Dedekind real numbers, yielding that $^*\mathbb{R}$ is an extension of $\mathbb{R}$. Indeed, the inclusion

$$\mathbb{R} \rightarrow {}^*\mathbb{R}$$

is an isomorphism exactly if De Morgan's Law is validated in the topos under consideration. In the absence of this non-constructive condition, the set $^*\mathbb{R}$ is the natural completion of $\mathbb{R}$ to a partially ordered field in which both infima and suprema may be constructed.

Consideration of the set $^*\mathbb{R}$ of Dedekind *real numbers provides an immediate solution to one of the problems already noted in connection with proving the Hahn-Banach theorem, namely the need, in establishing that a maximal extension chosen by Zorn's Lemma is necessarily defined on the seminormed space itself, to consider an extension obtained by assigning to an element lying outside the domain of the maximal extension a value satisfying

$$\sup_{a\in A}\{\varphi(a) - \|a-b\|\} \leq \beta \leq \inf_{a\in A}\{-\varphi(a) + \|a+b\|\}$$

In the Dedekind *reals, these suprema and infima may be calculated, although it may then be necessary for the value chosen to be itself an element $\beta \in {}^*\mathbb{R}$. Taking the lead from this observation that, in the constructive context, it may be necessary to consider *linear *functionals*

$$\psi : B \rightarrow {}^*\mathbb{R}$$

on a seminormed linear space, instead of linear functionals, and applying the existence of an internal form of Zorn's Lemma in the category of sheaves on a topological space X, a delicate argument allowed Burden [10] to prove the following extension of the Hahn-Banach theorem to this particular case:

Theorem 2.2 *For any bounded linear *functional φ on a linear subspace A of a seminormed linear space B in the category of sheaves on a topological space X, there exists a norm-preserving extension to a linear *functional ψ defined on the seminormed space B itself.*

From any kind of constructive philosophical standpoint, this is still highly unsatisfactory, since it depends entirely on the Axiom of Choice in the context within which the sheaves are being defined. From a pragmatic standpoint, there are also serious drawbacks, notably that inherently there exist normed spaces that are non-zero, yet on which only the zero *functional is defined. Hence, any hope of applying this version of the Hahn-Banach theorem to establish a canonical embedding of any normed space in its double dual will be shortlived. For a discussion of these ideas the reader is nevertheless referred to Burden and Mulvey [11], since these linear *functionals occupy an interesting place in the investigation of seminormed bundles over topological spaces.

The import of this failed attempt at overcoming the difficulties introduced by such things as the absence of De Morgan's Law from the logic of constructive mathematics is to identify more definitely the problems caused by absence of an Axiom of Choice as those which must be tackled. That this is indeed the case becomes further evident on considering a second context within which limitations can be seen in trying to extend the naive Hahn-Banach theorem, namely that of the Hahn-Banach theorem in the category of M-sets for any monoid M. This is very different from the context just discussed in that the Law of the Excluded Middle, and hence, *a fortiori*, De Morgan's Law, is always validated in a topos of this kind, implying that the difficulties arise simply due to the invalidity of the Axiom of Choice in such a topos. The result, again due to Banaschewski [1], is somewhat technical, although an examination of the proof shows clearly the reason why the condition concerned arises:

Theorem 2.3 *In the category of M-sets for a monoid M, the naive Hahn-Banach theorem is valid if, and only if, the monoid M admits an invariant mean.*

It may be recalled that by an invariant mean on a monoid M is meant a linear functional

$$\mu : \mathbb{R}^b(M) \to \mathbb{R}$$

on the Banach space $\mathbb{R}^b(M)$ of bounded real functions on the monoid M, satisfying the conditions that

$$\mu(\mathbf{1}) = 1 \quad \text{and} \quad \mu(m\varphi) = \mu(\varphi) \quad \text{for all } m \in M \text{ and } \varphi \in \mathbb{R}^b(M)$$

in which $\mathbf{1} \in \mathbb{R}^b(M)$ denotes the constant function with value 1, and the action of $m \in M$ on $\varphi \in \mathbb{R}^b(M)$ is defined by $m\varphi(x) = \varphi(xm)$ for each $x \in M$.

In each of these cases, the problems introduced by applying the Axiom of Choice are clearly revealed. In the case of the Hahn-Banach theorem in the category of sheaves over a topological space X, although the classical Hahn-Banach theorem allows an extension of the linear functional induced on the fibre of the subspace at each point $x \in X$ to be chosen in a norm-preserving manner, the necessary arbitrariness of this choice, implied by the Axiom of Choice being applied to do the choosing, means that one has no control over the coherence of these choices across the topological space X. The consequence is that although the linear functional originally given varies continuously across the space X, there is no chance of showing this to be the case for the extending functional, unless the topological space is extremally disconnected. It will be seen later that the extension of the Hahn-Banach theorem contained in the result of Burden may in fact be recovered by a purely geometric argument from that provable over an extremally disconnected space.

In the case of the Hahn-Banach theorem in the category of M-sets, the equivariant linear functional on the subspace admits, by the classical Hahn-Banach theorem, a multitude of norm-preserving extensions, chosen, by application of the Axiom of Choice, without concern for equivariance. The consequence is that there is no means of guaranteeing the existence amongst these of an equivariant extension, except in the case that there exists an invariant mean, in which case it may be applied to average out the action of the monoid on an arbitrarily chosen extension, allowing an equivariant extension to be constructed.

To tackle these situations properly, and to gain insight into the geometric aspects of avoiding applying the Axiom of Choice, requires a radically different approach to the problem.

3 Globalising the Hahn-Banach theorem

One may make a start by considering objectively what exactly the Axiom of Choice contributes through allowing the proof of a theorem like the Hahn-Banach. Certainly, it allows the choice of a norm-preserving extension of a particular linear functional. In so doing, it establishes the existence of such an extension. All it tells you about the extension to be chosen is that it is a linear functional, that it is an extension of the given functional, and that it has the same norm. In fact, any particular choice contributes no more, and no less, than any other, or indeed the totality of all possible choices. Since such a choice may be made for any bounded linear functional on the given subspace, one is left in reality only with a statement that the mapping from the collection of all functionals defined on the seminormed space B to the collection of all functionals defined on the subspace A, given by restriction of domain of functionals, is surjective. Including the condition that the extension is norm-preserving is achieved by constraining this assertion to apply not to all linear functionals, but only to those functionals of norm not exceeding 1.

With this realisation comes another, that the set of linear functionals of norm not exceeding 1 on a seminormed linear space B is not just a set, but actually a topological space once endowed with the weak* topology. Indeed, by

Alaoglu's theorem, the unit ball $\operatorname{Fn} B$ of the dual of a seminormed space B is a compact, Hausdorff topological space. Since the mapping given by restriction is indeed continuous, the Hahn-Banach theorem may be expressed exactly by saying that the restriction mapping

$$\operatorname{Fn} B \to \operatorname{Fn} A$$

from the unit ball of the dual space of the seminormed space B to the unit ball of the dual space of any subspace A is a quotient map of compact, Hausdorff spaces. Moreover, it may be remarked that this is frequently the form in which the Hahn-Banach theorem is applied, for instance to show that any normed linear space may be embedded isometrically in the Banach algebra of continuous real functions on a compact, Hausdorff topological space, namely the unit ball of its dual.

There is an immediate further connection with the Axiom of Choice to which these ideas point. One of its important applications is to establish the existence of the Stone-Čech compactification of any topological space. By work of Banaschewski and Mulvey [2, 3], and independently by Johnstone [15], the rôle of the Axiom of Choice in proving this theorem is known precisely. The context for these constructive approaches to Stone-Čech compactification is that within which topological spaces generally are discussed in the absence of the Axiom of Choice, namely the theory of locales [16]. The idea is that constructively it is generally straightforward to create the lattice of open subsets of a topological space, even if the Axiom of Choice is needed to construct the set of points that are to be topologised. For instance, one approach to Stone-Čech compactification is to consider the maximal ideal space $\operatorname{Max} \mathbb{R}^b(M)$ of the algebra of bounded continuous real functions on the topological space X. The Axiom of Choice, once more in the form of Zorn's Lemma, is needed to establish the existence of maximal ideals of this algebra, although the lattice of open subsets of the maximal ideal space is known to be that of ideals of the algebra that are closed with respect to the canonical sup-norm [3]. In this and other situations, one is led to give prominence to the lattice of open subsets of the topological space, rather than to the set of points being topologised, a concept which is abstracted in the following:

Definition 3.1 *By a* locale L *is meant a complete lattice in which the operations of finite meet and arbitrary join satisfy the condition that*

$$u \wedge \bigvee_i v_i = \bigvee_i u \wedge v_i$$

for any $u \in L$ and any elements $v_i \in L$.

Of course, any topological space will give rise to a locale, by considering its lattice of open subsets, but, constructively speaking, there are many situations in which one may define a locale, having the same properties as the locale of the topological space that would classically have been considered, but which is not necessarily the locale of a topological space, simply because in the absence of the Axiom of Choice the existence of its points cannot be established. Thus, it may be proved constructively that for any locale L there is a compact, completely regular locale βL together with a natural map of locales $L \to \beta L$ such that for any map of locales $L \to M$ to a compact, completely regular locale M there is a unique factorisation

$$\begin{array}{ccc} L & \longrightarrow & \beta L \\ & \searrow & \downarrow \\ & & M \end{array}$$

through the canonical map. It may be proved that in the presence of the Axiom of Choice every compact, completely regular locale is the lattice of open subsets of a compact, completely regular topological space [2]. Hence, classically this result is exactly the statement of the Stone-Čech compactification theorem, whilst constructively it is the appropriate formulation of that theorem. The concept of a map of locales met with here is that given by the following:

Definition 3.2 *By a* map of locales

$$\varphi : L \to M$$

is meant a mapping, referred to as the inverse image mapping,

$$\varphi^* : M \to L$$

that preserves finite meets and arbitrary joins.

The importance of locales constructively is that there is an extremely intuitive way of constructing them in practical situations. This arises from the observation that the lattices that underlie locales are, within constructive logic, exactly the Lindenbaum algebras of *propositional geometric theories* [20]. These are propositional theories generated starting from certain given primitive propositions by applying only the logical connectives of finite conjunction $\wedge$ and arbitrary disjunction $\bigvee$, of which the empty instances provide respectively the constants true $\top$ and false $\bot$. The axioms of such a theory are required to be entailments of the form

$$\varphi_1 \wedge \cdots \wedge \varphi_n \vdash \bigvee_i \psi_{i1} \wedge \cdots \wedge \psi_{in_i}$$

in which the symbols φ_i and ψ_{ij} denote primitive propositions of the theory. The Lindenbaum algebra of the theory is then the set of propositions

$$\bigvee_i \varphi_{i1} \wedge \cdots \wedge \varphi_{in_i}$$

of the theory, partially ordered by provable entailment in the theory, modulo provable equivalence in the theory. The fact that entailments are determined within constructive logic means precisely that the Lindenbaum algebra is a locale. Moreover, any locale is canonically isomorphic to the Lindenbaum algebra of an appropriate theory. It is perhaps significant that moving to constructive logic yields this abstraction of the concept of topological space, whilst remaining within classical logic gives only the much less interesting notion of a complete Boolean algebra.

The remark that any locale is canonically isomorphic to the Lindenbaum algebra of a propositional geometric theory is also of significance for our present purpose. In a quite precise sense, which may be found discussed at length elsewhere [20], the theory that is needed in order to obtain a particular locale is always the theory of its points. Presently, it will suffice to see this in the particular case of the locale that we wish to consider, namely that of the unit ball of the dual of a seminormed space B . To construct this locale, we need only find a propositional geometric theory of which the models within the constructive context are the linear functionals on B of norm not exceeding 1. To find the primitive propositions, we may take a lead from the way in which classically we define the open subsets of the weak* topology that we place on this set of linear functionals. This topology is defined by taking a subbasis of open subsets given by subsets

$$U\,(a, r, s) \;=\; \{\, \varphi \in \mathrm{Fn}\ B \mid \varphi\,(a) \in (r, s)\,\}$$

consisting of all linear functionals of norm ≤ 1 which map a given $a \in B$ into the rational open interval (r, s). Observe that taking these to be a subbasis of open subsets means that every open subset is an arbitrary join of finite meets of open subsets of this form.

Consider now the propositional geometric theory $\mathbb{F}\mathrm{n}\, B$ determined by taking a primitive proposition

$$a \in (r, s)$$

for each $a \in B$ and for each pair $r, s \in \mathbb{Q}$ of rational numbers, together with the following axioms:

(F1) $true \vdash 0 \in (r, s)$ whenever $r < 0 < s$;

(F2) $0 \in (r, s) \vdash false$ otherwise;

(F3) $a \in (r, s) \vdash -a \in (-s, -r)$;

(F4) $a \in (r, s) \vdash ta \in (tr, ts)$ whenever $t > 0$;

(F5) $a \in (r, s) \wedge a' \in (r', s') \vdash a + a' \in (r + r', s + s')$;

(F6) $a \in (r, s) \vdash a \in (r, s') \vee a \in (r', s)$ whenever $r < r' < s' < s$;

(F7) $true \vdash a \in (-1, 1)$ whenever $a \in N(1)$;

(F8) $a \in (r, s) \dashv\vdash \bigvee_{r<r'<s'<s} a \in (r', s')$.

Denoting the locale obtained from this theory by $\mathrm{Fn}\, B$, it may be verified [20] that its models are indeed exactly the linear functionals on B of norm ≤ 1. The correspondence between models Φ and functionals φ is given by

$$\Phi \models a \in (r, s) \quad \text{if and only if} \quad \varphi(a) \in (r, s)$$

In particular, the linear functional determined by a particular model of the theory assigns to each $a \in B$ the Dedekind real number of which the lower and upper cuts consist respectively of those rational numbers $r, s \in \mathbb{Q}$ for which the primitive proposition $a \in (r, s)$ is validated in the model. The axioms of the theory are exactly those required to prove that these are indeed the lower and upper cuts of a Dedekind real number, and that the assignment to each $a \in B$ of this Dedekind real number yields a linear functional on the seminormed space B of norm not exceeding 1. In this way, the axiomatisation of the theory is almost self-evident once one recalls the relationship classically between the points of the space to be constructed and the description of the topology that classically is placed upon them.

Recalling the motivating remarks concerning the expression of the Hahn-Banach theorem in terms of the continuous mapping from the unit ball of the dual of the seminormed space B to that of the dual of the subspace A, the Hahn-Banach theorem may be stated constructively in the following way:

Theorem 3.1 *For any linear subspace A of a seminormed linear space B, the map*

$$\mathrm{Fn}\ B \to \mathrm{Fn}\ A$$

canonically induced by restriction of functionals is a quotient map of compact, completely regular locales.

For the moment, we shall indicate a geometric proof of that theorem in a context that is not entirely constructive. Explicitly, we shall show that the theorem may be proved in any Grothendieck topos, that is, a topos that is bounded over the topos of sets [13], by a geometric argument that globalises the theorem from its classical counterpart in the topos of sets. The details of the proof may be found in the paper by Mulvey and Pelletier [20], in which the construction used is explained with some care. The view taken is that the technique applied is one which exactly shows how questions of continuity in parameters and of equivariance should be addressed constructively. In a sense, the implication is that this captures the constructive aspects of globalisation, even in the case where it is unknown whether the theorem to be globalised is indeed constructively provable. In the present case, we shall see later that the Hahn-Banach theorem may in fact be proved constructively in the form now introduced, thereby establishing a stronger assertion than that which we now consider. In each case, however, the map of locales about which the assertion is made is that given by interpreting the theory obtained from the linear subspace A in that obtained from the seminormed linear space B, by assigning to each primitive proposition

$$a \in (r, s)$$

of the theory $\mathbb{F}\mathrm{n}\ A$ the corresponding proposition of the theory $\mathbb{F}\mathrm{n}\ B$. The assignment determines a map of locales

$$\mathrm{Fn}\ B \rightarrow \mathrm{Fn}\ A$$

provided that each of the axioms of the theory $\mathbb{F}\mathrm{n}\,A$ is validated in the locale $\mathrm{Fn}\,B$, which is trivially the case since each such axiom is indeed one of the axioms of the theory $\mathbb{F}\mathrm{n}\,B$. The condition for the map of locales thereby defined to be a quotient map is exactly that its inverse image mapping is an embedding. In terms of the theories, this exactly states that the theory $\mathbb{F}\mathrm{n}\,B$ is a conservative extension of the theory $\mathbb{F}\mathrm{n}\ A$, an assertion which can be summarised by saying that one can learn no more in terms of linear functionals about the subspace A by considering it embedded in the seminormed space B than one learn by considering it as a seminormed space in its own right, an observation which is frequently used to summarise the content of the classical Hahn-Banach theorem.

The proof that we shall outline depends critically on the fact that the locales concerned are indeed compact, completely regular locales. That this is the case may indeed be proved constructively, and the details of one approach may be found in Mulvey and Pelletier [20]. The complete regularity is deduced from that of the locale of real numbers, itself defined by introducing a propositional geometric theory of the Dedekind real numbers with completely regularity deduced from the geometry of the set of rational numbers. The compactness is established by a syntactic argument, based on the compactness of the locale of any propositional geometric theory containing only finitary disjunctions. Although in this context this approach is extremely unintuitive, the technique is borrowed from another context in which the intuitive content of the method is both apparent and meaningful [3]. The statement that the locale

$$\mathrm{Fn}\,B$$

is compact, completely regular for any seminormed linear space B is exactly Alaoglu's theorem, which is thereby established to be constructively valid in this localic form [20].

The Hahn-Banach theorem is now proved in any Grothendieck topos $\mathbb{E}$ by constructing another Grothendieck topos $\mathbb{B}$, in which the Axiom of Choice is validated, together with a geometric map

$$\gamma : \mathbb{B} \to \mathbb{E}$$

which is a covering of the topos $\mathbb{E}$. The topos $\mathbb{B}$, known as the Barr covering [7] of the topos $\mathbb{E}$, is the category of sheaves on a complete Boolean algebra constructed from the topos $\mathbb{E}$. In such a topos, the mathematics is classical due to the presence of the Axiom of Choice. In particular, the dual locale of any seminormed space in the topos is actually the lattice of open subsets of a compact, completely regular topological space, hence necessarily that of the unit ball of the dual of the seminormed space in the classical sense. As a consequence, the classical Hahn-Banach theorem indeed implies that its constructive form is also valid for seminormed spaces in the topos $\mathbb{B}$. This may now be used to infer that the constructive form of the Hahn-Banach theorem is also valid in the topos $\mathbb{E}$ in which we are seeking to prove the theorem. That this is the case is argued in the following way.

Consider the inverse image along the geometric covering

$$\gamma : \mathbb{B} \to \mathbb{E}$$

of the inclusion map from the linear subspace A to the seminormed linear space B. This yields an inclusion map

$$\gamma^* A \to \gamma^* B$$

of seminormed spaces in the topos B. It should be noted in passing that it is to facilitate this step that we have considered linear spaces over the field $\mathbb{Q}$ of rational numbers, since the field of rational numbers is preserved under inverse image, whereas that of Dedekind real numbers is not. By the functoriality, straightforwardly verified, of the construction of the dual locale of a seminormed space, this yields a map

$$\text{Fn}\,\gamma^* B \to \text{Fn}\,\gamma^* A$$

of locales in the topos $\mathbb{B}$. Denoting the toposes of sheaves on each of the locales constructed by the corresponding open faced symbols, the functoriality of the construction of toposes of sheaves yields the following commutative diagram of toposes and geometric maps:

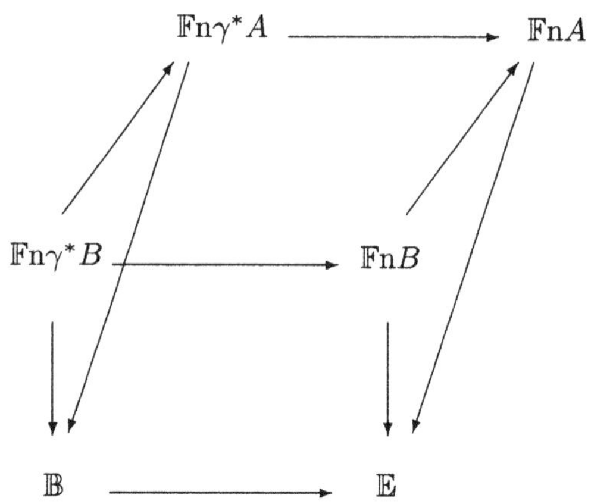

Since the axiomatisation of the theory of the dual locale is geometric, it is preserved under inverse image along a geometric map, hence the topos $\mathbb{F}\mathrm{n}\,\gamma^*B$ is actually the pullback of the topos $\mathbb{F}\mathrm{n}\,B$ along the geometric map

$$\gamma : \mathbb{B} \to \mathbb{E}.$$

Moreover, by the compact, complete regularity of the locale Fn B, the fact that this geometric map is actually a covering implies that its inverse image

$$\mathbb{F}\mathrm{n}\,\gamma^*B \to \mathbb{F}\mathrm{n}\,B$$

along the canonical map from $\mathbb{F}\mathrm{n}\,B$ to the topos $\mathbb{E}$ is again a covering. Furthermore, similar remarks apply to the geometric maps between the toposes determined by the linear subspace A. Finally, by the remark that the Hahn-Banach theorem applies classically in the topos $\mathbb{B}$, the map

$$\mathrm{Fn}\ \gamma^*B \ \to\ \mathrm{Fn}\ \gamma^*A$$

considered above is actually a quotient map of locales, hence determines a covering map of the corresponding toposes of sheaves. Hence, considering the commutative square that sits on top of the diagram of toposes, we observe that each of the maps concerned, with the exception of the map

$$\mathbb{F}\mathrm{n}\,B \to \mathbb{F}\mathrm{n}\,A$$

has already been shown to be a geometric covering, from which it follows that this map of toposes is also a covering. However, the geometric map determined by a map of locales is a covering exactly if the map of locales is a quotient map.

Hence, the canonical map

$$\mathrm{Fn}\, B \to \mathrm{Fn}\, A$$

is a quotient map of locales, establishing [20] the Hahn-Banach theorem in the topos $\mathbb{E}$.

Although this account has depended on a considerable amount of background from the theory of toposes, it has served to make the point that, once the Hahn-Banach theorem has been presented in a form appropriate to the constructive context, globalising the theorem to situations which include those of categories of sheaves on topological spaces, and of categories of sets acted upon by a monoid, is a matter of straightforward geometric argument involving covering toposes. The next stage of the development of these ideas involves in some sense simplifying these ideas to the point at which the argument may be carried out within the topos in which the Hahn-Banach theorem is needed. In other words, making the geometric argument self-contained, and thereby entirely constructive.

Before doing so, however, it may be of interest to apply the theorem just proved, and the technique used to prove it, to retrieving the extension of the naive Hahn-Banach theorem considered in the preceding section. So then, suppose that we are given a bounded linear *functional

$$\varphi : A \to {}^*\mathbb{R}_{\mathbb{E}}$$

on a linear subspace A of a seminormed space B in the topos $\mathbb{E}$ of sheaves on a topological space X. Recall that there exists another covering

$$\alpha : \mathbb{G} \to \mathbb{E}$$

of any topos $\mathbb{E}$, referred to as the Gleason covering [14], by a topos $\mathbb{G}$ in which De Morgan's Law is validated. In so doing, we should also remark that, unlike that of the covering considered above, the construction of this covering is entirely constructive. It has already been remarked that in any topos satisfying De Morgan's Law, the Dedekind real numbers are already order-complete, hence coincide with the Dedekind *real numbers. One of the properties of the Gleason covering [14] is that the direct image $\alpha_*\mathbb{R}_{\mathbb{G}}$ of the set of Dedekind real numbers in the topos $\mathbb{G}$ is exactly the set of Dedekind *real numbers ${}^*\mathbb{R}_{\mathbb{E}}$ in the topos $\mathbb{E}$. Hence, the bounded linear *functional considered above may in fact be written in the form

$$\varphi : A \to \alpha_*\mathbb{R}_{\mathbb{G}}$$

By the adjointness of inverse image and direct image functors, this therefore determines a bounded linear functional

$${}^*\varphi : \alpha^* A \to \mathbb{R}_{\mathbb{G}}$$

on the inverse image $\alpha^* A$ of the seminormed space A, hence a point of the locale $\mathrm{Fn}\, \alpha^* A$ in the topos $\mathbb{G}$. By the constructive form of the Hahn-Banach theorem, just proved to be valid in the topos $\mathbb{G}$, the canonical map

$$\mathrm{Fn}\ \alpha^*B \rightarrow \mathrm{Fn}\ \alpha^*A$$

is a quotient map of locales. By a result of Mulvey and Pelletier [19], for any quotient map of compact, completely regular locales in a topos satisfying De Morgan's Law, any point of the locale $\mathrm{Fn}\,\alpha^*A$ may then be lifted to a point

$$^*\psi : \alpha^*B \rightarrow \mathbb{R}_{\mathbb{G}}$$

of the locale $\mathrm{Fn}\,\alpha^*B$, which by adjointness yields the required extension

$$\psi : B \rightarrow {}^*\mathbb{R}_{\mathbb{E}}$$

of the given bounded linear *functional.

4 Constructivising the Hahn-Banach theorem

The approach to the constructive Hahn-Banach theorem described above relies on an application of geometric constructions at a contextual level to globalise the Hahn-Banach theorem to situations in which the Axiom of Choice cannot be applied. This technique may be extremely valuable in certain situations, of which the recovery of the extended version of the naive Hahn-Banach theorem is an instance. Although it gives some metamathematical credence to the view that the constructive Hahn-Banach theorem may indeed be provable constructively, it does not in itself lead to a constructive proof. To achieve this, one needs to be more careful in analysing the classical proof of the theorem, and the way in which its apparent dependence on the Axiom of Choice may be replaced by more subtle geometric arguments. Certainly, one aspect of the proof just outlined that is likely to need to be incorporated in arriving at such a constructive proof is that of the rôle played by the compact, complete regularity of the dual locale of a seminormed linear space.

This analysis, of which we shall give only an outline, involves working geometrically with locales, both in order to construct the locales that appear in the Hahn-Banach theorem itself, but also those needed to establish a constructive proof of the theorem. The principle that may be drawn from this is that careful geometrical argument can be used to avoid applications of the Axiom of Choice, except any that we may unwisely build into our mathematics by choosing to state a theorem in a hopelessly unconstructive way. For the Hahn-Banach theorem, this means that expressing the theorem in the form of a statement that the canonical map

$$\mathrm{Fn}\ B \rightarrow \mathrm{Fn}\ A$$

of locales is a quotient is a constructively appropriate statement that admits a constructive proof. In the event that the locales concerned are indeed topological spaces, then the Hahn-Banach theorem in its classical form will also be valid. However, the statement that any compact, completely regular locale is indeed spatial is a statement almost equivalent to the Axiom of Choice. The details of the proof to be outlined will appear in a paper by Mulvey and Vermeulen [22], depending heavily on earlier work by Vermeulen [25].

The concept behind the proof is straightforward. The classical proof of the Hahn-Banach theorem provides an inductive step, albeit for the moment not in a constructive form, by which it is shown that a bounded linear functional defined on a subspace of a seminormed space may be extended in a norm-preserving manner to the subspace obtained by adjoining a single element of the seminormed space. The classical proof applies this to show that a maximal norm-preserving extension is necessarily defined on the seminormed space itself. In the constructive context, instead of applying Zorn's Lemma to choose such a maximal extension, we must instead use this proof of the existence of an extension to a subspace obtained by adjoining a single element, adapted to a constructive form, as the inductive step of an inductive limit argument carried out over the finitary extensions of the given functional. The inductive step involves showing that the corresponding map of dual locales is indeed not just a quotient, but a kind of map known as a *proper surjection* [25]. That this is the case stems from the compact, complete regularity of the dual locales concerned, allowing a geometric result of Vermeulen concerning filtered limits of proper surjections [25] to be applied to prove that the canonical map

$$\mathrm{Fn}\ B \to \mathrm{Fn}\ A$$

is again a proper surjection, and hence a quotient map of locales. The subtler points of the proof are those involved in making the inductive step constructive, to which we shall turn after outlining the inductive framework within which that step will be applied.

Consider, then, a linear subspace A of a seminormed linear space B. Observe firstly that it is classically, and constructively, the case that the extension

$$A \to B$$

may be obtained as a filtered inductive limit of finitely generated extensions

$$A \to A[b_1, \cdots, b_n]$$

of the subspace A to the subspace of B obtained by adjoining finitely many elements $b_1, \cdots, b_n \in B$. This is evidently the case algebraically, and carries through without problem to the seminormed case. Intuitively, the idea of the proof would now be to establish, by applying the inductive step of the classical Hahn-Banach theorem, that, for the locales obtained by applying the construction of the dual locale to these seminormed spaces, the canonical map

$$\mathrm{Fn}\ A[b_1, \cdots, b_n] \to \mathrm{Fn}\ A$$

of locales is indeed a proper surjection. The canonical map

$$\mathrm{Fn}\, B \rightarrow \mathrm{Fn}\, A$$

would then be the filtered limit of these proper surjections, hence again a proper surjection, and in particular a quotient map, yielding the constructive Hahn-Banach theorem.

Constructively, in order to apply the inductive step of the classical Hahn-Banach theorem straightforwardly, we have to be slightly more subtle. The problem is one that is often met with constructively, namely that for a given element $b \in B$ by which we might wish to extend a subspace A, it may not be decidable whether $b \in A$ or $\neg (b \in A)$. To avoid this difficulty, which can be surmounted in a number of ways, we instead modify the inductive system considered by presenting the extension

$$A \rightarrow B$$

as the inductive limit of a filtered diagram of finitely freely generated extensions of the subspace A. Explicitly, consider for each finite family $b_1, \cdots, b_n \in B$ the extension

$$A \rightarrow A \langle b_1, \cdots, b_n \rangle$$

of the seminormed linear space A obtained by freely adjoining indeterminates representing these elements to the seminormed space A, together with the map of extensions

$$\begin{array}{ccc} A \langle b_1, ..., b_n \rangle & \longrightarrow & B \\ \uparrow & \nearrow & \\ A & & \end{array}$$

obtained by mapping each of these indeterminates to the corresponding element of the seminormed space B. Now consider the filtered diagram of extensions of the seminormed space A obtained by taking as maps those maps of extensions for which the diagram

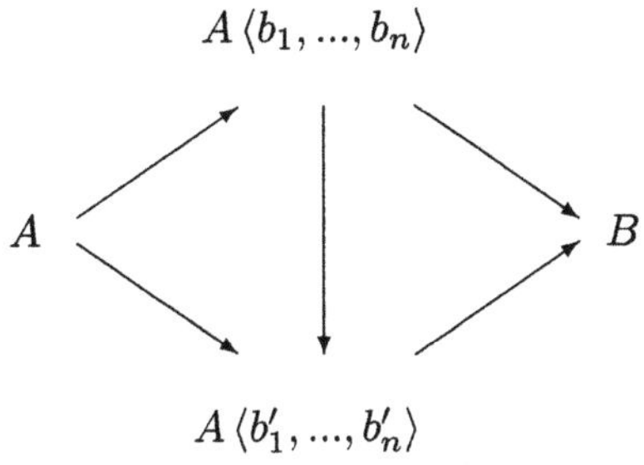

is the filtered inductive limit of this diagram in the category of extensions of the seminormed space A. Applying the syntactic construction of the dual locale of a seminormed space, the functoriality of the construction yields a corresponding filtered diagram in the category of locales over the locale $\mathrm{Fn}\, A$, of which it may formally be verified that the limit is the canonical map

$$\mathrm{Fn}\ B \rightarrow \mathrm{Fn}\ A$$

Applying the theorem mentioned above concerning filtered limits of proper surjections of locales [25], it follows that this is again a proper surjection, hence in particular a quotient map, provided that it may be shown that the canonical map

$$\mathrm{Fn}\ A\langle b_1, \cdots, b_n\rangle \rightarrow \mathrm{Fn}\ A$$

is a proper surjection for each finite family $b_1, \cdots, b_n \in B$. Since the composite of proper surjections is a proper surjection, it suffices to prove this in the case of an extension freely generated by a single element $b \in B$, which brings us to the inductive step in the theorem.

In the classical proof of the Hahn-Banach theorem, the inductive step involves showing that for any bounded linear functional

$$\varphi : A \rightarrow \mathbb{R}$$

on a subspace A of a seminormed space B and for any element $b \in B$ the inequality

$$\sup_{a\in A}\{\varphi(a) - \|a-b\|\} \leq \inf_{a\in A}\{-\varphi(a) + \|a+b\|\}$$

is satisfied, allowing a norm-preserving extension of the functional to be defined on the subspace obtained by adjoining the element $b \in B$ to the subspace A. In the constructive case, the same ideas are used to show that the canonical map

$$\mathrm{Fn}\ B \rightarrow \mathrm{Fn}\ A$$

is a proper surjection whenever the seminormed space B is that obtained by freely adjoining an indeterminate $b \in B$ to the seminormed space A. To prove this, we observe firstly that one may construct a locale $\mathbb{R}$ of real numbers by taking a propositional geometric theory of which the primitive propositions are of the form:

$$x \in (r, s)$$

for each pair of rational numbers $r, s \in \mathbb{Q}$, together with axioms that describe that these rationals lie respectively in the lower and upper cuts of a Dedekind real number.

One may also construct locales $\mathbb{R}_L$ and $\mathbb{R}_U$ of lower and upper real numbers by taking propositional geometric theories describing respectively lower and upper cuts on the rationals, together with a sublocale

$$\sqsubseteq \rightarrow \mathbb{R}_L \times \mathbb{R}_U$$

of the product $\mathbb{R}_L \times \mathbb{R}_U$ obtained by adjoining the axiom

$$p < \sigma \wedge \tau < q \vdash p < q$$

to the axioms for the generic lower and upper cuts σ and τ. Equally, within the product $\mathbb{R}_L \times \mathbb{R} \times \mathbb{R}_U$ one may construct a sublocale

$$I \rightarrow \mathbb{R}_L \times \mathbb{R} \times \mathbb{R}_U$$

which classifies triples (σ, α, τ) having the property that (σ, τ) lies in the sublocale defined above, and that $\sigma \leq L(\alpha)$ and $U(\alpha) \leq \tau$, in which $L(\alpha)$ and $U(\alpha)$ denote respectively the lower and the upper cuts of a Dedekind real number α. It must be remembered throughout that these discussions are considered at a syntactic level, rather than in terms of actual Dedekind cuts on the rationals.

There is a canonical map

$$I \rightarrow \sqsubseteq$$

obtained by factoring the projection, and this map of locales may be shown to be a proper surjection. It is this fact, derived from the compactness properties of closed bounded intervals of the locale of reals, that lies at the heart of the Hahn-Banach theorem. Given this observation, of which the proof may be found elsewhere [22], the required result may be proved by remarking that there is a canonical diagram

$$\begin{array}{ccc} \mathrm{Fn}\, B & \longrightarrow & I \\ \downarrow & & \downarrow \\ \mathrm{Fn}\, A & \longrightarrow & \sqsubseteq \end{array}$$

defined by encoding syntactically the inequality between the suprema and infima involved in the construction of an extension, described above in the case of the classical Hahn-Banach theorem. Moreover, this diagram is actually a pullback in the category of locales. In consequence, the fact that the right hand side is a proper surjection implies that the canonical map determined by the extension of seminormed spaces is also a proper surjection, which gives the required inductive step, and with it a constructive proof of the Hahn-Banach theorem stated in the preceding section.

In conclusion, it should be noted that the constructive proof has depended on simple manipulation of geometric objects, taken in conjunction with the deeper geometric observations of Vermeulen concerning the foundations of descent theory in the context of closed maps of locales [25]. In a sense, together these provide the geometric situation in which dependence on the Axiom of Choice is relegated to the question of whether the dual locales considered are in fact topological spaces, in which case the Hahn-Banach theorem can be recovered in its classical form. However, as may be found in papers by Mulvey and Pelletier [20, 21], there is no need to have the Hahn-Banach theorem in the classical form, when it may be applied directly in the constructive form to yield the same consequences. Nevertheless, in the presence of the Axiom of Choice, the constructive form immediately yields the classical one, just by the observation that the locales considered are then exactly the lattices of open subsets of their topological spaces of points.

5 Conclusions

In this chapter, we have considered one particular application of a technique that seems to have quite widespread relevance. The Hahn-Banach theorem classically has been seen to be a spatialisation of a constructive theorem of which the proof is closely linked to the geometric content of the classical theorem. The geometry of the Axiom of Choice has been seen to be that which is founded on the notion of point as a primitive construct. Instead, the constructive geometry of locales, in which points have little relevance, with locality taking the primitive place, has appeared naturally to signify the extent to which we consider points as an ideal realisation of an otherwise syntactic expression of mathematical relationships. The geometry of constructive logic has been seen as a fundamental component of the constructivisation of mathematical concepts. In particular, through the link between locales and constructive logic, the geometry of locales provides a mathematical expression of metamathematical constructs.

The ideas behind these techniques have already been applied in a number of other situations. The construction of the Stone-Čech compactification of a locale is one instance that predates the work described here, with a variety of approaches leading to the existence of compactifications of locales both to compact, regular and to compact, completely regular locales. One of the approaches taken, establishing the existence of the Stone-Čech compactification by developing a constructive version of the Tychonoff theorem [15], has led to further work extending the constructive form of the Tychonoff theorem to compact locales

[24]. The constructive form of Gelfand duality has been investigated, thereby stimulating the development of an extension towards the non-commutative case through the introduction of the theory of quantales [18]. The Fundamental Theorem of Galois theory has also been shown to have a constructive expression [12].

In each of these cases, the rôle played by the Axiom of Choice has been marginalised to one of recovering a familiar, but not necessarily more advantageous, form of the theorem from a version that gives constructive insight into the mathematics that lies behind it. In achieving this, constructive mathematics itself has moved forward from being involved with managing to achieve a proof against the odds, to describing positively the fundamentals of the mathematics that is being investigated. The importance of this step is also measured in terms of the applications of these ideas to situations within mathematics in which the formalism of constructive mathematics is needed to extend a classical theorem in another context, whether involving continuity across a space or equivariance with respect to an action of some kind.

The importance of the Axiom of Choice in allowing the rapid development of mathematics would appear to be justified by exactly the arguments being advanced here, pointing to the extent to which meaningful results within classical mathematics may be shown to carry through to the constructive context. But the message is also that with techniques that allow this more meaningful examination of mathematical structures increasingly at hand, whether one moves towards greater constructive insight or languishes in the perhaps less meaningful power of an approach to mathematics that may actually obscure insight into what is really happening is now at least largely a matter of choice.

References

[1] B. Banaschewski, *Extensions of invariant linear functionals: Hahn-Banach in the topos of M-sets*, J. Pure and Appl. Alg. **17**(1980), 227–248.

[2] B. Banaschewski, C. J. Mulvey, *The Stone-Čech compactification of locales, I*, Houston J. Math. **6**(1980), 301–312.

[3] B. Banaschewski, C. J. Mulvey, *The Stone-Čech compactification of locales, II*, J. Pure Applied Algebra, **33**(1984), 107–122.

[4] B. Banaschewski, C. J. Mulvey, *The spectral theory of commutative C*-algebras: the constructive spectrum*, Quaestiones Math., **25**(2000), 425–464.

[5] B. Banaschewski, C. J. Mulvey, *A constructive proof of the Stone-Weierstrass theorem*, J. Pure Applied Algebra **116**(1977), 25–40.

[6] B. Banaschewski, C. J. Mulvey, *A globalisation of the Gelfand duality theorem*, to appear.

[7] M. Barr, *Toposes without points*, J. Pure and Appl. Alg. **5**(1974), 265–280.

[8] E. A. Bishop, Foundations Of Constructive Analysis, McGraw-Hill (New York), 1967.

[9] D. S. Bridges, Constructive Functional Analysis, Pitman (London), 1979.

[10] C. W. Burden, *The Hahn-Banach Theorem in a category of sheaves*, J. Pure Applied Algebra **17**(1980), 25–34.

[11] C. W. Burden, C. J. Mulvey, *Banach spaces in categories of sheaves*, in *Applications of Sheaves: Proceedings of the Durham Symposium* (July 1977), Lecture Notes in Mathematics **753**, 169–196, Springer, 1979.

[12] G. M. French, *Syntactic and semantic aspects of locales and quantales*, Thesis, University of Sussex, 1992.

[13] P. T. Johnstone, Topos Theory, Academic Press (London), 1977.

[14] P. T. Johnstone, *The Gleason cover of a topos II*, J. Pure and Appl. Alg. **22**(1981), 229–247.

[15] P. T. Johnstone, *Tychonoff's theorem without the axiom of choice*, Fund. Math. **113**(1981), 21–35.

[16] P. T. Johnstone, Stone Spaces, Cambridge University Press (Cambridge), 1982.

[17] C. J. Mulvey, *Intuitionistic algebra and representation of rings*, Memoirs Amer. Math. Soc. **148**(1974), 3–57.

[18] C. J. Mulvey, *&*, Rendiconti Circ. Mat. Palermo **12**(1986), 99–104.

[19] C. J. Mulvey, J. W. Pelletier, *On the points of locales in a De Morgan topos*, in *Categorical Topology*, Toledo, 1983, Sigma Series in Pure Mathematics **5**(1984), 392–407, Heldermann.

[20] C. J. Mulvey, J. W. Pelletier, *A globalisation of the Hahn-Banach theorem*, Advances in Mathematics **89**(1991), 1–59.

[21] C. J. Mulvey, J. W. Pelletier, *A geometric characterization concerning compact convex sets*, Math. Proc. Camb. Phil. Soc. **109**(1991), 351–361.

[22] C. J. Mulvey, J. J. C. Vermeulen, *A constructive proof of the Hahn-Banach theorem*, to appear.

[23] J. Staples, *On constructive fields*, Proc. London Math. Soc. **23**(1971), 753–768.

[24] J. J. C. Vermeulen, *Constructive Techniques In Functional Analysis*, Dissertation, University of Sussex, 1986.

[25] J. J. C. Vermeulen, *Proper maps of locales*, J. Pure and Appl. Alg. **92**(1994), 79–107.

CHAPTER 12

On Some Fuzzy Categories Related To Category L-TOP Of L-Topological Spaces

A. P. ŠOSTAK

Introduction

Since the inception of fuzzy set theory (see [73], [51], [16]), the efforts of many researchers were directed to find fuzzy analogues of basic concepts of classical mathematics and to develop corresponding theories. General topology was one of the first fields to which the interest of "fuzzy mathematicians" was focused. Already as soon as in 1968 topological structures on lattices of [0,1]-fuzzy sets were considered by C. L. Chang [5]; C. L. Chang's ideas for more general lattices were developed by J.A. Goguen [17], B. Hutton [34], [35], S. E. Rodabaugh [46], [48], U. Höhle [21] , P. Eklund [9], [10], P. Eklund and W. Gähler [11] *et al.* However in all these works the fuzziness revealed itself only on the level of the powerset, while the topological structure still remained crisp. In particular, in the Chang-Goguen approach an L-fuzzy topology on set X is a family $\tau \subset L^X$ satisfying natural analogues of the usual axioms of topology. In the sequel, following [30], such family τ will be referred to as *an L-topology* on a set X, see 3.1.

Noticing this feature of the Chang-Goguen approach, some authors turned their attention to the study of fuzzy topological type structures on crisp or, more generally, on fuzzy powersets. For example, according to [53], [56], cf also [22], [36], [55], [29], [38], [30], [49], *et al*, an L-fuzzy topology $\mathcal{T}$ on a set X is an L-subset of the fuzzy powerset L^X, i.e. a mapping $\mathcal{T} : L^X \to L$ satisfying certain analogues of axioms of an ordinary topology.

Another feature of the Chang-Goguen theory of L-topological spaces is that when studying topological-type structures in the fuzzy context, the division of all functions into continuous and non-continuous seems to be too straightforward

S.E. Rodabaugh and E.P. Klement (eds.), Topological and Algebraic Structures in Fuzzy Sets, 337–372.
© 2003 *Kluwer Academic Publishers.*

and coarse: one could anticipate a more subtle classification. Such an idea was discussed in [56] where we introduced the *continuity defect* for mappings of L-topological spaces (in case $L = [0, 1]$). Later the continuity defect and other similar values for mappings were considered in [3], [4], [63], etc.

The idea to evaluate the measure to which a mapping between two L-(fuzzy) topological spaces is continuous leads us to the concept of a fuzzy category. As far as we know this idea for the first time was explicitly discussed in [57]; later in [61] and [63] we started to develop a general theory of fuzzy categories; some concrete fuzzy categories were briefly discussed there, too.

The main initial aim of this chapter was to define and to study a concrete fuzzy category based on the category L-**TOP** of Chang-Goguen L-topological spaces which would allow to develop a sufficiently subtle classification of *the measure of continuity* for mappings of such spaces. However, to do this consistently we had to enlarge the original category L-**TOP** (in the direction of the closure spaces in the sense of Čech [6]). Therefore we first introduce the category L-Ker of L-kernel spaces and the category L-$PInt$ of L-preinterior. For mappings of such spaces we define the *measure of continuity* thus coming to fuzzy categories $\mathcal{F}L$-Ker and $\mathcal{F}L$-$PInt$ (Subsection 4.C). On the other hand, in this situation it seems natural to evaluate also the degree to which a given L-kernel or L-preinterior space, i.e. an object of L-Ker or L-$PInt$, differs from a "normal" L-topological space, i.e. from an object of L-**TOP**. This leads to fuzzy categories of another type FL-Ker and FL-$PInt$ where "fuzziness" is applied to the class of objects (Subsection 4.A), and finally to fuzzy categories $\mathcal{F}FL$-Ker and $\mathcal{F}FL$-$PInt$ where both morphisms and objects are "fuzzy" (Subsection 4.E).

The structure of the chapter is as follows. We start with a short introduction into the theory of GL-monoids—complete lattices enriched with a further binary operation $*$ ([25], [27]; see Section 1). GL-monoids are constantly used in our work. In particular, a GL-monoid L plays the role of a standard range for fuzzy sets. Further, in Section 2, the concept of a fuzzy category and some related concepts are discussed. Starting with Section 3 we essentially turn to the basic subject of the chapter. In particular, in Section 3 the notions of an L-kernel and an L-preinterior operators are introduced and corresponding categories L-Ker and L-$PInt$ are discussed. Section 4 deals with fuzzification of these categories: here we introduce fuzzy categories $\mathcal{F}L$-Ker, $\mathcal{F}L$-$PInt$, FL-Ker, and FL-$PInt$ which are the central subject of the chapter. In Section 5 two modifications of L-preinterior structure Int are studied: α-weakening $Int^{\check{\alpha}}$ and α-strengthening $Int^{\hat{\alpha}}$; these modifications lead to self-functors $\Phi_\alpha : \mathcal{F}L\text{-}PInt \to \mathcal{F}L\text{-}PInt$ and $\Psi_\alpha : \mathcal{F}L\text{-}PInt \to \mathcal{F}L\text{-}PInt$, respectively. The subject of Section 6 are initial and final structures in fuzzy categories $\mathcal{F}L$-Ker and $\mathcal{F}L$-$PInt$; these structures in particular, allow to describe products and coproducts in these categories.

1 Prerequisities: GL-monoids

Let $(L, \leq)$ be a complete infinitely distributive lattice, i.e. $(L, \leq)$ is a partially ordered set such that for every subset $A \subset L$ the join $\bigvee A$ and the meet $\bigwedge A$ are defined and $(\bigvee A) \wedge \alpha = \bigvee\{a \wedge \alpha) \mid a \in A\}$ and $(\bigwedge A) \vee \alpha = \bigwedge\{a \vee \alpha) \mid a \in A\}$ for every $\alpha \in L$. In particular, $\bigvee L =: \top$ and $\bigwedge L =: \bot$ are respectively the universal upper and the universal lower bounds in L. We assume that $\bot \neq \top$, i.e. L has at least two elements.

A GL-monoid is a complete lattice enriched with a further binary operation $*$, i.e. a triple $(L, \leq, *)$ such that:

(1) $*$ is monotone, i.e. $\alpha \leq \beta$ implies $\alpha * \gamma \leq \beta * \gamma$, $\forall \alpha, \beta, \gamma \in L$;

(2) $*$ is commutative, i.e. $\alpha * \beta = \beta * \alpha$, $\forall \alpha, \beta \in L$;

(3) $(L, \leq, *)$ is integral, i.e. $\top$ acts as the unity: $\alpha * \top = \alpha$, $\forall \alpha \in L$;

(4) $\bot$ acts as the zero element in $(L, \leq, *)$, i.e. $\alpha * \bot = \bot$, $\forall \alpha \in L$;

(5) $*$ is distributive over arbitrary joins, i.e. $\alpha * (\bigvee_j \beta_j) = \bigvee_j (\alpha * \beta_j)$, $\forall \alpha \in L$, $\forall \{\beta_j : j \in J\} \subset L$;

(6) $(L, \leq, *)$ is divisible, i.e. $\alpha \leq \beta$ implies the existence of $\gamma \in L$ such that $\alpha = \beta * \gamma$.

(See [25], [26], [27]. Note that GL stands for *generalized logics.*)

It is known that every GL-monoid is residuated, i.e. there exists a further binary operation "$\longmapsto$" (implication) on L satisfying the following condition:

$$\beta * \alpha \leq \gamma \Longleftrightarrow \alpha \leq (\beta \longmapsto \gamma) , \quad \forall \alpha, \beta, \gamma \in L$$

Explicitly the implication is given by $\alpha \longmapsto \beta = \bigvee\{\lambda \in L \mid \alpha * \lambda \leq \beta\}$.

Below we list some properties of GL-monoids which will be useful in the sequel (see e.g. [25], [26], [27]):

(i) $\alpha \longmapsto \beta = \top \Longleftrightarrow \alpha \leq \beta$;

(ii) $\alpha \longmapsto (\bigwedge_i \beta_i) = \bigwedge_i (\alpha \longmapsto \beta_i)$;

(iii) $(\bigvee_i \alpha_i) \longmapsto \beta = \bigwedge_i (\alpha_i \longmapsto \beta)$;

(iv) $\alpha * (\bigwedge_i \beta_i) = \bigwedge_i (\alpha * \beta_i)$;

(v) $(\alpha \longmapsto \gamma) * (\gamma \longmapsto \beta) \leq \alpha \longmapsto \beta$;

(vi) $\alpha \leq \beta \quad \Longrightarrow \quad \gamma \longmapsto \alpha \leq \gamma \longmapsto \beta$.

Important examples of GL-monoids are Heyting algebras and MV-algebras. Namely, a *Heyting algebra* is GL-monoid of the type $(L, \leq, \wedge, \vee, \wedge)$ (i.e. in case of a Heyting algebra $\wedge = *$). A GL-monoid is called an *MV-algebra* if $(\alpha \longmapsto 0) \longmapsto 0 = \alpha$, $\forall \alpha \in L$. Thus in an MV-algebra an order reversing involution $^c : L \to L$ can be naturally defined by setting $\alpha^c := \alpha \longmapsto 0$, $\forall \alpha \in L$.

If X is a set and L is a GL-monoid, then the fuzzy powerset L^X in an obvious way can be pointwise endowed with a structure of a GL-monoid. In particular the L-sets 1_X and 0_X defined by $1_X(x) := \top$ and $0_X(x) := \bot$, $\forall x \in X$ are respectively the universal upper and lower bounds in L^X.

In the sequel L denotes an arbitrary GL-monoid.

2 L-fuzzy categories: basic concepts

In this section we introduce the concepts of a fuzzy category, a fuzzy functor and some related notions. Other needed concepts and facts of the theory of fuzzy categories will More information about fuzzy categories can be found in [61], [63], [62], cf. also [57], [58].

2.A: L-fuzzy categories

Definition 2.1 *An L-fuzzy category is a quituple $\mathcal{C} = (\mathcal{O}b(\mathcal{C}), \omega, \mathcal{M}(\mathcal{C}), \mu, \circ)$ where $\mathcal{C}_\bot = (\mathcal{O}b(\mathcal{C}), \mathcal{M}(\mathcal{C}), \circ)$ is a usual (classical) category called* the bottom frame *of the fuzzy category $\mathcal{C}$; $\omega : \mathcal{O}b(\mathcal{C}) \longrightarrow L$ is an L-subclass of the class of objects $\mathcal{O}b(\mathcal{C})$ of $\mathcal{C}_\bot$ and $\mu : \mathcal{M}(\mathcal{C}) \longrightarrow L$ is an L-subclass of the class of morphisms $\mathcal{M}(\mathcal{C})$ of $\mathcal{C}_\bot$. Besides ω and μ must satisfy the following conditions:*

(1) *if $f : X \to Y$, then $\mu(f) \leq \omega(X) \wedge \omega(Y)$;*

(2) *$\mu(g \circ f) \geq \mu(g) * \mu(f)$ whenever composition $g \circ f$ is defined;*

(3) *if $e_X : X \to X$ is the identity morphism, then $\mu(e_X) = \omega(X)$.*

Given an L-fuzzy category $\mathcal{C} = (\mathcal{O}b(\mathcal{C}), \omega, \mathcal{M}(\mathcal{C}), \mu, \circ)$ and $X \in \mathcal{O}b(\mathcal{C})$, the intuitive meaning of the value $\omega(X)$ is the *degree* to which a potential object X of the L-fuzzy category $\mathcal{C}$ is indeed its object; similarly, for $f \in \mathcal{M}(\mathcal{C})$ the intuitive meaning of $\mu(f)$ is the degree to which a potential morphism f of $\mathcal{C}$ is indeed its morphism.

Let $\mathcal{O}b_\alpha(\mathcal{C}) = \{X \in \mathcal{O}b(\mathcal{C}) \mid \omega(X) \geq \alpha\}$, $\mathcal{M}_\alpha(\mathcal{C}) = \{f \in \mathcal{M}(\mathcal{C}) \mid \mu(f) \geq \alpha\}$. The elements of $\mathcal{O}b_\alpha(\mathcal{C})$ will be referred to as α-objects, and the elements of $\mathcal{M}_\alpha(\mathcal{C})$ as $\alpha-$morphisms of the L-fuzzy category $\mathcal{C}$.

Obviously a usual category $(\mathcal{O}b(\mathcal{C}), \mathcal{M}(\mathcal{C}), \circ)$ can be identified with the L-fuzzy category

$$(\mathcal{O}b(\mathcal{C}), \omega, \mathcal{M}(\mathcal{C}), \mu, \circ), \text{ where } \omega \equiv 1_{\mathcal{O}b(\mathcal{C})} \text{ and } \mu \equiv 1_{\mathcal{M}(\mathcal{C})}$$

Given an L-fuzzy category $\mathcal{C} = (\mathcal{O}b(\mathcal{C}), \omega, \mathcal{M}(\mathcal{C}), \mu, \circ)$ one can obtain a crisp category $(\mathcal{O}b(\mathcal{C}), \mathcal{M}(\mathcal{C}), \circ)$ by taking all potential objects and potential morphisms as respective objects and morphisms of $\mathcal{C}_\perp$ and leaving the composition law unchanged. Obviously, this category is just the bottom frame $\mathcal{C}_\perp$ of $\mathcal{C}$.

This construction can be generalized: let ι be an idempotent element of L and let $\mathcal{C}_\iota = (\mathcal{O}b_\iota(\mathcal{C}), \mathcal{M}_\iota(\mathcal{C}), \circ)$. Then $\mathcal{C}_\iota$ is a crisp category having $\mathcal{O}b_\iota(\mathcal{C})$ as objects and $\mathcal{M}_\iota(\mathcal{C})$ as morphisms. In particular, the category $\mathcal{C}_\top$ is called the *top frame* of $\mathcal{C}$. (Of course, the case when $\mathcal{C}_\top$ is empty is not excluded.)

2.B: Fuzzy subcategories of fuzzy categories

Definition 2.2 *Let $\mathcal{C} = (\mathcal{O}b(\mathcal{C}), \omega, \mathcal{M}(\mathcal{C}), \mu, \circ)$ be an L-fuzzy category. An (L-fuzzy) subcategory of $\mathcal{C}$ is an L-fuzzy category $\mathcal{C}' = (\mathcal{O}b(\mathcal{C}), \omega', \mathcal{M}(\mathcal{C}), \mu', \circ)$ where $\omega' \le \omega$ and $\mu' \le \mu$. A subcategory $\mathcal{C}'$ of the category $\mathcal{C}$ is called full if $\mu'(f) = \mu(f) \wedge \omega'(X) \wedge \omega'(Y)$ for every $f \in \mathcal{M}_\mathcal{C}(X,Y)$, $X, Y \in \mathcal{O}b(\mathcal{C})$.*

Thus an L-fuzzy category and its subcategory have the same classes of potential objects and morphisms. The only difference of a subcategory from the whole category is in L-fuzzy classes of objects and morphisms, i.e. in the belongness degrees of potential objects and morphisms.

Let $\mathcal{C} = (\mathcal{O}b(\mathcal{C}), \mathcal{M}(\mathcal{C}), \circ)$ be a crisp category and $\mathcal{D} = (\mathcal{O}b(\mathcal{D}), \mathcal{M}(\mathcal{D}), \circ)$ be its subcategory. Then for every GL-monoid L the category $\mathcal{D}$ can be identified with the L-fuzzy subcategory $\tilde{\mathcal{D}} = (\mathcal{O}b(\mathcal{C}), \omega', \mathcal{M}(\mathcal{C}), \mu', \circ)$ of $\mathcal{C}$ such that $\omega'(X) = \top$ if $X \in \mathcal{O}b(\mathcal{D})$ and $\omega'(X) = \perp$ otherwise; $\mu'(f) = \top$ if $f \in \mathcal{M}(\mathcal{D})$ and $\mu'(f) = \perp$ otherwise. In particular, $\tilde{\mathcal{D}}_\top = \mathcal{D}$.

On the other hand sometimes it is convenient to identify a fuzzy subcategory $\mathcal{C}' = (\mathcal{O}b(\mathcal{C}), \omega', \mathcal{M}(\mathcal{C}), \mu', \circ)$ of the fuzzy category $\mathcal{C} = (\mathcal{O}b(\mathcal{C}), \omega, \mathcal{M}(\mathcal{C}), \mu, \circ)$ with the fuzzy category $\mathcal{D} = (\mathcal{O}b(\mathcal{D}), \omega_\mathcal{D}, \mathcal{M}(\mathcal{D}), \mu_\mathcal{D}, \circ)$ where $\mathcal{O}b(\mathcal{D}) := \{X \in \mathcal{O}b(\mathcal{C}) \mid \omega'(X) \neq \perp\}$ and $\mathcal{M}(\mathcal{D}) := \{f \in \mathcal{M}(\mathcal{C}) \mid \mu'(f) \neq \perp\}$ and $\omega_\mathcal{D}$ and $\mu_\mathcal{D}$ are restrictions of ω' and μ' to $\mathcal{O}b(\mathcal{D})$ and $\mathcal{M}(\mathcal{D})$, respectively.

2.C: Functors between fuzzy categories

Definition 2.3 *Let*

$$\mathcal{C} = (\mathcal{O}b(\mathcal{C}), \omega_\mathcal{C}, \mathcal{M}(\mathcal{C}), \mu_\mathcal{C}, \circ), \;\; \mathcal{D} = (\mathcal{O}b(\mathcal{D}), \omega_\mathcal{D}, \mathcal{M}(\mathcal{D}), \mu_\mathcal{D}, \circ)$$

be fuzzy categories, and let $F : \mathcal{C}_\perp \longrightarrow \mathcal{D}_\perp$ be a functor between their bottom frame categories. Then F is called a γ-functor *($\gamma \in L$) from a fuzzy category $\mathcal{C}$ to a fuzzy category $\mathcal{D}$ (in notations $F : \mathcal{C} \longrightarrow \mathcal{D}$) if $\mu(f) * \gamma \le \mu(F(f))$ for any $f \in \mathcal{M}(\mathcal{C})$.*

The term *a functor* is used as a common name for any γ-functor, $\gamma \in L$, in particular for a $\perp$-functor.

Remark 2.4 (a) Obviously, if $\gamma' \leq \gamma$, then a γ-functor is also an γ'-functor. In particular, a $\top$-functor is an γ-functor for any γ.

(b) Since $\omega(X) = \mu(e_X)$ we conclude that $\omega(X) * \gamma \leq \omega(F(X))$. In particular, $\omega(X) \leq \omega(F(X))$ if F is a $\top$-functor. (c) Let $\mathcal{C}$ and $\mathcal{D}$ be fuzzy categories and $\mathcal{C}^\perp$ and $\mathcal{D}^\perp$ be the corresponding bottom frame categories. Then $F : \mathcal{C} \longrightarrow \mathcal{D}$ is a $\perp$-functor if and only if $F : \mathcal{C}^\perp \longrightarrow \mathcal{D}^\perp$ is a functor in the "classical" sense.

3 L-kernel and L-preinterior spaces and some related categories

In order to make appropriate references we start with recalling and specifying some well-known concepts and facts.

Definition 3.1 ([5], [17], cf also [30]) An L-topology *on a set X is a subset $\tau \subset L^X$ such that*

(1O) $0_X, 1_X \in \tau$;

(2O) $U, V \in \tau \Longrightarrow U \wedge V \in \tau$;

(3O) $\{U_i \mid i \in I\} \subset L^X \Longrightarrow \bigvee_i U_i \in L^X$.

The pair (X, τ) is called an L-topological space. *An L-topology is called stratified if*

(SO) $U \in \tau,\ \alpha \in L \implies \alpha * U \in \tau$.

Given two L-topological spaces (X, τ_X) and (Y, τ_Y), a mapping $f : X \to Y$ is called continuous if

$$V \in \tau_Y \Longrightarrow f^{-1}(V) \in \tau_X$$ [1]

*Let L-**TOP** stand for the category of L-topological spaces and L-continuous mappings.*

Definition 3.2 ([64], cf also [30, p.153, p.165].) *A mapping $K : L^X \to L^X$ is called an L-interior operator on a set X if it satisfies the following conditions:*

(1K) $K(A) \leq A$, $\forall \in L^X$ *(kernel condition)* ;

(2K) $A \leq B \Longrightarrow K(A) \leq K(B)$, $\forall A, B \in L^X$ *(monotonicity)* ;

(3K) $K(A) \wedge K(B) \leq K(A \wedge B)$, $\forall A, B \in L^X$ *(subadditivity)* ;

[1] This chapter uses the traditional notation for the Zadeh preimage operator rather than the notation given in Chapters 2,4,6,7,8,10 of U. Höhle, S. E. Rodabaugh, eds, Mathematics Of Fuzzy Sets: Logic, Topology, And Measure Theory, The Handbooks of Fuzzy Sets Series, Volume **3**(1999), Kluwer Academic Publishers (Dordrecht/Boston/London).

(4K) $K(K(A)) = K(A)$ *(idempotency)* ;

(5K) $K(1_X) = 1_X$ *(globality condition)* .

An L-interior operator K is called stratified if

(SK) $\alpha * K(A) \leq K(\alpha * A)$ *for all* $\alpha \in L$ *and* $A \in L^X$.

Similar as in the crisp case there is an intimate important relation between L-topologies and L-interior operators.

Proposition 3.3 (cf e.g. [64], [30, p. 172]) *Given an L-topology τ on X let the mapping* $Int : L^X \to L^X$ *be defined by* $IntA = \bigvee\{U \mid U \leq A, U \in \tau\}$. *Then* $Int : L^X \to L^X$ *is an L-interior operator on X. Conversely, given an L-interior operator* $Int : L^X \to L^X$ *on X, by setting* $\tau_{Int} = \{A \in L^X \mid A = IntA\}$ *an L-topology is defined. Besides,* $\tau_{Int_\tau} = \tau$ *and* $Int_{\tau_{Int}} = Int$. *An L-interior operator Int is stratified if and only if the corresponding L-topology* τ_{Int} *is stratified. A mapping* $f : (X, \tau_X) \to (Y, \tau_Y)$ *is continuous if and only if* $Int_X(f^{-1}(V)) \geq f^{-1}(Int_Y(V))$, $\forall V \in L^Y$.

There are known several generalizations of (L-)topological structures—some of them are done in terms of axioms of (L-)topology, others are based on the axioms of (L-)interior or (L-)closure operators: these generalizations are obtained by omitting certain axioms in the definitions or replacing them by weaker axioms. Unfortunately, generalizations obtained by modifying axioms of (L-) topology and generalizations obtained from the axioms of (L-)interior operator lead often to different, non-equivalent concepts. For our merits more appropriate are generalizations obtained from the notion of an L-interior operator. In the next definition we specify terminilogy which is used throuhout the chapter.

Definition 3.4 . *A mapping* $K : L^X \to L^X$ *will be called*

- *an L-kernel operator on X if it satisfies conditions (1K) and (2K) in 3.2;*
- *an L-preinterior operator on X if it satisfies conditions (1K) - (3K) in 3.2;*
- *an idempotent L-preinterior operator on X if it satisfies conditions (1K), (2K) and (4K) in 3.2;*
- *a global L-kernel operator on X if it satisfies conditions (1K), (2K) and (5K) in 3.2.*

An L-kernel operator $K : L^X \to L^X$ *is called stratified, if* $\alpha * K(A) \leq K(\alpha * A)$ *for all* $\alpha \in L$ *and* $A \in L^X$. *A pair* (X, K) *is called an L-kernel (resp. L-preinterior, L-interior) space if* $K : L^X \to L^X$ *is an L-kernel (resp. L-preinterior, L-interior) operator on X. A mapping* $f : (X, K_X) \to (Y, K_Y)$, *where* (X, K_X) *and* (Y, K_Y) *are L-kernel spaces, is called continuous if* $K_X(f^{-1}(V)) \geq f^{-1}(K_Y(V))$, $\forall V \in L^Y$.

Let L-Ker denote the category of L-kernel spaces and their continuous mappings and let L-$PInt$ and L-Int denote, resp., its full subcategories consisting of all L-preinterior spaces and all L-interior spaces. Obviously, L-Int up to an isomorphism is just the category L-**TOP** of L-topological spaces.

Our main interest in this chapter are fuzzy categories based on categories L-Ker and L-$PInt$ as crisp bottom frames.

Remark 3.5 Note that L-kernel operators appear in [13] under the name of *extended fuzzy topologies*. In its turn, in crisp case, i.e. when $L := \{0, 1\}$, extended topologies, i.e. mappings $Int : P(X) \to P(X)$ such that (1) $IntA \subset A$ and (2) $A \subset B \implies IntA \subset IntB$, were considered in [19] and [44]. A global L-kernel operator appear in [13] under the name of a *fuzzy supratopology*. Note also that in case $L = \{0, 1\}$ the concept of an $\{0, 1\}$-preinterior operator is equivalent to the concept of a *closure operator in the sense of E.Čech* [6]. Closure spaces in fuzzy context (i.e. pairs (X, cl) where X is a set and $cl : L^X \to L^X$ is a closure operator) appear also in [42] and [14]; in case when the lattice L is endowed with an order reversing involution L-closure operators are in a natural one-to-one correspondence with L-preinterior operators on the given set.

Remark 3.6 When speaking about an L-kernel operator on a set X we shall use both notations $K : L^X \to L^X$ and $Int : L^X \to L^X$. Usually notation K will be used to denote a general L-kernel operator, while by notation Int it will be usually emphasized that we are dealing with an L-preinterior (in particular an L-interior) operator.

Remark 3.7 Given an L-preinterior operator $Int : L^X \to L^X$ let $\tau_{Int} = \{A \in L^X \mid A = IntA\}$ (cf 3.2). Then it is easy to verify that τ_{Int} satisfies intersection and union axioms of the definition of an L-topology, and hence, if additionally $Int 1_X = 1_X$, then τ_{Int} is an L-topology. Besides, $Int_{\tau_{Int}} \leq Int$, but the equality generally does not hold.

Proposition 3.8 *Let $K : L^X \to L^X$ be an L-kernel operator. Then the mapping $\tilde{K} : L^X \to L^X$ defined by the formula:*

$$\tilde{K}(A) = \bigvee_{\substack{A_1, \ldots, A_n \in L^X \\ A_1 \wedge \ldots \wedge A_n = A}} \Big(K(A_1) \wedge \ldots \wedge K(A_n) \Big)$$

is an L-preinterior operator. Besides it is the smallest one among all L-preinterior operators which dominate K. If a mapping $f : (X, K_X) \to (Y, K_Y)$ is continuous, then $f : (X, \tilde{K}_X) \to (Y, \tilde{K}_Y)$ is continuous, too. In particular, if Int_X is an L-preinterior operator, then continuity of a mapping $f : (X, Int_X) \to (Y, K_Y)$ implies the continuity of the mapping $f : (X, Int_X) \to (Y, \tilde{K}_Y)$.

Proof. The properties

$$\tilde{K}(A) \leq A, \quad A \leq B \Longrightarrow \tilde{K}(A) \leq \tilde{K}(B) \text{ and } \tilde{K}(A \wedge B) \geq \tilde{K}(A) \wedge \tilde{K}(B)$$

for all $A, B \in L^X$ can be easily established directly and hence $\tilde{K}$ is an L-preinterior operator on X. Besides, from its construction it is clear that $\tilde{K}$ is the smallest one ($\leq$) of all L-preinterior operators which are larger than or equal to K. Finally, it easy to verify derectly that continuity of a mapping $f : (X, K_X) \to (Y, K_Y)$ implies continuity of the mapping $f : (X, \tilde{K}_X) \to (Y, \tilde{K}_Y)$. □

Let $\mathcal{K}(L, X)$ stand for the set of all L-kernel operators on a set X and let $\mathcal{I}(L, X)$ denote its subset consisting of all L-preinterior operators on X. Then introducing the partial order $\preceq$ on these sets by setting:

$$K_1 \preceq K_2 \text{ if and only if } K_1(U) \leq K_2(U) \text{ for every } U \in L^X$$

$(\mathcal{K}(L, X), \preceq)$ and $(\mathcal{I}(L, X), \preceq)$ become complete lattices. The details are given in the following two propositions the proofs of which are straightforward and therefore omitted:

Proposition 3.9 *$\mathcal{K}(L, X)$ is a complete lattice. Its bottom and top elements are respectively $Int_\perp$ and $Int_\top$ defined by $Int_\perp(U) = 0_X$ and $Int_\top(U) = U$ for every $U \in L^X$. If $\mathcal{K} = \{K_j \mid j \in J\}$ is a family of L-kernel operators on a set X, then their meet $\bigwedge\{K_j \mid j \in J\} =: \bigwedge \mathcal{K}$ and join $\bigvee\{K_j \mid j \in J\} =: \bigvee \mathcal{K}$ are L-kernel operators, too and hence* $\inf \mathcal{K} := \bigwedge \mathcal{K}$ *and* $\sup \mathcal{K} := \bigvee \mathcal{K}$ *are respectively the infimum and the supremum of $\mathcal{K}$ in $(\mathcal{K}(L, X), \preceq)$.*

Proposition 3.10 *$\mathcal{I}(L, X)$ is a complete lattice. Its bottom and top elements are $Int_\perp$ and $Int_\top$ resp. (cf 3.9). If $\mathcal{I} = \{Int_j \mid j \in J\}$ is a family of L-preinterior operators on a set X, then their meet $\bigwedge\{Int_j \mid j \in J\}$ is an L-preinterior operator, too and hence* $\inf \mathcal{I} := \bigwedge\{Int_j \mid j \in J\}$ *is the infimum of $\mathcal{I}$ in $\mathcal{I}(L, X)$. On the other hand the join of L-preinterior operators $\bigvee\{Int_j \mid j \in J\} =: \bigvee \mathcal{I}$ generally may fail to be subadditive. Therefeore the supremum of $\mathcal{I}$ in $\mathcal{I}(L, X)$ is defined by* $\sup \mathcal{I} = \inf\{Int_l \mid Int_l \geq Int_j\ ,\ \forall Int_j \in \mathcal{I}\}$.

From 3.9 and 3.10 it follows that $\mathcal{I}(L, X)$ is a complete $\wedge$-subsemilattice of $\mathcal{K}(L, X)$. The lattice $\mathcal{I}(L, X)$ fails to be a complete sublattice of $\mathcal{K}(L, X)$ because generally $\sup \mathcal{I} \neq \bigvee \mathcal{I} := \bigvee\{Int_\lambda \mid Int_\lambda \in \mathcal{I}\}$. Note also that the (lattice-theretic) infimum of a family of idempotent L-kernel or L-preinterior operators may fail to be idempotent, while the (lattice-theoretic) supremum of idempotent kernel operators is idempotent; see also Theorem 4.3 below.

Remark 3.11 $Int_\top$ is also the maximal element in the lattice $\mathcal{T}(L, K)$ of all L-interior operators on X. The minimal element in $\mathcal{T}(L, K)$ is the L-interior operator Int_{ind} corresponding to the indiscrete L-topology: $Int_{ind}(U) = 0_X$ if $U \neq 1_X$ and $Int_{ind}(1_X) = 1_X$.

By applying Proposition 3.8 one can easily establish the following efficient description of the L-preinterior operator $\sup \mathcal{I}$, the supremum of a family of L-preinterior operators in $(\mathcal{I}(L,X), \preceq)$:

Proposition 3.12 *Let $\mathcal{I} := \{Int_\lambda \mid \lambda \in \Lambda\}$ be a family of L-preinterior operators on a set X and let $K := \bigvee\{Int_\lambda \mid Int_\lambda \in \mathcal{I}\}$ be the (pointwise) supremum of this family. Then the mapping $\tilde{K} : L^X \to L^X$ defined as in 3.8 is the supremum of the family $\mathcal{I}$ in $(\mathcal{I}(L,X), \preceq)$.*

Proposition 3.13 *Let X and Y be sets and K_λ^X, K_λ^Y where $\lambda \in \Lambda$ be two families of L-kernel operators on X and Y respectively. If a mapping $f : (X, K_\lambda^X) \to (Y, K_\lambda^Y)$ is continuous for every $\lambda \in \Lambda$, then the mappings $f : (X, \inf_\lambda K_\lambda^X) \to (Y, \inf_\lambda K_\lambda^Y)$ and $f : (X, \sup_\lambda K_\lambda^X) \to (Y, \sup_\lambda K_\lambda^Y)$ are continuous, too.*

Proof. From the assumptions it follows that $f^{-1}(K_\lambda^Y(V)) \le K_\lambda^X(f^{-1}(V))$ for every $V \in L^Y$ and every $\lambda \in \Lambda$; hence

$$f^{-1}\Big(\bigwedge_{\lambda\in\Lambda} K_\lambda^Y(V)\Big) = \bigwedge_{\lambda\in\Lambda} f^{-1}(K_\lambda^Y(V)) \le \bigwedge_{\lambda\in\Lambda} K_\lambda^X\big(f^{-1}(V)\big) \text{ for every } V \in L^Y,$$

$$f^{-1}\Big(\bigvee_{\lambda\in\Lambda} K_\lambda^Y(V)\Big) = \bigvee_{\lambda\in\Lambda} f^{-1}(K_\lambda^Y(V) \le \bigvee_{\lambda\in\Lambda} K_\lambda^X\big(f^{-1}(V)\big) \text{ for every } V \in L^Y$$

The first one of these inequalities just means continuity of the mapping $f : (X, \inf_\lambda K_\lambda^X) \to (Y, \inf_\lambda K_\lambda^Y)$ (both if we consider inf in the lattice $(\mathcal{K}(L,X), \preceq)$ or if we consider inf in the lattice $(\mathcal{I}(L,X), \preceq)$). The second one of the inequalities means continuity of the mapping $f : (X, \sup_\lambda K_\lambda^X) \to (Y, \sup_\lambda K_\lambda^Y)$ in case when we consider supremum in the lattice $(\mathcal{K}(L,X), \preceq)$; to complete the proof in case when we deal with L-preinterior operators and the supremum is considered in the lattice $(\mathcal{I}(L,X), \preceq)$ we must additionally apply 3.8 and 3.12.
□

3.14 (Case of MV-algebra: L-hull and L-closure operators) In case when $L = (L, \le, \wedge, \vee, *)$ is an MV-algebra, one can give an alternative description of L-kernel spaces by an operator which we call *an L-hull operator.* Namely, given an L-kernel space (X, K) we define the corresponding L-hull operator $H := H_K : L^X \to L^X$ by setting

$$H(A) := \big(K(A) \longmapsto 0\big) \longmapsto 0\,, \quad \forall A \in L^X$$

One can easily see that axioms (1K), (2K), (3K), (4K), (5K) of an L-kernel operator (the last three whenever they are valid for K) are respectively reflected as axioms

(1H) $H(A) \geq A$, $\forall A \in L^X$;

(2H) $A \leq B \Longrightarrow H(A) \leq H(B)$, $\forall A, B \in L^X$;

(3H) $H(A) \vee H(B) \geq H(A \vee B)$, $\forall A, B \in L^X$;

(4H) $H(H(A)) = H(A)$, $\forall A \in L^X$;

(5H) $H(0_X) = 0_X$

for the corresponding L-kernel operator $H := H_K$. On the other hand, given an L-kernel operator H (i.e. $H : L^X \to L^X$ satisfying (1H) and (2H)) by setting

$$K_H(A) := \big(H(A) \longmapsto 0\big) \longmapsto 0\ , \quad \forall A \in L^X$$

an L-kernel operator $K_H : L^X \to L^X$ is defined and besides $K_{H_K} = K$ and $H_{K_H} = H$. Note that an L-hull operator satisfying properties (3H) and (5H) is obviously just the L-valued version of a Čech closure operator, [6], see also [14].

4 Fuzzy categories of L-kernel and L-preinterior spaces

4.A: Evaluation of "topologiness degree" of L-kernel spaces

As it was mentioned (and is well-known), L-topological spaces and L-interior spaces (i.e. L-kernel spaces (X, K) whose kernel operator K satisfies all five conditions of 3.2), are actually the same things. Thus one can say that an L-preinterior space fails to be an L-topological space to the degree, to which it lacks the last two properties - of idempotency and globality; on its turn, an L-kernel space fails to be an L-preinterior space to the degree to which it lacks subadditivity. The aim of this subsection is to develop an approach, how such degree can be evaluated. Given an L-kernel operator $K : L^X \to L^X$ let

$$\omega_{sa}(K) := \bigwedge_{A_1, \dots, A_n \in L^X} \big(K(A_1) \wedge \dots \wedge K(A_n) \longmapsto K(A_1 \wedge \dots \wedge A_n)\big),$$

$$\omega_{id}(K) := \bigwedge_{A \in L^X} \big(K(A) \longmapsto K(K(A))\big),$$

$$\omega_{gl}(K) := 1_X \longmapsto K(1_X)$$

Obviously, ω_{sa}, ω_{id} and ω_{gl} measure the degree to which K is subadditive, idempotent and global, respectively. An L-kernel operator K is an L-preinterior operator if and only if $\omega_{sa}(K) = \top$; an L-kernel operator is an L-interior operator (and thus, actually is an L-topology), if and only if $\omega(K) = \top$, where

$$\omega(K) := \omega_{id}(K) \wedge \omega_{gl}(K) \wedge \omega_{sa}(K)$$

Remark 4.1 Having defined measures of subadditivity, of idempotency and of globality for a kernel operator K, we can evaluate also the deficiency δ of these properties for K, e.g. by the formulae $\delta_{sa}(K) := \omega_{sa}(K) \longmapsto \bot$; $\delta_{id}(K) := \omega_{id}(K) \longmapsto \bot$ and $\delta_{gl}(K) := \omega_{gl}(K) \longmapsto \bot$, and finally $\delta(K) = \omega(K) \longmapsto \bot$. Obviously, $\delta(K) = \delta_{sa}(K) \vee \delta_{gl}(K) \vee \delta_{id}(K)$. In case L is an MV-algebra, $\omega(K) = \delta(K) \longmapsto \bot$, and hence $\omega(K)$ and $\delta(K)$ mutually determine one another.

Our next aim is to establish how the operations of taking infima and suprema influence the "degree of topologiness" of the corresponding operators. First we prove the following

Lemma 4.2 *Let $K : L^X \to L^X$ be an L-kernel operator and $\tilde{K} : L^X \to L^X$ be defined as in 3.8. Then $\omega_{id}(\tilde{K}) \geq \omega_{id}(K)$, $\omega_{gl}(\tilde{K}) \geq \omega_{gl}(K)$ and $\omega_{sa}(\tilde{K}) = \top$.*

Proof. To prove the first inequality take $A \in L^X$ and let $A_1, \ldots A_n \in L^X$ be such that $A = A_1 \wedge \ldots \wedge A_n$. Then obviously $K(A_1) \wedge \ldots \wedge K(A_n) \longmapsto K(K(A_1)) \wedge \ldots \wedge K(K(A_n)) \geq \bigwedge_{i=1}^{n} (K(A_i) \longmapsto K(K(A_i))) \geq \xi := \omega_{id}(K)$. Since $\tilde{K}$ is subadditive and $\tilde{K} \geq K$ it follows from here that $K(A_1) \wedge \ldots \wedge K(A_n) \longmapsto \tilde{K}(\tilde{K}(A)) \geq K(A_1) \wedge \ldots \wedge K(A_n) \longmapsto \tilde{K}\big(K(A_1) \wedge \ldots \wedge K(A_n)\big) = K(A_1) \wedge \ldots \wedge K(A_n) \longmapsto \tilde{K}(K(A_1)) \wedge \ldots \wedge \tilde{K}(K(A_n)) \geq \xi$. Now taking infimum over all decompositions $A = A_1 \wedge \ldots \wedge A_n$, we get $\tilde{K}(A) \longmapsto \tilde{K}\big(\tilde{K}(A)\big) \geq \xi$, and hence $\omega_{id}(\tilde{K}) \geq \omega_{id}(K)$.

The proof of the second inequality is obvious. Finally, $\omega_{sa}(\tilde{K}) = \top$ since, according to 3.8, $\tilde{K}$ is an L-preinterior operator. □

Theorem 4.3 *Let $\mathcal{K} = \{K_j \mid j \in J\}$ be a family of L-kernel operators (resp. L-preinterior operators) on a set X and let $\inf \mathcal{K}$ and $\sup \mathcal{K}$ be its infimum and supremum in the lattice $\mathcal{K}(L,X)$ (in the lattice $\mathcal{I}(L,X)$ resp.) (see 3.9, 3.10). Then*

$$\omega_{gl}(\inf \mathcal{K}) = \bigwedge_{j \in J} \omega_{gl}(K_j), \;\; \omega_{sa}(\inf \mathcal{K}) \geq \bigwedge_{j \in J} \omega_{sa}(K_j),$$

$$\omega_{gl}(\sup \mathcal{K}) \geq \bigvee_{j \in J} \omega_{gl}(K_j), \; \omega_{id}(\sup \mathcal{K}) \geq \bigwedge_{j \in J} \omega_{id}(K_j), \; \omega_{sa}(\sup \mathcal{K}) \geq \bigwedge_{j \in J} \omega_{sa}(K_j)$$

Proof. First we consider the case of L-kernel operators and, respectively, the lattice $\mathcal{K}(L,X)$. The equality $\omega_{gl}(\inf \mathcal{K}) = \bigwedge_{j \in J} \omega_{gl}(K_j)$ is obvious since $1_X \longmapsto \bigwedge_{j \in J} K_j(1_X) = \bigwedge_{j \in J} (1_X \longmapsto K_j(1_X))$. To show that $\omega_{sa}(\inf \mathcal{K}) \geq \bigwedge_{j \in J} \omega_{sa}(K_j)$ fix $A_1, \ldots, A_n \in L^X$, and $j_0 \in J$ and notice that

$$(\inf \mathcal{K})(A_1) \wedge \ldots \wedge (\inf \mathcal{K})(A_n) \longmapsto K_{j_0}(A_1 \wedge \ldots \wedge A_n) \geq$$

$$K_{j_0}(A_1) \wedge \ldots \wedge K_{j_0}(A_n) \longmapsto K_{j_0}(A_1 \wedge \ldots \wedge A_n) \geq \omega_{sa}(K_{j_0})$$

Now, taking infimum over all $j_0 \in J$ and over all finite families $A_1, \dots A_n \in L^X$, we get the demanded inequality.

The first one of the inequalities characterizing $\sup \mathcal{K}$, namely, $\omega_{gl}(\sup \mathcal{K}) \geq \bigvee_{j\in J} \omega_{gl}(K_j)$, is obvious. To show that $\omega_{id}(\sup \mathcal{K}) \geq \bigwedge_{j\in J} \omega_{id}(K_j)$, notice first that since $K_{j_0}(A) \longmapsto K_{j_0}(K_{j_0}(A)) \geq \omega_{id}(K_{j_0})$ for some $j_0 \in J$ and $A \in L^X$, then moreover $K_{j_0}(A) \longmapsto \bigvee_{j\in J} K_j(\bigvee_{j\in J} K_j(A)) \geq \omega_{id}(K_{j_0})$, and hence, taking inf over all $j_0 \in J$ we get $\bigvee_{j\in J} K_j(A) \longmapsto \bigvee_{j\in J} K_j(\bigvee_{j\in J} K_j(A)) \geq \bigwedge_{j\in J} \omega_{id}(K_j)$. In view of Proposition 3.9, we can conclude from here that $\omega_{id}(\sup \mathcal{K}) \geq \bigwedge_{j\in J} \omega_{id}(K_j)$

To show that $\omega_{sa}(\sup \mathcal{K}) \geq \bigwedge_{j\in J} \omega_{sa}(K_j)$ fix some $A_1, \dots, A_n \in L^X$, $j_0 \in J$ and notice that

$$K_{j_0}(A_1) \wedge \dots \wedge K_{j_0}(A_n) \longmapsto \bigvee_{j\in J} K_j(A_1 \wedge \dots \wedge A_n) \geq$$

$$K_{j_0}(A_1) \wedge \dots \wedge K_{j_0}(A_n) \longmapsto K_{j_0}(A_1 \wedge \dots \wedge A_n) \geq \omega_{sa}(K_{j_0})$$

Now, taking inf over all $j_0 \in J$ we get

$$(\bigvee_{j\in J} K_j(A_1)) \wedge \dots \wedge (\bigvee_{j\in J} K_j(A_n)) \longmapsto \bigvee_{j\in J} K_j(A_1 \wedge \dots \wedge A_n) \geq \bigwedge_{j\in J} \omega_{sa}(K_j)$$

and hence, taking into account Proposition 3.9, the demanded inequality follows.

Now we pass to the case of L-preinterior operators and resp. consider the lattice $\mathcal{I}(L,X)$. Since $\mathcal{I}(L,X)$ is a complete $\wedge$-subsemilattice of $\mathcal{K}(L,X)$, the inequalities estimating $\inf \mathcal{K}$ follow immediately from the corresponding inequalities estimating $\inf \mathcal{K}$ in $\mathcal{K}(L,X)$. On the other hand the inequalities estimating $\sup \mathcal{K}$ in $\mathcal{I}(L,X)$ follow from the corresponding inequalities estimating $\sup \mathcal{K}$ in $\mathcal{K}(L,X)$ by applying Lemma 4.2 and 3.12. □

The evaluation ω of an L-kernel (resp. L-interior) operator suits as one of the tools for introducing fuzzy categories on the base of the categories of L-kernel and L-preinterior spaces as crisp bottom frames. Here are the details:

We view ω as a mapping defined on the class $\mathcal{O}b(L\text{-}Ker)$ of all objects of L-Ker and taking values in L, i.e. ω is an L-fuzzy subclass of $\mathcal{O}b(L\text{-}Ker)$. Further, given a mapping $f : (X, K_X) \to (Y, K_Y)$ we define $\nu(f) := \omega(K_X) \wedge \omega(K_Y)$ and thus come to an L-fuzzy subclass ν of the class $\mathcal{M}or(L\text{-}Ker)$. In the result we obtain a fuzzy category

$$FL\text{-}Ker := (\mathcal{O}b(L\text{-}Ker), \omega, \mathcal{M}or(L\text{-}Ker), \nu)$$

of L-kernel spaces and in an obvious way defined its full fuzzy subcategory

$$FL\text{-}PInt := (\mathcal{O}b(L\text{-}PInt), \omega, \mathcal{M}or(L\text{-}PInt), \nu)$$

of L-preinterior spaces.

4.B: Characterization of "topologiness degree" of L-kernel spaces by means of their neighbourhood structure

Given an L-kernel operator $K : L^X \to L^X$ we define a mapping $N := N_K : X \times L^X \to L^X$ by

$$N_K(x, A) = K(A)(x) \ , \quad \forall x \in X, \ \forall A \in L^X$$

Then

(N1) $N(x, A) \leq A(x)$;

(N2) $A \leq B \implies N(x, A) \leq N(x, B)$.

We refer to a mapping N_K as the L-neighbourhood structure generated by the L-kernel operator K or just as the L-neighbourhood structure of the L-kernel space (X, K). Conversely, given a mapping $N : X \times L^X \to L^X$ satisfying conditions (N1) and (N2), by setting $K_N(A)(x) = N(x, A)$ we obtain an L-kernel operator $K_N : L^X \to L^X$. Moreover, $N_{K_N} = N$ and K_{N_K} for every L-neighbourhood structure N and for every L-kernel operator K (cf. e.g. [30]).

L-neighbourhood structures allow us to obtain an alternative description of "topologiness degree" of a space (X, K) as follows. Given an L-neighbourhood structure $N : X \times L^X \to L^X$ (i.e. N satisfies conditions (N1) and (N2)), let

$$\omega_{sa}(N) := \bigwedge_{A_1,\ldots,A_n \in L^X} \Big(\bigwedge_{x \in X} (N(x, A_1) \wedge \ldots \wedge N(x, A_n) \longmapsto N(x, A_1 \wedge \ldots \wedge A_n)) \Big);$$

$$\omega_{gl}(N) := 1_X \longmapsto N(x, 1_X);$$

$$\omega_{id}(N) := \bigwedge_{A \in L^X} (\bigwedge_{x \in X} (N(x, A) \longmapsto \bigvee \{ N(y, B) \mid B(y) \leq N(y, A) \ , \quad \forall y \in X \})$$

Then $\omega_{sa}(N_K) = \omega_{sa}(K)$, $\omega_{gl}(N_K) = \omega_{gl}(K)$ and $\omega_{id}(N_K) = \omega_{id}(K)$.

The validity of the first two equalities can be established by an easy verification. To show the last equality take some $A \in L^X$ and $x \in X$ and notice first that

$$N_K(x, K(A)) = \bigvee \{ N_K(x, B) \mid B(y) \leq N_K(y, A) \ , \quad \forall y \in X \}$$

Indeed, to see the inequality $N_K(x, K(A)) \leq \bigvee \{ N_K(x, B) \mid B(y) \leq N_K(y, A)$, $\forall y \in X \}$, it is sufficient to take $B = K(A)$. To get the converse inequality note that $N_K(y, B) = K(B)(y) \leq K(K(A))(y) = N_K(y, K(A))$ for every $y \in X$ and hence, in particular, for $y = x$. From here it follows that

$$K(A)(x) \longmapsto K(K(A))(x) = N_K(x, A) \longmapsto N_K(x, K(A)) =$$

$$N_K(x, A) \longmapsto \bigvee \{ N_K(x, B) \mid B(y) \leq N(y, A) \ , \quad \forall y \in X \}$$

Now, taking infimum on the both sides over all $A \in L^X$ and all $x \in X$ we obtain the required inequality.

4.C: Evaluation of continuity degree of potential morphisms

In the previous subsections the fuzzification of categories L-Ker and L-$PInt$ was caused by estimation of the degree to which an L-kernel space differs from the "nearest" L-topological space. The attraction of the fuzzy subset ν of the class of morphisms was forced in order to satisfy the axioms of the L-fuzzy category. As different from the previous subsections, here for us the primacy is the "original" fuzziness of a mapping: we are interested to measure "to what degree" a mapping between L-kernel (in particular, between L-preinterior or L-topological) spaces is continuous. Here are the details of one possible approach to this problem:

Let L-Ker_{SET}, L-$PInt_{SET}$ and L-$\mathbf{TOP}_{SET}$ be the categories having the same objects as L-Ker, L-$PInt$ and L-**TOP** respectively, and all mappings between the corresponding sets $f : X \to Y$ as potential morphisms from (X, K_X) to (Y, K_Y) respectively. Given a mapping $f : X \to Y$ we set

$$\mu(f) := \bigwedge_{A \in L^Y} (f^{-1}(K_Y A) \longmapsto K_X(f^{-1}(A)))$$

We call a mapping f α-continuous if $\mu(f) \geq \alpha, \ \ \alpha \in L$.

A direct verification shows that $\mu(g \circ f) \geq \mu(g) * \mu(f)$ whenever composition $g \circ f$ is defined and that $\mu(id_X) = \top$ for the identity mapping $id_X : (X, K) \to (X, K)$. Thus we come to L-fuzzy categories:

$$\left(\mathcal{O}b\,(L\text{-}Ker)\,, 1_{\mathcal{O}b(L\text{-}Ker)}, \mathcal{M}or\,(L\text{-}Ker_{SET})\,, \mu\right),$$
$$\left(\mathcal{O}b\,(L\text{-}PInt)\,, 1_{\mathcal{O}b(L\text{-}PInt)}, \mathcal{M}or\,(L\text{-}PInt_{SET})\,, \mu\right),$$
$$\left(\mathcal{O}b\,(L\text{-}\mathbf{TOP})\,, 1_{\mathcal{O}b(L\text{-}\mathbf{TOP})}, \mathcal{M}or\,(L\text{-}\mathbf{TOP}_{SET})\,, \mu\right)$$

Proposition 4.4 *Let X, Y be sets, K_X and K_Y be L-kernel operators on X and Y respectively and let $f : X \to Y$ be a mapping. Further, let $\tilde{K}_X$ and $\tilde{K}_Y$ have the same meaning as in Proposition 3.8. Then*

$$f^{-1}(\tilde{K}_Y(A)) \longmapsto \tilde{K}_X\big(f^{-1}(A)\big) \geq f^{-1}(K_Y(A)) \longmapsto K_X\big(f^{-1}(A)\big)\,, \quad \forall A \in L^Y$$

In particular, if K_1 and K_2 are two L-kernel operators on a set X, then

$$\tilde{K}_1(A) \longmapsto \tilde{K}_2(A) \geq K_1(A) \longmapsto K_2(A)\,, \quad \forall A \in L^X$$

Proof. Straightforward and therefore omitted. □

Proposition 4.5 *Let X and Y be sets, $\{K_\lambda^X \mid \lambda \in \Lambda\} \subset \mathcal{K}(L,X)$ (resp. $\subset \mathcal{I}(L,X)$) and $\{K_\lambda^Y \mid \lambda \in \Lambda\} \subset \mathcal{K}(L,Y)$ (resp. $\subset \mathcal{I}(L,X)$) be two families of L-kernel operators (resp. L-preinterior operators) on X and Y respectively. If a mapping $f : (X, K_\lambda^X) \to (Y, K_\lambda^Y)$ is α-continuous for every $\lambda \in \Lambda$, then the mappings $f : (X, \inf_{\lambda\in\Lambda} K_\lambda^X) \to (Y, \inf_{\lambda\in\Lambda} K_\lambda^Y)$ and $f : (X, \sup_{\lambda\in\Lambda} K_\lambda^X) \to (Y, \sup_{\lambda\in\Lambda} K_\lambda^Y)$ are also α-continuous.*

Proof. First we consider the case of L-kernel operators and hence the infimum and supremum are taken in the lattices $\mathcal{K}(L,X)$ and $\mathcal{K}(L,Y)$.

From the assumptions it follows that for every $A \in L^Y$ and every $\lambda \in \Lambda$ it holds $f^{-1}(K_Y^\lambda A) \longmapsto K_X^\lambda(f^{-1}(A)) \geq \alpha$. Hence for any fixed λ_0 we have $f^{-1}(\bigwedge_{\lambda\in\Lambda} K_Y^\lambda(A)) \longmapsto K_X^{\lambda_0}(f^{-1}(A)) \geq \alpha$. Now, taking infimum over all $\lambda_0 \in \Lambda$ we get

$$f^{-1}\Big(\bigwedge_{\lambda\in\Lambda} K_Y^\lambda(A)\Big) \longmapsto \bigwedge_{\lambda_0\in\Lambda} K_X^{\lambda_0}(f^{-1}(A)) \geq \alpha$$

and hence $f : (X, \inf_\lambda K_\lambda^X) \to (Y, \inf_\lambda K_\lambda^Y)$ is α-continuous.

To prove α-continuity of the mapping in case of supremum fix again $A \in L^Y$ and $\lambda_0 \in \Lambda$; then $f^{-1}(K_Y^{\lambda_0}(A)) \longmapsto \bigvee_{\lambda\in\Lambda} K_X^\lambda(f^{-1}(A)) \geq \alpha$; and hence, by taking infimum over all $\lambda_0 \in \Lambda$ on the left side of the inequality, we get $f^{-1}((\bigvee_{\lambda\in\Lambda} K_Y^\lambda(A)) \longmapsto \bigvee_{\lambda\in\Lambda} K_X^\lambda(f^{-1}(A)) \geq \alpha$, therefore by Proposition 3.9, we conclude that $f : (X, \sup K_\lambda^X) \to (Y, \sup K_\lambda^Y)$ is α-continuous.

Now we pass to the case of L-preinterior operators, and resp. work in the lattices $\mathcal{I}(L,X)$ and $\mathcal{I}(L,Y)$. The α-continuity of the mapping $f : (X, \inf_{\lambda\in\Lambda} K_\lambda^X) \to (Y, \inf_{\lambda\in\Lambda} K_\lambda^Y)$ was established above (since the lattice of L-preinterior operators is an $\wedge$-subsemilattice of the lattice of L-kernel operators). The α-continuity of the mapping $f : (X, \sup_{\lambda\in\Lambda} K_\lambda^X) \to (Y, \sup_{\lambda\in\Lambda} K_\lambda^Y)$ follows from the above established α-continuity of the mapping in case of L-kernel operators, Proposition 3.12 and Remark 4.4. □

4.D: Case of an MV-algebra

It is easy to see that in case L is an MV-algebra, the evaluation of "topologiness degree" ω and "measure of continuity" μ can be characterised by means of the corresponding L-hull operator. Namely, let (X, K) be an L-kernel space and $H = H_K$ the corresponding L-hull operator (see 3.14). Then

$$\omega_{sa}(K) = \bigwedge_{A_1,\ldots,A_n\in L^X} \Big(H(A_1 \vee \ldots \vee A_n) \longmapsto (H(A_1) \vee \ldots \vee H(A_n))\Big),$$

$$\omega_{id}(K) = \bigwedge_{A\in L^X} (H(H(A)) \longmapsto H(A)),$$

$$\omega_{gl}(K) = H(0_X) \longmapsto 0_X$$

Further, given a mapping $f : (X, K_X) \to (Y, K_Y)$, its measure of continuity is:

$$\mu(f) = \bigwedge_{A \in L^Y} \left(H_X(f^{-1}(A)) \longmapsto f^{-1}(H_Y(A)) \right)$$

4.E: Fuzzy categories of L-kernel and L-preinterior spaces and related functors

In the previous subsections we have already introduced L-fuzzy categories of L-kernel spaces and L-preinterior spaces:

$$\begin{aligned}
FL\text{-}Ker &:= \big(\mathcal{O}b(L\text{-}Ker), \omega, \mathcal{M}or(L\text{-}Ker), \nu\big);\\
FL\text{-}PInt &:= \big(\mathcal{O}b(L\text{-}PInt), \omega, \mathcal{M}or(L\text{-}PInt), \nu\big);\\
\mathcal{F}L\text{-}Ker &:= \big(\mathcal{O}b(L\text{-}Ker), 1_{\mathcal{O}b(L\text{-}Ker)}, \mathcal{M}or(L\text{-}Ker_{SET}), \mu\big);\\
\mathcal{F}L\text{-}PInt &:= \big(\mathcal{O}b(L\text{-}PInt), 1_{\mathcal{O}b(L\text{-}PInt)}, \mathcal{M}or(L\text{-}PInt_{SET}), \mu\big)
\end{aligned}$$

As it was mentioned above, the first two of these categories are mainly dealing with estimation of fuzziness of objects, while in the second pair the fuzziness of morphisms is of paramount concern. In a natural way we can unite both of these aspects in the following fuzzy categories where $\overline{\mu} := \mu \wedge \nu$:

$$\begin{aligned}
\mathcal{F}FL\text{-}Ker &:= \big(\mathcal{O}b(L\text{-}Ker), \omega, \mathcal{M}or(L\text{-}Ker_{SET}), \overline{\mu}\big)\\
\mathcal{F}FL\text{-}PInt &:= \big(\mathcal{O}b(L\text{-}PInt), \omega, \mathcal{M}or(L\text{-}PInt_{SET}), \overline{\mu}\big)
\end{aligned}$$

By restricting the class of potential objects to L-topological spaces we obtain the full fuzzy subcategory $\mathcal{F}L$-**TOP** of $\mathcal{F}L$-$PInt$, while the full fuzzy subcategory FL-**TOP** of FL-$PInt$ is redundant, since it is just the category L-**TOP**.

Another way to obtain interesting (in our opinion) subcategories of these fuzzy categories is to put restrictions on its potential objects. For example, given $\xi \in L$ let $\mathcal{F}L\text{-}Ker_\xi$ be the full subcategory of $\mathcal{F}L$-Ker whose objects (X, K) satisfy condition $\omega(K) \geq \xi$, let $\mathcal{F}L\text{-}Ker_\xi^{sa}$ be the full subcategory of $\mathcal{F}L$-Ker whose objects (X, K) satisfy condition $\omega_{sa}(K) \geq \xi$, etc. In particular, $FL\text{-}Ker_\top^{sa} = FL\text{-}PInt$ and $FL\text{-}Ker_\top$ is just the category of L-**TOP**.

The next theorem establishes one typical relation between these categories:

Theorem 4.6 *Let $\mathcal{E}_\xi : \mathcal{F}L\text{-}PInt \longrightarrow \mathcal{F}L\text{-}Ker_\xi^{sa}$ be the embedding functor and let the functor $\Upsilon_\xi : \mathcal{F}L\text{-}Ker_\xi^{sa} \longrightarrow \mathcal{F}L\text{-}PInt$ assign to an L-kernel space (X, K) the L-preinterior space $(X, \tilde{K})$ and leaving potential morphisms unchanged. Then $\mathcal{E}_\xi$, and Υ_ξ are $\top$-functors. Besides the functor $\mathcal{E}_\xi$ is a $(\xi, \top)$-reflection (actually a $(\xi, \top)$-epi-mono-reflection) and hence $\mathcal{F}L\text{-}PInt$ is $(\xi, \top)$-reflective in $\mathcal{F}L\text{-}Ker_\xi^{sa}$.[2] In particular, $\mathcal{F}L\text{-}PInt$ is $(\bot, \top)$ reflective in $\mathcal{F}L$-Ker.*

[2] (ξ, γ)-reflections in fuzzy categories, where $\xi, \gamma \in L$, are defined in the same way as

Proof. Since $\mathcal{E}_\xi : \mathcal{F}L\text{-}PInt \longrightarrow \mathcal{F}L\text{-}Ker_\xi$ is an embedding functor, it is clear that it is a $\top$-functor. Further, if $(X,K) \in \mathcal{O}b(L-Ker)$, then the space $(X,\tilde{K})$ belongs to $\mathcal{O}b(L-PInt)$ and $\omega(X,\tilde{K}) \geq \omega(X,K)$, cf. 4.2. If $f:(X,K_X) \to (Y,K_Y)$ is a potential morphism in $\mathcal{F}L\text{-}Ker$, then if considered as a mapping $f:(X,\tilde{K}_X) \to (Y,\tilde{K}_Y)$ it is a potential morphism in $\mathcal{F}L\text{-}PInt$ and from Remark 4.4 it easily follows that $\mu_{\mathcal{F}L\text{-}Ker}(f) \leq \mu_{\mathcal{F}L\text{-}PInt}(f)$. However, this means that Υ_ξ is a $\top$-functor.

Further, if $\eta_{(X,K)} : (X,K) \to (X,\tilde{K})$ is the identity mapping, then $\mu(\eta_{(X,K)}) = \omega_{sa}(K) \geq \xi$. This easily follows from the inequality $K(A) \longmapsto \tilde{K}(A) \geq K(A) \longmapsto K(A_1 \wedge \ldots \wedge A_n) \geq \omega_{sa}(K) \geq \xi$ which holds for every $A \in L^X$ and each of its decompositions $A = A_1 \wedge \ldots \wedge A_n$.

Let now $f:(X,K) \to (Z,Int)$ be a potential morphism in $\mathcal{F}L\text{-}Ker$, and let $\tilde{f}:(X,\tilde{K}) \to (Z,Int)$ be the same mapping considered as a potential morphism in $\mathcal{F}L_PInt$; then according to Proposition 4.4 $\mu(f) = \mu(\tilde{f})$. Thus $\eta_{(X,K)}$ is the universal ξ-arrow from (X,K) to the functor $\mathcal{E}_\xi$, i.e. $\mathcal{E}_\xi$ is a $(\xi,\top)$-reflection. Moreover, since $\eta_{(X,K)}$ is a bijection, it is an epi-mono-$(\xi,\top)$-reflection, see [62]. □

Similar results also hold for fuzzy categories $\mathcal{F}\mathcal{F}L\text{-}Ker_\xi$ and $\mathcal{F}\mathcal{F}L\text{-}PInt$.

5 α-weakening and α-strengthening of L-kernel and L-preinterior operators.

5.A: α-weakening of L-kernel and L-preinterior operators

Given an L-kernel space (X,K) let $\mathcal{K}_K = \{K_\lambda \mid \lambda \in \Lambda\}$ be the family of all L-kernel operators on X such that the identity mapping $id:(X,K_\lambda) \to (X,K)$ is α-continuous. Obviously, $\mathcal{K}_K \not\models \emptyset$. Let $K^{\breve{\alpha}} := \bigwedge_{\lambda\in\Lambda} K_\lambda$. Then $K^{\breve{\alpha}} \in \mathcal{K}_K$. Indeed, $K^{\breve{\alpha}}$ is an L-kernel operator and for every $V \in L^X$

$$K(V) \longmapsto K^{\breve{\alpha}}(V) = K(V) \longmapsto \wedge_\lambda K_\lambda(V) = \wedge_\lambda (K(V) \longmapsto K_\lambda(V)) \geq \alpha$$

and hence $id:(X,K^{\breve{\alpha}}) \longmapsto (X,K)$ is α-continuous. Therefore $K^{\breve{\alpha}}$ is the minimal element in $\mathcal{K}_K$. Since obviously $K^{\breve{\alpha}} \leq K$, the inverse mapping $id:(X,K) \to (X,K^{\breve{\alpha}})$ is continuous and hence the objects (X,K) and $(X,K^{\breve{\alpha}})$ are α-isomorphic [61].

In case Int is an L-preinterior operator $Int^{\breve{\alpha}}$ is defined as the meet of all L-preinterior operators for which the identity mapping is α-continuous. According

reflections in the usual category theory (more precisely, as reflections of the corresponding bottom frame categories) with a special demand that the reflection functor must be a γ-functor (in our case it is a $\top$-functor) and universal arrows must be morphisms at least to degree ξ—see [62] for the details.

to Proposition 3.10, $Int^{\check{\alpha}}$ is the minimal one among all L-preinterior operators having this property. In the sequel the constructions

$$K \Longrightarrow K^{\check{\alpha}} \quad \text{and} \quad Int \Longrightarrow Int^{\check{\alpha}}$$

will be referred to as the *α-weakening* of the L-kernel operator K and the L-preinterior operator Int respectively.

It is easy to give also an effective description of the operators $K^{\check{\alpha}}$ and $Int^{\check{\alpha}}$. Indeed, a direct verification shows that they can be defined by the equalities:

5.1

$$K^{\check{\alpha}}(V) = \alpha * K(V) \quad \textit{and} \quad Int^{\check{\alpha}}(V) = \alpha * Int(V)\ , \quad \forall V \in L^X$$

Remark 5.2 From here, in particular, it is clear that even if the original operator Int was idempotent or global (i.e. $Int 1_X = 1_X$) its α-weakening $Int^{\check{\alpha}}$ may loose these properties and hence the construction $Int \Longrightarrow Int^{\check{\alpha}}$ when applied to an L-topological space leads out of the category L-**TOP**.

Next we shall study some properties of the operation of α-weakening. Since the results in case of L-kernel operators and in case of L-preinterior operators are analogous, we restrict our consideration to the case of L-preinterior operators.

Theorem 5.3 1^0. *A mapping $f : (X, Int_X) \to (Y, Int_Y)$ is α-continuous iff the mapping $f : (X, Int_X) \to (Y, Int_Y^{\check{\alpha}})$ is continuous. In particular, if $f : (X, Int_X^{\check{\alpha}}) \to (Y, Int_Y^{\check{\alpha}})$ is continuous, then $f : (X, Int_X) \to (Y, Int_Y)$ is α-continuous.*

2^0. *If a mapping $f : (X, Int_X) \to (Y, Int_Y)$ is β-continuous, then the mapping $f : (X, Int_X^{\check{\alpha}}) \to (Y, Int_Y)$ is $\alpha * \beta$-continuous. In particular, if $f : (X, Int_X) \to (Y, Int_Y)$ is continuous, then $f : (X, Int_X^{\check{\alpha}}) \to (Y, Int_Y)$ is α-continuous.*

Proof. 1^0. α-continuity of a mapping $f : (X, Int_X) \to (Y, Int_Y)$ means that $f^{-1}(Int_Y(V)) * \alpha \leq Int_X\big(f^{-1}(V)\big)$ for all $V \in L^Y$,or, equivalently, that $f^{-1}(Int^{\check{\alpha}}(V)) \leq Int_X\big(f^{-1}(V)\big)$ for all $V \in L^Y$,i.e. $f : (X, Int_X) \to (Y, Int_Y^{\check{\alpha}})$is continuous.

2^0. β-continuity of $f : (X, Int_X) \to (Y, Int_Y)$means that $f^{-1}(Int_Y V) * \beta \leq Int_X(f^{-1}(V))$ and hence

$$f^{-1}(Int_Y V) * \beta * \alpha \leq Int_X(f^{-1}(V)) * \alpha = Int_X^{\check{\alpha}}(f^{-1}(V))$$

i.e. $f^{-1}(Int_Y V) \longmapsto Int_X^{\check{\alpha}}(f^{-1}(V)) \geq \alpha * \beta$. □

Proposition 5.4 *Let $\mathcal{I} = \{Int_\lambda \mid \lambda \in \Lambda\}$ be a family of L-preinterior operators on X and let $\inf \mathcal{I} = \inf\{Int_\lambda \mid \lambda \in \Lambda\}$ and $\sup \mathcal{I} = \sup\{Int_\lambda \mid \lambda \in \Lambda\}$ be respectively the infimum and supremum of this family in the lattice $\mathcal{I}(X, L)$ of all L-preinterior operators on X. Then $\left(\inf \mathcal{I}\right)^{\breve{\alpha}} = \inf\{Int_\lambda^{\breve{\alpha}} \mid \lambda \in \Lambda\}$ and $\left(\sup \mathcal{I}\right)^{\breve{\alpha}} = \sup\{Int_\lambda^{\breve{\alpha}} \mid \lambda \in \Lambda\}$*

Proof. The equality $\inf\{Int_\lambda^{\breve{\alpha}} \mid \lambda \in \Lambda\} = (\inf \mathcal{I})^{\breve{\alpha}}$ follows from 5.1, 3.10, and the distributivity of $*$over arbitrary meets. To show the second equality let $K = \bigvee\{Int_\lambda \mid \lambda \in \Lambda\}$.Then, obviously, $K * \alpha = \bigvee\{Int_\lambda * \alpha \mid \lambda \in \Lambda\} = \bigvee\{Int^{\breve{\alpha}}{}_\lambda \mid \lambda \in \Lambda\}$and $\tilde{K} = \sup \mathcal{I}$ (cf 3.8). Applying Proposition 3.12 and distributivity of $*$-operation over arbitrary joins and meets, it is easy to conclude from here that $\left(\sup \mathcal{I}\right)^{\breve{\alpha}} = \tilde{K} * \alpha = \widetilde{K * \alpha} = \sup\{Int_\lambda^{\breve{\alpha}} \mid \lambda \in \Lambda\}$. □

Next we study the problem how the operation of α-weakening of an L-preinterior operator can influence the degree to which it is "topological-like" (see Subsection 4.A). The answer is given in the following:

Proposition 5.5 *Let $Int : L^X \to L^X$ be an L-preinterior operator on X and let $Int^{\breve{\alpha}} : L^X \to L^X$ be its α-weakening. Then $\omega_{gl}(Int^{\breve{\alpha}}) \geq \omega_{gl}(Int) * \alpha$. If moreover, Int is stratified, then $\omega_{id}(Int^{\breve{\alpha}}) \geq \omega_{id}(Int) * \alpha$. Thus operation of α-weakening can diminish the degree of idempotency (in case of a stratified operator) and globality of an L-preinterior operator at most α-times.*

Proof. Since $1_X \longmapsto Int^{\breve{\alpha}}(V) = 1_X \longmapsto Int(V) * \alpha \geq \left(1_X \longmapsto Int(V)\right) * \alpha$ for any $V \in L^X$, it is clear that $\omega_{gl}(Int^{\breve{\alpha}}) \geq \omega_{gl}(Int) * \alpha$.To prove the second relation, let $V \in L^X$and assume that $Int(V) \longmapsto Int(Int(V)) \geq \xi$.Then for $Int^{\breve{\alpha}}$, under assumption of stratifiability, we get

$$Int^{\breve{\alpha}}(V) \longmapsto Int^{\breve{\alpha}}(Int^{\breve{\alpha}}(V)) = \alpha * Int(V) \longmapsto \alpha * Int\left[\alpha * Int(V)\right] \geq$$

$$\alpha * Int(V) \longmapsto \alpha^2 * Int\left(Int(V)\right) \geq \left(\alpha * (Int(V) \longmapsto \alpha * (Int(IntV))\right) * \alpha \geq \xi * \alpha$$

□

Remark 5.6 It is easy to show that for an L-kernel operator K it holds: $\omega_{sa}(K^{\breve{\alpha}}) \geq \omega_{sa}(K)$; i.e., operation of α-weakening does not diminish the degree of subadditivity of an L kernel operator.

5.B α-strengthening of an L-preinterior operator

Given an L-preinterior space (X, Int) we are interested to describe the strongest (i.e. largest) L-preinterior operator $Int^{\hat{\alpha}}$ on X such that the identity mapping $id : (X, Int) \to (X, Int^{\hat{\alpha}})$ is α-continuous. This L-preinterior operator $Int^{\hat{\alpha}}$ will be referred to as α-strengthening of Int. In a similar way, given an L-kernel space (X, K) the strongest L-kernel operator $K^{\hat{\alpha}}$ on X for which $id : (X, K) \to (X, K^{\hat{\alpha}})$ is α-continuous will be referred to as α-strengthening of K.

As in the previous subsection, when studying the operation of α-strengthening, we shall consider the case of L-preinterior operators.

Theorem 5.7 *Let Int be an L-preinterior operator on a set X and let $\mathcal{I} = \{Int_\lambda \mid \lambda \in \Lambda\}$ be the family of L-preinterior operators on X for which the identity mapping $id : (X, Int) \to (X, Int_\lambda)$ is α-continuous. Then* $\sup \mathcal{I} = Int^{\hat{\alpha}}$ [3].

Proof. Let $K = \bigvee\{Int_\lambda \mid \lambda \in \Lambda\}$ and let $\tilde{K} : L^X \to L^X$ be defined as in Proposition 3.8 (i.e. $\tilde{K} = \sup \mathcal{I}$ is the supremum of the family $\mathcal{I} = \{Int_\lambda \mid \lambda \in \Lambda\}$ in $\mathcal{I}(L, X)$, see Proposition 3.10). Then $\tilde{K}$ is the required L-interior operator. To show this notice first that, by definition, $Int_\lambda(A) \longmapsto Int(A) \geq \alpha$ for all $\lambda \in \Lambda$ and for all $A \in L^X$, and hence also $K(A) \longmapsto Int(A) \geq \alpha$ for all $A \in L^X$. Now, applying Proposition 4.4, we get $\tilde{K}(A) \longmapsto Int(A) \geq \alpha$. Since this holds for every $A \in L^X$ we conclude that $\tilde{K}$ belongs to the family $\mathcal{I}$ and hence it is its maximal element, i.e. $\tilde{K} = Int^{\hat{\alpha}}$. □

Theorem 5.8 *Let (X, Int_X) and (Y, Int_Y) be L-preinterior spaces, and let $f : (X, Int_X) \to (Y, Int_Y)$ be a mapping. Then:*

1^0 *if $f : (X, Int_X) \to (Y, Int_Y)$ is β-continuous, then $f : (X, Int_X) \to (Y, Int^{\hat{\alpha}})$ is $\alpha * \beta$-continuous, and hence in particular, $f : (X, Int_X^{\hat{\alpha}}) \to (Y, Int_Y^{\hat{\alpha}})$ is also $\alpha * \beta$-continuous;*

2^0 *if $f : (X, Int_X^{\hat{\alpha}}) \to (Y, Int_Y^{\hat{\alpha}})$ is β-continuous, then $f : (X, Int_X) \to (Y, Int_Y^{\hat{\alpha}})$ is $\alpha * \beta$-continuous, and hence $f : (X, Int_X) \to (Y, Int_Y)$ is also $\alpha * \beta$-continuous.*

Proof. 1^0 $f : (X, Int_X) \to (Y, Int_Y^{\hat{\alpha}})$ is α-continuous as the composition of a continuous mapping $f : (X, Int_X) \to (Y, Int_Y)$ and an α-continuous identity mapping $id : (Y, Int_Y) \to (Y, Int_Y^{\hat{\alpha}})$.

2^0 $f : (X, Int_X) \to (Y, Int_Y^{\hat{\alpha}})$ is $\alpha * \beta$-continuous as the composition of an α-continuous identity mapping $id : (X, Int_X) \to (X, Int_X^{\hat{\alpha}})$ and a β-continuous mapping $f : (X, Int_X^{\hat{\alpha}}) \to (Y, Int_Y^{\hat{\alpha}})$. □

---

[3]The supremum obviously is taken in $\mathcal{I}(L, X)$ in this case

5.C: Functors $\Phi_\alpha : \mathcal{F}L\text{-}PInt \longmapsto \mathcal{F}L\text{-}PInt$ and $\Psi_\alpha : \mathcal{F}L\text{-}PInt \longmapsto \mathcal{F}L\text{-}PInt$

From Theorem 5.3 it follows that replacing an L-preinterior space (X, Int_X) by $(X, Int_X^{\check{\alpha}})$ and leaving morphisms unchanged, we obtain an α-functor $\Phi_\alpha : \mathcal{F}L\text{-}PInt \to \mathcal{F}L\text{-}PInt$. On the other hand from Theorem 5.8 it follows that replacing an L-preinterior space (X, Int_X) by $(X, Int_X^{\hat{\alpha}})$ and leaving morphisms unchanged, we obtain an α-functor $\Psi_\alpha : \mathcal{F}L\text{-}PInt \to \mathcal{F}L\text{-}PInt$. Moreover, both Φ_α and Ψ_α are α-isomorphisms, and their inverses are $\top$-functors.

Let $Int^{\hat{\alpha}\check{\alpha}}$ ($Int^{\check{\alpha}\hat{\alpha}}$ resp.) denote the L-preinterior operator obtained from Int by applying consequently operations of α-strengthening and α-weakening (resp. α-weakening and α-strengthening). In other words, $(X, Int^{\check{\alpha}\hat{\alpha}}) = \Psi_\alpha\Phi_\alpha(X, Int)$ and $(X, Int^{\hat{\alpha}\check{\alpha}}) = \Phi_\alpha\Psi_\alpha(X, Int)$.

Proposition 5.9 *L-preinterior spaces (X, Int), $(X, Int^{\hat{\alpha}})$, and $(X, Int^{\hat{\alpha}\check{\alpha}})$ are α-isomorphic; L-preinterior spaces (X, Int), $(X, Int^{\check{\alpha}})$ and $(X, Int^{\check{\alpha}\hat{\alpha}})$ are α-isomorphic. Functors $\Phi_\alpha \circ \Psi_\alpha : \mathcal{F}L\text{-}PInt \to \mathcal{F}L\text{-}PInt$ and $\Psi_\alpha \circ \Phi_\alpha : \mathcal{F}L\text{-}PInt \to \mathcal{F}L\text{-}PInt$ are α-isomorphisms; besides the inverse of $\Phi_\alpha \circ \Psi_\alpha$ is a $\top$-functor.*

Proof. The α-isomorphism of (X, Int) and $(X, Int^{\check{\alpha}})$ and the α-isomorphism of (X, Int) and $(X, Int^{\hat{\alpha}})$ was established above. Hence also the spaces $(X, Int^{\check{\alpha}})$ and $(X, Int^{\check{\alpha}\hat{\alpha}})$ and the spaces $(X, Int^{\hat{\alpha}})$ and $(X, Int^{\hat{\alpha}\check{\alpha}})$ are α-isomorphic. Thus, to complete the proof we have to show only that the compositions $\Phi_\alpha \circ \Psi_\alpha : \mathcal{F}L\text{-}PInt \to \mathcal{F}L\text{-}PInt$ and $\Psi_\alpha \circ \Phi_\alpha : \mathcal{F}L\text{-}PInt \to \mathcal{F}L\text{-}PInt$ are α-isomorphisms. To see this notice first that the identity mappings $id : (X, Int) \to (X, Int^{\check{\alpha}\hat{\alpha}})$, $id : (X, Int) \to (X, Int^{\hat{\alpha}\check{\alpha}})$ and $id : (X, Int^{\hat{\alpha}\check{\alpha}}) \to (X, Int)$ are α-continuous as compositions of a continuous and an α-continuous mappings. On the other hand, from 5.3 and 5.8 it easily follows that $Int^{\check{\alpha}\hat{\alpha}} \leq Int$ and hence the identity mapping $id : (X, Int^{\check{\alpha}\hat{\alpha}}) \to (X, Int)$ is continuous. □

Note that from the above it does not generally follow that $(X, Int^{\hat{\alpha}\check{\alpha}})$ and $(X, Int^{\check{\alpha}\hat{\alpha}})$ are α-isomorphic.

Remark 5.10 It is easy to define similar functors of α-strengthening and α-weakening in case of the fuzzy categories $\mathcal{F}L\text{-}Ker$, $\mathcal{F}\mathcal{F}L\text{-}Ker$ and $\mathcal{F}\mathcal{F}L\text{-}PInt$, too. We shall not go into details here.

6 Initial and final structures in fuzzy categories of L-kernel and L-preinterior spaces

6.A: Initial structures in fuzzy categories of L-kernel and L-preinterior spaces

Proposition 6.1 *Let (Y, K_Y) be an L-kernel space. Given a structured map $f : X \to (Y, K_Y)$ let $f^{\leftarrow}(K_Y) : L^X \to L^X$ be defined by the equality*

$$f^{\leftarrow}(K_Y)(A) = \bigvee_{f^{-1}(B) \leq A, B \in L^Y} f^{-1}(K_Y(B)) \quad \text{for every } A \in L^X$$

Then

1^0 *$f^{\leftarrow}(K_Y)$ is an L-kernel operator on X; moreover, if $K_Y = Int_Y$ is an an L-preinterior operator, then $f^{\leftarrow}(Int_Y)$ is an L-preinterior operator on X.*

2^0 *$\omega(f^{\leftarrow}(K_Y)) \geq \omega(K_Y)$ ($\omega_{id}(f^{\leftarrow}(K_Y)) \geq \omega_{id}(K_Y)$, $\omega_{gl}(f^{\leftarrow}(K_Y)) = \omega_{gl}(K_Y)$, $\omega_{sa}(f^{\leftarrow}(K_Y)) \geq \omega_{sa}(K_Y)$).*

3^0 *$f : (X, K_X) \to (Y, K_Y)$ is continuous. Moreover, if composition $f \circ g : (Z, K_Z) \to (Y, K_Y)$ is β-continuous for a mapping $g : (Z, K_Z) \to (X, K_X)$, then $\mu(g) \geq \beta$. In particular, continuity of $g \circ f$ implies continuity of g.*

Thus $f^{\leftarrow}(K_Y)$ is the (unique) initial structure on X for the map $f : X \to (Y, K_Y)$ in the fuzzy categories FL-Ker and $\mathcal{F}L$-Ker.

Proof. 1^0. The validity of properties (1K) and (2K) of the L-kernel operator for $f^{\leftarrow}(K_Y)$ is obvious. Therefore in the sequel we shall use also notation $f^{\leftarrow}(K_Y) =: K_X$. The validity of Property 3 for K_X (in case $K_Y = Int_Y$ is an L-preinterior operator) follows from the inequality $\omega_{sa}(f^{\leftarrow}(K_Y)) \geq \omega_{sa}(K_Y)$ established below.

2^0. To show that $\omega_{sa}(f^{\leftarrow}(K_Y)) \geq \omega_{sa}(K_Y)$ take $A_1, \ldots, A_n \in L^X$ and notice that (since L is infinitely distributive)

$$\begin{array}{rl} f^{\leftarrow}(K_Y)(A_1) \wedge \ldots \wedge f^{\leftarrow}(K_Y)(A_n) & \longmapsto \\ f^{\leftarrow}(K_Y)(A_1 \wedge \ldots \wedge A_n) & = \\ \bigwedge_{k=1}^{n} \left(\bigvee_{\substack{f^{-k}(B_k) \leq A_k \\ B_k \in L^Y}} f^{-1}(K_Y(B_k) \right) & \longmapsto \end{array}$$

$$\bigvee_{\substack{f^{-1}(C)\le A_1\wedge\ldots\wedge A_n,\\ C\in L^Y}} f^{-1}(K_Y(C)) \quad =$$

$$\bigvee_{\substack{f^{-1}(B_k)\le A_k\\ k=1,\ldots,n}} (f^{-1}(K_Y(B_1))\wedge\ldots\wedge f^{-1}(K_Y(B_n)) \quad \longmapsto$$

$$\bigvee_{f^{-1}(C)\le A_1\wedge\ldots\wedge A_n,} f^{-1}(K_Y(C)) \quad \geq$$

$$\bigvee_{\substack{f^{-1}(B_k)\le A_k\\ k=1,\ldots,n}} (f^{-1}(K_Y(B_1))\wedge\ldots\wedge f^{-1}(K_Y(B_n))) \quad \longmapsto$$

$$\bigvee_{\substack{f^{-1}(B'_k)\le A_k\\ k=1,\ldots,n}} f^{-1}(K_Y(B'_1\wedge\ldots B'_n)) \quad \geq \quad \omega_{sa}(K_Y)$$

and hence $\omega_{sa}(f^{\leftarrow}(K_Y)) \geq \omega_{sa}(K_Y)$.

The verification of the equality $\omega_{gl}(f^{\leftarrow}(K_Y)) = \omega_{gl}(K_Y)$ is straightforward and therefore omitted. To show that $\omega_{id}(f^{\leftarrow}(K_Y)) \geq \omega_{id}(K_Y)$, take some $A \in L^X$ and let $B \in L^Y$ be such that $f^{-1}(B) \leq A$. Then

$$\begin{aligned}
K_X(f^{-1}(B)) &\longmapsto K_X(K_X(A)) = \\
K_X(f^{-1}(B)) &\longmapsto \bigvee_{f^{-1}(C)\le K_X(A)} K_X(f^{-1}(C)) \geq \\
K_X(f^{-1}(B)) &\longmapsto \bigvee_{f^{-1}(C)\le K_X(f^{-1}(B))} K_X(f^{-1}(C)) = \\
K_X(f^{-1}(B)) &\longmapsto K_X(f^{-1}(K_Y(B))) = \\
f^{-1}(K_Y(B)) &\longmapsto f^{-1}(K_Y(K_Y(B))) \geq \\
K_Y(B) &\longmapsto K_Y(K_Y(B)) \geq \omega_{id}(K_Y)
\end{aligned}$$

Now, taking the infimum over all $B \in L^Y$ such that $f^{-1}(B) \leq A$ we get

$$\begin{aligned}
K_X(A) &\longmapsto K_X(K_X(A)) = \\
\bigvee_{f^{-1}(B)\le A} K_X(f^{-1}(B)) &\longmapsto \bigvee_{f^{-1}(C)\le K_X(A)} K_X(f^{-1}(C)) \\
&\geq \omega_{id}(K_Y)
\end{aligned}$$

and hence $\omega_{id}(K_X) \geq \omega_{id}(K_Y)$.

3^0. The continuity of the mapping $f : (X, K_X) \to (Y, K_Y)$ is obvious since from the definition of K_X it is clear that for every $V \in L^Y$ it holds $K_X(f^{-1}(V)) = f^{-1}(K_Y(V))$. Further, let $g : (Z, K_Z) \to (X, K_X)$ be such that the composition $f \circ g : (Z, Int_Z) \to (Y, K_Y)$ is β-continuous and let $U \in L^X$

Then

$$K_Z(g^{-1}(f^{-1}(A))) \geq g^{-1}(f^{-1}(K_Y(A))) * \beta$$

for every $A \in L^Y$ such that $f^{-1}(A) \leq U$. It follows from here that

$$\begin{array}{lcl} g^{-1}(f^{-1}(K_Y(A))) & \longmapsto & K_Z(g^{-1}(U)) \geq \\ g^{-1}(f^{-1}(K_Y(A))) & \longmapsto & K_Z(g^{-1}(f^{-1}(A))) \geq \beta \end{array}$$

and hence, by taking the infimum over all $A \in L^Y$ such that $f^{-1}(A) \leq U$, we conclude that

$$g^{-1}(K_X(U) \longmapsto K_Z(g^{-1}(U)) \geq \beta$$

Since this holds for all $U \in L^X$ it follows that $\mu(g) \geq \beta$. In particular, continuity of $g \circ f$ implies the continuity of g. □

Theorem 6.2 *Let $\mathcal{S} = \{f_i : X \to (Y_i, K_{Y_i}) \mid i \in I\}$ be a nonempty source in the fuzzy category FL-Ker (in the fuzzy category FL-PInt resp.) and for each $i \in I$ let $f_i^{\leftarrow}(K_{Y_i}) := K_i^X$ be the initial structure for the corresponding f_i. Further, let $\sup\{f_i^{\leftarrow}(K_{Y_i}) \mid i \in I\} =: K_X$ be the supremum of this family in $\mathcal{K}(L,X)$ (in $\mathcal{I}(L,X)$ resp.) (see 3.9, 3.10). Then K_X is an L-kernel operator (resp. is an L-preinterior operator) on X. Besides,*

1. $\omega(K_X) \geq \bigwedge_{i \in I} \omega(K_{Y_i})$. *Actually,*

$$\omega_{id}(K_X) \geq \bigwedge_{i\in I} \omega_{id}(K_{Y_i}), \omega_{gl}(K_X) \geq \bigwedge_{i\in I} \omega_{gl}(K_{Y_i}), \omega_{sa}(K_X) \geq \bigwedge_{i\in I} \omega_{sa}(K_{Y_i})$$

2. *Given a mapping $g : (Z, K_Z) \to (X, K_X)$, if all compositions $f_i \circ g : (Z, K_Z) \to (Y_i, K_{Y_i})$ are β-continuous, then the g is β-continuous, too.*

Thus the L-kernel operator K_X is the initial structure[4] for the source $\mathcal{S}$ in the fuzzy categry FL-Ker (in the fuzzy category FL-PInt, respectively)

Proof. Follows from Proposition 6.1 by applying 3.9, 3.10, 4.3 and 3.13. □

From the above it is clear that in $\mathcal{F}L$-$PInt$ direct products exist. Here is an effective description of the product:

Let $\mathcal{K} = \{(X_\lambda, K_\lambda) : \lambda \in \Lambda\}$ be a family of L-kernel spaces (resp., of L-preinterior spaces) and let $\alpha \in L$. Further, let $X = \prod_\lambda X_\lambda$ be the product of the corresponding sets and let $p_\lambda : X \to (X_\lambda, K_\lambda)$ be the projection. Then the product L-kernel operator on X (resp., the product L-preinterior operator) is defined as $K := \sup\{p_\lambda^{\leftarrow}(K_\lambda) \mid \lambda \in \Lambda\}$ and the supremum is taken in $\mathcal{K}(L,X)$ (in $\mathcal{I}(L,X)$, resp.). The pair (X, K) is the product of the family in the fuzzy category FL-Ker (in the fuzzy category FL-$PInt$, resp.).

[4] actually T-initial, cf [62]

Applying proposition 6.1 to the case when $f : X \to (Y, K_Y)$ is injective, we obtain description of subobjects in the fuzzy categories FL-Ker and FL-$PInt$.

Next we consider how the operation of α-weakening influence the construction of the initial structure.

Proposition 6.3 *Let $\mathcal{S} = \{f_i : X \to (Y_i, K_{Y_i}) \mid i \in I\}$ be a nonempty source in the fuzzy category FL-Ker where all (Y_i, K_{Y_i}) are stratified, let $\sup\{f_i^{\leftarrow}(K_{Y_i}) \mid i \in I\} =: K_X$ (cf 6.2) and let $K_X^{\breve{\alpha}}$ be its α-weakening. Then $\omega(K_X^{\breve{\alpha}}) \geq \wedge_{i \in I}\omega(K_{Y_i}) * \alpha$. Besides, if $g : (Z, K_Z) \to (X, K_X^{\breve{\alpha}})$ is a mapping such that all compositions $f_i \circ g : (Z, K_Z) \to (Y_i, K_{Y_i})$ are $\alpha * \beta$-continous, then g is β-continuous.*[5]

Proof. The inequality $\omega(K_X^{\breve{\alpha}}) \geq \wedge_{i \in I}\omega(K_{Y_i}) * \alpha$ follows from propositions 5.4 and 5.5. In view of 4.5 to complete the proof it is sufficient to consider the case when the source § consists of a single mapping $f : X \to (Y, K_Y)$ and hence $K_X = f^{\leftarrow}(K_Y)$ and $K_X^{\breve{\alpha}} := f^{\leftarrow}(K_Y)^{\breve{\alpha}}$. Let $g : (Z, K_Z) \to (X, K_X^{\breve{\alpha}})$ be such that $\mu(f \circ g) \geq \alpha * \beta$ (we can view f as a mapping $f : (X, K_X^{\breve{\alpha}}) \to (Y, K_Y)$). Thus

$$(f \circ g)^{-1}(K_Y(A)) \longmapsto K_Z((f \circ g)^{-1}(A)) \geq \alpha * \beta$$

for every $A \in L^Y$. The above inequality can be rewritten as

$$g^{-1}(f^{-1}((K_Y(A) * \alpha * \beta)) \leq K_Z(g^{-1}(f^{-1}(A)))$$

Now, fixing some $U \in L^X$ and taking the supremum on the both sides of the above inequality over all $A \in L^Y$ such that $f^{-1}(A) \leq U$ we obtain:

$$g^{-1}(K_X(U)) * \alpha * \beta \leq K_Z(g^{-1}(U))$$

and hence

$$g^{-1}(K_X^{\breve{\alpha}}(U) * \beta) \leq K_Z(g^{-1}(U))$$

Thus $g : (Z, K_Z) \to (X, K_X^{\breve{\alpha}})$ is β-continuous. We complete the proof by applying 5.5. □

Question. Is the α-initial structure $f^{\leftarrow}(K_Y)^{\breve{\alpha}}$ on X in the fuzzy category $\mathcal{F}L$-Ker in some natural sense unique?

[5] In the terminology from [62] this means that $K_X^{\breve{\alpha}}$ is the $(\alpha, \top)$-initial structure for the source $\mathcal{S}$ in the fuzzy categories $\mathcal{F}L$-Ker and $\mathcal{F}L$-$PInt$

6.B: Final structures in fuzzy categories of L-kernel and L-preinerior spaces

In the sequel we need the following construction allowing to extend L-kernel operators from a subset to the whole set.

Lemma 6.4 (Construction of the maximal extension) *Let Y be a set, Y_0 be its subset and let $K_0 : L^{Y_0} \to L$ be an L-kernel operator on Y_0. Define a mapping $K : L^Y \to L$ by $K(A) = K_0(A_0) \vee A_1$ where A_0 and A_1 are restrictions of A to Y_0 and $Y \setminus Y_0$, respetively.*[6] *Then K is an L-kernel operator on Y and besides $\omega_{id}(K) = \omega_{id}(K_0)$; $\omega_{gl}(K) = \omega_{gl}(K_0)$ and $\omega_{sa}(K) = \omega_{sa}(K_0)$. Moreover, a mapping $g : (Y, K) \to (Z, K_Z)$ is α-continuous if and only if its restriction $g_0 : (Y_0, K_0) \to (Z, K_Z)$ is α-continuous.*

Proof. It is obvious that K is an L-kernel operator. To show the equality $\omega_{id}(K) = \omega_{id}(K_0)$, fix $A \in L^Y$; then

$$\begin{array}{rcl} K(A) & \longmapsto & K(K(A)) \ = \\ K_0(A_0) \vee A_1 & \longmapsto & K\big(K_0(A_0) \ \vee \ A_1\big) \ = \\ K_0(A_0) \vee A_1 & \longmapsto & K_0\big(K_0(A_0)\big) \ \vee \ A_1 \ \geq \\ K_0(A_0) & \longmapsto & K_0(K_0(A_0)) \ \geq \ \omega(K_0) \end{array}$$

and hence $\omega_{id}(K) \geq \omega_{id}(K_0)$. Since the converse inequality is obvious, we get $\omega_{id}(K) = \omega_{id}(K_0)$.

Since $1_Y \longmapsto K(1_Y) = 1_{Y_0} \vee 1_{Y_1} \longmapsto K(1_{Y_0}) \vee 1_{Y_1} = 1_{Y_0} \longmapsto K(1_{Y_0})$ it follows that $\omega_{gl}(K) = \omega_{gl}(K_0)$.

To establish the last equality fix $A^1, \dots, A^n \in L^Y$. Then:

$$\begin{array}{rc} K(A^1) \wedge \dots \wedge K(A^n) & \longmapsto \\ K(A^1 \wedge \dots \wedge A^n) & = \\ \big[K_0(A_0^1) \vee A_1^1\big] \wedge \dots \wedge \big[K_0(A_0^n) \vee A_1^n\big] & \longmapsto \\ \big[K_0(A_0^1 \wedge \dots \wedge A_0^n) \vee (A_1^1 \wedge \dots \wedge A_1^n)\big] & = \\ \big(K_0(A_0^1) \wedge \dots \wedge K_0(A_0^n)\big) \bigvee (A_1^1 \wedge \dots A_1^n) & \longmapsto \\ K_0(A_0^1 \wedge \dots \wedge A_0^n) \bigvee (A_1^1 \vee \dots \vee A_1^n) & = \\ K_0(A_0^1) \wedge \dots \wedge K_0(A_0^n) & \longmapsto \\ K_0(A_0^1 \wedge \dots \wedge A_0^n) & \end{array}$$

Taking the infimum on the the both sides of this equality over all $A^1, \dots, A^n \in L^Y$ (and hence, respectively, over all $A_0^1, \dots, A_0^n \in L^{Y_0}$), we get $\omega_{sa}(K_0) =$

[6] Note that, when it is convenient, we identify the restriction $A_0 \in L^{Y_0}$ of an L-set $A \in L^Y$ to Y_0 with the L-set $A_0 \in L^Y$ defined by $A_0(y) = A(y)$ if $y \in Y_0$ and $A_0(y) = 0$ otherwise

$\omega_{sa}(K)$. Since obviously $\mu(g_0) \leq \mu(g)$, to complete the proof we have to establish the converse inequality. Let $B \in L^Z$. Then

$$g^{-1}(K_Z(B)) \longmapsto K(g^{-1}(B)) =$$

$$\left(g^{-1}(K_Z(B))\right)_0 \bigvee \left(g^{-1}(K_Z(B))\right)_1 \longmapsto \left(K(g^{-1}(B)\right)_0 \bigvee \left(g^{-1}(B)\right)_1 \geq$$

$$\left(g^{-1}(K_Z(B))\right)_0 \longmapsto \left(K(g^{-1}(B))\right)_0 \bigwedge \left(g^{-1}(K_Z(B)) \longmapsto g^{-1}(B)\right) \geq \mu(g_0)$$

and hence, taking the infimum over all $B \in L^Z$, we get $\mu(g) \geq \mu(g_0)$. □

Proposition 6.5 *Let (X, K_X) be an L-kernel space, $f : (X, K_X) \to Y$ be a mapping, and let $Y_0 = f(X)$. Define $K_0 : L^{Y_0} \to L^{Y_0}$ by the equality*

$$K_0(V) = f(K_X(f^{-1}(V))) \ , \quad \forall V \in L^{Y_0}$$

and let $f^{\to}(K_X) : L^Y \to L^Y$ be the maximal extension of K_0 to Y. Then

1^0 *$f^{\to}(K_X) =: K_Y$ is an L-kernel operator.*

2^0 *Moreover, $\omega(f^{\to}(K_X)) \geq \omega(K_X)$: actually $\omega_{sa}(f^{\to}(K_X)) \geq \omega_{sa}(K_X)$; $\omega_{gl}(f^{\to}(K_X)) \geq \omega_{gl}(K_X)$, $\omega_{id}(f^{\to}(K_X)) \geq \omega_{id}(K_X)$, and hence , if K_X is an L-preinterior operator, then $f^{\to}(K_X)$ is an L-preinterior operator, too.*

3^0 *Further, given an L-kernel space (Z, K_Z) and a mapping $g : (Y, K_Y) \to (Z, K_Z)$, composition $g \circ f : (X, K_X) \to (Z, K_Z)$ is β-continuous if and only if g is β-continuous.*

Thus K_Y is the (unique) final structure on Y for the map $f : (X, K_X) \to Y$ in the fuzzy categories FL-Ker and $\mathcal{F}L$-Ker.

Proof. Taking into account Lemma 6.4 it is sufficient to prove proposition under additional assumption that f is surjective, i.e. that $f(X) = Y$, and this will be assumed henceforth.

1^0 From the definition of K_Y it is obvious that it satisfies conditions (1K) and (2K) of Definition 3.4 and hence it is a kernel operator.

2^0 The inequality $\omega_{gl}(K_Y) \geq \omega_{gl}(K_X)$ is obvious. To show that $\omega_{sa}(K_Y) \geq \omega_{sa}(K_X)$ take some $V_1, \ldots, V_n \in L^Y$; then (taking into account that $f(A) \longmapsto f(B) \geq A \longmapsto B$ for any $A, B \in L^X$):

$$K_Y(V_1) \wedge \ldots \wedge K_Y(V_n) \longmapsto K_Y(V_1 \wedge \ldots \wedge V_n) =$$

$$f(K_X(f^{-1}(V_1))) \wedge \ldots \wedge f(K_X(f^{-1}(V_n))) \longmapsto f(K_X(f^{-1}(V_1 \wedge \ldots \wedge V_n))) \geq$$

$$(K_X(f^{-1}(V_1)) \wedge \ldots \wedge K_X(f^{-1}(V_n))) \longmapsto (K_X(f^{-1}(V_1 \wedge \ldots V_n))) \geq$$

$$K_X(f^{-1}(V_1))\wedge\ldots\wedge K_X(f^{-1}(V_n)) \longmapsto K_X(f^{-1}(V_1)\wedge\ldots\wedge f^{-1}(V_n)) \geq \omega_{sa}(K_X)$$

and hence, taking inf over all $V_1,\ldots,V_n \in L^Y$, we get $\omega_{sa}(K_Y) \geq \omega_{sa}(K_X)$. To show the inequality $\omega_{id}(K_Y) \geq \omega_{id}(K_X)$ take $B \in L^Y$ and notice that, in view of the surjectivity of the mapping f

$$K_Y(B) \longmapsto K_Y(K_Y(B)) = f(K_X(f^{-1}(B))) \longmapsto f(K_X(f^{-1}(K_Y(B)))) =$$

$$f(K_X(f^{-1}(B))) \longmapsto f(K_X(f^{-1}(f(K_X(f^{-1}(B)))))) \geq K_X(f^{-1}(B) \longmapsto$$

$$K_X(f^{-1}(f(K_X(f^{-1}(B))))) \geq K_X(f^{-1}(B)) \longmapsto K_X(K_X(f^{-1}(B))) \geq \omega_{id}(K_X)$$

and hence $\omega_{id}(K_Y) \geq \omega_{id}(K_X)$.

From the definition of K_Y it clear that it is the strongest L-kernel operator on Y for which the mapping $f : (X, K_X) \to (Y, K_Y)$ is continuous.

3^0 Let an L-kernel space (Z, K_Z) be given and assume that $g : (Y, K_Y) \to (Z, K_Z)$ be a mapping such that $\mu(g \circ f) \geq \beta$. Then $f^{-1}(g^{-1}(K_Z(V))) \longmapsto f^{-1}(K_Y(g^{-1}(V))) \geq f^{-1}(g^{-1}(K_Z(V))) \longmapsto K_X(g \circ f)^{-1}(V) \geq \beta$, and hence, in view of surjectivity of f,

$$g^{-1}(K_Z(V)) \longmapsto K_Y(g^{-1}(V)) \geq \beta$$

for every $V \in L^Z$. Now, taking infimum over all $V \in L^Z$ we obtain that $\mu(g) \geq \beta$, and this completes the proof of the proposition. □

Question: Is the α-strengthening $\left(f^{\rightarrow}(K_X)\right)^{\hat{\alpha}}$ of $f^{\rightarrow}(K_X)$ the α-final L-kernel operator for the mapping $f : (X, K_X) \to Y$ in the categories $\mathcal{F}L$-Ker and $\mathcal{FF}L$-Ker?

Theorem 6.6 *Let $\S = \{f_i : (X_i, K_{X_i}) \to Y \mid i \in I\}$, be a sink in FL-Ker and let $K_Y^i := f_i^{\rightarrow}(K_{X_i})$ and $K_Y := \inf_{i\in I} K_Y^i$. Then*

1^0 *$K_Y : L^Y \to L^Y$ is an L-kernel operator on Y. In particular, K_Y is an L-preinterior operator whenever all K_{X_i} are L-preinterior operators.*

2^0 *$\omega(K_Y) \geq \bigwedge_{i\in I} \omega(K_{X_i})$, (actually $\omega_{id}(K_Y) \geq \bigwedge_{i\in I} \omega_{id}(K_{X_i})$, $\omega_{gl}(K_Y) \geq \bigwedge_{i\in I} \omega_{gl}(K_{X_i})$, and $\omega_{sa}(K_Y) \geq \bigwedge_{i\in I} \omega_{sa}(K_{X_i})$).*

3^0 *If $g : (Y, K_Y) \to (Z, K_Z)$ is a mapping, then $\mu(g) \geq \wedge_{i\in I}\mu(g \circ f_i)$*

Thus K_Y is the (unique) final structure on Y for the sink $\S = \{f_i : (X_i, K_{X_i}) \to Y \mid i \in I\}$ in the category FL-Ker.

To prove the theorem, first we establish the following lemma:

Lemma 6.7 *Let $(X_i, K_i), i \in I$ be a family of L-kernel spaces and let $\oplus_{i\in I} X_i$ be the direct sum of these spaces. Define a mapping $K : L^X \to L^X$ by setting $K(A) = \bigvee_{i\in I} K_i(A_i)$, where A_i is the restriction of A to X_i. Further, given a family of mappings $f_i : (X_i, K_i) \to (Y, K_Y)$, let $f : (X, K) \to (Y, K_Y)$ be defined by $f(x) = f_i(x)$ whenever $x \in X_i$. Then*

1^0 *K is an L-kernel operator on X.*

2^0 *$\omega(K) = \bigwedge_{i\in I} \omega(K_i)$. Actually,*

$$\omega_{sa}(K) = \bigwedge_{i\in I} \omega_{sa}(K_i), \omega_{id}(K) = \bigwedge_{i\in I} \omega_{id}(K_i), \omega_{gl}(K) = \bigwedge_{i\in I} \omega_{gl}(K_i)$$

3^0 *If $g : (Y, K_Y) \to (Z, K_Z)$ is a mapping, then $\mu(g \circ f) = \bigwedge_{i\in I} \mu(g \circ f_i)$*

Proof. The proof of the first two statements can be obtained by a direct verification applying 3.9, 3.10, 3.13, 4.3. To show the third statement take $A \in L^Z$ and notice that

$$\begin{array}{rcl}
(g \circ f)^{-1}(K_Z(A)) & \longmapsto & K_X(g \circ f)^{-1}(A) = \\
f^{-1}(g^{-1}(K_Z(A))) & \longmapsto & K_X(f^{-1}(g^{-1}(A))) = \\
\bigvee_{i\in I} (f_i^{-1}(g^{-1}(K_Z(A)))) & \longmapsto & \bigvee_{i\in I} (K_i(f_i^{-1}(g^{-1}(A)))) = \\
\bigwedge_{i\in I} (f_i^{-1}(g^{-1}(K_Z(A)))) & \longmapsto & K_i(f_i^{-1}(g^{-1}(A)))
\end{array}$$

This completes the proof of the Lemma. □

To prove the theorem let X, K_X, and f be defined as in the Lemma 6.7. Then it is easy to notice that $f^{\to}(K_X) =: K_Y = \inf_{i\in I} K_Y^i$, is the final L-kernel operator for the sink $f : (X, K_X) \to Y$. Now the statement of the Theorem follows from Lemma 6.7 and Proposition 6.5. □

Remark 6.8 The existence of final structures in fuzzy categories FL-Ker, FL-$PInt$, $\mathcal{F}L$-Ker and $\mathcal{F}L$-$PInt$ mean, in particular, that coproducts and quotients exist in these fuzzy categories. Actually, the effective description of coproducts is given already by 6.7, while the description of quotients is contained in Proposition 6.5 (in case $Y - f(X)$).

References

[1] R. Artico, G. Moresco, *Fuzzy proximities according with Lowen fuzzy uniformities*, Fuzzy Sets and Systems **21**(1987), 85–98.

[2] G. Birkhoff, Lattice Theory, Amer. Math. Soc. Colloquium Publication XXXV, 3rd ed., American Mathematical Society (Providence, RI), 1967.

[3] M. Burgin, A. Šostak, *Towards the theory of continuity defect and continuity measure for mappings of metric spaces*, Acta Univ. Latv. **576**(1992), 45–62.

[4] M. Burgin, A. Šostak, *Fuzzification of the theory of continuous functions*, Fuzzy Sets and Systems **62**(1994), 71–81.

[5] C. L. Chang, Fuzzy topological spaces, *J. Math. Anal. Appl.*, **24**(1968), pp. 182–190.

[6] E. Čech, Topological Spaces, English translation, Interscience-Wiley (London-New York), 1966.

[7] J. Coulon, J. L. Coulon, *Une notion de proximite floue*, Busefal **10**(1990), 17–34.

[8] D. Dzhajanbajev, A. Šostak, *On a fuzzy uniform structure*, Acta et Comm. Univ. Tartu. **940**(1992), 31–36.

[9] P. Eklund, *Category theoretic properties of fuzzy topological spaces*, Fuzzy Sets and Systems **19**(1984), 303–310.

[10] P. Eklund, *Categorical fuzzy topology*, Acta Acad. Aboensis Ser. B **46:1**(1986), 1–13.

[11] P. Eklund, W. Gähler, *Basic notions for fuzzy topology, I*, Fuzzy Sets and Systems **26**(1988), 333–356.

[12] D. H. Foster, *Fuzzy topological groups*, J. Math. Anal. Appl. **67**(1979), 549–564.

[13] W. Gähler, A. S. Abd-Allah, A. Kandil, *On extended fuzzy topologies*, Fuzzy Sets and Systems **109**(2000), 149–172.

[14] M. H. Ghanim, H. M. Hasan, *L-closure spaces*, Fuzzy Sets and Systems **33**(1989), 383–391.

[15] G. Gierz, K. H. Hofmann, K. Keimel, J. D. Lawson, M. Mislove, D. S. Scott, A Compendium Of Continuous Lattices, Springer Verlag (Berlin/Heidelberg/New York), 1980.

[16] J. A. Goguen, *L-fuzzy sets*, J. Math. Anal. Appl. **18**(1967), 145–174.

[17] J. A. Goguen, *The fuzzy Tychonoff theorem*, J. Math. Anal. Appl. **43**(1973) 737–742.

[18] R. Goldblatt, Topoi: The Categorical Analysis Of Logic, revised ed., North-Holland (Amsterdam), 1981.

[19] P. C. Hammer, *Extended topology: set-valued set-functions*, Nieuw Archief voor Wiskunde **3:10**(1962), 55–77.

[20] H. Herrlich, G. E. Strecker, Category Theory, Heldermann Verlag (Berlin), 1979.

[21] U. Höhle, *Probabilistische topologies*, Manuscripta Math. **26**(1978), 223–245.

[22] U. Höhle, *Upper semicontinuous fuzzy sets and applications*, J. Math. Anal. Appl. **78**(1980) 659–673.

[23] U. Höhle, *G-fuzzy topologies on algebraic structures*, J. Math. Anal. Appl. **108**(1985), 113–150.

[24] U. Höhle, *Fuzzy topologies and topological space objects in a topos*, Fuzzy Sets and Systems **19**(1986), 299–304.

[25] U. Höhle, *Monoidal closed categories, weak topoi and generalized logics*, Fuzzy Sets and Systems **42**(1991), 15–35.

[26] U. Höhle, *M-valued sets and sheaves over integral commutative cl-monoids*, Chapter 2 in: S. E. Rodabaugh, E. P. Klement, U. Höhle, eds, Application Of Category To Fuzzy Sets, Theory and Decision Library—Series B: Mathematical and Statistical Methods, Volume **14**, Kluwer Academic Publishers (Boston/Dordrecht/London), 1992, pp. 33–72.

[27] U. Höhle, *Commutative residuated l-monoids*, Chapter IV in: U. Höhle, E. P. Klement, eds, Non-Classical Logics And Their Applications To Fuzzy Subsets—A Handbook of the Mathematical Foundations of Fuzzy Set Theory, Theory and Decision Library—Series B: Mathematical and Statistical Methods, Volume **32**, Kluwer Academic Publishers (Boston/Dordrecht/London), 1995, pp. 53–106.

[28] U. Höhle, *On the foundations of fuzzy set theory*, preprint, 1996.

[29] U. Höhle, A. Šostak, *A general theory of fuzzy topological spaces*, Fuzzy Sets and Systems **73**(1995), 131–149.

[30] U. Höhle, A. Šostak, *Axiomatics of fixed-basis fuzzy topology*, Chapter 3 in: U. Höhle, S. E. Rodabaugh, eds, Mathematics Of Fuzzy Sets: Logic, Topology, And Measure Theory, The Handbooks of Fuzzy Sets Series, Volume **3**(1999), Kluwer Academic Publishers (Dordrecht/Boston/London), pp. 123–273.

[31] U. Höhle, L. N. Stout, *Foundations of fuzzy sets*, Fuzzy Sets and Systems **40** (1991), 257–296.

[32] B. Hutton, *Normality in fuzzy topological spaces*, J. Math. Anal. Appl. **50**(1975), 74–79.

[33] B. Hutton, *Uniformities on fuzzy topological spaces*, J. Math. Anal. Appl. **58**(1977), 559–571.

[34] B. Hutton, *Products of fuzzy topological spaces*, Top. and Appl. **11**(1980), 59–67.

[35] B. Hutton, I. Reilly, *Separation axioms in fuzzy topological spaces*, Fuzzy Sets and Systems **3**(1980), 93–104.

[36] T. Kubiak, *On Fuzzy Topologies*, Doctoral Dissertation, Adam Mickiewicz University (Poznan, Poland), 1985).

[37] T. Kubiak, *The topological modification of the L-fuzzy unit interval*, Chapter 11 in: S. E. Rodabaugh, E. P. Klement, U. Höhle, eds, Application Of Category To Fuzzy Sets, Theory and Decision Library—Series B: Mathematical and Statistical Methods, Volume **14**, Kluwer Academic Publishers (Boston/Dordrecht/London), 1992, pp. 275–305.

[38] T. Kubiak, A. Šostak, *Lower set-valued fuzzy topologies*, Quaestiones Mathematicae **20:3**(1997), 423–429.

[39] Liu Wangjing, *Fuzzy invariant subgroups and fuzzy ideals*, Fuzzy Sets and Systems **8**(1982), 133–139.

[40] R. Lowen, *Fuzzy topological spaces and fuzzy compactness*, J. Math. Anal. Appl. **56**(1976), 621–633.

[41] R. Lowen, *Initial and final fuzzy topologies and the fuzzy Tychonoff theorem*, J. Math. Anal. Appl. **58**(1977), 621–633.

[42] A. S. Mashhour, M. H. Ghanim, *On closure spaces*, Indian J. Pure Appl. Math. **14**(1983), 680–691.

[43] C. V. Negoita, D. A. Ralescu, Application Of Fuzzy Sets To System Analysis, John Wiley & Sons (New York), 1975.

[44] N. Netzer, *Verallgemeinerte topologische Strukturen*, Jahrbuch Überblicke Math.(1978), 87–196.

[45] H. E. Porst, A. Šostak, *On the concept of a fuzzy function*, in preparation.

[46] S. E. Rodabaugh, *A categorical accomodation of various notions of fuzzy topology*, Fuzzy Sets and Systems **9** (1983), 241–265.

[47] S. E. Rodabaugh, *A theory of fuzzy uniformities with applications to the fuzzy real lines*, J. Math. Anal. Appl. **129**(1988), 37–70.

[48] S. E. Rodabaugh, *Categorical frameworks for Stone representation theories*, Chapter 7 in: S. E. Rodabaugh, E. P. Klement, U. Höhle, eds, Applications Of Category Theory To Fuzzy Subsets, Kluwer Academic Publishers (Boston/Dordrecht/London), 1992, pp. 177–232.

[49] S. E. Rodabaugh, *Categorical foundations of variable-basis fuzzy topology*, Chapter 4 in: U. Höhle, S. E. Rodabaugh, eds, Mathematics of Fuzzy Sets: Logic, Topology, and Measure Theory, The Handbooks of Fuzzy Sets Series, Volume **3**(1999), Kluwer Academic Publishers (Dordrecht/Boston/London), pp. 273–388.

[50] A. Rosenfeld, *Fuzzy groups*, J. Math. Anal. Appl. **35**(1971), 512–517.

[51] V. Salij, *Binary $\mathcal{L}$-relations*, Izv. Vysh. Ucheb. Zaved. (Ser. Math.) **1:44**(1965), 133–145 (in Russian).

[52] B. Schweitzer, A. Sklar, Probabilistic Metric Spaces, North Holland (Amsterdam), 1983.

[53] A. Šostak, *On a fuzzy topological structure*, Suppl. Rend. Circ. Matem. Palermo (Ser. II) **11**(1985), 89–103.

[54] A. Šostak, *Coreflectivity in categories of fuzzy topology*, in: Continuous Functions on Topological Spaces, Latv. Univ. Publ. (Riga) (1986), 159–165 (in Russian).

[55] A. Šostak, *On a category for fuzzy topology*, Zb. Radova Univ. u Nišu (Ser. Matem.) **2**(1988), 61–67.

[56] A. Šostak, *Two decades of fuzzy topology: basic ideas, notions, and results*, Russian Math. Surveys **44:6**(1989), 125–186.

[57] A. Šostak, *Towards the concept of a fuzzy category*, Acta Univ. Latviensis (Ser. Math.) **562**(1991), 85–94.

[58] A. Šostak, *On a concept of a fuzzy category*, in: 14th Linz Seminar on Fuzzy Set Theory: Non-Classical Logics and Their Applications, Linz, Austria, 1992, 63–66.

[59] A. Šostak, *Basic structures of fuzzy topology*, J. Math. Sci. textbf78:6(1996), 662–701.

[60] A. Šostak, *On a fuzzy syntopogeneous structure*, Quaestiones Mathematicae **20:3**(1997), 431–463.

[61] A. Šostak, *Fuzzy categories versus categories of fuzzily structured sets: Elements of the theory of fuzzy categories*, Mathematik-Arbeitspapiere **48**: Categorical Methods In Algebra And Topology (A collection of papers in honor of Horst Herrlich), Hans-E. Porst ed., pp. 407–438, Bremen, August 1997.

[62] A. Šostak, *Fuzzy categories versus categories of fuzzily structured sets: Elements of the theory of fuzzy categories, Part II*, Acta Univ. Latv. (Riga, Latvia) (2001), to appear.

[63] A. Šostak, *Fuzzy categories related to algebra and topology*, Tatra Mount. Math. Publ. **16**(1999), 159–186.

[64] R. H. Warren, *Neighbourhoods, bases and continuity in fuzzy topological spaces*, Rocky Mount. J. Math. **8**(1978), 459–470.

[65] Wang Guojun, *Topological molecular lattices*, Fuzzy Sets and Systems **47**(1992), 351–376.

[66] Yuan Bo, Wu Wangming, *Fuzzy ideals on a distributive lattice*, Fuzzy Sets and Systems **35**(1990), 231–240.

[67] Ying Mingsheng, *A new approach to fuzzy topology, Part I*, Fuzzy Sets and Systems **39**(1991), 303–321.

[68] Ying Mingsheng, *A new approach to fuzzy topology, Part II*, Fuzzy Sets and Systems **47**(1992), 221–232.

[69] Ying Mingsheng, *A new approach to fuzzy topology, Part III*, Fuzzy Sets and Systems **55**(1993), 193–207.

[70] Ying Mingsheng, *Fuzzifying uniform spaces*, Fuzzy Sets and Systems **53**(1993), 93–104.

[71] Ying Mingsheng, *Fuzzifying topology based on a residuated lattice-valued logic, I*, Fuzzy Sets and Systems **56**(1993), 337–373.

[72] Ying Mingsheng, *Fuzzy topology based on a residuated lattice-valued logic*, preprint.

[73] L. A. Zadeh, *Fuzzy sets*, Inform. and Control **8**(1965), 338–365.

Part III:

Related Topics In Topological And Algebraic Structures

CHAPTER 13

Fuzzy Compactness Via Categorical Closure Operators

I. W. ALDERTON [1]

Introduction

The Kuratowski-Mrówka result [12, 15] that a topological space X is compact if and only if the second projection map $\pi_Y : X \times Y \to Y$ is closed for each topological space Y, has led to a number of categorical definitions of the notion of compactness.

One, introduced by Herrlich, Salicrup, and Strecker [10], is based upon factorization structures. In [2], the α-compactness of Gantner, Steinlage, and Warren [9] was identified as the categorical compactness notion arising from a certain factorization structure on the category $\mathbb{I}$-**TOP** of $\mathbb{I}$-topological spaces in the sense of Chang [6] and $\mathbb{I}$-continuous maps, where $\mathbb{I}$ is the complete lattice $[0, 1]$ with the usual ordering.

Another more recent categorical definition of compactness is due to Clementino, Giuli, and Tholen [7], and is based upon categorical closure operators. Various categorical closure operators on **S**$\mathbb{I}$-**TOP**, the category of stratified $\mathbb{I}$-topological spaces in the sense of Lowen [13] and $\mathbb{I}$-continuous maps, were studied in [4]. It is the purpose of this chapter to identify the categorical compactness notions (in **S**$\mathbb{I}$-**TOP** and/or $\mathbb{I}$-**TOP**, where appropriate) associated with three of these closure operators. One of them gives rise to α-compactness, another gives rise to α^*-compactness, whilst the third gives rise to a compactness notion which we shall term semi-α-compactness—the characterization of the latter proceeds via the closure spaces of Čech.

Throughout this chapter, categories of lattice-valued topological spaces and associated continuous mappings are denoted according to the recently standardized terminology and notation of Höhle-Šostak [11] and Rodabaugh [16], according to which the category of Lowen is denoted **S**$\mathbb{I}$-**TOP** or $\mathbb{I}$**S**-**TOP** and the category of Chang is denoted $\mathbb{I}$-**TOP**.

[1] Professor L. Stout is thanked for remarks which led to improvements in this chapter.

S.E. Rodabaugh and E.P. Klement (eds.), Topological and Algebraic Structures in Fuzzy Sets, 375–387.
© 2003 *Kluwer Academic Publishers.*

1 Preliminaries

If X is a set, then $\mathcal{P}X$ denotes the powerset of X. The characteristic function of a subset M of X is denoted by χ_M. For any number $\gamma \in [0,1]$, the symbol γ will also denote denote the constant map with value γ, regardless of the domain. Given a map $u : X \to [0,1]$ the notation supp(u) means the support of u, i.e. the set $\{x \in X \mid u(x) > 0\}$. If (X, τ) and (Y, σ) are fuzzy topological spaces, then their product is denoted by $(X, \tau) \times (Y, \sigma)$ or $(X \times Y, \tau \times \sigma)$.

Suppose $\mathcal{X}$ is a category with finite products and X, Y are $\mathcal{X}$-objects. Then $\pi_X : X \times Y \to X$ and $\pi_Y : X \times Y \to Y$ denote the projection morphisms. In case $\mathcal{X}$ is a concrete category the subscripts will be the underlying sets of the corresponding $\mathcal{X}$-objects. **TOP** denotes the category of topological spaces and continuous maps. The reader may consult [1] for any undefined categorical notions.

The following definitions and information regarding categorical closure operators are taken from [7], where further information and references can be found. Let $\mathcal{X}$ be a complete category with a proper $(\mathcal{E}, \mathcal{M})$-factorization system for morphisms, so that $\mathcal{E}$ is a class of epimorphisms and $\mathcal{M}$ is a class of monomorphisms in $\mathcal{X}$, both containing the isomorphisms of $\mathcal{X}$, such that every $\mathcal{X}$-morphism has an $(\mathcal{E}, \mathcal{M})$-factorization and the $(\mathcal{E}, \mathcal{M})$-diagonalization property holds. For every object X, the class $sub(X)$ of $\mathcal{M}$-morphisms with codomain X is preordered as follows: $m \leq n$ if and only if there exists j with $n \circ j = m$. If $m \leq n$ and $n \leq m$, then we write $m \cong n$. If $f : X \to Y$ is a morphism, then $f(m)$ denotes the $\mathcal{M}$-part of the $(\mathcal{E}, \mathcal{M})$-factorization of $f \circ m$, for $m \in sub(X)$.

1.1 Definition. A **closure operator** c of $\mathcal{X}$ with respect to $\mathcal{M}$ is given by a family of maps $c_X : sub(X) \to sub(X)$ $(X \in \mathcal{X})$ such that:

(i) $m \leq c_X(m)$ for all $m \in sub(X)$;

(ii) $m \leq n \Rightarrow c_X(m) \leq c_X(n)$ for all m, $n \in sub(X)$;

(iii) every morphism $f : X \to Y$ satisfies $f(c_X(m)) \leq c_Y(f(m))$ for all $m \in sub(X)$.

Due to (i) every $m : M \to X$ in $\mathcal{M}$ factors as

$$M \overset{\gamma_X(m)}{\to} c_X(M) \overset{c_X(m)}{\to} X$$

A subobject $m \in sub(X)$ is: c-**closed** if $m \cong c_X(m)$ (i.e. $\gamma_X(m)$ is an isomorphism); c-**dense** if $c_X(m)$ is an isomorphism. A morphism $f : X \to Y$ is c-**dense** if $f(1_X)$ is c-dense. Let $\mathcal{M}^c$ and $\mathcal{E}^c$ denote the classes of all c-closed ($\mathcal{M}$-)subobjects and of all c-dense morphisms respectively. Although the $(\mathcal{E}^c, \mathcal{M}^c)$-diagonalization property holds in $\mathcal{X}$, morphisms in $\mathcal{X}$ may not have $(\mathcal{E}^c, \mathcal{M}^c)$-factorizations. A closure operator c is: **idempotent** if $c_X(m)$ is c-closed for every $m \in sub(X)$; **weakly hereditary** if $\gamma_X(m)$ is c-dense for every $m \in sub(X)$. $(\mathcal{E}^c, \mathcal{M}^c)$ is a factorization structure if and only if c is idempotent and weakly hereditary. A morphism $f : X \to Y$ in $\mathcal{X}$ is c-**preserving** if $f(c_X(m) \cong c_Y(f(m))$ for all $m \in sub(X)$.

1.2 Definition. An object $X \in \mathcal{X}$ is c-**compact** if $\pi_Y : X \times Y \to Y$ is c-preserving for every object $Y \in \mathcal{X}$.

In the sequel we take $\mathcal{X}$ to be $\mathcal{F}$, where $\mathcal{F}$ can denote either **S$\mathbb{I}$-TOP** or **$\mathbb{I}$-TOP**; $\mathcal{E}$ to be the class of $\mathbb{I}$-continuous surjections; and $\mathcal{M}$ to be the class of embeddings. Because initial fuzzy topologies in $\mathcal{F}$ are unique, given a fuzzy topological space (X, τ) and an embedding $(Z, \sigma) \hookrightarrow (X, \tau)$, we shall identify the latter simply with the subset Z of X. The corresponding image under c will then be denoted by $c_{(X,\tau)}(Z)$, or just $c(Z)$ if no confusion can result. With all these conventions, the previous definitions unpack in $\mathcal{F}$ as follows:

1.3 Definition. A **closure operator** c on $\mathcal{F}$ is a map defined on all subsets of all $\mathbb{I}$-topological spaces so that the following hold for all $(X, \tau) \in \mathcal{F}$:

(i) $M \subseteq c(M)$ for all $M \subseteq X$;

(ii) $M \subseteq N \Rightarrow c(M) \subseteq c(N)$ for all M, $N \subseteq X$;

(iii) for all $(Y, \sigma) \in \mathcal{F}$, all $\mathbb{I}$-continuous maps $f : (X, \tau) \to (Y, \sigma)$, and all $M \subseteq X$, it holds that $f(c(M)) \subseteq c(f(M))$.

If $(X, \tau) \in \mathcal{F}$, then $M \subseteq X$ is: c-**closed** if $M = c(M)$; c-**dense** if $c(M) = X$. An $\mathbb{I}$-continuous map $f : (X, \tau) \to (Y, \sigma)$ is c-**dense** if $c(f(X)) = Y$. The subscript notation is used for the next two definitions. A closure operator c on $\mathcal{F}$ is: **weakly hereditary** if for all $(X, \tau) \in \mathcal{F}$ and $M \subseteq X$, the embedding $M \hookrightarrow c_{(X,\tau)}(M)$ is c-dense, i.e. $c_{c_{(X,\tau)}(M)}(M) = c_{(X,\tau)}(M)$; **idempotent** if for all $(X, \tau) \in \mathcal{F}$ and $M \subseteq X$, we have $c_{(X,\tau)}(c_{(X,\tau)}(M)) = c_{(X,\tau)}(M)$. An $\mathbb{I}$-continuous map $f : (X, \tau) \to (Y, \sigma)$ is c-**preserving** if $f(c(M) = c(f(M))$ for all $M \subseteq X$.

1.4 Definition. $(X, \tau) \in \mathcal{F}$ is c-**compact** if $\pi_Y : (X, \tau) \times (Y, \sigma) \to (Y, \sigma)$ is c-preserving for all $(Y, \sigma) \in \mathcal{F}$.

2 Closure operator l_α

The results in this section are valid for both **𝕀-TOP** and **S𝕀-TOP**, but they are all presented for the latter category.

Given $\alpha \in [0,1)$, the closure operator l_α on **S𝕀-TOP** is defined as follows: if $(X,\tau) \in$ **S𝕀-TOP** and $M \subseteq X$, then

$$l_\alpha(M) = (\overline{\chi_M})^{-1}[1-\alpha, 1]$$

where the horizontal bar denotes ordinary 𝕀-topological closure. (Note that the definition of l_α, as it appears in [4], has been modified here so that the results are terminologically consistent, i.e. l_α-compactness turns out to be semi-α-compactness (defined in 2.12) rather than semi-$(1-\alpha)$-compactness.)

To show that l_α-compactness with respect to $\mathcal{M}$ (the class of embeddings in **S𝕀-TOP**) is the same as semi-α-compactness, we shall proceed via **CCS**, the category of Čech closure spaces and continuous maps, the relevant definitions being as follows:

2.1 Definition (cf. 14 A.1 of [5]). A **Čech closure space** is a pair (X, C), where X is a set and $C : \mathcal{P}X \to \mathcal{P}X$ is a map (called a **Čech closure operation** for X) such that the following axioms are satisfied:

(i) $C(\emptyset) = \emptyset$;

(ii) $M \subseteq C(M)$ for each $M \subseteq X$;

(iii) $C(M \cup N) = C(M) \cup C(N)$ for each $M \subseteq X$ and $N \subseteq X$.

It should help the reader to note that a categorical closure operator in this chapter is denoted by a lower case letter whilst a Čech closure operation on a set is denoted by an upper case letter.

2.2 Definition (cf. pp. 269 - 270 of [5]). Let (X, C) and (Y, D) be Čech closure spaces, and let $f : X \to Y$ be a map. Then f is said to be **continuous** provided $M \subseteq X \Rightarrow f(C(M)) \subseteq D(f(M))$.

The categorical closure operator l_α on **S𝕀-TOP** induces a concrete functor L_α : **S𝕀-TOP** $\to$ **CCS** in the obvious way [8]: if $(X,\tau) \in$ **S𝕀-TOP**, then $L_\alpha(X,\tau) = (X, L_\alpha^\tau)$, where the Čech closure operation L_α^τ for X is given as follows: if $M \subseteq X$, then $L_\alpha^\tau(M) = l_\alpha(M)$.

2.3 Definition (14 B.1 of [5]). A **neighbourhood** (nbd for short) of a point x in a Čech closure space (X, C) is a subset U of X which contains x in its interior, $int_C(U)$, where $int_C(U) = X - C(X - U)$.

2.4 Lemma. *Given an $\mathbb{I}$-topological space (X,τ), a subset U of X is a nbd of $x \in X$ in the Čech closure space (X, L_α^τ) if and only if $\chi_U^\circ(x) > \alpha$.*

The category **CCS** has products (Section 17 C of [5]). In particular, if (X,C) and (Y,D) are Čech closure spaces, and $(X \times Y, C \times D)$ denotes their product, then $C \times D$ is given as follows: if $M \subseteq X \times Y$, then $(C \times D)(M)$ is the set of all (x,y) in $X \times Y$ such that whenever U is a nbd of x in (X,C) and V is a nbd of y in (Y,D), then $(U \times V) \cap M \neq \emptyset$.

2.5 Proposition. *The functor L_α :* **S$\mathbb{I}$-TOP** $\to$ **CCS** *preserves binary products, i.e. if (X,τ), $(Y,\sigma) \in$* **S$\mathbb{I}$-TOP** *then $L_\alpha(X \times Y, \tau \times \sigma) = (X \times Y, L_\alpha^\tau \times L_\alpha^\sigma)$.*

Proof. As a result of the product property in **CCS** we obtain that the identity $1 : (X \times Y, L_\alpha^{\tau\times\sigma}) \to (X \times Y, L_\alpha^\tau \times L_\alpha^\sigma)$ is continuous. It just needs to be shown that $1 : (X \times Y, L_\alpha^\tau \times L_\alpha^\sigma) \to (X \times Y, L_\alpha^{\tau\times\sigma})$ is continuous, i.e. if $M \subseteq X \times Y$, then $(L_\alpha^\tau \times L_\alpha^\sigma)(M) \subseteq L_\alpha^{\tau\times\sigma}(M)$. So let $(x,y) \in (L_\alpha^\tau \times L_\alpha^\sigma)(M)$ and let W be a nbd of (x,y) in $(X \times Y, L_\alpha^{\tau\times\sigma})$. By 14 B.6 of [5] it should be shown that W intersects M. Now $\chi_W^\circ(x,y) > \alpha$ by 2.4. Since $\chi_W^\circ = \bigvee\{w \mid a \in \tau \times \sigma \text{ and } a(z,w) = 0$ for all $(z,w) \notin W\} = \bigvee\{u \times v \mid u \in \tau,\ v \in \sigma; \text{ and } (z,w) \notin W \Rightarrow u(z) = 0 \text{ or } v(w) = 0\}$ it follows that there exist $u \in \tau$ and $v \in \sigma$ with the properties that $(z,w) \notin W \Rightarrow u(z) = 0$ or $v(w) = 0$; and $(u \times v)(x,y) > \alpha$ (so that $u(x) > \alpha$ and $v(y) > \alpha$). Now $u \leq \chi_{\text{supp}(u)}$ so $u \leq \chi_{\text{supp}(u)}^\circ$. Hence $\chi_{\text{supp}(u)}^\circ(x) > \alpha$. Similarly $\chi_{\text{supp}(v)}^\circ(y) > \alpha$. Thus supp$(u)$ is a nbd of x in (X, L_α^τ) and supp(v) is a nbd of y in (Y, L_α^σ). It follows that $(\text{supp}(u) \times \text{supp}(v)) \cap M \neq \emptyset$. That $W \cap M \neq \emptyset$ follows from the fact that supp$(u) \times$ supp$(v) \subseteq W$. □

2.6 Definition (17 A.17 of [5]). An **interior cover** of a Čech closure space (X,C) is a cover $\{M_i \mid i \in I\}$ of X such that $\{int_C(M_i) \mid i \in I\}$ is a cover of X.

2.7 Definition (cf. 41 A.9 of [5]). A Čech closure space (X,C) is said to be **Čech-compact** provided that every interior cover of (X,C) has a finite subcover.

2.8 Definition. Let (X,C) and (Y,D) be Čech closure spaces, and let $f : X \to Y$ be a map. Then f is said to be **closure-preserving** if for all $M \subseteq X$ we have $f((C(M)) = D(f(M))$.

2.9 Proposition. *Let (X, C) be a Čech closure space. The following three statements are equivalent:*

(i) *(X, C) is Čech-compact.*

(ii) *$\pi_Y : (X, C) \times (Y, D) \to (Y, D)$ is closure-preserving for each $(Y, D) \in$* **CCS**.

(iii) *$\pi_Y : (X, C) \times (Y, D) \to (Y, D)$ is closure-preserving for each$(Y, D) \in$* **TOP** *(where, of course, the topology is identified with its unique associated Kuratowski closure operation).*

Proof. *(i)* $\Rightarrow$ *(ii)*: Let (Y, D) be a Čech closure space and let $M \subseteq X \times Y$. By continuity $\pi_Y((C \times D)(M)) \subseteq D(\pi_Y(M))$. If $y_0 \notin \pi_Y((C \times D)(M))$, then for each $x \in X$ there exists a nbd W_x of (x, y_0) in $(X \times Y, C \times D)$ so that $W_x \cap M = \emptyset$. Now $\pi_X(W_x)$ is a nbd of x in (X, C) (17 C.7 of [5]), and $\{ \pi_X(W_x) \mid x \in X\}$ is an interior cover of (X, C). Let

$$\{\pi_X(W_{x_1})\, \pi_X(W_{x_2}), ..., \pi_X(W_{x_n})\}$$

be a finite subcover, and put $V = \bigcap_{i=1}^{n} (W_{x_i})$. Then V is a nbd of y_0 in (Y, D) which does not meet $\pi_Y(M)$. Consequently $y_0 \notin D(\pi_Y(M))$.

(ii) $\Rightarrow$ *(iii)*: Clear.

(iii) $\Rightarrow$ *(i)*: This part of the proof is based upon Manes' proof [14] of the Kuratowski-Mrówka result. Let $(\beta X, \delta)$ be the Stone-Čech compactification of X, considered as a discrete topological space, i.e. βX is the set of all ultrafilters on X, and the topology δ is given by the following procedure: for each $A \subseteq X$, let $A^* = \{\mathcal{U} \in \beta X \mid A \in \mathcal{U}\}$. Then $\{A^* \mid A \subseteq X\}$ is a base for the topology δ on βX. Let the corresponding Čech closure space be denoted by $(\beta X, D)$. Let $\xi \subseteq X \times \beta X$ be given by $\{(x, \mathcal{U}) \mid \mathcal{U}$ converges to x in $(X, C)\}$. Then ξ is closed in the product Čech closure space $(X, C) \times (\beta X, D)$, i.e. $(C \times D)(\xi) = \xi$. Now $D(S) = \beta X$, where S denotes the set of all ultrafilters in (X, C) which converge to at least one point. The projection $\pi_{\beta X} : (X, C) \times (\beta X, D) \to (\beta X, D)$ is closure-preserving so $\pi_{\beta X}(\xi) = \pi_{\beta X}((C \times D)(\xi)) = D(\pi_{\beta X}(\xi)) = D(S) = \beta X$. Consequently all ultrafilters on X converge. Hence (X, C) is Čech-compact by 41 A.6 of [5]. $\square$

2.10 Remarks. The equivalence of conditions (i) and (ii) in the preceding proposition is already in [8]. However, the proof presented here is different, and proceeds via condition (iii), which is not in [8]. Moreover, the equivalence of conditions (i) and (iii) is a vital ingredient in the proof of Proposition 2.13, the main result in this section.

2.11 Lemma. *Given a topological space* (Y, σ), *let* $(Y, \overline{\sigma})$ *be the fuzzy topological space which has as a base of closed sets the collection* $\{\chi_K \vee \gamma \mid K$ *is closed in* (Y, σ), *and* $\gamma \in [0,1]\}$. *Then the Čech closure space* $(Y, L_\alpha^{\overline{\sigma}})$ *is topological, and* $L_\alpha^{\overline{\sigma}}$ *is exactly the Kuratowski closure operation* Cl_σ *corresponding to* (Y, σ).

Proof. Let $M \subseteq Y$. It will be shown that $Cl_{\overline{\sigma}}(\chi_M) = \chi_{Cl_\sigma(M)}$, for then $L_\alpha^{\overline{\sigma}}(M) = (Cl_{\overline{\sigma}}(\chi_M))^{-1}[1-\alpha, 1] = (\chi_{Cl_\sigma(M)})^{-1}[1-\alpha, 1] = Cl_\alpha(M)$. Now $\chi_{Cl_\sigma(M)}$ is closed in $(Y, \overline{\sigma})$ and $\chi_{Cl_\sigma(M)}(m) = 1$ for all $m \in M$, so $Cl_{\overline{\sigma}}(\chi_M) \leq \chi_{Cl_\sigma(M)}$. Suppose that $Cl_{\overline{\sigma}}(\chi_M) < \chi_{Cl_\sigma(M)}$. Then for some $y \in Y$ it holds that $\chi_{Cl_\sigma(M)}(y) = 1$ (i.e. $y \in Cl_\alpha(M)$) and $Cl_{\overline{\sigma}}(\chi_M)(y) < 1$. Thus there exists a closed set N in (Y, σ) and $\gamma \in [0,1]$ such that $(\chi_N \vee \gamma)(y) < 1$ and $(\chi_N \vee \gamma)(m) = 1$ for all $m \in M$. Hence $y \notin N$ and $M \subseteq N$. Since N is closed in (Y, σ), we have $y \notin Cl_\sigma(M)$, a contradiction. □

The next proposition shows that l_α-compactness with respect to $\mathcal{M}$ is the same as semi-α-compactness, which is defined as follows:

2.12 Definition. Let $\alpha \in [0, 1)$. An $\mathbb{I}$-topological space (X, τ) is called **semi-α-compact** if, whenever $\{u_i \mid i \in I\} \subseteq \tau$ is such that $\bigvee u_i > \alpha$, then there exists a finite subset $\{u_{i_1}, u_{i_2}, ..., u_{i_n}\}$ of $\{u_i \mid i \in I\}$ such that $\bigvee_{k=1}^{n} u_{i_k} > 0$.

2.13 Proposition. *Let* $(X, \tau) \in$ **S$\mathbb{I}$-TOP**. *The following three statements are equivalent:*

(i) (X, τ) *is* l_α*-compact with respect to* $\mathcal{M}$.

(ii) (X, L_α^τ) *is Čech-compact.*

(iii) (X, τ) *is semi-*α*-compact.*

Proof. *(i)* $\Rightarrow$ *(ii)*:
(X, τ) is l_α-compact

$\Rightarrow$ for all $(Y, \sigma) \in$ **S$\mathbb{I}$-TOP** and all $M \subseteq X \times Y$ it holds that $\pi_Y(l_\alpha(M)) = l_\alpha(\pi_Y(M))$

$\Rightarrow$ $\pi_Y : (X \times Y, L_\alpha^{\tau\times\sigma}) \to (Y, L_\alpha^\sigma)$ is closure-preserving for all $(Y, \sigma) \in$ **S$\mathbb{I}$-TOP**

$\Rightarrow$ $\pi_Y : (X, L_\alpha^\tau) \times (Y, L_\alpha^\sigma) \to (Y, L_\alpha^\sigma)$ is closure-preserving for all $(Y, \sigma) \in$ **S$\mathbb{I}$-TOP** (by 2.5)

$\Rightarrow$ $\pi_Y : (X, L_\alpha^\tau) \times (Y, D) \to (Y, D)$ is closure-preserving for all $(Y, D) \in$ **TOP** (by 2.11)

$\Rightarrow$ (X, L_α^τ) is Čech-compact (by 2.9).

$(ii) \Rightarrow (i)$:

(X, L_α^τ) is Čech-compact

$\Rightarrow \pi_Y : (X, L_\alpha^\tau) \times (Y, D) \to (Y, D)$ is closure-preserving for all $(Y, D) \in$ **CCS**

$\Rightarrow \pi_Y : (X, L_\alpha^\tau) \times (Y, L_\alpha^\sigma) \to (Y, L_\alpha^\sigma)$ is closure-preserving for all $(Y, \sigma) \in$ **S𝕀-TOP**

$\Rightarrow \pi_Y : (X \times Y, L_\alpha^{\tau\times\sigma}) \to (Y, L_\alpha^\sigma)$ is closure-preserving for all $(Y, \sigma) \in$ **S𝕀-TOP**

$\Rightarrow (X, \tau)$ is l_α-compact.

$(ii) \Rightarrow (iii)$: Suppose $\{u_i \mid i \in I\} \subseteq \tau$ is such that $\bigvee_{i\in I} u_i > \alpha$. For each $i \in I$, put $M_i = \text{supp}(u_i)$. Since each $u_i \in \tau$ and $u_i \leq \chi_{M_i}$ we have that $u_i \leq \chi_{M_i}^\circ$. Now $\{M_i \mid i \in I\}$ is an interior cover of (X, L_α^τ) so we can obtain a finite subcover, say $\{M_{i_1}, M_{i_2}, \ldots , M_{i_n}\}$. Consequently $\bigvee_{j=1}^n u_{i_j} > 0$.

$(iii) \Rightarrow (ii)$: Let $\{M_i \mid i \in I\}$ be an interior cover of (X, L_α^τ). Note that $int_{L_\alpha^\tau}(M_i) = X - (\overline{\chi_{X-M_i}})^{-1}[1-\alpha]$. For each $i \in I$, put $u_i = \chi_{M_i}^\circ$. Then $\{u_i \mid i \in I\} \subseteq \tau$ and $\vee_{i\in I} u_i > \alpha$. Hence there exists a finite subset $\{u_{i_1}, u_{i_2}, \ldots, u_{i_n}\} \subseteq \{u_i \mid i \in I\}$ with $\bigvee_{j=1}^n u_{i_j} > 0$. This means that $\{M_{i_1}, M_{i_2}, \ldots , M_{i_n}\}$ is a cover of (X, L_α^τ). □

3 Closure operator k_α

In this section we consider only the category **𝕀-TOP**. Although [4] was concerned just with **S𝕀-TOP**, any results from that chapter which are used here are valid for both categories.

Given $\alpha \in [0, 1)$ the closure operator k_α on **𝕀-TOP** is defined as follows: if $(X, \tau) \in$ **𝕀-TOP** and $M \subseteq X$, then $k_\alpha(M) = \bigcap\{w^{-1}[1-\alpha, 1] \mid w$ is fuzzy closed in (X, τ) and $M \subseteq w^{-1}[1-\alpha, 1]\}$.

Now in certain fairly commonly occurring situations, such as in this section, the categorical compactness notions of [7] and [10] coincide. This is made explicit by the following general result, which is not difficult to prove: Let $\mathcal{X}$ be a concrete category over the category of sets and maps. Suppose that $\mathcal{X}$ has finite products, and the property that each embedding of a subset into the underlying set of any $\mathcal{X}$-object has a unique initial lift. Let c be an idempotent weakly hereditary closure operator on $\mathcal{X}$ with respect to $\mathcal{M}$, the class of $\mathcal{X}$-embeddings. Then c-compactness with respect to $\mathcal{M}$ is precisely the same as $\mathcal{M}^c$-compactness (in the sense of [10]). (Here $\mathcal{M}$ is part of the factorization structure $(\mathcal{E}, \mathcal{M})$ on $\mathcal{X}$, where $\mathcal{E}$ is the class of surjective $\mathcal{X}$-morphisms.)

The closure operator k_α is idempotent (2.14 of [4]), and it is easy to show that it is also weakly hereditary. Because of this, $(\mathcal{E}^{k_\alpha}, \mathcal{M}^{k_\alpha})$ is a factorization structure on $\mathbb{I}$-**TOP**, and, by the general result of the previous paragraph, k_α-compactness with respect to $\mathcal{M}$ is the same as $\mathcal{M}^{k_\alpha}$-compactness. The remainder of this section is devoted to identifying the class $\mathcal{M}^{k_\alpha}$, and the notion of k_α-compactness with respect to $\mathcal{M}$.

The factorization structure on $\mathbb{I}$-**TOP** considered in [2] is (α^*-dense $\mathbb{I}$-continuous map, α^*-closed embedding). The required definitions are as follows:

3.1 Definitions (cf. 2.1 and 2.2 of [2]). Let (X,τ), $(Y,\sigma) \in \mathbb{I}$-**TOP**, and let $f : X \to Y$ be a map. Then $f : (X,\tau) \to (Y,\sigma)$ is said to be:

(i) α^*-**dense** provided that whenever v is a fuzzy closed set in (Y,σ) such that $f(X) \subseteq v^{-1}[1-\alpha, 1]$, it follows that $Y = v^{-1}[1-\alpha, 1]$;

(ii) α^*-**closed** provided that for each fuzzy closed set w in (X,τ) there exists a fuzzy closed set v in (Y,σ) such that $f(w^{-1}[1-\alpha,1]) = v^{-1}[1-\alpha,1]$.

The required characterization of k_α-compactness with respect to $\mathcal{M}$ is a consequence of the next result.

3.2 Proposition. *Let $f : (X,\tau) \to (Y,\sigma)$ be an embedding in $\mathbb{I}$-**TOP**. The following are equivalent:*

(i) *f is α^*-closed;*

(ii) *$f(X) = v^{-1}[1-\alpha,1]$ for some fuzzy closed set v in (Y,σ);*

(iii) *f is k_α-closed.*

Proof. *(i)* $\Leftrightarrow$ *(ii)*: See 2.5 of [2].

(iii) $\Leftrightarrow$ *(ii)*: f is k_α-closed if and only if $k_\alpha(f(X)) = f(X)$ if and only if there exists a fuzzy closed set v in (X,τ) such that $f(X) = v^{-1}[1-\alpha,1]$ (by the remark in 2.14 of [4]). □

3.3 Corollary. *k_α-compactness with respect to $\mathcal{M}$ is precisely α-compactness.*

Proof. The proposition shows that $\mathcal{M}^{k_\alpha}$ is precisely the class of α^*-closed embeddings, so by the general result mentioned above, k_α-compactness with respect to $\mathcal{M}$ coincides with (α^*-closed embedding)-compactness, which is just α-compactness by 3.12 of [2]. □

3.4 Remark. The closure operator l_α of Section 2, although weakly hereditary, is not idempotent (2.32 of [4]). Hence the general result mentioned above was not used there.

4 Closure operator e_α

We again consider only the category **$\mathbb{I}$-TOP**. The remark made in Section 3 about the use of results from [4] applies here as well.

Given $\alpha \in (0,1]$ the closure operator e_α is defined as follows: if $(X,\tau) \in$ **$\mathbb{I}$-TOP** and $M \subseteq X$, then $e_\alpha(M) = \bigcap\{w^{-1}(1-\alpha,1] \mid w$ is fuzzy closed in (X,τ) and $M \subseteq w^{-1}(1-\alpha,1]\}$.

The definition of e_α, as it appears in 2.26 of [4] has been altered for the sake of terminological consistency. Although this closure operator appears to be a minor modification of k_α, and it is easy to guess that e_α-compactness with respect to $\mathcal{M}$ (the class of embeddings in **$\mathbb{I}$-TOP**) is α^*-compactness, the results needed to prove this are more than merely trivial analogues of those in the preceding section, and are therefore presented in some detail.

4.1 Definitions. Let (X,τ), $(Y,\sigma) \in$ **$\mathbb{I}$-TOP**, and let $f : X \to Y$ be a map. Then $f : (X,\tau) \to (Y,\sigma)$ is said to be:

(i) **α-dense** provided that whenever $v_i (i \in I)$ are fuzzy closed sets in (Y,σ) such that $f(X) \subseteq \bigcap_{i\in I} v_i^{-1}(1-\alpha,1]$, it follows that

$$Y = \bigcap_{i\in I} v_i^{-1}(1-\alpha,1]$$

(ii) **α-closed** provided that for each fuzzy closed set w in (X,τ) there exist fuzzy closed sets v_i $(i \in I)$ in (Y,σ) such that

$$f(w^{-1}(1-\alpha,1] = \bigcap_{i\in I} v_i^{-1}(1-\alpha,1]$$

For each $\alpha \in (0,1]$, define a concrete functor $\lambda_\alpha :$ **$\mathbb{I}$-TOP** $\to$ **TOP** as follows (2.2(iii) of [3] and 2.28 of [4]): if $(X,\tau) \in$ **$\mathbb{I}$-TOP**, then $\lambda_\alpha(X,\tau) = (X, \lambda_\alpha\tau)$, where the topology $\lambda_\alpha\tau$ has as a base the collection $\{u^{-1}[\alpha,1] \mid u \in \tau\}$. In fact, $\lambda_\alpha\tau$ is the topology on X which is initial with respect to the collection $(X \xrightarrow{u} ([0,1], \{\emptyset, [0,1], [\alpha,1]\}))_{u\in\tau}$. By 2.2(iii) of [3], λ_α preserves initial morphisms, in particular, embeddings.

4.2 Proposition. *Let $f : (X,\tau) \to (Y,\sigma)$ be $\mathbb{I}$-continuous. The following are equivalent:*

(i) $f : (X,\tau) \to (Y,\sigma)$ *is α-dense.*

(ii) $f : (X,\lambda_\alpha\tau) \to (Y,\lambda_\alpha\sigma)$ *is dense.*

4.3 Proposition. *Let $f : (X, \tau) \to (Y, \sigma)$ be an embedding in* $\mathbb{I}$**-TOP**. *The following are equivalent:*

(i) $f : (X, \tau) \to (Y, \sigma)$ *is α-closed.*

(ii) $f(X) = \bigcap_{i \in I} v_i^{-1}(1 - \alpha, 1]$, *where each v_i $(i \in I)$ is a fuzzy closed set in (Y, σ).*

(iii) $f : (X, \tau) \to (Y, \sigma)$ *is e_α-closed.*

(iv) $f : (X, \lambda_\alpha \tau) \to (Y, \lambda_\alpha \sigma)$ *is a closed embedding.*

Proof. $(i) \Rightarrow (ii)$: This follows since $X = 1^{-1}(1 - \alpha, 1]$.

$(ii) \Rightarrow (i)$: Let w be a fuzzy closed set in (X, τ). Then there exists $u \in \sigma$ such that $w = (1 - u) \circ f$. Then $f(w^{-1}(1 - \alpha]) = ((1 - u)^{-1}\ (1 - \alpha]) \cap (\bigcap_{i \in I} v_i^{-1}(1 - \alpha, 1])$.

$(ii) \Rightarrow (iii) \Rightarrow (iv) \Rightarrow (ii)$: These implications are easily shown. □

4.4 Proposition. *(α-dense $\mathbb{I}$-continuous map, α-closed embedding) is a factorization structure on* $\mathbb{I}$**-TOP**.

Proof. This result follows from 4.2 and 4.3 together with the comment just preceding 2.2 in [3]. □

Since e_α is weakly hereditary and idempotent, it is a consequence of 4.3 and the general result in Section 3 that e_α-compactness with respect to $\mathcal{M}$ coincides with (α-closed embedding)-compactness. It just remains to show that (α-closed embedding)-compactness is the same as α^*-compactness.

4.5 Proposition. *The functor λ_α :* $\mathbb{I}$**-TOP** $\to$ **TOP** *preserves finite products.*

4.6 Proposition. *Let $(X, \tau) \in$* $\mathbb{I}$**-TOP**. *The following are equivalent:*

(i) (X, τ) *is (α-closed embedding)-compact.*

(ii) $(X, \lambda_\alpha \tau)$ *is compact.*

(iii) (X, τ) *is α^*-compact.*

Proof. $(i) \Leftrightarrow (ii)$: Define a concrete functor μ : **TOP** $\to$ $\mathbb{I}$**-TOP** as follows: if $(X, \tau) \in$ $\mathbb{I}$**-TOP**, then $\mu(X, \tau) = (X, \mu(\tau))$, where $\mu(\tau) = \{\chi_G \mid G \in \tau\}$. Then $\lambda_\alpha \circ \mu = 1_{\mathbf{TOP}}$. Now apply 3.6 and 3.7 of [3].

$(ii) \Leftrightarrow (iii)$: These implications are easy to prove. □

4.7 Corollary. *e_α-compactness with respect to $\mathcal{M}$ is precisely α^*-compactness.*

4.8 Remarks. In each of the three sections 2 to 4 there has been defined a closure operator d, or rather, two closure operators, one for each of the categories $\mathbb{I}$-**TOP** and $\mathbf{S}\mathbb{I}$-**TOP**, where we use the obvious notation $d^{\mathbb{I}-\mathbf{TOP}}$ and $d^{\mathbf{S}\mathbb{I}-\mathbf{TOP}}$ to distinguish the two. However, there is no essential difference between $d^{\mathbb{I}-\mathbf{TOP}}$ and $d^{\mathbf{S}\mathbb{I}-\mathbf{TOP}}$: For suppose $(X, \tau) \in \mathbb{I}$-**TOP**, and $(X, \tau') \in$ $\mathbf{S}\mathbb{I}$-**TOP** is such that τ' has as a subbase the collection which is the union of τ and all constant fuzzy maps on X. If $M \subseteq X$, then $d^{\mathbb{I}-\mathbf{TOP}}(M) = d^{\mathbf{S}\mathbb{I}-\mathbf{TOP}}(M)$.

References

[1] J. Adámek, H. Herrlich, G. E. Strecker, Abstract and Concrete Categories: The Joy of Cats, Wiley Interscience Pure and Applied Mathematics, John Wiley & Sons (Brisbane/Chicester/New York/Singapore/Toronto), 1990.

[2] I. W. Alderton, *α-Compactness of fuzzy topological spaces: a categorical approach*, Quaestiones Math. **20**(1997), 269–282.

[3] I. W. Alderton, *Compactness in topological categories*, in: B. Banaschewski, C. R. A. Gilmour, H. Herrlich, eds, *Festschrift (On the Occasion of the 60th Birthday of Guillaume Brümmer) incorporating Proceedings, Symposium on Categorical Topology*, University of Cape Town, 1994, 9–16, Department of Mathematics and Applied Mathematics, University of Cape Town, 1999.

[4] I. W. Alderton, G. Castellini, *Closure operators and fuzzy connectedness*, Fuzzy Sets and Systems **104**(1999), 397–405.

[5] E. Čech, Topological Spaces, rev. ed. by Z. Frolík and M. Katětov, Publishing House of the Czechoslovak Academy of Sciences, Prague, Wiley-Interscience (London/New York), 1966.

[6] C. L. Chang, *Fuzzy topological spaces*, J. Math. Anal. Appl. **24**(1968), 182–190.

[7] M. M. Clementino, E. Giuli, W. Tholen, *Topology in a category*, Portugal. Math. **53**(1996), 397–433.

[8] D. Dikranjan, E. Giuli, *Compactness, minimality and closedness with respect to a closure operator*, in: J. Adámek, S. MacLane, eds, Categorical Topology And Its Relations To Analysis, Algebra, And Combinatorics, Prague, Czechoslovakia, 1988, 284–296, World Scientific (Singapore et al.), 1989.

[9] T. E. Gantner, R. C. Steinlage, R. H. Warren, *Compactness in fuzzy topological spaces*, J. Math. Anal. Appl. **62**(1978), 547–562.

[10] H. Herrlich, G. Salicrup, G. E, Strecker, *Factorizations, denseness, separation, and relatively compact objects*, Topology Appl. **27**(1987), 157–169.

[11] U. Höhle, A. Šostak, *Axiomatic foundations of fixed-basis fuzzy topology*, Chapter 3 in: U. Höhle, S. E. Rodabaugh, eds, Mathematics Of Fuzzy Sets: Logic, Topology, And Measure Theory, The Handbooks of Fuzzy Sets Series, Volume **3** (1999), Kluwer Academic Publishers (Boston/Dordrecht/London), pp. 123–272.

[12] K. Kuratowski, *Evaluation de la class borélienne d'un ensemble de points à l'aide des symboles logiques*, Fund. Math. **17**(1931), 249–272.

[13] R. Lowen, *Fuzzy topological spaces and fuzzy compactness*, J. Math. Anal. Appl. **56**(1976), 621–633.

[14] E. G. Manes, *Compact Hausdorff objects*, Gen. Topology Appl. **4**(1974), 341–360.

[15] S. Mrówka, *Compactness and product spaces*, Colloq. Math. **7**(1959), 19–22.

[16] S. E. Rodabaugh, *Categorical foundations of variable-basis fuzzy topology*, Chapter 4 in: U. Höhle, S. E. Rodabaugh, eds, Mathematics Of Fuzzy Sets: Logic, Topology, And Measure Theory, The Handbooks of Fuzzy Sets Series, Volume **3** (1999), Kluwer Academic Publishers (Boston/Dordrecht/London), pp. 273–388.

CHAPTER 14

Discrete Triangular Norms

BERNARD DE BAETS [1] AND RADKO MESIAR [2]

Introduction

In this chapter, we study the relationships between discrete t-norms, i.e. t-norms on a finite chain, and t-norms on the unit interval. Firstly, we investigate when and how a discrete t-norm can be extended to a (continuous) t-norm on the unit interval. Secondly, we investigate when a discretization of a t-norm on the unit interval yields a discrete t-norm. Moreover, several classes of discrete t-norms are discussed.

1 T-norms on $[0, 1]$

Applications in various fields such as probabilistic metric spaces, fuzzy logic, fuzzy control, generalized measure theory, multi-criteria decision making, etc., have increased the interest in triangular norms. Triangular norms (t-norms, for short) have been introduced in the sixties by Schweizer and Sklar [12] as commutative, associative and increasing $[0,1]^2 \to [0,1]$ mappings with neutral element 1. The strongest (continuous) t-norm is the minimum operator $T_{\mathrm{M}}(x,y) = \min(x,y)$, the weakest (right-continuous) t-norm T_{W} is defined by $T_{\mathrm{W}}(x,y) = \min(x,y)$ if $\max(x,y) = 1$ and $T_{\mathrm{W}}(x,y) = 0$ otherwise. For an in-depth analysis of various aspects of t-norms, we refer to a recent monograph [9].

In this paper, we consider t-norms in a discrete setting. In Section 2, we discuss the structure of divisible (better known as smooth) t-norms and list the number of discrete t-norms, divisible discrete t-norms, and discrete t-norms inducing an involutive residual negator on a finite chain. In Section 3, we are

[1] Post-Doctoral Fellow of the Fund for Scientific Research—Flanders(Belgium).
[2] Partially Supported by Grant VEGA 1/4064/97.

S.E. Rodabaugh and E.P. Klement (eds.), Topological and Algebraic Structures in Fuzzy Sets, 389–400.
© 2003 *Kluwer Academic Publishers.*

concerned with the possible extension of a given discrete t-norm to the unit interval. The inverse problem, the restriction of a given t-norm to a given finite subchain, is treated in Section 4. Section 5 concludes with some remarks on t-norms on non-usual chains.

Before we proceed with the discussion of discrete t-norms, we first briefly recall some basic facts concerning the usual continuous t-norms. Continuous t-norms have been completely characterized as ordinal sums with continuous Archimedean summands [10, 12]. Recall that continuous Archimedean t-norms are characterized by the diagonal inequality $(\forall x \in]0,1[)(T(x,x) < x)$.

Continuous Archimedean t-norms are representable by means of additive generators [10]: a t-norm T is continuous and Archimedean if and only if there exists a continuous, strictly decreasing $[0,1] \to [0,\infty]$ mapping f with $f(1) = 0$ (called an additive generator of T) such that

$$T(x,y) = f^{-1}(\min(f(0), f(x) + f(y)))$$

for any $(x,y) \in [0,1]^2$. Note that an additive generator f of a continuous Archimedean t-norm T is unique up to a positive multiplicative constant.

Any continuous Archimedean t-norm T is either a strict t-norm (i.e. continuous and strictly increasing on $]0,1]^2$) or a nilpotent (i.e. non-strict) t-norm. Strict t-norms are characterized by unbounded additive generators ($f(0) = +\infty$) and are isomorphic to the product t-norm $T_{\mathrm{P}}(x,y) = xy$. On the other hand, nilpotent t-norms are characterized by bounded additive generators ($f(0) < +\infty$) and are isomorphic to the Łukasiewicz t-norm $T_{\mathrm{L}}(x,y) = \max(0, x+y-1)$; if $f(0) = 1$, then f is called the normed additive generator.

2 Discrete t-norms

Practical applications supported by computer implementations are often based on arguments taken from a finite scale, e.g. a finite subchain $C_n = \{x_1, \ldots, x_n\}$ of $[0,1]$, $n \in \mathbb{N}_0$, with $x_1 = 0 < x_2 < \ldots < x_n = 1$. Since there is no problem with introducing the concept of a t-norm on an arbitrary bounded poset [3] (for an in-depth study, in particular on product lattices, see [2]), we can introduce t-norms on a finite chain as well.

Definition 1 [5]
Consider a finite chain $C_n = \{x_1, \ldots, x_n\}$, $n \in \mathbb{N}_0$, with $x_1 < x_2 < \ldots < x_n$. A ${C_n}^2 \to C_n$ mapping D is called a discrete t-norm (on C_n) if it is commutative, associative, increasing and has x_n as neutral element, i.e.

$$(\forall i \in \{1, \ldots, n\})(D(x_i, x_n) = x_i)$$

All algebraic notions, such as the Archimedean property, strict monotonicity, nilpotent elements, etc., related to t-norms (on $[0,1]$) can be introduced for discrete t-norms in a straightforward way. The particular structure of a finite chain, however, leads to additional observations [2]: the Archimedean property is equivalent to the diagonal inequality, i.e. $(\forall i \in \{2,\ldots,n-1\})(T(x_i,x_i) < x_i)$, and also to the fact that x_{n-1} is a nilpotent element, there exists no strictly increasing discrete t-norm, etc. Recall that on the unit interval $[0,1]$ the diagonal inequality is only a consequence of the Archimedean property. The strongest discrete t-norm D_{M} is again the minimum operator and also the weakest discrete t-norm D_{W} is defined in the same way.

An important subclass is the class of divisible discrete t-norms characterized by Mayor and Torrens as a counterpart of continuous t-norms on $[0,1]$ [11]. In fact, the continuity of a t-norm T on $[0,1]$ is equivalent with

$$(\forall(x,y) \in [0,1]^2)(x \leq y \Leftrightarrow (\exists z \in [0,1])(x = T(y,z))), \tag{1}$$

a property also known under the name *divisibility* [6]. It is nothing else but the intermediate value theorem.

However, for a discrete t-norm D on a finite chain C_n, the above property is equivalent with the following: for any $(i,j) \in \{2,\ldots,n\}^2$ it holds that

$$\text{if } D(x_i,x_j) = x_r, \text{ then } D(x_{i-1},x_j) = x_p \text{ and } D(x_i,x_{j-1}) = x_q$$

with $r-1 \leq p,q \leq r$. This property is called the smoothness property by Godo and Sierra [5]. It can be rewritten as follows:

$$\text{if } D(x_i,x_j) = x_p \text{ and } D(x_k,x_l) = x_q, \text{ then } |p-q| \leq |i-k| + |j-l|,$$

and corresponds to the Lipschitz condition on the indices (with parameter value 1). From here one, we prefer to talk about divisible discrete t-norms, because of the non-standard intuition behind the term *smooth*.

The class of divisible discrete t-norms on a finite chain C_n has been characterized completely by Mayor and Torrens [11]. Firstly, there exists a unique divisible Archimedean discrete t-norm D_{L} on it, defined by

$$D_{\mathrm{L}}(x_i,x_j) = x_{\max(1,i+j-n)}$$

i.e. nothing else but a Łukasiewicz t-norm like operation on the indices. Secondly, for any given subset I of $\{x_2,\ldots,x_{n-1}\}$, there exists a unique divisible discrete t-norm D that has I as set of non-trivial idempotent elements. As a consequence, divisible discrete t-norms show an ordinal sum structure similar to that of continuous t-norms on $[0,1]$. In between two consecutive idempotent elements x_i and x_j it holds that

$$(\forall(x_k,x_l) \in [x_i,x_j]^2)(D(x_k,x_l) = x_{\max(i,k+l-j)})$$

i.e. D behaves as the unique divisible Archimedean discrete t-norm on $[x_i, x_j]$. It then also follows that there exist 2^{n-2} divisible discrete t-norms on C_n. Note that in the discrete case the ordinal sum construction reads as follows:

Definition 2

Let $(]a_\alpha, e_\alpha[)_{\alpha\in A}$ be a family of non-empty, pairwise disjoint open subintervals of C_n and let D_α be a discrete t-norm on $[a_\alpha, e_\alpha]$, for any $\alpha \in A$. The ordinal sum

$$D \approx (\langle a_\alpha, e_\alpha, D_\alpha\rangle)_{\alpha\in A}$$

is the t-norm on C_n defined by

$$D(x,y) = \begin{cases} D_\alpha(x,y) & , \text{ if } (x,y) \in [a_\alpha, e_\alpha]^2 \\ \min(x,y) & , \text{ otherwise} \end{cases}$$

Since the partial mappings of a discrete t-norm D on a finite chain C_n are always supremum preserving (a property equivalent to left-continuity of the partial mappings in the case of the unit interval), it makes sense to consider its residual implicator I_D defined by

$$I_D(x,y) = \sup\{z \in C_n \mid D(x,z) \leq y\}$$

Similarly, there is no discrete counterpart of the notion of right-continuity. Recently, there has been an increasing interest in the class of left-continuous t-norms T that induce an involutive residual negator [7, 8], i.e. the mapping $N = I_D(\cdot, 0)$ is an involution: $N(N(x)) = x$. In the discrete setting, the same role is played by discrete t-norms of which the residual negator $I_D(\cdot, x_1)$ coincides with the unique decreasing involution N of the chain C_n given by $N(x_i) = x_{n-i+1}$. Several techniques exist for the construction of such t-norms, but a full characterization is still lacking. These techniques can be carried over to the discrete setting without problem. A well-known member of this class of t-norms is the nilpotent minimum T^{nM} (see e.g. [4]) defined by

$$T^{\mathrm{nM}}(x,y) = \begin{cases} 0 & , \text{if } x+y \leq 1 \\ \min(x,y) & , \text{elsewhere} \end{cases}.$$

Remarkably, in the discrete case, there exist both a strongest discrete t-norm D^{nM}, a discrete version of the nilpotent minimum, and a weakest discrete t-norm G_{W} of this type, given by:

$$D^{\mathrm{nM}}(x_i, x_j) = \begin{cases} x_1 & , \text{if } i+j \leq n+1 \\ \min(x_i, x_j) & , \text{elsewhere} \end{cases}$$

and

$$G_{\mathrm{W}}(x_i, x_j) = \begin{cases} x_1 & , \text{ if } i+j \leq n+1 \\ x_2 & , \text{ if } i+j > n+1 \wedge \max(i,j) < n \\ \min(x_i, x_j) & , \text{ elsewhere} \end{cases}$$

The only divisible such t-norm is D_{L}. In case of $n \leq 4$, it holds that $G_{\mathrm{W}} = D_{\mathrm{L}}$; for $n > 4$, we have $G_{\mathrm{W}} < D_{\mathrm{L}} < D^{\mathrm{nM}}$.

The number of discrete t-norms and the number of discrete t-norms that induce an involutive residual negator on a finite chain C_n are known (by computation) for $n \leq 14$ (see [1] for the former). They are listed in Table 1 and compared to the number of divisible t-norms.

Table 1: Countings of Discrete T-Norms

n	t-norms	divisible	with involutive residual negator
2	1	1	1
3	2	2	1
4	6	4	2
5	22	8	3
6	94	16	7
7	451	32	12
8	2,386	64	31
9	13,775	128	59
10	86,417	256	161
11	590,489	512	329
12	4,446,029	1024	944
13	37,869,449	2048	2,067
14	382,549,464	4096	6,148

3 Extensions of discrete t-norms

From an applicational point of view, an interesting problem is that of the possible (continuous) extension(s) of a discrete t-norm D on a finite subchain C of $[0,1]$ to a t-norm T (on $[0,1]$) in the sense that $T|_{C^2} = D$. The following results hold.

Proposition 1
Consider a discrete t-norm D on a finite subchain $C_n = \{x_1, \ldots, x_n\}$ of $[0,1]$, $n \in \mathbb{N}_0$, with $x_1 = 0 < x_2 < \ldots < x_n = 1$. Then there exists a right-continuous extension T of D.

Proof. Indeed, consider the $[0,1] \to C_n$ mapping q defined by

$$q(x) = \max\{x_i \in C_n \mid x_i \le x\}$$

Next, we define the $[0,1]^2 \to [0,1]$ mapping T by

$$T(x,y) = \begin{cases} D(q(x),q(y)) & , \text{ if } (x,y) \in [0,1[^2 \\ \min(x,y) & , \text{ elsewhere} \end{cases}$$

Clearly, $T|_{C_n{}^2} = D$, the partial mappings of T are right-continuous, T is commutative (due to the commutativity of D), T is increasing (due to the fact that q and D are increasing) and 1 is its neutral element (by definition). Moreover, for any $(x,y,z) \in [0,1[^3$ it holds that

$$T(x,T(y,z)) = D(q(x),q(D(q(y),q(z)))) = D(q(x),D(q(y),q(z)))$$

and the associativity of D implies the associativity of T. For $(x,y,z) \in [0,1]^3 \setminus [0,1[^3$, the associativity of T is obvious. Hence, T is indeed a t-norm. □

Example 1 Let $C = \{0, 1/4, 1/2, 3/4, 1\}$ and consider the following $C^2 \to C$ mapping D:

$$D(x,y) = \begin{cases} \min(x,y) & , \text{ if } \max(x,y) = 1 \\ 1/4 & , \text{ if } (x,y) \in \{1/2, 3/4\}^2 \\ 0 & , \text{ elsewhere} \end{cases}$$

Then D is a discrete t-norm (the associativity of D follows from the equality $D(x, D(y,z)) = 0$ whenever $\max(x,y,z) < 1$). Moreover, D is Archimedean.

(i) The right-continuous extension T_r of D looks as follows

$$T_r(x,y) = \begin{cases} \min(x,y) & , \text{ if } \max(x,y) = 1 \\ 1/4 & , \text{ if } (x,y) \in [1/2, 1[^2 \\ 0 & , \text{ elsewhere} \end{cases}$$

(ii) Not only a right-continuous extension exists, also a left-continuous one can be constructed. Indeed, the $[0,1]^2 \to [0,1]$ mapping T_l defined by

$$T_l(x,y) = \begin{cases} 0 & , \text{ if } (x,y) \in [0,3/4]^2 \setminus]1/4, 3/4]^2 \\ 1/4 & , \text{ if } (x,y) \in]1/4, 3/4]^2 \\ \min(x,y) & , \text{ elsewhere} \end{cases}$$

is a left-continuous t-norm that extends D.

(iii) Although both a right-continuous extension and left-continuous extension exist, the discrete t-norm D does not have a continuous extension. We note first that the discrete counterpart of the cancellation property of continuous Archimedean t-norms, i.e.

$$T(x,y) = T(x,z) \Leftrightarrow (T(x,y) = 0 \vee y = z)$$

is violated by D. Indeed, $D(1/2, 1/2) = D(1/2, 3/4) = 1/4 \neq 0$. Now suppose that there exists a continuous extension T of D. Then $3/4$ and hence all elements belonging to the interval $]0, 3/4]$ are nilpotent elements of T. But then T should fulfill the above cancellation property for all $(x, y, z) \in [0, 3/4]^3$ and hence it cannot be an extension of D, which yields a contradiction.

The above example shows that there exist Archimedean discrete t-norms that cannot be extended in a continuous way. Therefore, Archimedean discrete t-norms are not necessarily representable by means of a continuous additive generator. However, divisible discrete t-norms can always be extended in a continuous way as is shown in the following proposition.

Proposition 2
Consider a ***divisible*** *discrete t-norm D on a finite subchain $C_n = \{x_1, \ldots, x_n\}$ of $[0,1]$, $n \in \mathbb{N}_0$, with $x_1 = 0 < x_2 < \ldots < x_n = 1$. Then there exists a continuous extension T of D that has the same idempotent elements as D.*

Proof. We distinguish two cases.

(i) If D has no non-trivial idempotent elements, then it is the unique Archimedean divisible discrete t-norm $D_{\mathbf{L}}$. Hence, any nilpotent t-norm T generated by a continuous normed additive generator that satisfies $f(x_i) = 1 - i/n$ is a continuous extension of D.

(ii) In case D has some non-trivial idempotent elements, it is an ordinal sum and the above procedure can be applied to each of its summands. More explicitly, let $\{x_0, x_{n_1}, \ldots, x_{n_k}, x_n\}$ be the set of all idempotent elements of D, and suppose that $(n_i)_{i=1}^k$ is an increasing sequence. Then any ordinal sum

$$T \approx (\langle x_0, x_{n_1}, T_1\rangle, \ldots, \langle x_{n_k}, x_n, T_{k+1}\rangle)$$

with nilpotent summands T_i generated by a normed additive generator f_i, $i = 1, \ldots, k+1$, such that

$$f_i\left(\frac{x_j - x_{n_{i-1}}}{x_{n_i} - x_{n_{i-1}}}\right) = 1 - \frac{j - n_{i-1} + 1}{n_i - n_{i-1}}$$

whenever $n_{i-1} \leq j \leq n_i$, is a continuous extension of D. □

Example 2 The existence of a continuous extension is not reserved for divisible t-norms! There also exist non-divisible Archimedean discrete t-norms admitting a continuous extension. This is the case for the weakest discrete t-norm D_W. Indeed, any nilpotent t-norm T generated by a normed additive generator f such that $f(x_{n-1}) \geq 0.5$ is a continuous extension of D_W. Note that the right-continuous extension of D_W as in Proposition 1 leads to the drastic product T_W.

The above example leads to the formulation of the following *open problem*: characterize the discrete t-norms admitting a continuous extension.

4 Discretizations of t-norms

The converse problem of the one dealt with in the previous section is the following: when does a discretization of a t-norm T yield a discrete t-norm? In other words, determine the finite subchains $C = \{x_1, \ldots, x_n\}$ of $[0,1]$, $n \in \mathbb{N}_0$, with $x_1 = 0 < x_2 < \ldots < x_n = 1$, on which $D = T|_{C^2}$ is a discrete t-norm. The main concern is, of course, that the discretization should yield an operation which is internal on the selected subchain, i.e. $\text{Ran}(D) \subseteq C$.

The first proposition below describes the t-norms that can be discretized on any finite subchain. It applies in particular to the minimum operator. Then we discuss the case of continuous Archimedean t-norms. These results are summarized into a single theorem describing the discretizability of an arbitrary continuous t-norm.

Proposition 3
*Consider a t-norm T. Then for **any** finite subchain $C = \{x_1, \ldots, x_n\}$, $n \in \mathbb{N}_0$, with $x_1 = 0 < x_2 < \ldots < x_n = 1$, the restriction $D = T|_{C^2}$ is a discrete t-norm on C if and only if T satisfies*

$$(\forall (x,y) \in [0,1]^2)(T(x,y) \in \{0, \min(x,y)\}), \tag{2}$$

or, in other words, if and only if there exists a symmmetric down-set (order ideal) K of $([0,1[^2, \leq)$ such that T is given by

$$T(x,y) = \begin{cases} 0 & , \text{ if } (x,y) \in K \\ \min(x,y) & , \text{ elsewhere} \end{cases} \tag{3}$$

Proof. It is immediately clear that Eq. 2 characterizes the desired t-norms. The formulation in terms of down-sets is a direct consequence. The demonstration that Eq. 3 indeed describes a t-norm requires the verification of the associativity by a simple method of cases. □

Well-known examples of the t-norms introduced in the above proposition are the minimum operator T_{M} (the only continuous one), the drastic product T_{W} (the only Archimedean one) and the nilpotent minimum T^{nM}.

Proposition 4
Consider a continuous Archimedean t-norm T.

(i) *If T is a strict t-norm, then only the trivial restriction $D = T|_{\{0,1\}^2}$ yields a discrete t-norm, namely the Boolean conjunction.*

(ii) *If T is a nilpotent t-norm with additive generator f, then $D = T|_{C^2}$ is a discrete t-norm on C if and only if $f(C) = \{f(x) \mid x \in C\}$ is relatively closed under addition, i.e. for any $(x, y) \in C^2$ there either exists $z \in C$ such that $f(x) + f(y) = f(z)$ or $f(x) + f(y) > f(0)$. If we consider the normed additive generator t, then the formulation is as follows: for any $(x, y) \in C^2$ it holds that $t(x) + t(y) \in t(C) \cup [1, 2]$.*

Proof. The strict case is immediately clear in view of the strict monotonicity of T. The nilpotent case is a simple consequence of the representation in terms of the additive generator f. □

As a consequence of the above proposition (the strict case), the only subchain on which any t-norm can be discretized is the trivial chain $\{0, 1\}$. Taking into account the ordinal sum representation of continuous t-norms, the above propositions can be summarized into the following theorem characterizing the subchains on which an arbitrary continuous t-norm can be discretized.

Theorem 5
Consider a continuous t-norm T with ordinal sum representation

$$T \approx (\langle a_\alpha, e_\alpha, T_\alpha \rangle)_{\alpha \in A}$$

Any subchain C of $[0, 1]$ on which T can be discretized is characterized by:

(i) *C can contain any of the idempotent elements of T;*

(ii) *if T_α is a strict t-norm, then $C \cap]a_\alpha, e_\alpha[= \emptyset$;*

(iii) *if T_α is a nilpotent t-norm, then either $C \cap]a_\alpha, e_\alpha[= \emptyset$ or $a_\alpha \in C$ and $f_\alpha(C_\alpha)$ is relatively closed under addition, with f_α an additive generator of T_α and*

$$C_\alpha = \left\{ \frac{x - a_\alpha}{e_\alpha - a_\alpha} \mid x \in C \cap [a_\alpha, e_\alpha] \right\}$$

Clearly, discretization preserves idempotent elements, nilpotent elements and the Archimedean property. However, the following examples show that divisibility and zero divisors are not necessarily preserved.

Example 3

(i) It is immediately clear that the Łukasiewicz t-norm T_L can be discretized on any equidistant subchain $E_n = \{0, \frac{1}{n}, \frac{2}{n}, \ldots, \frac{n-1}{n}, 1\}$ yielding a divisible Archimedean discrete t-norm for any $n \in \mathbb{N}_0$. Another suitable subchain on which T_L (with normed additive generator $t(x) = 1 - x$) can be discretized is $C = \{0, 0.2, 0.4, 1\}$ leading to the weakest t-norm D_W on C. This shows that the discretization of a continuous t-norm does not necessarily yield a divisible t-norm.

(ii) Due to Proposition 3, the nilpotent minimum T^{nM} can be discretized on any finite subchain C. Note that it has $]0,1[$ as set of zero divisors. However, if we discretize T^{nM} on a subchain $C \subseteq \{0\} \cup]1/2, 1]$, then the resulting discrete t-norm is the minimum operator D_M and has no zero divisors.

5 T-norms on non-usual chains

Divisible t-norms on the unit interval and on finite chains are well understood. From a mathematical point of view, nothing prevents us from considering (divisible) t-norms on other types of bounded subchains of the real line (or even general bounded posets). However, the particular structure of the chain in question may cause some peculiarities.

Example 4 The particular structure of the chain may force certain elements to be idempotent elements of any divisible t-norm on it. Consider for instance the chain $C = [0, 1/2] \cup \{1\}$. Then $1/2$ is an idempotent element of any divisible t-norm T on it. Indeed, suppose that $a = T(1/2, 1/2) < 1/2$. The divisibility (as formulated in Eq. (1), replacing $[0,1]$ by C) means that for $a < b < 1/2$, there exists some $c \in C$ such that $b = T(1/2, c)$. This is clearly impossible. The above also implies that there exists no Archimedean divisible t-norm on C.

For divisible t-norms on an arbitrary complete chain, we can rephrase the proof of the ordinal sum decomposition (as on $[0,1]$ or on a finite chain), leading to the following proposition.

Proposition 6
Consider a divisible t-norm T on a complete chain C with smallest element 0 and greatest element 1. Let $a \in C \setminus \{0,1\}$ be an idempotent element of T, i.e. $T(a,a) = a$. Then the restrictions $T|_{(\downarrow a)^2}$ and $T|_{(\uparrow a)^2}$ are divisible t-norms on $\downarrow a$ and $\uparrow a$ respectively, and $T(x,y) = \min(x,y)$ whenever $x \leq a \leq y$.

References

[1] B. De Baets, *Database entry A030453*, Sloane's On-Line Encyclopedia of Integer Sequences, URL:

http://www.research.att.com/~njas/sequences/index.html

[2] B. De Baets, R. Mesiar, *Triangular norms on product lattices*, Fuzzy Sets and Systems, Special Issue "Triangular norms" (B. De Baets, R. Mesiar, eds) **104**(1999), 61–75.

[3] G. De Cooman, E. Kerre, *Order norms on bounded partially ordered sets*, J. Fuzzy Math. **2**(1994), 281–310.

[4] J. Fodor, *Contrapositive symmetry of fuzzy implications*, Fuzzy Sets and Systems **69**(1995), 141–156.

[5] L. Godo, C. Sierra, *A new approach to connective generation in the framework of expert systems using fuzzy logic*, Proc. Eighteenth Internat. Symposium on Multiple-Valued Logic (Palma de Mallorca, Spain), IEEE Computer Society Press, 1988, pp. 157–162.

[6] U. Höhle, *Commutative, residuated l-monoids*, Chapter IV in: U. Höhle, E. P. Klement, eds, Non-Classical Logics And Their Applications To Fuzzy Subsets—A Handbook of the Mathematical Foundations of Fuzzy Set Theory, Theory and Decision Library—Series B: Mathematical and Statistical Methods, Volume **32**, Kluwer Academic Publishers (Boston/Dordrecht/London), 1995, pp. 53–106.

[7] S. Jenei, *Geometry of left-continuous triangular norms with strong induced negations*, JORBEL **38**(1998), 5–16.

[8] S. Jenei, *The structure of Girard monoids on* $[0,1]$, in E. P. Klement, S. E. Rodabaugh, eds, Abstracts of the Twentieth International Seminar on Fuzzy Set Theory: Topological and Algebraic Structures (Linz, Austria), 1999, pp. 21–23.

[9] E. P. Klement, R. Mesiar, E. Pap, Triangular Norms, Trends in Logic, Studia Logica Library, Volume **8**, Kluwer Academic Publishers (Boston/Dordrecht/London), 2000.

[10] C. Ling, *Representation of associative functions*, Publ. Math. Debrecen **12**(1965), 189–212.

[11] G. Mayor, J. Torrens, *On a class of operators for expert systems*, Internat. J. Intell. Sys. **8**(1993), 771–778.

[12] B. Schweizer, A. Sklar, Probabilistic Metric Spaces, North-Holland Elsevier (Amsterdam/New York), 1983.

CHAPTER 15

Powerset Operators Based Approach To Fuzzy Topologies On Fuzzy Sets

C. Guido [1]

Introduction

Topological structures have been extensively studied in the context of fuzzy set theories, and many and well organized approaches to this matter have been developed (see [7]).

This chapter merges and extends two notions of fuzzy topological spaces, each of which extends the Chang-Goguen approach to fuzzy topology: the first notion considers a fuzzy topological structure as a suitable fuzzy set on some set (see [10,11,15]); the second notion introduces fuzzy topologies on (not necessarily crisp) fuzzy sets (see [2,3,5,9]). Indeed we can speak here of fuzzy topological spaces on fuzzy sets since both the carrier and the topology of the space are fuzzy sets.

As a ground category we consider the category of fuzzy sets on a fixed base, already introduced in [3,4,5]. The category of fuzzy topological spaces described here is topological over its ground. In such a category the fundamental concepts of subspaces and product spaces are also introduced.

But the main character of the present chapter lies in considering and using fuzzy powerset operators in order to give a description of the presented matter. Such operators were introduced and studied, in somehow different context, in [12,13] as well as in [3,5]. Here fuzzy powerset operators allow us to simplify notation and to enlighten quite well the role of distributivity conditions in the used lattices.

[1]Thanks are due to Prof. S. E. Rodabaugh, whose help allowed an improvement of this chapter, and to Italian CNR (GNSAGA) and the Department of Mathematics (Pure & Applied) of Rhodes University (Grahamstown, South Africa) which supported this work.

S.E. Rodabaugh and E.P. Klement (eds.), Topological and Algebraic Structures in Fuzzy Sets, 401–413.
© 2003 *Kluwer Academic Publishers.*

1 Preliminaries

We shall follow the most used and well known notation in fuzzy set theory and fuzzy topology and in particular we shall refer to [3,5,12,13] for notation and terminology concerning fuzzy powersets and fuzzy powerset operators, even with some slight modification.

Here we only recall some fundamental or non-standard concepts. L-**fuzzy sets** on X are, of course, elements of the complete lattice L^X: in fact L will be in any case a complete lattice with further properties when specified.

We shall identify subsets of X, $T \subseteq X$, with their characteristic function and call them **crisp** L-fuzzy sets on X too. Hence L-fuzzy sets on X will be also called L-fuzzy subsets of X. The **support** of $Y \in L^X$ is the set $Y_o = \{x \in X : Y(x) > 0\}$.

We shall sometimes omit writing L when dealing with L-fuzzy sets: in any case a fuzzy set Y is a function; so it is meant that its domain (some set) and its range (some complete lattice) have to be known.

We shall call L^X the L-**powerset** of X and, if $Y \in L^X$, the complete lattice $[O, Y] = \{A \in L^X | A \leq Y\}$ will be the **powerset** of Y.

For any given function $f : T \to X$ and any fixed complete lattice l, we shall denote by $f_L^{\to} : L^T \to L^X$ and $f_L^{\leftarrow} : L^X \to L^T$, or more simply by $f^{\to}$ and $f^{\leftarrow}$, the Zadeh powerset operators of f (see [3,13]). The latter symbols will also denote the traditional powerset operators $f^{\to} : 2^T \to 2^X$ and $f^{\leftarrow} : 2^X \to 2^T$.

A **frame** is a complete lattice L that satisfies the infinite distributive law

$$a \wedge (\bigvee_{j \in J} b_j) = \bigvee_{j \in J} (a \wedge b_j) \quad \text{if } a \in L \quad \text{and} \quad b_j \in L,\ \forall j \in J$$

We shall denote by **CLAT** the category of complete lattices and maps preserving arbitrary infima and suprema, by **DCLAT** its full subcategory of completely distributive complete lattices, by **FRM** the category of frames and frame maps e.g. maps preserving finite infima and arbitrary suprema. $\bigvee$-**CSLAT** ($\bigwedge$-**CSLAT** respectively) is the category of complete lattices and maps preserving arbitrary suprema (infima respectively).

2 Ground category

Let $Z \in L^T$ and $Y \in L^X$ be fuzzy sets.

An L-**fuzzy function**, or more simply a fuzzy function, from Z to Y,

$$f : Z \to Y$$

is any function $f : T \to X$ such that $Y \circ f \geq Z$, i.e. $f^{\leftarrow}(Y) \geq Z$.

Clearly a fuzzy function $f : Z \to Y$ satisfies the condition $f^{\to}(Z_o) \subseteq Y_o$, so it determines uniquely, by restriction, a function $f' : Z_o \to Y_o$. Conversely, any extension $g'' : T \to X$ of any function $g : Z_o \to Y_o$ that is an L-fuzzy function from $Z_{|Z_o}$ to $Y_{|Y_o}$ is a fuzzy function from Z to Y. Moreover we note that all sets can be considered as (crisp of course) L-fuzzy sets and every function between sets can be considered as an L-fuzzy function between (crisp of course) L-fuzzy sets, for any complete lattice L.

The identity fuzzy function $i_Y : Y \to Y$, $Y \in L^X$, is the identity function from X onto itself.

If $R \in L^S, Z \in L^T, Y \in L^X$ and $g : R \to Z$, $f : Z \to Y$ are fuzzy functions, then the composition $f \circ g : S \to X$ is clearly a fuzzy function from R to Y.

So it is easily seen that we have a category, first considered in [4] and also recently used in [3,5,9], that we shall denote

$$\mathcal{A}L\text{-}\mathbf{SET}$$

whose objects are L-fuzzy sets and whose morphisms are L-fuzzy functions with the above composition and identities.

Following [3,5] we also describe the powerset operators in $\mathcal{A}L$-**SET**.

Given any L-fuzzy function $f : Z \to Y$, where $Z \in L^T$ and $Y \in L^X$, the **forward powerset operator** of f is the map

$$f_{\mathcal{A}}^{\to} : [O, Z] \to [O, Y], \ A \mapsto f_{\mathcal{A}}^{\to}(A)$$

defined by

$$f_{\mathcal{A}}^{\to}(A)(x) = \bigvee\{A(t) \mid f(t) = x, t \in T\}, \ \forall x \in X$$

$f_{\mathcal{A}}^{\to}$ is a $\bigvee$-complete semilattice morphism; hence it has a right adjoint (see [3,13])

$$f_{\mathcal{A}}^{\leftarrow} : [O, Y] \to [O, Z], \ B \mapsto f_{\mathcal{A}}^{\leftarrow}(B)$$

defined by

$$f_{\mathcal{A}}^{\leftarrow}(B) = \bigvee\{A \in [O, Z] : f_{\mathcal{A}}^{\to}(A) \leq B\}$$

which is the **backward powerset operator** of f.

With the above notation the following statements are evident (see also [3,9]).

Proposition 2.1 *For any fuzzy function $f : Z \to Y$:*

1. $f_{\mathcal{A}}^{\to}(A) = f^{\to}(A)$, $\forall A \in [O, Z]$.
2. $f_{\mathcal{A}}^{\leftarrow}(B) = (B \circ f) \wedge Z = f^{\leftarrow}(B) \wedge Z$, $\forall B \in [O, Y]$.
3. $f_{\mathcal{A}}^{\leftarrow}$ *is a $\bigwedge$-complete semilattice morphism.*
4. *The adjunction inequalities hold:*

$$f_{\mathcal{A}}^{\to}(f_{\mathcal{A}}^{\leftarrow}(B)) \leq B, \ \ \forall B \in [O, Y], \quad f_{\mathcal{A}}^{\leftarrow}(f_{\mathcal{A}}^{\to}(A)) \geq A, \ \ \forall A \in [O, Z]$$

5. $f_{\mathcal{A}}^{\leftarrow}$ *is a complete lattice morphism if L is a frame.*

Furthermore, the following can be easily proved.

Proposition 2.2 *The correspondences*

$$Y \in L^X \mapsto \rightarrow_{\mathcal{A}L}(Y) = [O, Y] \quad \text{and} \quad f : Z \to Y \mapsto \rightarrow_{\mathcal{A}L}(f) = f_{\mathcal{A}}^{\rightarrow}$$

give a covariant functor

$$\rightarrow_{\mathcal{A}L}: \mathcal{A}L\text{-}\mathbf{SET} \longrightarrow \bigvee\text{-}\mathbf{CSLAT}$$

The correspondences

$$Y \in L^X \mapsto \leftarrow_{\mathcal{A}L}(Y) = [O, Y] \quad \text{and} \quad f : Z \to Y \mapsto \leftarrow_{\mathcal{A}L}(f) = f_{\mathcal{A}}^{\leftarrow}$$

give a contravariant functor

$$\leftarrow_{\mathcal{A}L}: \mathcal{A}L\text{-}\mathbf{SET} \longrightarrow \bigwedge\text{-}\mathbf{CSLAT}$$

We remark that if we restrict this last functor to **SET**, we obtain in particular a contravariant functor

$$\leftarrow_L: \mathbf{SET} \longrightarrow \mathbf{CLAT}$$

Also, $\leftarrow_{\mathcal{A}L}: \mathcal{A}L$-**SET** $\longrightarrow$ **CLAT** is a contravariant functor whenever L is a frame.

We use different notation for our powerset operators and Zadeh powerset operators for an easier reading of the text. Indeed, if sets are considered as (crisp) fuzzy sets and functions between sets are considered as fuzzy functions between (crisp) fuzzy sets, then the Zadeh and the traditional powerset operators are special cases of our powerset operators.

We note that the name $\mathcal{A}L$-**SET** used in this chapter merges notation of [3,4,14]. In particular, the symbol $\mathcal{A}$ is justified, as in [3,5], for distinguishing the category used here from other, possible grounds that are or may be considered with the same purpose of building categories of fuzzy topological spaces.

Definition 2.1 *Let $(Y^j)_{j\in J}$ be any family of L-fuzzy sets, $Y^j \in L^{X^j}$, $\forall j \in J$. The **product** L-fuzzy set*

$$\prod_{j\in J} Y^j \in L^{\prod_{j\in J} X^j}$$

is defined by

$$x \in \prod_{j\in J} X^j \mapsto \prod_{j\in J} Y^j(x) = \bigwedge_{j\in J} Y^j(x(j))$$

The projections on the factor L-fuzzy sets Y^k are the functions

$$p_k : \prod_{j\in J} X^j \to X^k, \quad x \mapsto p_k(x) = x(k)$$

It is easily seen that the projections p_k are morphisms in $\mathcal{A}L$-**SET**. In fact $x \in \prod_{j\in J} X^j \Rightarrow Y^k(p_k(x)) = Y^k(x(k)) \geq (\prod_{j\in J} Y^j)(x)$.

Moreover $\prod_{j\in J} Y^j$ is a product in the category $\mathcal{A}L$-**SET**.

In fact let $Z \in L^T$ and $f_j : Z \to Y^j$ be fuzzy functions, i.e. $Y^j(f^j(t)) \geq Z(t),\ \forall t \in T,\ \forall j \in J$.

Then the function

$$f : T \to \prod_{j\in J} X^j, \quad t \mapsto f(t)$$

defined by

$$f(t)(j) = f_j(t)$$

is a fuzzy function, as the following holds:

$$(\prod Y^j)(f(t)) = \wedge_{j\in J} Y^j(f(t)(j)) = \wedge_{j\in J} Y^j(f_j(t)) \geq Z(t),\ \forall t \in T$$

Moreover, $p_k(f(t)) = f(t)(k) = f_k(t),\ \forall k \in J$, and any fuzzy function $f' : T \to \prod_{j\in J} X^j$ that satisfies the similar condition $f_k = p_k \circ f',\ \forall k \in J$, must coincide with f, since

$$f(t)(k) = p_k(f(t)) = f_k(t) = p_k(f'(t)) = f'(t)(k),\ \forall t \in T,\ \forall k \in J$$

The following proposition states some useful properties of the powerset operators of suitable maps between complete lattices.

Proposition 2.3 *Let T, S, M be complete lattices, $\tau \in M^T$, $\sigma \in M^S$ and let $F : T \to S$ be a map.*
Then the following statements hold

1) *If F preserves the upper bounds, then $F^{\to}(\tau)(1) \geq \tau(1)$ and $F^{\leftarrow}(\sigma)(1) = \sigma(1)$.*

1′) *If F preserves the lower bounds, then $F^{\to}(\tau)(0) \geq \tau(0)$ and $F^{\leftarrow}(\sigma)(0) = \sigma(0)$.*

2) *If M is a frame, F preserves finite meets and $\tau(a \wedge a') \geq \tau(a) \wedge \tau(a')$, $\forall a, a' \in T$, then $F^{\to}(\tau)(b \wedge b') \geq F^{\to}(\tau)(b) \wedge F^{\to}(\tau)(b')$, $\forall b, b' \in S$.*

2′) *If F preserves finite meets and $\sigma(b \wedge b') \geq \sigma(b) \wedge \sigma(b')$, $\forall b, b' \in S$, then $F^{\leftarrow}(\sigma)(a \wedge a') \geq F^{\leftarrow}(\sigma)(a) \wedge F^{\leftarrow}(\sigma)(a')$, $\forall a, a' \in T$.*

3) *If M is completely distributive and F preserves arbitrary joins and*

$$\tau(\bigvee_{j\in J} a_j) \geq \bigwedge_{j\in J} \tau(a_j)$$

whenever $a_j \in T$, $\forall j \in J$, *then*

$$F^{\rightarrow}(\tau)(\bigvee_{k\in K} b_k) \geq \bigwedge_{k\in K} F^{\rightarrow}(\tau)(b_k)$$

whenever $b_k \in S$, $\forall k \in K$.

3′) *If* F *preserves arbitrary joins and* $\sigma(\bigvee_{j\in J} b_j) \geq \bigwedge_{j\in J} \sigma(b_j)$ *whenever* $b_j \in S$, $\forall j \in J$, *then* $F^{\leftarrow}(\sigma)(\bigvee_{k\in K} a_k) \geq \bigwedge_{k\in K} F^{\leftarrow}(\sigma)(a_k)$ *whenever* $a_k \in T$, $\forall k \in K$.

Proof. 1), 1′), 2′), 3′) are easily verified; we only recall that $F^{\rightarrow}(\tau)(b) = \bigvee\{\tau(a) : F(a) = b\}$ and $F^{\leftarrow}(\sigma)(a) = \sigma(F(a))$, $\forall a \in T$, $\forall b \in S$. We prove the remaining statements.

2). If $b, b' \in S$, $a, a' \in T$ and $F(a) = b$, $F(a') = b'$, then $F(a \wedge a') = b \wedge b'$. Hence

$$F^{\rightarrow}(\tau)(b \wedge b') \geq \tau(a \wedge a') \geq \tau(a) \wedge \tau(a')$$

whenever $a, a' \in T$ and $F(a) = b$, $F(a') = b'$. Since M is a frame we have:

$$\begin{aligned} F^{\rightarrow}(\tau)(b) \wedge F^{\rightarrow}(\tau)(b') &= \bigvee\{\tau(a) : F(a) = b\} \wedge \bigvee\{\tau(a') : F(a') = b'\} \\ &= \bigvee\{\tau(a) \wedge \tau(a') : F(a) = b,\ F(a') = b'\} \\ &\leq F^{\rightarrow}(\tau)(b \wedge b') \end{aligned}$$

3). Let $b_k \in S$, $\forall k \in K$. Denote $\mathcal{A}_k = \{a \in T : F(a) = b_k\}$, $\forall k \in K$, $\mathcal{A} = \prod_{k\in K} \mathcal{A}_k$, and $s(k) = a_{s_k} \in \mathcal{A}_k$, $\forall s \in \mathcal{A}$, $\forall k \in K$. Then $F(\bigvee_{k\in K} a_{s_k}) = \bigvee_{k\in K} b_k$, $= \forall s \in \mathcal{A}$. Hence

$$F^{\rightarrow}(\tau)(\bigvee_{k\in K} b_k) \geq \tau(\bigvee_{k\in K} a_{s_k}) \geq \bigwedge_{k\in K} \tau(a_{s_k}),\ \forall s \in \mathcal{A}$$

By using complete distributivity of M we have

$$\begin{aligned} \bigwedge_{k\in K} F^{\rightarrow}(\tau)(b_k) &= \bigwedge_{k\in K}(\bigvee_{a\in\mathcal{A}_k} \tau(a)) \\ &= \bigvee_{s\in\mathcal{A}}(\bigwedge_{k\in K} \tau(a_{s_k})) \\ &\leq F^{\rightarrow}(\tau)(\bigvee_{k\in K} b_k) \quad \square \end{aligned}$$

We remark that all the hypotheses required for F in the statements 1)...3′) above are satisfied iff F is a frame map. As for M, complete distributivity is the strongest required condition.

3 Fuzzy topology

All along this section L will be a fixed frame and M a fixed completely distributive complete lattice.

Definition 3.1 *An M-**fuzzy topology** τ, on an L-fuzzy set Y, is an M-fuzzy set on $[O, Y]$, namely a function*

$$\tau : [O, Y] \to M$$

that satisfies the following conditions:

(i) $\tau(O) = \tau(Y) = 1$;

(ii) $A, A' \in [O, Y] \Rightarrow \tau(A \wedge A') \geq \tau(A) \wedge \tau(A')$;

(iii) $A^j \in [O, Y],\ \forall j \in J \Rightarrow \tau(\bigvee_{j \in J} A^j) \geq \bigwedge_{j \in J} A^j$.

The above definition evidently extends previous definitions given by Höhle (see [6], where $L = 2$ and Y is crisp), Šostak and Ying (see [15,16], where $L = M = I$ and Y is crisp), Kubiak (see [10], where Y is crisp), De Mitri and Guido (see [2], where $L = I,\ M = 2$).

If τ is an M-fuzzy topology on the L-fuzzy set Y, the pair

$$(Y, \tau)$$

is an **M-fuzzy topological L-fuzzy space.**

More simply we can call (Y, τ) fuzzy topological space, without specifying either L or M as they should be known to be the range lattices for the fuzzy sets Y and τ, respectively.

If $M = 2$ or τ is crisp we call (Y, τ) a topological L-space and we identify τ with its support.

The L-fuzzy set Y is the carrier and, if $Y \in L^X$, the set X is the support of the M-fuzzy topological L-fuzzy space.

We remark that for every $\beta \in M$, the subset $\tau_\beta = \tau^{\leftarrow}([\beta, 1])$ gives $[O, Y]$ a structure of topological L-space (Y, τ_β).

Definition 3.2 *Given two M-fuzzy topological L-fuzzy spaces (Z, τ), (Y, σ) and an L-fuzzy function from Z to Y, namely for $Z \in L^T$ and $Y \in L^X$ a function $f : T \to X$ s.t. $f^{\leftarrow}(Y) \geq Z$, we say that f is M-**fuzzy continuous** if its backward powerset operator is an L-fuzzy function from σ to τ*

$$f^{\leftarrow}_{\overline{A}} : \sigma \to \tau$$

which means that the condition $(f^{\leftarrow}_{\overline{A}})^{\leftarrow}(\tau) \geq \sigma$ is satisfied. We shall write

$$f : (Z, \tau) \to (Y, \sigma)$$

to denote such M-fuzzy continuous L-fuzzy function.

We shall denote by $\mathcal{A}(L,M)$-**TOP** the category whose objects are the M-fuzzy topological L-fuzzy spaces and whose morphisms are the M-fuzzy continuous L-fuzzy functions, with the same composition law as in $\mathcal{A}L$-**SET** and the same identity morphisms.

In fact the following can be easily verified.

Proposition 3.1 *Let (Y,τ), (Y',τ'), (Y'',τ'') be M-fuzzy topological L-fuzzy spaces, $f : Y \to Y'$, $g : Y' \to Y''$ M-fuzzy continuous L-fuzzy functions and $i_{Y'} : Y' \to Y'$ the identity fuzzy function. Then:*

1. *$i_{Y'}$ is M-fuzzy continuous and, of course, $i_{Y'} \circ f = f$, $g \circ i_{Y'} = g$.*
2. *$g \circ f$ is M-fuzzy continuous.*

Remark 3.1 *If we define, for every object (Y,τ) of $\mathcal{A}(L,M)$-**TOP**, $U(Y,\tau) = Y$ and, for every morphism $f : (Z,\tau) \to (Y,\sigma)$, $U(f) = f$, then we have a faithful functor, hence*

*($\mathcal{A}(L,M)$-**TOP**, **U**) is a concrete category over $\mathcal{A}L$-**SET**.*

Clearly such a category is fibre-small.

Definition 3.3 *Let τ, τ' be M-fuzzy topologies on the L-fuzzy set Y. We say that τ is coarser than τ' or τ' is finer than τ if $\tau \le \tau'$, i.e. $\tau(A) \le \tau'(A)$, $\forall A \in [O,Y]$.*

Of course $\tau \le \tau'$ iff the identity $i_Y : Y \to Y$ is a morphism, $i_Y : (Y,\tau') \to (Y,\tau)$, in $\mathcal{A}(L,M)$-**TOP**.

The order relation $\le$ gives the set $\Omega_M(Y)$ of all M-fuzzy topologies on a fixed L-fuzzy set Y the structure of a complete lattice. More precisely $(\Omega_M(Y),\le)$ is a complete $\bigwedge$-subsemilattice of the complete lattice $(M^{[O,Y]},\le)$ even if it is not a complete sublattice. In fact the following holds.

Proposition 3.2 *Let $(\tau^j)_{j\in J}$ be any family of M-fuzzy topologies on $Y \in L^X$. Then*

$$\tau = \bigwedge_{j\in J} \tau^j : [O,Y] \to M, \qquad A \mapsto \bigwedge_{j\in J} \tau^j(A)$$

is the infimum of the given family in the ordering $\le$.

Proof. Clearly $\tau(O) = \tau(Y) = 1$. $A, A' \in [O,Y] \Rightarrow$

$$\begin{aligned}
\tau(A \wedge A') &= \bigwedge_{j\in J} \tau^j(A \wedge A') \ge \bigwedge_{j\in J} (\tau^j(A) \wedge \tau^j(A')) \\
&\ge \bigwedge_{i,j\in J} (\tau^i(A) \wedge \tau^j(A')) = \bigwedge_{i\in J} \tau^i(A) \wedge \bigwedge_{j\in J} \tau^j(A') \\
&= \tau(A) \wedge \tau(A')
\end{aligned}$$

$A^k \in [O,Y], \forall k \in K \Rightarrow$

$$\begin{aligned}\tau(\bigvee_{k\in K} A^k) &= \bigwedge_{j\in J} \tau^j(\bigvee_{k\in K} A^k) \geq \bigwedge_{j\in J}\bigwedge_{k\in K} \tau^j(A^k)\\ &= \bigwedge_{k\in K}\bigwedge_{j\in J} \tau^j(A^k) = \bigwedge_{k\in K} \tau(A^k)\end{aligned}$$

Hence τ is an M-fuzzy topology. Moreover τ is clearly coarser than each τ^j and, if τ' is a coarser M-fuzzy topology on Y than each τ^j, then for every $A \in [O,Y]$ we have

$$\tau'(A) \leq \tau^j(A), \quad \forall j \in J \Rightarrow \tau'(A) \leq \bigwedge_{j\in J} \tau^j(A) = \tau(A) \quad \square$$

Now let $f : Z \to Y$ be any morphism in $\mathcal{A}L$-**SET** and let M be, as usual, a completely distributive complete lattice. Let τ, σ be M-fuzzy topologies on Z and Y respectively. Then if we denote

$$(f_{\mathcal{A}}^{\leftarrow})^{\rightarrow}(\sigma) = \tau' \in M^{[O,Z]} \quad \text{and} \quad (f_{\mathcal{A}}^{\leftarrow})^{\leftarrow}(\tau) = \sigma' \in M^{[O,Y]}$$

we clearly have the equalities

$$\tau'(A) = \bigvee\{\sigma(B) : B \in [O,Y],\ (B \circ f) \wedge Z = A\}, \quad \forall A \in [O,Z]$$

$$\sigma'(B) = \tau((B \circ f) \wedge Z), \quad \forall B \in [O,Y]$$

Moreover we prove the following.

Proposition 3.3 *With the above notation we have:*

1. *τ' is the coarsest M-fuzzy topology on Z making $f : (Z,\tau') \to (Y,\sigma)$ M-fuzzy continuous (namely τ' is the initial M-fuzzy topology of σ by f).*

2. *σ' is the finest M-fuzzy topology on Y making $f : (Z,\tau) \to (Y,\sigma')$ M-fuzzy continuous (namely σ' is the final M-fuzzy topology of τ by f).*

Proof. Since L is a frame, $f_{\mathcal{A}}^{\leftarrow}$ is a complete lattice morphism (see Proposition 2.1), hence it follows from Proposition 2.3 that τ' is an M-fuzzy topology on Z, as M is completely distributive, and σ' is an M-fuzzy topology on Y.

By using the adjunction inequalities of the powerset operators related to the function $f_{\mathcal{A}}^{\leftarrow}$ we have

$$(f_{\mathcal{A}}^{\leftarrow})^{\leftarrow}(\tau') = (f_{\mathcal{A}}^{\leftarrow})^{\leftarrow}((f_{\mathcal{A}}^{\leftarrow})^{\rightarrow}(\sigma)) \geq \sigma$$

and, more simply,

$$(f_{\mathcal{A}}^{\leftarrow})^{\leftarrow}(\tau) = \sigma'$$

Hence the required continuity conditions for f are satisfied.

Now let

$$f : (Z, \tau'') \to (Y, \sigma) \quad \text{and} \quad f : (Z, \tau) \to (Y, \sigma'')$$

be M-fuzzy continuous. Then by also using monotonicity of the forward powerset operators, we have

$$\tau' = (f_{\mathcal{A}}^{\leftarrow})^{\rightarrow}(\sigma) \leq (f_{\mathcal{A}}^{\leftarrow})^{\rightarrow}((f_{\mathcal{A}}^{\leftarrow})^{\leftarrow}(\tau'')) \leq \tau''$$

and

$$\sigma'' \leq (f_{\mathcal{A}}^{\leftarrow})^{\leftarrow}(\tau) = \sigma'. \quad \square$$

Corollary 3.4 *Every* **U***-structured* 1*-source has a unique* **U***-initial lift in the concrete category* ($\mathcal{A}(L, M)$*-**TOP**,* **U***).*

Proof. Let Z be an L-fuzzy set on T, (Y, σ) an M-fuzzy topological L-fuzzy space, $Y \in L^X$, and let $F : Z \to Y$ be any morphism in $\mathcal{A}L$-**SET**. Let $\tau = (f_{\mathcal{A}}^{\leftarrow})^{\rightarrow}(\sigma)$ be the initial M-fuzzy topology of τ by f in Z. Now let $g : S \to Z$ be an L-fuzzy function, $S \in L^R$, τ' an M-fuzzy topology on S and assume $f \circ g : (S, \tau') \to (Y, \sigma)$ be M-fuzzy continuous. Then, by using the adjoint inequalities and functoriality of the powerset operators we have

$$\begin{aligned}(f_{\mathcal{A}}^{\leftarrow})^{\leftarrow}((g_{\mathcal{A}}^{\leftarrow})^{\leftarrow}(\tau')) &= ((f \circ g)_{\mathcal{A}}^{\leftarrow})^{\leftarrow}(\tau') \geq \sigma \\ &\Rightarrow \tau = (f_{\mathcal{A}}^{\leftarrow})^{\rightarrow}(\sigma) \leq (g_{\mathcal{A}}^{\leftarrow})^{\leftarrow}(\tau') \\ &\Rightarrow (g_{\mathcal{A}}^{\leftarrow})^{\rightarrow}(\tau) \leq \tau'\end{aligned}$$

Hence $g : (S, \tau') \to (Z, \tau)$ is M-fuzzy continuous, since $(g_{\mathcal{A}}^{\leftarrow})^{\rightarrow}(\tau)$ is the initial M-fuzzy topology of τ by g on S, and so $f : (Z, \tau) \to (Y, \sigma)$ is a **U**-initial lift of the source $f : Z \to (Y, \sigma)$.

Given any other **U**-initial lift $f : (Z, \bar{\tau}) \to (Y, \sigma)$, by using initiality of both lifts with respect to the identity function $i : T \to T$ we have $\tau \leq \bar{\tau}$ and $\bar{\tau} \leq \tau$, thence the uniqueness of the **U**-initial lift. $\square$

Proposition 3.5 *Let $g : Z \to T$, $f : T \to Y$ be L-fuzzy functions; let τ be any M-fuzzy topology on Z, τ' its final M-fuzzy topology on T by g and τ'' the final M-fuzzy topology of τ' by f. Moreover, let σ be any M-fuzzy topology on Y, σ' its initial M-fuzzy topology by f and σ'' the initial M-fuzzy topology of σ' by g. Then τ''(σ'', respectively) is the final (initial, respectively) M-fuzzy topology of τ (σ, respectively) by $f \circ g$.*

Proof. Straightforward. $\square$

Definition 3.4 *Let Z, $Y \in L^X$, $Z \leq Y$ and let $i : Z \to Y$ be the L-fuzzy function related to the identity function of X. Moreover let σ be any M-fuzzy* *topology on Y. Then the* ***induced M-fuzzy topology*** *on Z by σ is defined to be the initial M-fuzzy topology*

$$\sigma_{|Z} = (i_{\mathcal{A}}^{\leftarrow})^{\rightarrow}(\sigma)$$

*of σ by i. $(Z, \sigma_{|Z})$ is the L-**subspace** of (Y, σ) on the carrier Z.*

It is clear that

$$\sigma_{|Z}(A) = \bigvee\{\sigma(B) : B \in [O, Y],\ B \wedge Z = A\},\ \forall A \in [O, Z]$$

As a special case of Corollary 3.4 one can see that the above defined subspace produces an initial subobject $((S, \sigma_{|Z}), i)$. More generally, it follows from Corollary 3.4 that $\mathcal{A}(L, M)$-**TOP** has initial subobjects.

Since M-fuzzy topologies on a given L-fuzzy set form a complete lattice, initial (final respectively) M-fuzzy topologies by families of L-fuzzy functions from (to respectively) a given L-fuzzy set can be considered. In particular, the product of any family of M-fuzzy topological L-fuzzy spaces can be constructed.

In fact let (Y^j, σ^j) be an M-fuzzy topological L-space, $Y^j \in L^{X^j}$, $\forall j \in J$. For every $j \in J$ denote by τ^j the initial M-fuzzy topology of σ^j by p^j, on the product $Y = \prod_{j \in J} Y^j$ defined in Section 2 and consider the supremum

$$\tau = \bigvee_{j \in J} \tau^j$$

of the family $(\tau^j)_{j \in J}$ in the lattice of all M-fuzzy topologies on Y (note such a supremum is not necessarily the supremum in the lattice $(M^{[O,Y]}, \leq)$).

Then the following result holds.

Proposition 3.6 *With the above notation, $\mathcal{P} = (p_j : (Y, \tau) \rightarrow (Y^j, \sigma^j))_{j \in J}$ is a concrete product in $\mathcal{A}(L, M)$-**TOP**.*

Proof. The inequalities

$$\begin{aligned} ((p_j)_{\mathcal{A}}^{\leftarrow})^{\leftarrow}(\tau)(B) &= \tau(B \circ p_j) \geq \tau^j(B \circ p_j) \\ &= ((p_j)_{\mathcal{A}}^{\leftarrow})^{\leftarrow}(\tau^j)(B) \geq \sigma^j(B), \\ \forall j &\in J,\ \forall B \in [O, Y^j] \end{aligned}$$

show that all

$$p_j : (Y, \tau) \rightarrow (Y^j, \sigma^j)$$

are M-fuzzy continuous; hence $\mathcal{P}$ is a source in $\mathcal{A}(L, M)$-**TOP**.

Moreover $\mathcal{P}$ is initial. In fact let (A, α) be an M-fuzzy topological L-fuzzy space and $h : A \rightarrow Y$ any L-fuzzy function such that $p^j \circ h : (A, \alpha) \rightarrow (Y^j, \sigma^j)$

is M-fuzzy continuous, $\forall j \in J$. Then, by the adjunction inequalities, it follows from $(h^{\leftarrow}_{\mathcal{A}} \circ (p^j)^{\leftarrow}_{\mathcal{A}})^{\leftarrow}(\alpha) \geq \sigma^j$ that

$$(h^{\leftarrow}_{\mathcal{A}})^{\leftarrow}(\alpha) \geq ((p^j)^{\leftarrow}_{\mathcal{A}})^{\rightarrow}(\sigma^j) = \tau^j, \forall j \in J$$

Hence the continuity condition of h, $(h^{\leftarrow}_{\mathcal{A}})^{\leftarrow}(\alpha) \geq \tau$, follows.

Now the assertion is a consequence of Proposition 10.53 of [1], since $\mathbf{U}(\mathcal{P}) = (p^j : Y \to Y^j)_{j \in J}$ is a product in $\mathcal{A}L$-**SET** (see Section 2). □

Definition 3.5 *The M-fuzzy topological L-fuzzy space*

$$(\prod_{j \in J} Y^j, \tau)$$

*is the **product** of the M-fuzzy topological L-spaces (Y^j, τ^j), $j \in J$.*

Theorem 3.7 *The concrete category $(\mathcal{A}(L, M)$-**TOP**, **U**$)$ is topological over $\mathcal{A}L$-**SET***

Proof. We choose to verify the conditions (1), (2), (3) of Proposition 21.37 of [1]. Condition (1) follows from Proposition 3.6 and Condition (2) follows from Corollary 3.4.

As for condition (3), it is clear that for any given L-fuzzy set $Y \in L^X$, the fuzzy set $\nu : [O, Y] \to M$ such that

$$\nu(O) = \nu(Y) = 1 \text{ and } \nu(A) = 0, \forall O \neq A \neq Y$$

make (Y, σ) an indiscrete object in $\mathcal{A}(L, M)$-**TOP**. In fact for every M-fuzzy topological L-fuzzy space (Z, τ) and for every L-fuzzy function $f : Z \to Y$, it follows from

$$f^{\leftarrow}_{\mathcal{A}}(Y) = f^{\leftarrow}(Y) \wedge Z = Z \text{ and } f^{\leftarrow}_{\mathcal{A}}(O) = O$$

that $(f^{\leftarrow}_{\mathcal{A}})^{\leftarrow}(\tau) \geq \nu$. □

References

[1] J. Adámek, H. Herrlich, G. E. Strecker, Abstract And Concrete Categories: The Joy Of Cats, Wiley Interscience Pure and Applied Mathematics, John Wiley & Sons (Brisbane/Chicester/New York/Singapore/Toronto), 1990.

[2] C. De Mitri, C. Guido, *G-fuzzy topological spaces and subspaces*, Suppl. Rend. Circolo Matem. Palermo **29**(1992), 363–383.

[3] C. De Mitri, C. Guido, *Some remarks on fuzzy powerset operators*, Fuzzy Sets and Systems **126**(2002), 241–251.

[4] J. A. Goguen, *Categories of V-sets*, Bull. Amer. Math. Soc.**75**(1969), 622–624.

[5] C. Guido, *The subspace problem in the traditional point set context of fuzzy topology*, Quaestiones Mathematicae **20(3)**(1997), 351–372.

[6] U. Höhle, *Upper semicontinuous fuzzy sets and applications*, J. Math. Anal. Appl. **78**(1980), 659–673.

[7] U. Höhle, S. E. Rodabaugh, eds, Mathematics Of Fuzzy Sets: Logic, Topology, And Measure Theory, The Handbooks of Fuzzy Sets Series, Volume **3**, Kluwer Academic Publishers (Boston/Dordrecht/London), 1999.

[8] U. Höhle, A. Šostak, *Axiomatic foundation of fixed-basis fuzzy topology*, in [7], 123–272.

[9] G. Jäger, *Compactness and connectedness as absolute properties in fuzzy topological spaces*, Fuzzy Sets and Systems **94**(1998), 405–410.

[10] T. Kubiak, *On Fuzzy Topologies*, Doctoral Dissertation, A. Mickiewicz University (Poznan, Poland), 1985.

[11] P. P. Matutu, *(L,M)-Fuzzy Topological Spaces*, Master's Dissertation, Rhodes University (Grahamstown, South Africa), 1992.

[12] S. E. Rodabaugh, *Powerset operator based foundation for point-set lattice-theoretic (poslat) fuzzy set theories and topologies*, Quaestiones Mathematicae **20(3)**(1997), 463–530.

[13] S. E. Rodabaugh, *Powerset operator foundations for poslat fuzzy set theories and topologies*, Chapter 2 in [7], 91–116.

[14] S. E. Rodabaugh, *Categorical foundations of variable-basis fuzzy topology*, Chapter 4 in [7], 273–388.

[15] A. P. Šostak, *On a fuzzy topological structure*, Suppl. Rend. Circolo Matem. Palermo **11**(1985), 89–103.

[16] M. S. Ying, *A new approach to fuzzy topology (I)*, Fuzzy Sets and Systems **39**(1991), 303–321.

CHAPTER 16

Lifting Of Sobriety Concepts With Particular Reference To (L, M)-Topological Spaces

W. KOTZÉ [1]

Introduction

The concept of sobriety of ordinary topological spaces has been around since at least the 1970's. Sober topological spaces were introduced by Grothendieck et al ([1] IV 4.2.1) and independently by T. Blanksma [3]. Since then various authors studied this axiom, e.g. R-E. Hoffmann ([5] and [6]) and P. T. Johnstone [10]. It was pointed out by [21] and [25] that this axiom is also of importance to computer science, domain theory in particular. Smyth argues that "computationally reasonable spaces are sober". There are several reasons for this, one being the possibility of, in the case of sobriety, moving from frame maps between frames to continuous maps between topological spaces; in other words, continuous maps being categorically (dually) equivalent to frame maps requires sobriety (see e.g. [11], Theorems 3.3 and 3.4). Another is that the Scott topology on continuous posets are highly non-Hausdorff but sober. See e.g. [10]. (The Scott topology on a two point set is the Sierpinski space).

In more recent years, sobriety has been extended to L-topological spaces with L a suitable lattice. Such sobriety has been studied by Rodabaugh [16, 17, 18, 19], the present author [11, 12], and Höhle [8].

Furthermore, a weaker property, semi-sobriety, was established for L-topological spaces in [13]. This separation axiom has however not been extended to the (L, M)-topological spaces of Šostak [22, 23, 15]. We do so in this chapter.

[1]This work was supported by a grant from the NRF (South Africa). I wish to thank A. K. Srivastava for his most valuable input and stimulation.

S.E. Rodabaugh and E.P. Klement (eds.), Topological and Algebraic Structures in Fuzzy Sets, 415–426.
© 2003 *Kluwer Academic Publishers.*

1 Different concepts of topological spaces

1.1 Ordinary (crisp) topological case

If a topology $\mathcal{T}$ on a space X (i.e. a frame in $\mathcal{P}(X)$ with bottom $\emptyset$ and top X) is identified with a map

$$\mathcal{T} : \mathcal{P}(X) \to \{0,1\} \quad \text{where } \mathcal{T}(A) = 1 \text{ iff } A \text{ is open, then}$$

(*i*) $\mathcal{T}(\emptyset) = \mathcal{T}(X) = 1$;

(*ii*) $\mathcal{T}(U \cap V) \geq \mathcal{T}(U) \wedge \mathcal{T}(V)$ for all $U, V \in \mathcal{P}(X)$;

(*iii*) $\mathcal{T}(\cup U_i) \geq \wedge \mathcal{T}(U_i)$ for all $U_i \in \mathcal{P}(X)$.

$(X, \mathcal{T})$ is called a *topological space.*
Viewing a topology as an operator on $\mathcal{P}(X) = 2^X$ goes back to [7].

1.2 L-topological case

(L is a frame with 0 and 1 and order reversing involution.)

If an L-topology $\mathcal{T}$ on X (i.e. a frame in L^X with $\mathbf{0}$ and $\mathbf{1}$ where $\mathbf{0}(x) \equiv 0$ and $\mathbf{1}(x) \equiv 1$) is identified with a map

$$\mathcal{T} : L^X \to \{0,1\} \quad \text{where } \mathcal{T}(a) = 1 \text{ iff } a \text{ is } L\text{-open, then also}$$

(*i*) $\mathcal{T}(\mathbf{0}) = \mathcal{T}(\mathbf{1}) = 1$;

(*ii*) $\mathcal{T}(u \wedge v) \geq \mathcal{T}(u) \wedge \mathcal{T}(v)$ for all $u, v \in L^X$;

(*iii*) $\mathcal{T}(\bigvee u_i) \geq \bigwedge \mathcal{T}(u_i)$ for all $u_i \in L^X$.

$(X, L, \mathcal{T})$ is called an *L-topological space.*

In what follows, we identify the operator $\mathcal{T} : L^X \to \{0,1\}$ with the frame $\mathcal{F} = \{u \in L^X : \mathcal{T}(u) = 1\}$ in L^X, and the corresponding "closed sets" with the complements $u'(x) \equiv u(x)'$ where the second $'$ is the order reversing involution of L.

If we write $\mathcal{T} \subseteq \mathcal{T}^*$ for two L – topologies on X, it is in the sense of the frame inclusion: $\mathcal{F} \subseteq \mathcal{F}^*$.

1.3 (L, M)-topological case

(L as in **1.2** and M a frame with 0 and 1.)

This concept is due to Šostak (see e.g. [22] and [23] in case $L = M = [0,1]$) and Šostak and Kubiak [15] where these spaces are referred to as "(L, M)-fuzzy topologies" (cf. Chapters 3 and 4 of [9]). Mingsheng Ying [26] called the case $L = \{0,1\}$ and $M = [0,1]$ a "fuzzifying topology".

An (L, M)*-topology* is a map $\mathcal{T} : L^X \to M$ such that

(*i*) $\mathcal{T}(\mathbf{0}) = \mathcal{T}(\mathbf{1}) = 1$ (where $\mathbf{0}(x) \equiv 0$ and $\mathbf{1}(x) \equiv 1$);

(*ii*) $\mathcal{T}(u \wedge v) \geq \mathcal{T}(u) \wedge \mathcal{T}(v)$ for all $u, v \in L^X$;

(*iii*) $\mathcal{T}(\bigvee u_i) \geq \bigwedge \mathcal{T}(u_i)$ for all $u_i \in L^X$.

$(X, L, M, \mathcal{T})$ is called an (L, M)*-topological space.*

2 Sobriety

2.1 Sobriety of crisp topological spaces

(See [10] and [16].)

Given an ordinary topological space $(X, \mathcal{T})$, *pt* $\mathcal{T}$ (the *spectrum* of $\mathcal{T}$) denotes the collection of all frame morphisms p from $\mathcal{T}$ into the frame $\{0,1\}$ (the topology on a singleton space). Then X can be mapped into *pt* $\mathcal{T}$ by

$$\Psi : X \to pt\ \mathcal{T} \quad \text{given by} \quad \Psi(x)(u) \equiv \chi_u(x) \quad \text{for all } u \in \mathcal{T} \text{ and } x \in X.$$

Clearly Ψ is injective iff $(X, \mathcal{T})$ is T_0.

If $(X, \mathcal{T})$ is Hausdorff, then Ψ is a bijection (see [10], Chapter 2). We say $(X, \mathcal{T})$ is *sober* iff Ψ is a bijection.

So $(X, \mathcal{T})$ Hausdorff $\Rightarrow$ $(X, \mathcal{T})$ sober $\Rightarrow$ $(X, \mathcal{T})$ T_0.

There is no simple connection between sobriety and the T_1 separation axiom. See example **3.3** below.

From [10] we have:

A topological space $(X, \mathcal{T})$ is sober iff every non-empty irreducibly closed set is the closure of a unique singleton.

2.2 Sobriety of L-topological spaces

(See e.g. [16]–[19], [11])

Given an L-topological space $(X, L, \mathcal{T})$, $Lpt\ \mathcal{T}$ denotes the collection of all frame morphisms p from $\mathcal{T}$ into L. Then X is mapped into $Lpt\ \mathcal{T}$ by

$$\Psi : X \to Lpt\ \mathcal{T} \quad \text{given by} \quad \Psi(x)(u) \equiv u(x) \quad \text{for all } u \in \mathcal{T} \text{ and } x \in X.$$

Ψ is 1-1 iff $(X, L, \mathcal{T})$ is T_0 ($x_1 \neq x_2 \Rightarrow$ there exists a $u \in \mathcal{T}$ such that $u(x_1) \neq u(x_2)$).

We say $(X, L, \mathcal{T})$ is *sober* iff Ψ is a bijection. So $(X, L, \mathcal{T})$ is sober $\Rightarrow$ $(X, L, \mathcal{T})$ is T_0.

3 Semi-sobriety in L-topological case

3.1 Definition

An L-closed set a is irreducibly closed iff $b \vee c \geq a$, for b and c L-closed sets, implies that $b \geq a$ or $c \geq a$. Of course $b \geq a$ in L^X means $b(x) \geq a(x)$ in the lattice L for all $x \in X$.

3.2 Theorem (see [13])

Let 0 in L be prime and $(X, L, \mathcal{T})$ be a sober L-topological space. Then every non-zero irreducibly L-closed set is the β-closure of a singleton for some irreducible non-zero $\beta \in L$ (where the β-closure of singleton $x_1 \in X$ is the smallest L-closed set which has value at least β at x_1).

The necessary condition of the theorem is called *semi-sober.* So $(X, L, \mathcal{T})$ sober implies it is semi-sober. The discrete L–topological space (X, L, L^X) is clearly semi-sober for non-empty X.

In [13] we showed that $(X, L, \mathcal{T})$ (fuzzy) Hausdorff implies $(X, L, \mathcal{T})$ semi-sober (and T_0). It was also pointed out that in the crisp case ($L = \{0, 1\}$) sobriety and semi-sobriety do not always coincide, but that in this case, sobriety is equivalent to semi-sobriety plus T_0.

3.3 Example

$X = (-\infty, k]$ on the real line with the topology $\{X, \emptyset, (a, k];\ a \in X\}$ is sober and semi-sober but not T_1 (or T_2), whereas $\mathbb{R}$ with the cofinite topology is T_1 but not sober because $\mathbb{R}$ itself is irreducibly closed but not the closure of a singleton.

3.4 Example

It was pointed out in [24] and again in [13] that a difference between the sobriety of the crisp case ($L = \{0,1\}$) and sobriety or semi-sobriety of the general case is that in the first case we have a unique singleton but not necessarily in the second:

Consider the fuzzy Sierpinski space of Srivastava, viz. $X = [0,1]$, $L = [0,1]$, and $\mathcal{T}$ assigns value 1 to each of $\mathbf{0}$, $\mathbf{1}$, id (the identity map on $[0,1]$). So the non-zero irreducibly closed sets are $\mathbf{1}$ and $\mathbf{1}\text{-}id$. Now $\mathbf{1}\text{-}id$ is the $\beta = 1$ closure of 0 whereas $\mathbf{1}$ is the $\beta = 1$ closure of any $x \in (0,1]$; or alternatively, $\mathbf{1}\text{-}id$ is the β-closure, $\beta = 1 - x$, of any $x \in [0,1)$ and $\mathbf{1}$ is the $\beta = 1$ closure of $x = 1$.

Note that this example is also sober: $Lpt\mathcal{T} = \{\varphi_x : x \in [0,1]\}$ where $\varphi_x(\mathbf{0}) = 0$, $\varphi_x(\mathbf{1}) = 1$, $\varphi_x(id) = x$ and so $\Psi : X \to Lpt\mathcal{T}$ defined as $\Psi(x)(u) = u(x)$ is a surjection (e.g. $\Psi(x)(id) = id(x) = x = \varphi_x(id)$). $\mathcal{T}$ is clearly T_0 since id distinguishes between points, hence Ψ is injective.

4 Lifting of sobriety and semi-sobriety

A property P of a topological space (ordinary or L-valued) is said to *lift* if $(X, \mathcal{T}^*)$ has property P whenever $(X, \mathcal{T})$ has property P for $\mathcal{T} \subseteq \mathcal{T}^*$.

Clearly, many separation axioms lift, e.g. T_0, T_1, and T_2.

The following is an unpublished result by S. F. Barger (but see also [2] and [5]).

4.1 Theorem

If a T_0 topological space $(X, \mathcal{T})$ is sober and $\mathcal{T} \subseteq \mathcal{T}^*$, and if H is irreducibly closed in $(X, \mathcal{T}^*)$, then $H \subseteq cl_{\mathcal{T}}\{x_1\}$ for some $x_1 \in cl_{\mathcal{T}}H$. ($cl_{\mathcal{T}}$ means closure w.r.t. $\mathcal{T}$)

Proof. Firstly, $cl_{\mathcal{T}}H$ is irreducibly closed in $(X, \mathcal{T})$; because if not, $cl_{\mathcal{T}}H = F_1 \cup F_2$, F_1 and F_2 closed in $(X, \mathcal{T})$ with $cl_{\mathcal{T}}H$ not contained in either F_1 or F_2.

Now $H = (F_1 \cap H) \cup (F_2 \cap H)$ with both $F_1 \cap H$ and $F_2 \cap H$ closed in $(X, \mathcal{T}^*)$. So $H \subseteq F_1 \cap H$ or $H \subseteq F_2 \cap H$, i.e. $H \subseteq F_1$ or $H \subseteq F_2$. But then $cl_{\mathcal{T}}H \subseteq F_1$ or $cl_{\mathcal{T}}H \subseteq F_2$. Contradiction.

Since $cl_{\mathcal{T}}H$ is irreducibly closed in $(X, \mathcal{T})$ and $(X, \mathcal{T})$ is sober, by **2.1** above, $cl_{\mathcal{T}}H = cl_{\mathcal{T}}\{x_1\}$ for some $x_1 \in cl_{\mathcal{T}}H$. Thus $H \subseteq cl_{\mathcal{T}}\{x_1\}$. □

4.2 Corollary

Sobriety lifts for T_1 topological spaces.

4.3 Example

Here is a non-T_1 sober space where sobriety does not lift: Consider example **3.3** above where $X = (-\infty, k]$ and with the sober topology

$$\mathcal{T} = \{X, \emptyset, (a, k];\ a \in X\}$$

Enlarge this to the topology $\mathcal{T}^*$ on X generated by $\mathcal{T}$ together with the cofinite topology on X. Then the closed sets are: $(-\infty, a]$, $a \in X$; all finite sets; and X and $\emptyset$. The sets $(-\infty, a]$ are again irreducible, but are not the closures of singletons since the space is now T_1.

4.4 Definition

An L-topological space $(X, L, \mathcal{T})$ is T_1 iff the map $\mathcal{T} : L^X \to \{0, 1\}$ assigns to the complement of a "fuzzy point" the value 1.

A "fuzzy point" $u(x)$ in L^X (or on X) is a member of L^X which has a singleton of X as support. Its complement is $u'(x) \equiv u(x)'$ where the second $'$ is the order reversing involution of L. We denote a fuzzy point with support x_0 and value 1 at x_0 by 1_{x_0} and if it has value α at x_0, by $\alpha 1_{x_0}$.

So in a T_1 L-topological space, the fuzzy points are closed in view of the remarks under **1.2** above.

Theorem **4.1** can be directly extended to the semi-sober L-topological case so as to obtain:

4.5 Theorem

If $(X, L, \mathcal{T})$ is semi-sober and T_1 and $\mathcal{T} \subseteq \mathcal{T}^*$, then $(X, L, \mathcal{T}^*)$ is also semi-sober.

Proof. Here $\cup, \cap$ and $\subseteq$ in the proof of Theorem **4.1** should be read as $\vee, \wedge$ and $\leq$ respectively in L^X. Then $H \leq cl_{\mathcal{T}}\{x_1\}$, where $cl_{\mathcal{T}}\{x_1\}$ denotes the β-closure of x_1 referred to in **3.2**. Since $(X, L, \mathcal{T})$ is T_1, $cl_{\mathcal{T}}\{x_1\} = \beta 1_{x_1}$ (see notation in **4.4** above). Since H is irreducibly closed in $(X, L, \mathcal{T}^*)$, H is also a fuzzy point $\beta^* 1_{x_1}$ where β^* is irreducible and $0 < \beta^* \leq \beta$; hence H is the β^*-closure of a singleton, and so $(X, L, \mathcal{T}^*)$ is semi-sober. □

4.6 Note

The considerations above are also valid for L from the category **SFRM** of semi-frames and semi-frame morphisms, namely all complete lattices and mappings preserving arbitrary joins and finite meets (see Chapter 4 of [9]). A semi-frame with order-preserving involution is a complete de Morgan lattice or algebra. The frame distributive law of finite meets over arbitrary joins is in fact not used above.

5 Semi-sobriety in (L, M)-topological case

Let $(X, L, M, \mathcal{T})$ be an (L, M)-topological space and put $\mathcal{T}_\alpha = \mathcal{T}^{\leftarrow}\{[\alpha, 1]\}$ where $[\alpha, 1] = \{\beta \in M : \beta \geq \alpha\}$. $\mathcal{T}_\alpha$ is a frame in L^X (an L-topology on X) for each $\alpha \in M$ with $\mathcal{T}_0 = L^X$ and $\alpha \geq \beta \Rightarrow \mathcal{T}_\alpha \subseteq \mathcal{T}_\beta$.

5.1 Definition [4]

In a complete lattice L, x is *way below* y $(x << y)$ iff for directed subsets D of L, the relation $y \leq \sup D$ implies the existence of a $d \in D$ such that $x \leq d$. In L $x << y \Rightarrow x \leq y$.

A complete lattice L is a *continuous lattice* if for all $x \in L$, $x = \sup\{u \in L : u << x\}$.

L a frame $\Rightarrow$ L a continuous lattice $\Rightarrow$ L a semi-frame, but these implications are not reversible.

5.2 Proposition

If M is a continuous lattice, then $\mathcal{T}_\alpha = \bigcap_{\beta << \alpha} \mathcal{T}_\beta$.

Proof. Since $\alpha \geq \beta \Rightarrow \mathcal{T}_\alpha \subseteq \mathcal{T}_\beta$ it follows that $\mathcal{T}_\alpha \subseteq \bigcap_{\beta << \alpha} \mathcal{T}_\beta$. On the other hand, if $u \in \mathcal{T}_\beta$ for all $\beta << \alpha$, then since $\alpha = \sup\{\beta : \beta << \alpha\}$ we have

$$[\mathcal{T}(u) \geq \beta, \text{ for all } \beta << \alpha] \Rightarrow \mathcal{T}(u) \geq \alpha \Rightarrow u \in \mathcal{T}_a \quad \square$$

5.3 Definition

An (L, M)-topological space $(X, L, M, \mathcal{T})$ is T_1 iff the map $\mathcal{T} : L^X \to M$ assigns to the complement of a fuzzy point on X the value 1 in M.

If $(X, L, M, \mathcal{T})$ is T_1, then all the complements of fuzzy points in L^X are in $\mathcal{T}_a$. This means that for each $\alpha \in M$ we get a T_1 L-topological space $(X, L, \mathcal{T}_\alpha)$ i.t.o. Definition **4.4** after identification of the frame $\mathcal{T}_\alpha$ in L^X with $\mathcal{T}_\alpha : L^X \to \{0, 1\}$ through $u \in \mathcal{T}_a$ iff $\mathcal{T}_\alpha(u) = 1$.

5.4 Definition

A T_1 (L, M)-fuzzy topological space $(X, L, M, \mathcal{T})$ is *semi-sober of degree* $\geq m$ iff $(X, L, \mathcal{T}_m)$ is semi-sober (in the sense of an L-topological space (**3.2**)).

Then, in view of Theorem **4.5**, all the $\mathcal{T}_\alpha$ between $\mathcal{T}_m$ and $\mathcal{T}_0 = L^X$ are semi-sober.

In the definition above we cannot specify $(X, L, M, \mathcal{T})$ to be semi-sober of degree m if m is the supremum of members of M for which $(X, L, \mathcal{T}_m)$ is semi-sober, since the infimum (intersection) of semi-sober topologies is not necessarily semi-sober, as can be seen through the following example:

5.5 Example

Consider $\mathbb{R}$ with the topology $\mathcal{T} = \{\mathbb{R}; \emptyset; (a,\infty) : a \in \mathbb{R}\}$. Then $(\mathbb{R}, \mathcal{T})$ is not semi-sober (nor sober) since $\mathbb{R}$ is irreducibly closed but not the closure of a singleton. Now define $\mathcal{T}_x$ on $\mathbb{R}$ as the topology which contains $\mathcal{T}$ as well as sets of the form $\{(-\infty, b) : b > x\}$. The non-empty closed sets are then $\mathbb{R}$, $\{[b,\infty) : b > x\}$, $\{[b,b'] : b > x\}$, $\{(-\infty, a] : a \in \mathbb{R}\}$, and $\{\{b\} : b > x\}$. The first three types are reducibly closed whereas the fourth is reducible for $a > x$, but in the case $a \leq x$ is both irreducibly closed and the closure of $\{a\}$. The fifth type is irreducible and the closure of $\{b\}$. So $\mathcal{T}_x$ is semi-sober (and sober). But $\mathcal{T} = \bigcap_{x \in \mathbb{R}} \mathcal{T}_x$.

On the other hand, given L-topologies $\{\mathcal{T}_\alpha : \alpha \in M\}$ on X with $\alpha \geq \beta \Rightarrow \mathcal{T}_\alpha \subseteq \mathcal{T}_\beta$ and $\mathcal{T}_0 = L^X$, then $\mathcal{T}^* : L^X \to M$ defined by

$$\mathcal{T}^*(u) = \bigvee \{\alpha : u \in \mathcal{T}_\alpha\}$$

is an (L, M)-topology on X. Now consider $\mathcal{T}^*_\alpha = \mathcal{T}^{*\leftarrow}\{[\alpha, 1]\}, \quad \alpha \in M$. Then as in Proposition **5.2**, if M is a continuous lattice, then

$$\mathcal{T}^*_\alpha = \bigcap_{\beta << \alpha} \mathcal{T}^*_\beta .$$

Furthermore:

5.6 Proposition

If M is a continuous lattice and $\beta << \alpha$, then

$$\bigcap_{\gamma << \alpha} \mathcal{T}_\gamma \subseteq \mathcal{T}^*_\beta \quad \textit{and} \quad \bigcap_{\gamma << \alpha} \mathcal{T}^*_\gamma \subseteq \mathcal{T}_\beta .$$

Proof. Suppose there exists a u such that $u \in \mathcal{T}_\gamma$ for all $\gamma << \alpha$ but $u \not\in \mathcal{T}^*_\beta$. Since $\sup\{\gamma : \gamma << \alpha\} = \alpha$ and $\beta << \alpha$, there exists a $\gamma_0 << \alpha$ such that $\gamma_0 \geq \beta$. Hence $\mathcal{T}_\beta \subseteq \mathcal{T}_{\gamma_0}$. Now $u \in \mathcal{T}_{\gamma_0}$ and so $\mathcal{T}^*(u) \geq \gamma_0$. Hence $u \in \mathcal{T}^*_{\gamma_0}$ and so $u \in \mathcal{T}^*_\beta$. Contradiction.

The other case follows similarly. □

5.7 Corollary

If M is a continuous lattice, then $\mathcal{T}_\alpha = \mathcal{T}^*_\alpha$.

Proof.

$$\mathcal{T}^*_\alpha = \bigcap_{\beta << \alpha} \mathcal{T}^*_\beta = \bigcap_{\gamma << \alpha} \mathcal{T}_\gamma = \mathcal{T}_\alpha \quad \square$$

So if all the $\mathcal{T}_\alpha$, $\alpha \in [0, m]$ are semi-sober (and the rest not), then $(X, L, M, \mathcal{T}^*)$ obtained is an (L, M) topological space which is semi-sober of degree $\geq m$ $(= m)$.

6 Further questions

6.1 Converse of Theorem 4.5

Is the converse of Theorem **4.5** true; i.e., if $(X, L, \mathcal{T})$ is semi-sober and if for any $\mathcal{T}^* \supseteq \mathcal{T}$, $(X, L, \mathcal{T}^*)$ is semi-sober, is $(X, L, \mathcal{T})$ T_1 ? The answer is "no" as can be seen from the following two examples, the first a "crisp" case and second an L-topological one:

(i) The two-point Sierpinski space $(\{0,1\}, \mathcal{S})$ with $\mathcal{S}$ consisting of $\emptyset, \{0,1\}$ and $\{1\}$ is sober (hence semi-sober) by **3.1**, but all larger topologies on $\{0,1\}$—and there is only one such, namely the discrete topology—are sober (semi-sober). However $(\{0,1\}, \mathcal{S})$ is not T_1.

(ii) Take $X = \{0,1\}$ and $\mathcal{F} = \{\mathbf{0}, \iota, \mathbf{1}\}$ where ι is the inclusion map of $\{0,1\}$ into $[0,1]$. (Here we use the topology $\mathcal{F}$ as established in **1.2** as the L-topology where $L = [0,1]$.) Denote $\mu : X = \{0,1\} \to [0,1]$, defined by $\mu(0) = \alpha$, $\mu(1) = \beta$ by $< \alpha, \beta >$. (So $\iota =< 0, 1 >$.) The irreducibly non-zero L-closed sets are $\mathbf{1}$ and $\mathbf{1}\text{-}\iota = \ < 1, 0 >$ and $\mathbf{1} = cl_{\mathcal{F}} 1_1$ while $\mathbf{1}\text{-}\iota = cl_{\mathcal{F}} 1_0$. So by **3.2**, $(X, L, \mathcal{F})$ is semi-sober.

Now let $\mathcal{F}^*$ be another L-topology on X with $\mathcal{F} \subseteq \mathcal{F}^*$ and let $\mu = \ < \alpha, \beta >$ be a non-zero irreducibly $\mathcal{F}^*$ L-closed set.

Then as $< 1, 0 >$ is $\mathcal{F}$-closed, hence $\mathcal{F}^*$-closed,

$$< \alpha, \beta > \wedge < 1, 0 > \ = \ < \alpha, 0 > \ \text{ is } \mathcal{F}^* - \text{closed.}$$

If $\beta = 0$, then $< \alpha, \beta > \ = \ < \alpha, 0 >$ is clearly $cl_{\mathcal{F}^*} \alpha 1_0$.
If $\beta \neq 0$ and $\alpha = 0$, then $< \alpha, \beta > \ = \ < 0, \beta >$ is clearly $cl_{\mathcal{F}^*} \beta 1_1$.
If $\beta \neq 0$ and $\alpha \neq 0$, then also $< \alpha, \beta >= cl_{\mathcal{F}^*} \beta 1_1$, for if not we could have an $\mathcal{F}^*$-closed set $< \alpha', \beta >$ with $\alpha' < \alpha$, which would mean that $< \alpha, \beta >= \ < \alpha, 0 > \vee < \alpha', \beta >$, contradicting the $\mathcal{F}^*$-irreducibility of $< \alpha, \beta >$.

So by **3.2** again, $(X, L, \mathcal{F}^*)$ is semi-sober.

But $(X, L, \mathcal{F})$ is not T_1 since the fuzzy point $1_1 = \iota = \ < 0, 1 >$ is not $\mathcal{F}$-closed.

6.2 Crisp case

The question in **6.1** was answered for sobriety (but not semi-sobriety) in the crisp case by R-E. Hoffmann in [5] where he showed that the following two statements are equivalent:

(a) $(X, \mathcal{T})$ is sober and T_D;

(b) every space finer than $(X, \mathcal{T})$ is sober.

The T_D property is: every singleton is the intersection of an open and closed set, or equivalently, for any $A \in \mathcal{P}(X)$, *der* A is closed.

It is easy to see that $T_1 \Rightarrow T_D \Rightarrow T_0$. This begs the question to come up with a T_D axiom for L-topological spaces, weaker than the T_1 axiom, which will give an analogous result for semi-sobriety.

A further interesting question in this connection is, can we have two (crisp) topologies $\mathcal{T}$ and $\mathcal{T}^*$ on X, $\mathcal{T} \subseteq \mathcal{T}^*$ with $(X, \mathcal{T})$ sober, $|pt\,\mathcal{T}| = |pt\,\mathcal{T}^*|$ but $(X, \mathcal{T}^*)$ not sober ? The following example answers the question in the affirmative:

Consider $X = \mathbb{Z} \cap (-\infty, k]$, $k \in \mathbb{Z}$ (i.e. all integers up to and including k) with the topology $\mathcal{T}$ consisting of $X, \emptyset$, and all $(a, k] \cap \mathbb{Z}$, $a \in X$ (simply $(a, k]$ in what follows).

$(X, \mathcal{T})$ is sober since the irreducibly closed sets are $(-\infty, a] \cap \mathbb{Z}$ (simply $(-\infty, a]$ in what follows), each of which is the closure of $\{a\}$ (see also Example **3.3**). $pt\,\mathcal{T}$ consists of, for $a \in X$,

$$p_a(l, k] = \begin{cases} 1 \text{ if } l < a \\ 0 \text{ if } l \geq a \end{cases}$$

Note that $p_a(\emptyset) = p_a((k, k]) = 0$ and

$$p_k(l, k] = \begin{cases} 1 \text{ if } l < k \\ 0 \text{ if } l = a \end{cases}$$

So $|pt\,\mathcal{T}| = |X| = \aleph_0$. Furthermore $\Psi : X \to pt\,\mathcal{T}$ defined as $\Psi(x)(u) = \chi_u(x)$ (see **2.1**) gives

$$\begin{aligned} \Psi(a)(l, k] &= \begin{cases} 1 \text{ if } l < a \\ 0 \text{ if } l \geq a \end{cases} \\ &= p_a(l, k] \end{aligned}$$

Now define $\mathcal{T}^*$ on X as $X; \emptyset; (a, k]$, $a \in X; \{k\}$. So the closed sets are $\emptyset, X, (-\infty, a], (-\infty, k)$. $(X, \mathcal{T}^*)$ is not sober since $(-\infty, k)$ is irreducibly closed but is not the closure of a singleton. Further, $pt\,\mathcal{T}^*$ consists of the following: (i) for $a \in X$,

$$p_a\,(l, k) = \begin{cases} 1 \text{ if } l < a \\ 0 \text{ if } l \geq a \end{cases}$$

$$p_a\,\{k\} = 0$$

and (ii)

$$p'_k\,(l, k) = \begin{cases} 1 \text{ if } l < k \\ 0 \text{ if } l = a \end{cases}$$

$$p'_k\,\{k\} = 1$$

So $|pt\,\mathcal{T}^*| = \aleph_0 + 1 = \aleph_0$.

References

[1] M. Artin, A. Grothendieck, J. Verdier, *Théorie des topos et cohomologie étale des schémas*, Lecture Notes in Mathematics **269**, Springer-Verlag (Berlin/Heidelberg/New York), 1972.

[2] S. F. Barger, *Heredity sobriety and the cardinality of T_1 topologies on countable sets*, Quaestiones Mathematicae **20(3)**(1997), 117-126.

[3] T. Blanksma, *Lattice characterizations and compactifications*, Doctoral Dissertation, Rijksuniversiteit te Utrecht 1968, MR **37**, p. 5851.

[4] G. Gierz et al, A Compendium Of Continuous Lattices, Springer-Verlag (Berlin/Heidelberg/New York), 1980.

[5] R.-E. Hoffmann, *Irreducible filters and sober spaces*, Manuscripta Math. **22**(1977), 365-380.

[6] R.-E. Hoffmann, *Sobrification of partially ordered sets*, Semigroup Forum **17**(1979), 123-138.

[7] U. Höhle, *Uppersemicontinuous fuzzy sets and applications*, J. Math. Anal. Appl. **78**(1978), 659-673.

[8] U. Höhle, *Locales and L-topologies*, Mathematik-Arbeitspapiere **48**(1997), 223-250, Universität Bremen.

[9] U. Höhle, S. E. Rodabaugh, eds, Mathematics Of Fuzzy Sets: Logic, Topology, And Measure Theory, The Handbooks of Fuzzy Sets Series, Volume 3, Kluwer Academic Publishers (Boston/Dordrecht/London), 1999.

[10] P. T. Johnstone, Stone Spaces, Cambridge University Press (Cambridge), 1982.

[11] W. Kotzé, *Lattice morphisms, sobriety, and Urysohn lemmas*, Chapter 10 in [20], 257-274.

[12] W. Kotzé, *Fuzzy sobriety and fuzzy Hausdorff*, Quaestiones Mathematicae **20**(1997), 415-422.

[13] W. Kotzé, *Sobriety and semi-sobriety of L-topological spaces*, Quaestiones Mathematicae **24**(2001), 549–554.

[14] T. Kubiak, *On L-Tychonoff spaces*, Fuzzy Sets and Systems **73**(1995), 25-53.

[15] T. Kubiak, A. P. Šostak, *Lower set-valued fuzzy topologies*, Quaestiones Mathematicae **20(3)**(1997), 423-429.

[16] S. E. Rodabaugh, *A point-set lattice theoretic framework* $\mathbb{T}$ *for topology which contains LOC as a subcategory of singleton spaces and in which there are general classes of Stone Representations and compactification theorems*, Youngstown State University Central Printing Office (first edition February 1986, second edition April 1987), Youngstown, Ohio.

[17] S. E. Rodabaugh, *Point-set lattice-theoretic topology*, Fuzzy Sets and Systems **40**(1991), 297-345.

[18] S. E. Rodabaugh, *Applications of localic separation axioms, compactness axioms, representations, and compactifications to poslat topological spaces*, Fuzzy Sets and Systems **73**(1995), 55-87.

[19] S. E. Rodabaugh, *Separation axioms: representation theorems, compactness, and compactifications*, Chapter 7 in [9], 481-552.

[20] S. E. Rodabaugh, E. P. Klement, U. Höhle, eds, Application Of Category To Fuzzy Sets, Theory and Decision Library—Series B: Mathematical and Statistical Methods, Volume **14**, Kluwer Academic Publishers (Boston/Dordrecht/London), 1992.

[21] M. B. Smyth, *Topology*, Handbook of Logic in Computer Science, Vol. I, Clarendon Press (Oxford), 1992, pp. 641-761.

[22] A. P. Šostak, *On a fuzzy topological structure*, Suppl. Rend. Circ. Matem. Palermo, sr. II, **11**(1985), 89-103.

[23] A. P. Šostak, *On compactness and connectedness degrees of fuzzy sets in fuzzy topological spaces.* General Topology and related Modern Analysis and Algebra. Proceedings of the V-th Prague Topological Symposium, Prague 1986, 519-523; Heldermann Verlag (Berlin), 1988.

[24] A. K. Srivastava, A. S. Khastgir, *On fuzzy sobriety*, Information Sciences **110**(1998), 195-205.

[25] S. Vickers, Topology Via Logic, Cambridge Tracts in Theoretical Computer Science **5**, Cambridge University Press, 1989.

[26] Mingsheng Ying, *A new approach for fuzzy topology (I)*, Fuzzy Sets and Systems **39**(1991), 303-321.

CHAPTER 17

Examples For Different Sobrieties In Fixed-Basis Topology

A. PULTR [1] AND S. E. RODABAUGH

Introduction

In a previous chapter [21], the authors introduced a new approach to describing L-topological spaces using categorical constructs called fuzzy frames. This approach not only gives new decriptions of previously known types of sober spaces, but it also leads naturally to a new type of sober spaces not previously documented in the literature.

It is the purpose of this chapter is to give examples for this new class of sober spaces, along with examples for some previously known classes of sober spaces. Thus this chapter is a trailer to [21], providing pertinent examples for the theory of that chapter.

1 Preliminaries

Except as noted below, we generally follow the notation and conventions of [21], where are found needed ideas and references.

1.1 Summary of Types of Sobriety. Let L be a complete lattice and (X, τ) be an L-topological space.

[1]Support of the Grant Agency of the Czech Republic under Grant 201/99/0310 is gratefully acknowledged.

S.E. Rodabaugh and E.P. Klement (eds.), Topological and Algebraic Structures in Fuzzy Sets, 427–440.
© 2003 Kluwer Academic Publishers.

1. **L-Sobriety**. Recall $Lpt(\tau) = \mathbf{SFrm}(\tau, L)$ and the evaluation map $\Psi_L : X \to Lpt(\tau)$ by $\Psi_L(x)(u) = u(x)$. Then (X, τ) is **L-sober** if Ψ_L is a bijection. Recall that when $Lpt(\tau)$ is equipped with the canonical L-topology $\Phi_L^{\rightarrow}(\tau)$, then L-sobriety says that Ψ_L is an L-homeomorphism and L-T_0 says that Ψ_L is an L-embedding.

2. **Modified L-Sobriety**. Modify $Lpt(\tau)$ in (1) to be $Lpt_{\text{mod}}(\tau)$ [21] as follows:
$$Lpt_{\text{mod}}(\tau) = \{p \in Lpt(\tau) : \forall\, \underline{\alpha} \in \tau,\, p(\underline{\alpha}) = \alpha\}$$
where $\underline{\alpha}$ is the constant L-subset with value α. It is a fact that we always have
$$\Psi_L : X \to Lpt_{\text{mod}}(\tau) \subset Lpt(\tau)$$
We say (X, τ) is **modified L-sober** if Ψ_L is a bijection onto $Lpt_{\text{mod}}(\tau)$, in which case Ψ_L is a homeomorphism onto $Lpt_{\text{mod}}(\tau)$ equipped with the subspace topology inherited from $Lpt(\tau)$. See Lemma 2.2 below.

3. **Modified L-Sobriety to Degree D**. Let D be a complete sublattice of L, i.e. D is closed under the arbitrary $\bigvee$ and aribitrary $\bigwedge$ of L. It follows that $\{0, 1\} \subset D$. Assume that (X, τ) has the property that $\{\underline{\alpha} : \alpha \in D\} \subset \tau$, in which case it is said to be **stratified to degree** D. If (X, τ) is a space in the sense of [18], it is stratified to degree $D = L$ and is said to be **stratified** [20, 30]. We say (X, τ) is stratified to **nontrivial** degree D if $D \supsetneqq \{0, 1\}$. Assuming that (X, τ) is stratified to degree D, we modify $Lpt(\tau)$ in (1) to be $Lpt_D(\tau)$ as follows:
$$Lpt_D(\tau) = \{p \in Lpt(\tau) : \forall\, \alpha \in D,\, p(\underline{\alpha}) = \alpha\}$$
Note $\{\underline{\alpha} : \alpha \in D\} \subset \tau \Rightarrow \text{Im}(\Psi_L) \subset Lpt_D(\tau)$, so that Ψ_L in this case is well-defined into $Lpt_D(\tau)$. Now we say (X, τ) is **modified L-sober to degree** D if Ψ_L is a bijection onto $Lpt_D(\tau)$, and it is modified L-sober to **nontrivial** degree D if $D \supsetneqq \{0, 1\}$. See Lemma 2.2(1) below for relationships to the notions of (1) above.

4. **Ultra-L-Sobriety, [L-Sobriety, ι_L-Sobriety, ι-Sobriety.** The first term recognizes an early tradition for naming properties of L-topological spaces [16]; the second term arises naturally in [21]; and the third and fourth terms are the same—if L is understood in context—and are motivated by the "ι-P" naming template of [16]. All these terms refer to the same notion of [21]. Let L be a complete lattice, and consider the upper topology SUP(L) having subbase $\{L - \downarrow(a) : a \in L\}$. Then the functor $\iota_L : L\text{-}\mathbf{Top} \to \mathbf{Top}$ of [16], or ι if L is understood, is defined as follows, given $(X, \tau), f$ from L-**Top**:
$$\iota(\tau) = \langle\{u^{\leftarrow}(U) : u \in \tau,\, U \in \text{SUP}(L)\}\rangle,\ \iota(X, \tau) = (X, \iota(\tau)),\ \iota(f) = f$$

Alternatively, we may define $\iota(\tau) = \langle\{[u \nleq a] : u \in \tau,\, a \in L\}\rangle$, where $[u \nleq a]$ is the Halmos notation for the preimage $\{x \in X : u(x) \nleq a\}$. If L is a complete chain, as in much of [21], then $\iota(\tau) = \langle\{[u > a] : u \in \tau,\, a \in L\}\rangle$. When $L = \mathbb{I}$ (unit interval), ι is the ι functor originally defined in [18]. The ordinary topological space $(X, \iota(\tau))$ is called the **topological modification** of (X, τ). An L-space (X, τ) is **ι-sober** if its topological modification $(X, \iota(\tau))$ is sober as an ordinary topological space as in [8, 9, 10, 12, 23]. Now ι-sobriety is the sense of sobriety which arises naturally in the constructions of [21]. We refer below to the left adjoint ω of ι [16], which is the ω functor originally defined in [18] when $L = \mathbb{I}$.

1.2 Goal. The purpose of this chapter is to investigate and inventory canonical examples of various classes of L-topological spaces:

1. L-sober spaces which are not modified L-sober to any non-trivial degree.
2. Modified L-sober spaces to any desired degree induced from any L-topological space or any traditional topological space.
3. Non-generated (by ω) L-spaces which are modified L-sober to any desired degree and not to any other desired degree.
4. L-sober spaces or modified L-sober spaces or modified L-spaces to degree D which are not ι-sober.
5. ι-sober spaces which are modified L-sober to highest degree L and not L-sober.
6. ι-sober spaces which are not modified L-sober to any nontrivial degree.
7. ι-sober spaces which are modified L-sober to one degree but not another and are not L-sober.
8. ι-sober spaces which are not modified L-sober to any nontrivial degree and are not L-sober.

While the examples of 1.2(5–8) are the main point of this chapter, the examples of 1.2(1–3) are related to the constructions of 1.2(5–8), as well as being of interest in their own right. The rationale for 1.2(6), in light of 1.2(8), is: (i) there are many more examples for 1.2(6) than for 1.2(8), and (ii) 1.2(6) is used in 1.2(7).

In Remark 3.9, we state categorical ramifications of the examples for 1.2 (5–8).

To carry out Goal 1.2, we often appeal to: the L-fuzzy real line $\mathbb{R}(L)$ and the L-fuzzy unit interval $\mathbb{I}(L)$—see [11, 2, 16, 17, 32] and their references; the duals $\mathbb{R}_L$ and $\mathbb{I}_L$ of $\mathbb{R}(L)$ and $\mathbb{I}(L)$—see [25, 26, 32]; and the alternative L-fuzzy real line $\mathbb{R}^*(L)$ and the alternative L-fuzzy unit interval $\mathbb{I}^*(L)$—namely the L-2 soberifications of $\mathbb{R}$ and $\mathbb{I}$—see Subsections 2.15–2.16, 8.15–8.16 of [31] as well as [5].

In Remark 3.9, we outline how these classes of examples categorically justify the various kinds of sobriety defined in 1.1, including that of [21].

1.3 Discussion (Sobrieties and Semi-Sobriety). This chapter's survey of sobrieties in lattice-valued topology does not include the new and interesting semi-sobriety axioms of [14, 15], which in traditional topology are equivalent to the comparison map Ψ being surjective and hence to the S_0 axiom—Ψ_L is surjective—of [24] restricted to the crisp case. Indeed, a traditional space is sober iff $T_0 + S_0$ iff T_0 + semi-sober. However, semi-sobriety is not equivalent to S_0 in the setting of L-topology.

1. Let L satisfy the condition of Lemma 2.1(1) below and consider the discrete L-topological space (X, L^X) with $X \neq \emptyset$. By the first paragraph below Theorem 3.2 of [15], (X, L^X) is semi-sober. It is also T_0 (in the sense that $\forall\, x \neq y$ in X, $\exists\, u \in L$, $u(x) \neq u(y)$). But by the contrapositive of Lemma 2.1(1), (X, L^X) is not L-sober. Since $T_0 + S_0 = L$-sobriety, we have (X, L^X) is not S_0. (In fact, the proof of Lemma 2.1(1) given in [24] and repeated in [28], shows that it is the S_0 part of L-sobriety which implies that the topology of the space cannot contain the constant map $\underline{\alpha}$ referred to in Lemma 2.1(1).)

2. So (1) gives a class of spaces, indexed by lattices satisfying the condition of Lemma 2.1(1), which are T_0 + semi-sober, but not S_0. Thus T_0 + semisober $\not\Rightarrow$ L-sobriety.

3. Theorem 3.2 of [15] shows that $T_0 + S_0 = L$-sobriety $\Rightarrow T_0$ + semi-sober. Thus [15] suggests a new sobriety axiom for L-topological spaces with L a De Morgan algebra, logically distinct from L-sobriety, namely T_0 + semi-sober. Since semi-sobriety deals with the co-primes of L and the primes of τ, we for the moment call T_0 + semi-sobriety **prime sobriety**. Restated, we have shown that L-sobriety $\Rightarrow$ prime sobriety and that prime sobriety $\not\Rightarrow$ L-sobriety.

4. **Open question.** What is the relationship between prime sobriety and the other sobrieties summarized above?

5. **Open question.** Does $S_0 \Rightarrow$ semi-sober without the assumption of the space being T_0?

It is our conjecture that the above prime sobriety axiom is indeed distinct from the other sobrieties of Definition 1.1 and the examples of this chapter could be relevant to determining such relationships. However, such a study is beyond the scope of this chapter and must await future work.

2 Useful lemmas

Proofs of the following are either given, referenced, or straightforward.

2.1 Lemma [24, 28]. Let $L \in |\mathbf{SFrm}|$.

1. If L has an endomorphism ϕ other than the identity—$\exists\, \alpha \in L,\ \phi(\alpha) \neq \alpha$, then each nonempty L-sober space does not have the constant map $\underline{\alpha}$ as an open set and hence is not stratified for each degree D containing α.

2. If L satisfies the condition

$$\forall\, \alpha \in L - \{0, 1\}\,,\ \exists \phi \in Endo\,(L)\,,\ \phi\,(\alpha) \neq \alpha$$

(guaranteed if 0 is meet-irreducible (satisfied if L is linear or coherent), L is an atomic Boolean algebra, L is a generalized diamond [26] (and so is non-distributive when $|L| > 4$)), then each nonempty L-sober space has no constant open sets other than $\underline{0}, \underline{1}$ and hence is stratified only to the minimal degree $D = \{0, 1\}$.

Proof. See the statement of Lemma 6.2(2) of [28] for (2), and the proof of that lemma for (1). □

2.2 Lemma. Let $L \in |\mathbf{SFrm}|$. Given (X, τ), let

$$D_\tau = \{\alpha : \underline{\alpha} \in \tau\}$$

1. $Lpt_{\text{mod}}\,(\tau) = Lpt_{D_\tau}\,(\tau)$. Hence (X, τ) is modified L-sober $\Leftrightarrow$ (X, τ) is modified L-sober to degree D_τ.

2. If L satisfies any of the conditions of Lemma 2.1(2), (X, τ) is L-sober $\Leftrightarrow$ (X, τ) is modified L-sober $\Leftrightarrow$ (X, τ) is modified L-sober to degree $D_\tau = \{0, 1\}$, in which case (X, τ) has no constant open sets other than $\underline{0}, \underline{1}$.

3. (X, τ) is stratified to degree D and modified L-sober $\Rightarrow$ (X, τ) is modified L-sober to degree D $\Rightarrow$ (X, τ) is stratified to degree D. Hence if (X, τ) is stratified, then (X, τ) is modified L-sober $\Leftrightarrow$ (X, τ) is modified L-sober to degree $D_\tau = L$.

2.3 Lemma. Let $L \in |\mathbf{Frm} \cap \mathbf{Dmrg}|$.

1. $\mathbb{R}\,(L)$ and $\mathbb{I}\,(L)$ are L-sober $\Leftrightarrow$ L is a Boolean algebra [19].

2. $\mathbb{R}\,(L)$ $[\mathbb{I}\,(L)]$ is L-homeomorphic to $\mathbb{R}^*\,(L)$ $[\mathbb{I}^*\,(L)$, resp.$]$ $\Leftrightarrow$ L is a Boolean algebra [31].

Proof. (1) comes from Theorem of [19]. Sufficiency of (2) comes from Application 2.15.8 of [31], and necessity comes from the fact that L-2 soberifications are L-sober (e.g. Lemma 2.9 of [31]) and that L-sobriety is an L-topological property, itself a consequence of Ψ_L being a universal arrow (e.g. Theorem 2.4 of [31]). □

2.4 Lemma [16]. If L is completely distributive, then as fibre maps $\iota \circ \omega = id$, so that as functors $\omega \dashv \iota$ is an epicoreflection.

Proof. Proposition 3.6 of [16]. □

2.5 Lemma [16]. Let L be a completely distributive deMorgan algebra. Then $\mathbb{I}(L)$ is ι-Hausdorff; i.e. the topological modification $(\mathbb{I}(L), \iota(\tau(L)))$, is Hausdorff.

Proof. Proposition 6.11 of [16]. □

2.6 Lemma [26]. Let L be a complete deMorgan chain. Then $\iota(\mathbb{R}_L)$ and $\iota(\mathbb{I}_L)$ are $\mathbb{R}$ and $\mathbb{I}$, respectively.

Proof. Theorem 4.1(2) of [26]. □

2.7 Lemma [21]. Let $L \in |\mathbf{SFrm}|$ and (X, τ) be an L-topological space.

1. If L is completely distributive, then each sober traditional space $(X, \mathfrak{T})$ generates via ω an ι-sober L-space (X, τ).

2. If L is a complete chain, then each sober traditional space $(X, \mathfrak{T})$ generates via ω an ι-sober L-space (X, τ) which is modified L-sober to degree L.

3. If (X, τ) is L-sober and D is a subsemiframe of L, then $(X, \tau \vee \{\underline{\alpha} : \alpha \in D\})$ is modified L-sober to degree D, where $\tau \vee \{\underline{\alpha} : \alpha \in D\}$ is the smallest L-topology containing $\tau \cup \{\underline{\alpha} : \alpha \in D\}$. In particular, the stratification $G_k(X, \tau)$, where $G_k(\tau)$ is the smallest L-topology containing $\tau \cup \{\underline{\alpha} : \alpha \in L\}$ is modified L-sober to degree L (see [27, 28] and Section 6 of [30]).

Proof. Proposition 3.5.2 of [21]. □

2.8 Lemma. Let L be a complete chain and (X, τ) be an L-topological space. The following hold:

1. $\forall \{v_\gamma\}_\gamma \subset \tau$, $\forall \alpha \in L$,

$$\left[\bigvee_\gamma v_\gamma > \alpha\right] = \bigcup_\gamma [v_\gamma > \alpha], \quad \left[\bigwedge_\gamma v_\gamma > \alpha\right] = \bigcap_\gamma [v_\gamma > \alpha]$$

2. Let σ be another L-topology on X. Then

$$\iota(\tau \vee \sigma) = \iota(\tau) \vee \iota(\sigma)$$

3. Let D be a subsemiframe of L, and (X, τ) be an L-topological space. Then $\iota(X, \tau) = \iota(X, \tau \vee \{\underline{\alpha} : \alpha \in D\})$.

2.9 Lemma. Let $L \in |\mathbf{SFrm}|$, (X, τ) be an L-topological space, and $A \subset X$. The following hold:

1. $\forall u \in \tau, \forall \alpha \in L, [u_{|A} \not\leq \alpha] = [u \not\leq \alpha] \cap A$.

2. ι respects subspaces; i.e. $(\iota(\tau))_A = \iota(\tau_A)$, where the left-hand side is the ordinary subspace topology on A inherited from $(X, \iota(\tau))$ and the right-hand side is ι applied to the usual subspace L-topology $\{u_{|A} : u \in \tau\}$ on A inherited from $(X.\tau)$.

2.10 Lemma. Let $L \in |\mathbf{SFrm}|$. Then ι takes L-homeomorphisms to homeomorphisms.

3 Examples for different sobrieties

Unless stated otherwise, L is a semiframe, and, in the context of

$$\mathbb{R}(L), \mathbb{I}(L), \mathbb{R}_L, \mathbb{I}_L$$

a complete deMorgan algebra.

3.1 Examples for Goal 1.2(1).

1. Let $L \in |\mathbf{SFrm}|$ such that L satisfies any of the conditions of Lemma 2.1(2). Then the L-soberification $LPT(L\Omega(X, \tau))$ (2.2 of [21]) of any nonempty L-topological space (X, τ) and the L-2 soberification $LPT(\Omega(X, \mathfrak{T}))$ (1.7, 2.2 of [21]) of any nonempty ordinary topological space $(X, \mathfrak{T})$ are L-sober and not modified L-sober to any degree $D \neq \{0, 1\}$. This includes $\mathbb{R}^*(L)$ and $\mathbb{I}^*(L)$.

2. Let L be any complete atomic Boolean algebra. Then $\mathbb{R}(L)$ and $\mathbb{I}(L)$ are L-sober but are not modified L-sober to any degree $D \neq \{0, 1\}$.

Proof. For (1), see the cited sections of [21] for L-sobriety, and then apply Lemma 2.1(2) and Lemma 2.2(3). For (2), $\mathbb{R}(L)$ and $\mathbb{I}(L)$ are L-sober by Lemma 2.3; and then finish by Lemma 2.1(2) and Lemma 2.2 (3). □

3.2 Examples for Goal 1.2(2). Let L be a frame, and let (X, τ) be any L-topological space. Consider its L-soberification $(Y, \sigma) \equiv LPT(L\Omega(X, \tau))$ as in Examples 3.1. Now let D be any sub-semiframe of L and construct the space $(Y, \sigma \vee \{\underline{\alpha} : \alpha \in D\})$. By Lemma 2.7(3), this last space is modified L-sober to degree D. Given a topological space $(X, \mathfrak{T})$, consider its L-2 soberification $(Y, \sigma) \equiv LPT(\Omega(X, \mathfrak{T}))$ as in Examples 3.1, and proceed as before by adjoining $\{\underline{\alpha} : \alpha \in D\}$. The canonical examples in these classes include choosing (X, τ) in the above to be $\mathbb{R}(L)$ or $\mathbb{I}(L)$ and $(X, \mathfrak{T})$ in the above to be $\mathbb{R}$ or $\mathbb{I}$, in which case (Y, σ) is $\mathbb{R}^*(L)$ or $\mathbb{I}^*(L)$.

3.3 Examples for Goal 1.2(3). First, let $L \in |\mathbf{SFrm}|$ such that L satisfies the conditions of Lemma 2.1(2) and let (X, τ) be any L-topological space from Examples 3.1 which is L-sober but not modified L-sober to any nontrivial degree $D \hookrightarrow L$, Now let $D, E \hookrightarrow L$ with $D \neq E$. We are to construct an L-space which is modified L-sober to degree D but not E. We consider three cases.

1. *Case 1:* $E \not\subset D$. In this case, apply the procedure of Examples 3.2—the adjoining of the constant maps $\{\underline{\alpha} : \alpha \in D\}$—to get an L-space (Y, σ) which is modified L-space to degree D. It follows that $D_\sigma = D$ (Lemma 2.2(1)). Now $\exists\, \alpha \in E$, $\alpha \notin D$. But Lemma 2.2(3) now implies (Y, σ) is not modified L-sober to degree E.

2. *Case 2:* $E \subsetneqq D$. *In this case, we add the assumption*: $\forall \beta \in E - \{0, 1\}$, $\exists\, \phi_\beta \in Endo(L)$, ϕ fixes E and $\phi_\beta(\beta) \neq \beta$. Under the conditions of Lemma 2.1(2), such an assumption always holds if E is $\{0, 1\}$; such an assumption holds for $L = \mathbb{I}$ when $E \subsetneqq D$; such an assumption holds for the five-point diamond
$$L = \{0, a, b, c, 1\}$$
where a, b, c are all unrelated, where $E = \{0, a, 1\}$; etc. Now proceed as in Case 1 to obtain an L-space (Y, σ) which is a modified L-space to degree D. Since $E \subset D$, it follows that $Lpt_D(\sigma) \subset Lpt_E(\sigma)$. Now $\exists\, \alpha \in D$, $\alpha \notin E$. We claim that $Lpt_D(\sigma) \subsetneqq Lpt_E(\sigma)$, from which claim it will follow that (Y, σ) is not modified L-sober to degree E. To verify this claim, we must exhibit $p \in Lpt_E(\sigma)$ such that $p(\underline{\alpha}) \neq \alpha$. Now any space (X, τ) chosen from Examples 3.1 is nonempty, so therefore is the resulting space (Y, σ) from Examples 3.2. It follows that $Lpt_D(\sigma)$ is nonempty—each $y \in Y$ generates such a point through the map $\Psi_L : Y \to Lpt_D(\sigma)$ given by $\Psi_L(y)(u) = u(y)$. So let $p \in Lpt_D(\sigma)$. If $p(\underline{\alpha}) \neq \alpha$, then we are done. If not, let ϕ_α be given from our addiional assumption and put $q = \phi \circ p$. It follows $q \in Lpt_E(\sigma)$ and that $q(\underline{\alpha}) \neq \alpha$. This proves the claim (and is essentially the proof of Lemma 6.2 of [28] stated above as Lemma 2.1).

3. *Case 3:* $E \subsetneqq D$ with no additional assumption. We leave this as an *open question.* □

3.4 Examples for Goal 1.2(4). We have only open questions here:

1. *Open question*: do there exist $L \in |\mathbf{SFrm}|$ and L-sober space (X, τ) (or modified L-sober space (X, τ) or modified L-space (X, τ) to degree D) such that (X, τ) is not ι-sober?

2. *Open question*: is there $L \in |\mathbf{SFrm}|$ such that $\iota(\mathbb{R}^*(L))$ or $\iota(\mathbb{I}^*(L))$ is not sober (these are always L-sober)?

3. *Open question*: is there $L \in |\mathbf{Frm} \cap \mathbf{Dmrg}|$ such that $\mathbb{R}(L)$ or $\mathbb{I}(L)$ are L-sober and $\iota(\mathbb{R}(L))$ or $\iota(\mathbb{I}(L))$ is not sober? Because of Lemmas 2.3, 2.5, the only lattices to consider for this question are complete Boolean algebras which are not completely distributive (recall that each complete Boolean algebra is weakly completely distributive in the sense that both infinite distributive laws hold), standard examples of which are the regular open sets of $\mathbb{R}$ and the quotient algebra of the Lebésgue-measurable subsets of $\mathbb{I}$ with the measure 0 subsets identified and the measure 1 subsets identified: for if $\mathbb{R}(L)$ or $\mathbb{I}(L)$ is L-sober, then L is Boolean; but if L is completely distributive, then $\iota(\mathbb{I}(L))$ is Hausdorff, hence sober. It may be possible to approach this question through hereditary sobriety (each subspace is sober) and hereditary L-sobriety (each L-subspace based on a crisp subset is L-sober—see Section 5 of [30]) given that ι respects subspaces (Lemma 2.9(2)). □

3.5 Examples for Goal 1.2(5). Let L be completely distributive and satisfy any of the conditions of Lemma 2.1(2), and let $(X, \mathfrak{T})$ be any ordinary sober topological space. Then $\omega(X, \mathfrak{T}) \equiv (X, \omega(\mathfrak{T}))$ is an ι-sober space which is modified sober to the highest degree $D = L$ and is not L-sober. Canonical examples in this context include choosing $(X, \mathfrak{T})$ to be $\mathbb{R}$ or $\mathbb{I}$. When L is a chain, the resulting L-topological space is $\mathbb{R}_L$ or $\mathbb{I}_L$, the duals of $\mathbb{R}(L)$ and $\mathbb{I}(L)$.

Proof. Apply Lemma 2.7(2) followed by Lemma 2.1(2). The claim concerning $\mathbb{R}_L$ and $\mathbb{I}_L$ follows from Lemma 2.6. □

3.6 Examples for Goal 1.2(6). Let L satisfy any of the conditions of Lemma 2.1(2), and let $(X, \mathfrak{T})$ be any ordinary sober topological space. Take the image $G_\chi(X, \mathfrak{T}) \equiv (X, G_\chi(\mathfrak{T}))$ under the characteristic functor G_χ [27, 16, 30], where $G_\chi(X, \mathfrak{T}) = \{\chi_U : U \in \mathfrak{T}\}$. Then for each $\alpha \in L$ we have

$$[\chi_U \not\leq \alpha] = \begin{cases} U, & \alpha \in L - \{1\} \\ \emptyset, & \alpha = 1 \end{cases}$$

It follows that $\iota((X, G_\chi(\mathfrak{T}))) = (X, \mathfrak{T})$. Hence $(X, G_\chi(\mathfrak{T}))$ is an ι-sober space which by Lemma 2.2(3) is not modified L-sober to any degree—it has no non-trivial constant open sets.

3.7 Examples for Goal 1.2(7). We continue the construction of Examples 3.6, letting $(Y, \sigma) \equiv (X, G_X(\mathfrak{T}))$ of 3.6, but now impose the condition that L be a chain. Now let $D, E \hookrightarrow L$ with $D \neq E$ and apply the the second part of the procedures of Examples 3.2–3.3—the adjoining of the constant maps $\{\underline{\alpha} : \alpha \in D\}$—to obtain ι-sober spaces which are modified L-sober to degree D but not degree E. The proof that the adjoining of $\{\underline{\alpha} : \alpha \in D\}$ does not affect ι-sobriety follows from Lemma 2.8(3).

3.8 Examples for Goal 1.2(8). Let L be any lattice which satisfies each of the following conditions:

1. L is a non-Boolean frame;

2. L is a completely distributive deMorgan algebra.

Examples of such lattices include any non-trivial complete deMorgan chain such as $\mathbb{I}$, any complete atomic Boolean algebra with new upper and lower bounds adjoined (which algebra remains completely distributive and deMorgan and becomes non-Boolean—this includes the lattice L such that $\{\alpha \in L : \alpha < 1\}$ is isomorphic to the topology of the product of the Sierpinski space with itself), etc. Now consider $\mathbb{R}(L)$ and $\mathbb{I}(L)$. These spaces, by their construction, have no constant open sets other than $\underline{0}, \underline{1}$. Further, by Lemmas 2.2(3), and 2.3, these spaces are not L-sober and are not modified L-sober to any non-trivial degree. Now by Lemma 2.5, we have $\mathbb{I}(L)$ is ι-Hausdorff, i.e. $\iota(\mathbb{I}(L))$ is Hausdorff, hence sober. Thus $\mathbb{I}(L)$ is ι-sober. The ι-sobriety of $\mathbb{R}(L)$ now follows from that of $\mathbb{I}(L)$ by the following steps:

1. We have from Lemma 2.9(2) that ι respects subspaces, and the L-fuzzy open unit interval $(0,1)(L)$ of [2] is a subspace of $\mathbb{I}(L)$. Hence $\iota((0,1)(L))$ is a subspace of the Hausdorff space $\iota(\mathbb{I}(L))$ and so is Hausdorff, hence sober. Thus $(0,1)(L)$ is ι-sober.

2. By Lemma 2.10 we have that ι takes L-homeomorphisms to homeomorphisms. Now from [2] we have $\mathbb{R}(L)$ is L-homeomorphic to $(0,1)(L)$—the proof of [2] assumes that $L \in |\mathbf{Hut}|$, which is satisfied under the above conditions. (This L-homeomorphism may be valid under less stringent conditions.) Thus $\iota((0,1)(L))$ is homeomorphic to $\iota(\mathbb{R}(L))$.

3. Traditional sobriety is a topological property (as a consequence of the universal properties of $\Omega \dashv PT$), which implies that $\iota(\mathbb{R}(L))$ is sober. Hence $\mathbb{R}(L)$ is ι-sober.

3.9 Remark (Consequences of Examples 3.1–3.8 and Justifications of Spectum Adjunctions).

1. The examples of Examples 3.1 are not accommodated by the modified L-spectrum adjunctions of [27, 28] for each degree $D \subset L$ with $D \neq \{0, 1\}$, but are accommodated by, and hence justify the standard L-spectrum adjunctions of [3, 4, 24, 13, 27, 28, 29], as well as the variable-basis spectrum adjunctions of [31].

2. The examples of Examples 3.1–3.2 are not accommodated by the standard L-spectrum adjunction, but are accommodated by, and hence justify, the modified L-spectrum adjunction using modified sobriety for each degree $D \subset L$.

3. The examples of Examples 3.5 are not accommodated by the standard L-spectrum adjunction, but are accommodated by, and hence justify, both the modified L-spectrum adjunction using modified sobriety to degree $D = L$ as well as the spectrum adjunction of [21].

4. The examples of Examples 3.6–3.7 are not accommodated by the standard L-spectrum adjunction. Further, fixing L for Examples 3.6–3.7, we can state: *for each* $D \subset L$, *there exists* ι-sober spaces not accommodated by the modified L-spectrum adjunction for that degree D. For such fixed L, the only spectrum adjunction accommodating all the ι-sober spaces is that of [21]. This is sufficient to justify the spectrum adjunction for [21] *vis-a-vis* any previous L-spectrum and modified L-spectrum adjunctions in the literature. However, a much stronger justification for [21] is given our next remark for much more restricted L.

5. The examples of Examples 3.8 are not accommodated by the standard L-spectrum adjunction. Fixing L for Examples 3.8, we can state: *there exists* ι-sober spaces such that *for each* $D \subset L$, these spaces are not accommodated by the modified L-spectrum adjunction for that degree D. Restated, for such fixed L, there are ι-sober spaces which are *uniformly* non-accommodated by any previous L-spectrum and modified L-spectrum adjunctions. *A fortiori*, such spaces are only accommodated by the spectrum adjunction of [21]. So this remark strengthens the preceding remark for the L of Examples 3.8 due to an exchange of quantifiers.

References

[1] S. Barger, *Hereditary sobriety and cardinality of topologies on countable sets*, Quaestiones Mathematicae **20**(1997), 117–126.

[2] T. E. Gantner, R. C. Steinlage, R. H. Warren, *Compactness in fuzzy topological spaces*, J. Math. Anal. Appl. **62**(1978), 547–562.

[3] U. Höhle, *Fuzzy topologies and the topological space objects in a topos*, Fuzzy Sets and Systems **19**(1986) 299–304.

[4] U. Höhle, *Locales and L-topologies*, Mathematik-Arbeitspapiere **48**(1997), 233-250, Universität Bremen.

[5] U. Höhle, Many Valued Topology And Its Applications, Kluwer Academic Publishers (Boston/Dordrecht/London), 2001.

[6] U. Höhle, S. E. Rodabaugh, eds, Mathematics Of Fuzzy Sets: Logic, Topology, And Measure Theory, The Handbooks of Fuzzy Sets Series, Volume **3**, Kluwer Academic Publishers (Boston/Dordrecht/London), 1999.

[7] U. Höhle, S. E. Rodabaugh, *Weakening the requirement that L be a complete chain*, Appendix to Chapter 6 in this Volume.

[8] R.-E. Hoffmann, *Irreducible filters and sober spaces*, Manuscripta Math. **22**(1977), 365-380.

[9] R.-E. Hoffmann, *Soberification of partially ordered sets*, Semigroup Forum **17**(1979), 123-138.

[10] R.-E. Hoffmann, *On the soberification remainder $sX - X$*, Pacific J. Math. **83**(1979), 145–156.

[11] B. Hutton, *Normality in fuzzy topological spaces*, J. Math. Anal. Appl. **50**(1975), 74–79.

[12] P. T. Johnstone, Stone Spaces, Cambridge University Press (Cambridge) 1982.

[13] W. Kotzé, *Lattice morphisms, sobriety, and morphisms*, Chapter 10 in [33], 257–274.

[14] W. Kotzé, *Sobriety and semi-sobriety of L-topological spaces*, Quaestiones Mathematicae **24**(2001), 549–554.

[15] W. Kotzé, *Lifting of sobriety concepts with particular references to (L, M)-topological spaces*, Chapter 16 in this Volume.

[16] T. Kubiak, *The topological modification of the L-fuzzy unit interval*, Chapter 11 in [33], 275–305.

[17] T. Kubiak, *Separation axioms: extension of mappings and embedding of spaces*, Chapter 6 in [6], 433–479.

[18] R. Lowen, *Fuzzy topological spaces and fuzzy compactness*, J. Math. Anal. Appl. **56**(1976), 621–633.

[19] G. H. J. Meßner, *Sobriety and the fuzzy real lines*, Chapter in Ph.D. Thesis, Johannes Kepler Universität (Linz), 1987.

[20] Pu Bao-Ming, Liu Ying-Ming, *Fuzzy topology I*, J. Math. Anal. Appl. **76**(1980), 571–599; *Fuzzy topology II*, J. Math. Anal. Appl. **77**(1980), 20–37.

[21] A. Pultr, S. E. Rodabaugh, *Lattice-valued frames, functor categories, and classes of sober spaces*, Chapter 6 of this Volume.

[22] A. Pultr, A. Tozzi, *Separation facts and frame representation of some topological facts*, Applied Categorical Structures **2**(1994), 107–118.

[23] A. Pultr, A. Tozzi, *A note on reconstruction of spaces and maps from lattice data*, Quaestiones Mathematicae **24**(2001), 55–63.

[24] S. E. Rodabaugh, *A point set lattice-theoretic framework $\mathbb{T}$ which contains LOC as a subcategory of singleton spaces and in which there are general classes of Stone representation and compactification theorems*, first draft February 1986 / second draft April 1987, Youngstown State University Central Printing Office (Youngstown, Ohio).

[25] S. E. Rodabaugh, *Dynamic topologies and their applications to crisp topologies, fuzzifications of crisp topologies, and fuzzy topologies on the crisp real line*, J. Math. Anal. Appl. **131**(1988), 25–66.

[26] S. E. Rodabaugh, *Lowen, para-Lowen, and α-level functors and fuzzy topologies on the crisp real line*, J. Math. Anal. Appl. **131**(1988), 157–169.

[27] S. E. Rodabaugh, *Point-set lattice-theoretic topology*, Fuzzy Sets and Systems **40**(1991), 297–345.

[28] S. E. Rodabaugh, *Categorical frameworks for Stone representation theorems*, Chapter 7 in [33], 178-231.

[29] S. E. Rodabaugh, *Applications of localic separation axioms, compactness axioms, representations, and compactifications to poslat topological spaces*, Fuzzy Sets and Systems **73**(1995) 55–87.

[30] S. E. Rodabaugh, *Categorical foundations of variable-basis fuzzy topology*, Chapter 4 in [6], 273–388.

[31] S. E. Rodabaugh, *Separation axioms: representation theorems, compactness, and compactifications*, Chapter 7 in [6], 481–552.

[32] S. E. Rodabaugh, *Fuzzy real lines and dual real lines as poslat topological, uniform, and metric ordered semirings with unity*, Chapter 10 in [6], 607–631.

[33] S. E. Rodabaugh, E. P. Klement, U. Höhle, eds, Application Of Category To Fuzzy Sets, Theory and Decision Library—Series B: Mathematical and Statistical Methods, Volume **14**, Kluwer Academic Publishers (Boston/Dordrecht/London), 1992.

CHAPTER 18

Additive Generators Of Non-Continuous Triangular Norms

P. VICENÍK [1]

Introduction

This chapter discusses construction of t-norms by means of additive generators and introduces some necessary and sufficient conditions of associtivity of generated functions. The case of left-continuous additive generators of t-norms is completely analyzed. Examples of additive generators of t-norms whose ranges are not relatively closed under addition are given.

Since Abel [1] it is known that a strictly monotone surjective function $f : [a,b] \to [0,\infty]$ generates an associative two-place function $F : [a,b]^2 \to [a,b]$ via $F(x,y) = f^{-1}(f(x)+f(y))$. The function f is called an additive generator of F. Several generalizations of the concept of an additive generator have been introduced and investigated in the framework of triangular norms; e.g., see [3], [4], [6]. The main interest in this topic has been shaped by the relaxing of the requirements of the surjectivity and continuity of the additive generator f.

1 Preliminaries

We recall some known basic concepts.

1.1 Definition. A **triangular norm** (briefly **t-norm**) is a function $T : [0,1]^2 \to [0,1]$ such that for all $x,y,z \in [0,1]$ T satisfies:

$T(x,y) = T(y,x)$	(commutativity)
$T(T(x,y),z) = T(x,T(y,z))$	(associativity)
$x \leq y \Rightarrow T(x,z) \leq T(y,z)$	(monotonicity)
$T(x,1) = x$	(boundary conditon)

[1]This work was supported by grants VEGA 1/4064/97 and VEGA 2/6087/99.

S.E. Rodabaugh and E.P. Klement (eds.), Topological and Algebraic Structures in Fuzzy Sets, 441–454.
© 2003 *Kluwer Academic Publishers.*

1.2 Definition. Let $f : [a,b] \to [c,d]$ $(a,b,c,d \in [-\infty,\infty],\ a < b,\ c < d)$ be a non-increasing (non-decreasing) function. Then the function $f^{(-1)} : [c,d] \to [a,b]$ defined by

$$f^{(-1)}(y) = \sup\{x \in [a,b] : f(x) > y\}$$

$$(f^{(-1)}(y) = \sup\{x \in [a,b] : f(x) < y\})$$

is called the **pseudo-inverse of a non-increasing (non-decreasing) function** f.

Note that $\sup \emptyset = a$. This function $f^{(-1)}$ is a non-increasing (non-decreasing) function if f is a non-increasing (non-decreasing) function. Recall that the strict monotonicity of f ensures the continuity of its pseudo-inverse $f^{(-1)}$; see [3], [4], [6].

1.3 Definition [5]. Let $f : [0,1] \to [0,\infty]$ be a strictly decreasing function, $f(1) = 0$ and let the function $T : [0,1]^2 \to [0,1]$ be given by formula

$$T(x,y) = f^{(-1)}(f(x) + f(y)) \qquad \text{for } \forall x,y \in [0,1] \qquad (1)$$

where $f^{(-1)}$ is the pseudo-inverse of f. Then the function f is called a **conjunctive additive generator of** T.

1.4 Remark. Note that the value of a conjunctive additive generator f at point 0 has no influence on the generated function T.

A function T defined by (1) is always commutative, non-decreasing, fulfils the boundary condition but need not be associative.

1.5 Proposition [2]. *Let $f : [0,1] \to [0,\infty]$ be a conjunctive additive generator of T. Then T is a t-norm if and only if T is an associative function.*

1.6 Definition. Let $f : [0,1] \to [0,\infty]$ be a conjunctive additive generator of T. The function f is called an **additive generator of** T iff T is a t-norm.

We will say that T is a **generated t-norm** iff T is a t-norm and T has at least one conjunctive additve generator. The set of all generated t-norms is a proper subset of the set of all t-norms.

In general, if we have a given t-norm T, then we are not able to decide if T is a generated function or not; and on the other hand, if we have a conjunctive additive generator f of T, then we are not able to decide if the corresponding generated function T is an associative function or not. We can formulate the following two basic problems: How can we characterize the set of all additive generators? How can we characterize the set of all generated t-norms?

Let T be a t-norm, $x \in [0,1]$, and define

$$x_T^{(n)} = \begin{cases} 1 & \text{if } n = 0 \\ T(x_T^{(n-1)}, x) & \text{if } n \in \{1, 2 \dots\} \end{cases}$$

1.7 Definition. Let $T : [0,1]^2 \to [0,1]$ be a t-norm. We will say that T is **Archimedean** if $\lim_{n\to\infty} x_T^{(n)} = 0$ for all $x \in (0,1)$.

For $a \in (0,1]$, denote $f(a_-) = \lim_{x\to a^-} f(x)$ and for $a \in [0,1)$, $f(a_+) = \lim_{x\to a^+} f(x)$.

1.8 Remark. Let $f : [0,1] \to [0,\infty]$ be a conjunctive additive generator of T. Then any of the following conditions ensure the associativity of T:

1. The continuity of f on $(0,1]$.
2. The relatively closed range of f under addition (for all $x, y \in [0,1]$ we have $f(x) + f(y) \in Ran(f)$ or $f(x) + f(y) \geq f(0_+)$)—see [2].
3. If f satisfies (2), then T is an Archimedean t-norm; and if f satisfies (1), then T is a continuous Archimedean t-norm.

It is a well known result that T is a continuous Archimedean t-norm if and only if there exists a continuous conjunctive additive generator f of T. The set of all continuous Archimedean t-norms is a subset of the set of all generated t-norms. The relatively closed range of a conjunctive additive generator f under addition is a sufficient condition of the associativity of the corresponding generated function T and in this case, T is an Archimedean t-norm. Recent research shows that there are generated t-norms which are not Archimedean. The following problem is still open: What is a characterization of the set of all additive generators whose range is not relatively closed under addition?

1.9 Proposition. *Let a function $f : [0,1] \to [0,\infty]$ be a conjunctive additive generator of T and let f be left-continuous at point 1. Then the function f is continuous on interval $(0,1]$ if and only if $Ran(f)$ is relatively closed under addition.*

Proof. It is obvious that if a function f is continuous on interval $(0,1]$, then $Ran(f)$ is relatively closed under addition. We prove only assertion that if function f is not continuous on interval $(0,1]$ then $Ran(f)$ is not relatively closed under addition. Let f be discontinuous at point $a \in (0,1)$. Denote $\epsilon = f(a_-) - f(a_+) > 0$. As far as f is strictly decreasing and $f(1_-) = 0$, there exist points $x, y \in (a,1)$, $x \neq y$, such that $0 < f(x) < \epsilon$, $0 < f(y) < \epsilon$

and $f(x) \neq f(y)$. We can choose a point $z \in (a,1)$ such that $f(a_+) - f(x) < f(z) < f(a_+)$ and $f(a_+) - f(y) < f(z) < f(a_+)$. Hence, we obtain $f(a_+) < f(x)+f(z) < f(a_-)$, $f(a_+) < f(y)+f(z) < f(a_-)$ and $f(x)+f(z) \neq f(y)+f(z)$. Since $(f(a_+), f(a_-)) \cap Ran(f) \subseteq \{f(a)\}$, at least one of the values $f(x) + f(z)$ and $f(y) + f(z)$ does not belong to $Ran(f)$. From $f(x) + f(z) < f(0_+)$ and $f(y)+f(z) < f(0_+)$, we have that $Ran(f)$ is not relatively closed under addition. □

1.10 Remark. Note there are non-continuous additive generators on $(0,1]$ whose ranges are relatively closed under addition (see [2]) but having $f(1_-) > 0$.

2 Continuity of generated functions

In this section we discuss the relation between the continuity of a conjunctive additive generator and the continuity of the corresponding generated function.

Let $\mathcal{F}$ be the set of all conjunctive additive generators, namely

$$\mathcal{F} = \{\ f : [0,1] \to [0,\infty] \mid f \text{ is strictly decreasing and } f(1) = 0\}$$

2.1 Lemma. *Let $0 \leq c < d \leq \infty$, $f \in \mathcal{F}$. Then the following assertions are equivalent:*

1. $f^{(-1)}/_{[c,d]}$ *is constant.*
2. $f^{(-1)}/_{(c,d)}$ *is constant.*
3. $Ran(f) \cap [c,d]$ *is at most one-element set.*
4. $Ran(f) \cap (c,d)$ *is at most one-element set.*
5. $[c,d] \subset [0, f(1_-)]$ *or* $[c,d] \subset [f(0_+), \infty]$ *or* $\exists a \in (0,1)$ *such that* $[c,d] \subset [f(a_+), f(a_-)]$.

2.2 Remark. Let $f \in \mathcal{F}$, $a \in [0,1]$, and $u \in [0,\infty]$.

1. If $0 < a < 1$ and $f(a_+) \leq u \leq f(a_-)$, then $f^{(-1)}(u) = a$.
2. If $0 < a$ and $f(a_-) < u$, then $a > f^{(-1)}(u)$.
3. If $a < 1$ and $u < f(a_+)$, then $f^{(-1)}(u) > a$.

2.3 Proposition. *Let f be a conjunctive additive generator of T satisfying $f(0_-) = f(0)$ and $f(1_+) = f(1) = 0$.*

1. *T is continuous at point (x, y) if and only if $Ran(f) \cap [f(x_+) + f(y_+), f(x_-) + f(y_-)]$ is at most a one-element set.*
2. *T is left-continuous at point (x, y) if and only if $Ran(f) \cap [f(x) + f(y), f(x_-) + f(y_-)]$ is at most a one-element set.*
3. *T is right-continuous at point (x, y) if and only if $Ran(f) \cap [f(x_+) + f(y_+), f(x) + f(y)]$ is at most a one-element set.*

Proof. *Ad (1).* Since the function $f^{(-1)}$ is continuous and f is a strictly decreasing function, T is continuous at point (x, y) if and only if $f^{(-1)}(f(x_+) + f(y_+)) = f^{(-1)}(f(x) + f(y)) = f^{(-1)}(f(x_-) + f(y_-))$, which is equivalent to constantness of $f^{(-1)}$ on $[f(x_+) + f(y_+), f(x_-) + f(y_-)]$; and it is equivalent to the assertion that $Ran(f) \cap [f(x_+) + f(y_+), f(x_-) + f(y_-)]$ is at least a one-element set by (1) and (3) of Lemma 2.1. The proof of the other cases is very similar. □

2.4 Theorem. *Let f be a conjunctive additive generator of T.*

1. *T is left-continuous at $(1, 1)$ if and only if f is left-continuous at point 1.*
2. *T is continuous if and only if f is continuous on $(0, 1]$.*
3. *T is left-continuous if and only if f is left continuous.*
4. *If f is right-continuous on $(0, 1)$, then T is right-continuous.*

Proof. *Ad (1).* $f(1_-) > 0 = f(0)$ is equivalent with $T(1_-, 1_-) = f^{(-1)}(2f(1_-)) < 1 = T(1, 1)$, i.e., f is not left-continuous at point 1 if and only if T is not left-continuous at point $(1, 1)$.

Ad (2). If f is continuous function on $(0, 1]$, then T is also continuous by Proposition 2.3(1). We prove that if f is not continuous on $(0, 1]$, then T is not continuous. If $f(1_-) > 0$, then T is non-continuous at point $(1, 1)$ because of (1). If $f(1_-) = 0$, then there exists $x \in (0, 1)$ such that $f(x_-) - f(x_+) > 0$. Choose $y \in (0, 1)$ with $f(y_+) < f(x_-) - f(x_+)$. Then we obtain that $f(x_+) + f(y_+) < f(x_-) < f(x_-) + f(y_-)$. So we have that $Ran(f) \cap [f(x_+) + f(y_+), f(x_-) + f(y_-)]$ is an infinite set and T is non-continuous at point (x, y) by Proposition 2.3(1).

The proof of (3) is similar to the previous case and (4) is a simple consequence of Proposition 2.3. □

2.5 Remark.

1. Let f be a conjunctive additive generator of T. Then f is left-continuous at point 1 if and only if T is continuous on the boundary of the unit squere $[0,1]^2$. It follows from Propositon 2.3(1).

2. Note there are examples of right-continuous t-norms with additive generators which are not right-continuous. This example was introduced in [3]: Let $f \in \mathcal{F}$ and $\exists a \in (0,\infty)$ such that $\forall x \in (0,1)$, $a < f(x) < 2a$; then f is an additive generator of right-continuous t-norm T_D ($T_D(x,y) = 0$ if $(x,y) \in [0,1)^2$ and $T_D(x,y) = \min(x,y)$ if $\max(x,y) = 1$).

3 The case of left continuous additive generators

The left-continuous t-norms play important role in fuzzy logics. In this section we discuss the case of left-continuous generated t-norms. First we introduce some notation:

$$D_f = \{a \in (0,1) \mid f(a_-) > f(a_+)\},$$

$$L_-(f) = \{v \in [0,\infty) \mid \exists t \in (0,1],\, v = f(t_-)\},$$

$$H_-(f) = \{v \in [0,\infty) \mid \exists a \in D_f,\, v = f(a_-) - f(a)\}$$

The next theorem gives one of the necessary conditions of associativity of T.

3.1 Theorem. *Let $f : [0,1] \to [0,\infty]$ be a conjunctive additive generator of T. If T is a t-norm, then $H_-(f) \cap L_-(f) = \emptyset$.*

Proof. Suppose that f is a conjunctive additive generator of a t-norm T and $H_-(f) \cap L_-(f)$ is not empty set. We will consider two cases.

I. If $0 \in H_-(f) \cap L_-(f)$, then $\exists a \in (0,1)$ such that $f(a_-) = f(a) > f(a_+)$ and $f(1_-) = 0$. Denote $\epsilon = f(a_-) - f(a_+) > 0$ and $i = inf\{t \in [0,1] : f(t) < \epsilon\}$. It is obvious that $a \leq i < 1$. For arbitrary $x, z \in (i,1)$ we have $0 < f(x) < \epsilon$, $0 < f(z) < \epsilon$ and put $m = \min(f(x), f(z))$. Choose $y \in (a,1)$ such that $f(a_+) - m < f(y) < f(a_+)$. Then $f(a_+) < f(x) + f(y) < f(a_-)$ and $f(a_+) < f(y) + f(z) < f(a_-)$; and because of $f^{(-1)}(u) = a$, for all $u \in [f(a_+), f(a_-)]$ we have $f^{(-1)}(f(x) + f(y)) = f^{(-1)}(f(y) + f(z)) = a$. We proved that for arbitrary $x, z \in (i,1)$, there exists $y \in (a,1)$ such that $T(x,y) = T(y,z) = a$. By commutativity and associativity of T we obtain

$$T(x,a) = T(x, T(y,z)) = T(T(x,y), z) = T(a,z) = T(z,a)$$

This shows that $T(x,a) = T(z,a)$ for all $x, z \in (i,1)$.

Let $g(t) = T(t,a)$ for all $t \in (i,1]$. It is clear that $g(t) = c$ for all $t \in (i,1)$ and some $c \in [0,1]$, and $g(1) = T(1,a) = a$. Due to the left-continuity of f at point 1 ($f(1_-) = 0$) and continuity of $f^{(-1)}$ we have

$$c = \lim_{t\to 1^-} g(t) = \lim_{t\to 1^-} T(t,a) = f^{(-1)}(f(1_-) + f(a)) = f^{(-1)}(f(a)) = a$$

Thus $c = a$ and $g(t) = a$ for all $t \in (i,1]$.

Now let $s = \sup\{f(t) : t \in (i,1]\}$ and let $h \in [0,s)$. Then due to the properties of supremum s, there exists $x \in (i,1)$ such that $0 \leq h < f(x) < s$. Then $f(a) \leq f(a) + h < f(a) + f(x)$ and since the function $f^{(-1)}$ is non-increasing, we have

$$a = f^{(-1)}(f(a)) \geq f^{(-1)}(f(a)+h) \geq f^{(-1)}(f(a)+f(x)) = g(x) = a$$

We proved that $a = f^{(-1)}(f(a)+h) = \sup\{t \in [0,1] : f(t) > f(a)+h\}$ for all $h \in [0,s)$. This implies that $f(t) > f(a)+h$ for all $t \in [0,a)$ and for all $h \in [0,s)$. So we have $f(a_-) \geq f(a)+s > f(a)$, which is in contradiction with $f(a_-) = f(a)$.

II. If the set $H_-(f) \cap L_-(f)$ contains some positive number ϵ, then $\exists a \in (0,1)$ and $u \in (0,1]$ such that $\epsilon = f(a_-) - f(a) = f(u_-) > 0$. It is obvious that $a < u \leq 1$. Take $y \in (a,1)$ such that $f(a_+) - \epsilon < f(y) < f(a_+)$ and denote $i = inf\{t \in [0,1] : f(t) < \epsilon + (f(a_+) - f(y))\}$. It is obvious that $a \leq i < u$. Then for arbitrary $x, z \in (i,u)$ we have $\epsilon < f(x) < \epsilon + (f(a_+) - f(y))$ and $\epsilon < f(z) < \epsilon + (f(a_+) - f(y))$. Thus $f(a_+) < f(x) + f(y) < f(a_-)$ and $f(a_+) < f(y) + f(z) < f(a_-)$ and because of $f^{(-1)}(u) = a$ for all $u \in [f(a_+), f(a_-)]$ we have $f^{(-1)}(f(x)+f(y)) = f^{(-1)}(f(y)+f(z)) = a$, i.e., $T(x,y) = T(y,z) = a$. By commutativity and associativity of T we have

$$T(x,a) = T(x,T(y,z)) = T(T(x,y),z) = T(a,z) = T(z,a)$$

We have proved that $T(x,a) = T(z,a)$ for all $x, z \in (i,u)$.

Let $g(t) = T(t,a)$ for all $t \in (i,u]$. It is clear that $g(t) = c$ for all $t \in (i,u)$ and some $c \in [0,1]$. From inequality $f(a) \leq f(a) + f(u) \leq f(a) + \epsilon = f(a_-)$ and monotonicity of $f^{(-1)}$, we get $a = f^{(-1)}(f(a)) \geq f^{(-1)}(f(u) + f(a)) \geq f^{(-1)}(f(a_-)) = a$, i. e., $g(u) = f^{(-1)}(f(u) + f(a)) = a$. Due to continuity of $f^{(-1)}$ and $\epsilon = f(u_-) = f(a_-) - f(a)$ we have

$$c = \lim_{t\to u^-} g(t) = \lim_{t\to u^-} T(t,a) = f^{(-1)}(f(u_-) + f(a)) = f^{(-1)}(f(a_-)) = a$$

Thus $c = a$ and $g(t) = a$ for all $t \in (i,u]$.

Now let $s = \sup\{f(t) : t \in (i,u]\}$—obviously $\epsilon < s$—and let $h \in [\epsilon, s)$. Then due to properties of supremum s, there exists $x \in (i,u)$ such that $\epsilon \leq h < f(x) < s$. Then $f(a_-) = f(a) + \epsilon \leq f(a) + h < f(a) + f(x)$ and as since the function $f^{(-1)}$ is non-increasing, we have

$$a = f^{(-1)}(f(a_-)) \geq f^{(-1)}(f(a)+h) \geq f^{(-1)}(f(a)+f(x)) = g(x) = a$$

We proved that $a = f^{(-1)}(f(a)+h) = \sup\{t \in [0,1] : f(t) > f(a)+h\}$ for all $h \in [\epsilon, s)$. It implies that $f(t) > f(a)+h$ for all $t \in [0,a)$ and for all $h \in [\epsilon, s)$. Hence, obviously $f(a_-) \geq f(a)+s > f(a)+\epsilon$ which is in contradiction with $f(a_-) = f(a)+\epsilon$. We have proved that if T is a t-norm, then the set $H_-(f) \cap L_-(f)$ is empty. □

3.2 Corollary. *Let f be a conjunctive additive generator of T which is left-continuous at point 1 and suppose $D_f \neq \emptyset$. If T is a t-norm, then $f(a_-)-f(a) > 0$ for all $a \in D_f$.*

Proof. From $f(1_-) = 0$ we have $0 \in L_-(f)$, and because $H_-(f) \cap L_-(f) = \emptyset$ by Theorem 3.1, we obtain that $0 \notin H_-(f)$, which is equivalent to $f(a_-)-f(a) \neq 0$ for each $a \in D_f$. Due to the monotonicity of f we have $f(a_-)-f(a) > 0$ for all $a \in D_f$. □

3.3 Corollary. *Let f be a left-continuous conjunctive additive generator of T. Then T is a t-norm if and only if f is a continuous function on interval* $(0,1]$.

Proof. We prove only that if T is a t-norm, then f is a continuous function on $(0,1]$ since the converse is obvious. If f is a left-continuous function, then $H_-(f) \subset \{0\} \subset L_-(f)$ and the set $H_-(f) \cap L_-(f) = \emptyset$ by Theorem 3.1. Hence, $H_-(f) = \emptyset$. This is posible if and only if the set D_f is empty, which together with $f(1_-) = 0$, shows that f is continuous on $(0,1]$. □

3.4 Corollary. *Let f be a conjunctive additive generator of t-norm T. Then T is a left-continuous t-norm if and only if T is a continuous Archimedean t-norm.*

Proof. If T is a left-continuous generated t-norm with an additive generator f, then by Theorem 2.4(3), f is a left-continuous function, and by Corollary 3.3, f is a continuous function on $(0,1]$. The rest of the proof follows immediately from Remark 1.8(3). □

By the one-to-one correspondence between additive and multiplicative generators, all presented results also can be rewritten for multiplicative generators.

4 Some other necessary conditions

Let $f \in \mathcal{F}$. If $a \in D_f$, then $2f(a_+) \leq f(a_+)+f(a) \leq 2f(a)$. We devide the set D_f into four disjoint subsets.

$$\begin{aligned}
D_f^0 &= \{a \in (0,1) : 2f(a) \leq f(a_-)\}, \\
D_f^1 &= \{a \in (0,1) : f(a)+f(a_+) \leq f(a_-) < 2f(a)\}, \\
D_f^2 &= \{a \in (0,1) : 2f(a_+) \leq f(a_-) < f(a)+f(a_+)\}, \\
D_f^3 &= \{a \in (0,1) : a \in D_f \text{ and } f(a_-) < 2f(a_+)\}
\end{aligned}$$

Then $D_f = D_f^0 \cup D_f^1 \cup D_f^2 \cup D_f^3$ and $D_f^i \cap D_f^j = \emptyset$ for all $i, j \in \{0, 1, 2, 3\}$, $i \neq j$. Let $I_T = \{x \in (0,1) : T(x,x) = x\}$.

4.1 Theorem. *Let f be a conjunctive additive generator of T. Then $D_f^0 = I_T$.*

Proof. Let $a \in (0,1)$. Then $2f(a) \leq f(a_-)$ if and only if $f^{(-1)}(2f(a)) = a$. □

4.2 Remark. It is not difficult to prove that the set D_f^0 is countable and by Theorem 4.1 we know that the set I_T of all non-trivial idempotent elements of generated t-norm T is countable. It means that if T is a t-norm and the set I_T is uncountable, then T is not a generated t-norm.

4.3 Proposition. *Let f be a conjunctive additive generator of T. If T is a t-norm, then $D_f^1 = \emptyset$.*

Proof. If $D_f^1 \neq \emptyset$, then for $a \in D_f^1$ and

$$x, y \in f^{-1}\left(\left(\frac{f(a_+)}{2}, f(a_+)\right)\right) = \{t \in [0,1] : \frac{f(a_+)}{2} < f(t) < f(a_+)\}$$

we have

$$T(T(x,y),a) = T(a,a) = f^{(-1)}(2f(a)) < a = T(x,a) = T(x, T(y,a))$$

violating the associativity of T. □

4.4 Proposition. *Let f be a conjunctive additive generator of T which is left-continuous at point 1. If T is a t-norm, then $D_f = D_f^0 \cup D_f^3$.*

Proof. With respect to the assertion of Proposition 4.3, it is sufficient to prove that $D_f^2 = \emptyset$. If $D_f^2 \neq \emptyset$ and $a \in D_f^2$, then we have $f(a_-) - f(a) > 0$ by Corollary 3.2. Denote $\epsilon = f(a_-) - f(a) > 0$. Then for the triple $x, y, z \in (a, 1)$ such that

$$0 < f(x) < \epsilon, \max(f(a_+) - f(x), \frac{f(a_+)}{2}) < f(y) < f(a_+)$$

and

$$\max(f(a_-) - f(a), \frac{f(a_+)}{2}) < f(z) < f(a_+)$$

(note only that $f(a_-) - f(a) < f(a_+)$ for $a \in D_f^2$), we obtain

$$T(T(x,y),z) = T(a,z) = f^{(-1)}(f(a) + f(z)) < a = T(x,a) = T(x, T(y,z))$$

which contradicts to the associativity of T. □

5 Some sufficient conditions

Now we introduce some sufficient conditions of associativity of generated functions. Let $M \subset [0,\infty]$ and

$$C(M) \equiv \{t \in [0,\infty] : \exists x_n \in M, x_n \neq t, (n \in \mathbb{N}) \text{ such that } t = \lim_{n\to\infty} x_n\}$$

We assume that $f \in \mathcal{F}$ and $D_f = D_f^0$ for the rest of this chapter. In this case we have:

1. $C(D_f) \subset \{0,1\}$.
2. For $D_f \neq \emptyset$, $1 \notin C(D_f)$ if and only if $\exists c \in D_f$ such that $c = \max D_f$
3. If $0 \in C(D_f)$, then $f(0_+) = \infty$.

If $D_f = D_f^0 \neq \emptyset$ and $\exists c \in D_f$ such that $c = \max D_f$, then we can define $\bar{a} = \max\{D_f \cap [0,a)\}$ ($\max \emptyset = 0$) for all $a \in D_f \cup \{1\}$, where obviously $\bar{a} < a$. The class $\{[\bar{a}, a) : a \in D_f \cup \{1\}\}$ is a disjoint countable partition of the interval $[0,1)$ (if $0 \notin C(D_f)$) or interval (0,1) (if $0 \in C(D_f)$).

Note that $f(0_-) = \infty$, $f(1_+) = 0$ by convention for the rest of this chapter.

5.1 Lemma. *Let $f \in \mathcal{F}$. Then*

1. $f^{(-1)}(f(x)) = x$ *for* $\forall\, x \in [0,1]$.
2. $f(f^{(-1)}(f(x))) = f(x)$ *for* $\forall x \in [0,1]$.
3. $f(f^{(-1)}(u)) = u$ *for* $\forall\, u \in Ran(f)$.
4. *If* $a \in [0,1]$ *and* $f(a_+) < f(a_-)$, *then* $f^{(-1)}(y) = a$ *for* $\forall y \in [f(a_+), f(a_-)]$.

5.2 Lemma. *Let f be a conjunctive additive generator of T and $D_f = D_f^0 \neq \emptyset$. Then these assertions are equivalent:*

1. *There exists* $c = \max D_f$ *and* $f(\bar{a}_+) - f(a_-) \leq f(c_+)$ *for all* $a \in D_f$.
2. $f(\bar{a}_+) - f(a_-) \leq f(b_+)$ *for each* $a, b \in D_f$.

Proof. We prove (1) implies (2). Let there exist $c = \max D_f$ and let $f(\bar{a}_+) - f(a_-) \leq f(c_+)$ for each $a \in D_f$. Then for all $b \in D_f$ we have $b \leq c$; and due to the monotonicity of f, $f(c_+) \leq f(b_+)$. Hence $f(\bar{a}_+) - f(a_-) \leq f(c_+) \leq f(b_+)$ for all $a, b \in D_f$.

Now we prove (2) implies (1). First we prove there exists maximum c of D_f. If set D_f does not have the maximum, then $1 \in C(D_f)$ and there exists strictly increasing sequence $\{x_n\}$, $x_n \in D_f$, $(n \in N)$, such that $\lim_{n\to\infty} x_n = 1$. Since

$D_f = D_f^0$ and f is strictly decreasing, $\lim_{n\to\infty} f(x_n) = 0$. Fix an arbitrary $a \in D_f$. Then $\exists m \in N$ such that $f(\bar{a}_+) - f(a_-) > f(x_m) \geq f((x_m)_+)$, which contradicts (2). This means there exists $c = \max D_f$. If $b = c$, then $f(\bar{a}_+) - f(a_-) \leq f(c_+)$ for all $a \in D_f$. □

We frequently use the next two remarks, Remarks 5.3 and Remark 5.4, in the proof of Theorem 5.5.

5.3 Remark. Let $f \in \mathcal{F}$, $D_f = D_f^0 \neq \emptyset$, $1 \notin C(D_f)$, $a \in D_f \cup \{1\}$, $\bar{a} = \max\{[0,a) \cap D_f\}$, and $x, y, z \in [0,1]$. Then:

1. If $\bar{a} \leq x, y$ and $f(\bar{a}_+) \leq f(x) + f(y)$, then $f(\bar{a}_+) \leq f(x) + f(y) \leq f(\bar{a}_-)$ and $f^{(-1)}(f(x) + f(y)) = \bar{a}$.
2. If $\bar{a} \leq z$, then $f(\bar{a}_+) \leq f(z) + f(\bar{a}) \leq f(\bar{a}_-)$ and $f^{(-1)}(f(z) + f(\bar{a})) = \bar{a}$.

5.4 Remark. Let $f \in \mathcal{F}$, $D_f = D_f^0 \neq \emptyset$, $a, b \in D_f$, $a \leq b$, $\bar{a} = \max\{[0,a) \cap D_f\}$, $f(\bar{a}_+) - f(a_-) \leq f(b_+)$. The following holds: if $x \in [\bar{a}, a)$, then $f(\bar{a}_+) \leq f(x) + f(b_+) \leq f(x) + f(b) \leq f(\bar{a}_-)$ and $f^{(-1)}(f(x) + f(b_+)) = f^{(-1)}(f(x) + f(b)) = \bar{a}$.

5.5 Theorem. *Let f be a conjunctive additive generator of T and $D_f = D_f^0 \neq \emptyset$. If there exists $c = \max D_f$ and $f(\bar{a}_+) - f(a_-) \leq f(c_+)$ for all $a \in D_f$, then T is a t-norm.*

Proof. With respect to the assertion of Proposition 1.5, we have to prove the associativity of T, i.e., $T(T(x,y),z) = T(x,T(y,z))$ for all $x, y, z \in [0,1]$. Let $x, y, z \in [0,1]$, denote

$$L = T(T(x,y),z) = f^{(-1)}(f(f^{(-1)}(f(x) + f(y))) + f(z))$$

and

$$R = T(x,T(y,z)) = f^{(-1)}(f(x) + f(f^{(-1)}(f(y) + f(z))))$$

If $\min(x,y,z) = 0$,, then obviously $L = R = 0$. If $\max(x,y,z) = 1$, then by Lemma 5.1(1, 2), $L = R$. If $x, y, z \in (0,1)$, then either both values $f(x) + f(y)$, $f(y) + f(z)$ are in $Ran(f)$ or at least one of the values $f(x) + f(y)$, $f(y) + f(z)$ is not in $Ran(f)$. In the first case by Lemma 5.1(3), we get $L = R = f^{(-1)}(f(x) + f(y) + f(z))$. For the rest of this proof we will discuss the second case.

Let $f(x) + f(y) \notin Ran(f)$ or $f(y) + f(z) \notin Ran(f)$ and let $\min(x,y,z) \in [\bar{a}, a)$ for some $a \in D_f \cup \{1\}$. Obviously, $\bar{a} \leq x, y, z$. Now, we will consider the following cases:

1. If $f(\bar{a}_+) \leq f(x) + f(y)$ and $f(\bar{a}_+) \leq f(y) + f(z)$, then $\bar{a} = f^{(-1)}(f(x) + f(y)) = f^{(-1)}(f(y) + f(z))$ by Remark 5.3(1); and $L = f^{(-1)}(f(\bar{a}) + f(z))$ and $R = f^{(-1)}(f(x) + f(\bar{a}))$. By (2) of Remark 5.3, we obtain $L = \bar{a} = R$.

2. If $f(\bar{a}_+) \leq f(x) + f(y)$ and $f(y) + f(z) < f(\bar{a}_+)$, then we discuss the following two cases (a) and (b):

   (a) If $f(y) + f(z) \in Ran(f)$, then by (3) of Lemma 5.1, $f(f^{(-1)}(f(y) + f(z))) = f(y)+f(z)$ and $R = f^{(-1)}(f(x)+f(y)+f(z))$. From $f(\bar{a}_+) \leq f(x) + f(y) + f(z) < f(x) + f(\bar{a}_+) \leq 2f(\bar{a}) \leq f(\bar{a}_-)$ and Lemma 5.1(4), we obtain $R = \bar{a}$. From Remark 5.3(1) we have $f^{(-1)}(f(x) + f(y)) = \bar{a}$. Then $L = f^{(-1)}(f(\bar{a})+f(z))$ and $L = \bar{a}$ by Remark 5.3(2). So we have $L = R = \bar{a}$.

   (b) If $f(y) + f(z) \notin Ran(f)$, then $\exists b \in D_f$, $a \leq b$ such that $f(b_+) \leq f(y) + f(z) \leq f(b_-)$. It is obvious that $a \leq b \leq y, z$, $\bar{a} \leq x < a$ and $a \in D_f$. From Remark 5.3(1) we obtain $f^{(-1)}(f(x)+f(y)) = \bar{a}$, hence $L = f^{(-1)}(f(\bar{a}) + f(z))$ and $L = \bar{a}$ by Remark 5.3(2). We know that $f^{(-1)}(f(y) + f(z)) = b$ and $R = f^{(-1)}(f(x) + f(b))$. From Remark 5.4 we have $f^{(-1)}(f(x) + f(b)) = \bar{a}$. It shows that $R = \bar{a}$ and hence $L = R = \bar{a}$.

3. The case when $f(x)+f(y) < f(\bar{a}_+)$ and $f(\bar{a}_+) \leq f(y)+f(z)$ is completely analogous to case (2) above.

4. If $f(x) + f(y) < f(\bar{a}_+)$ and $f(y) + f(z) < f(\bar{a}_+)$, then we consider the following three cases (a), (b), and (c):

   (a) If $f(x)+f(y) \in Ran(f)$ and $f(y)+f(z) \notin Ran(f)$, then $\exists b \in D_f$, $a \leq b$ such that $f(b_+) \leq f(y)+f(z) \leq f(b_-)$. Obviously $a \leq b \leq y, z$ and $\bar{a} \leq x < a$ and $a \in D_f$. By Lemma 5.1(3) we obtain $f(f^{(-1)}(f(x) + f(y))) = f(x)+f(y)$ and $L = f^{(-1)}(f(x)+f(y)+f(z))$. We know that $f^{(-1)}(f(y) + f(z)) = b$ and $R = f^{(-1)}(f(x) + f(b))$. From Remark 5.4 we have $R = f^{(-1)}(f(x) + f(b)) = \bar{a}$ and $f(\bar{a}_+) \leq f(x) + f(b_+)$. From the inequality $f(\bar{a}_+) \leq f(x) + f(b_+) \leq f(x) + f(y) + f(z) \leq f(x) + f(b_-) \leq 2f(\bar{a}) \leq f(\bar{a}_-)$ and Lemma 5.1(4), we obtain $L = \bar{a}$. So we have $L = R = \bar{a}$.

   (b) The case $f(x) + f(y) \notin Ran(f)$ and $f(y) + f(z) \in Ran(f)$ is completely analogous to case (a).

   (c) If $f(x) + f(y) \notin Ran(f)$ and $f(y) + f(z) \notin Ran(f)$, then $a \leq x, y, z$, a contradiction to the inequality $\min(x, y, z) < a$.

This completes the proof that T is a t-norm. $\square$

5.6 Example. Let $f : [0, 1] \to [0, \infty]$ be defined by

$$f(x) = \begin{cases} 7 - 3x & \text{if } x \in [0, \frac{1}{3}) \\ 4 - 3x & \text{if } x \in [\frac{1}{3}, \frac{2}{3}) \\ 3 - 3x & \text{if } x \in [\frac{2}{3}, 1] \end{cases}$$

Then f is an additive generator of a t-norm by Theorem 5.5.

5.7 Corollary. *Let f be a conjunctive additive generator of T, $f(1_-) = 0$ and $D_f = D_f^0 \neq \emptyset$. Then T is a t-norm if and only if there exists $c = \max D_f$ and $f(\bar{a}_+) - f(a_-) \leq f(c_+)$ for each $a \in D_f$.*

Proof. With respect to the claim of Theorem 5.5 and the equivalence assertions (1) and (2) of Lemma 5.2, it is sufficient to prove that if $\exists a, b \in D_f$ such that $f(\bar{a}_+) - f(a_-) > f(b_+)$, then T is not a t-norm. To show this, denote $\epsilon = f(\bar{a}_+) - f(a_-) - f(b_+) > 0$ (note $\epsilon = \infty$ is possible), take $x \in (b, 1)$ such that $0 < f(x) < \frac{\epsilon}{2}$ and take $z \in (\bar{a}, a)$ such that $f(a_-) < f(z) < f(a_-) + \frac{\epsilon}{2}$. Choose $y \in (b, 1)$ such that $f(b_+) - f(x) < f(y) < f(b_+)$ and $f(x) + f(y) \neq f(b)$. Then

$$f(b_+) < f(x) + f(y) < 2f(b) \leq f(b_-),$$

$$f(a_-) < f(y) + f(z) < f(b_+) + f(a_-) + \frac{\epsilon}{2}$$

and hence

$$f(a_-) < f(y) + f(z) < f(x) + f(y) + f(z) < f(b_+) + f(a_-) + \epsilon = f(\bar{a}_+)$$

This implies that $f(y) + f(z) \in Ran(f)$, $f(f^{(-1)}(f(y) + f(z))) = f(y) + f(z)$, and $T(x, T(y, z)) = f^{(-1)}(f(x) + f(y) + f(z))$. We know that $f^{(-1)}(f(x) + f(y)) = b$ and $T(T(x, y), z) = f^{(-1)}(f(b) + f(z))$. Due to the strict monotonicity of f, continuity of f at point $f^{(-1)}(f(x) + f(y) + f(z))$, and $f(x) + f(y) + f(z) \neq f(b) + f(z)$, we have $f^{(-1)}(f(b) + f(z)) \neq f^{(-1)}(f(x) + f(y) + f(z))$. This proves that T is not an associative function. □

5.8 Corollary. *Let f be a conjunctive additive generator of T and let a be the only point of discontinuity of f and $a \in (0, 1)$. Then T is a t-norm if and only if $2f(a) \leq f(a_-)$ and $f(0_+) - f(a_-) \leq f(a_+)$.*

Proof. If $a \in (0, 1)$ is the only point of discontinuity of f in $[0, 1]$ such that $2f(a) \leq f(a_-)$ and $f(0_+) - f(a_-) \leq f(a_+)$, then T is a t-norm by Theorem 5.5.

If $a \in (0, 1)$ is the only point of discontinuity of f in $[0, 1]$ and T is a t-norm, then $f(1_-) = 0$ and by Proposition 4.4 we have that either $a \in D_f^0$ or $a \in D_f^3$. If $a \in D_f^3$, then $f(a_-) < 2f(a_+) \leq f(a_+) + f(a)$ and $f(a_-) - f(a) < f(a_+)$. This means that $\exists u \in (a, 1]$ such that $v = f(u_-) = f(a_-) - f(a)$, i.e., $v \in H_-(f) \cap L_-(f)$, which contradicts the assertion of Theorem 3.1. This shows that point a must be an element of the set D_f^0, and by Corollary 5.7, we have $f(0_+) - f(a_-) \leq f(a_+)$, completing the proof. □

References

[1] N. H. Abel, *Untersuchungen der Funktionen zweier unabhängigen veränderlichen Größen x und y wie f(x,y), welche die Eigenschaft haben, daß f(z,f(x,y)) eine symmetrische Funktionen von x,y und z ist,* J. Reine Angew. Math. **1**(1826) 11–15.

[2] E. P. Klement, R. Mesiar, E. Pap, *Additive generators of t-norms which are not necessarily continuous,* EUFIT'96, Vol. 1, Aachen 1996, 70–73.

[3] E. P. Klement, R. Mesiar, E. Pap, Triangular Norms, Trends in Logic, Studia Logica Library, Volume **8**(2000), Kluwer Academic Publishers (Boston/Dordrecht/London).

[4] C. M. Ling, *Representation of associative functions,* Publ. Math. Debrecen **12**(1965), 189–212.

[5] B. Schweizer, A. Sklar, *Espaces Métriques Aléatoires,* Comptes Rendus Acad. Sci. 247, Paris, 1958.

[6] B. Schweizer, A. Sklar, Probabilistic Metric Spaces, North-Holland Elsevier (Amsterdam/New York), 1983.

[7] J. Tkadlec, *Triangular norms with continuous diagonals,* Tatra Mt. Math. Publ. **16**, 187–196.

[8] P. Viceník, *A note on generators of t-norms,* Busefal **75**(1998), 33–38.

[9] P. Viceník, *Additive generators and discontinuity,* Busefal **76**(1998), 25–28.

[10] P. Viceník, *Non-continuous generated t-norms,* E. P. Klement, S. E. Rodabaugh, eds, Abstracts of the 20th Linz Seminar on Fuzzy Set Theory: Topological and Algebraic Structures (Bildungshaus St. Magdelena, Linz, Austria, February 1999), pp. 9–10.

CHAPTER 19

Groups, T-Norms, And Families Of De Morgan Systems

C. WALKER AND E. WALKER

Introduction

Continuous Archimedean t-norms are generated by automorphisms of the unit interval with its usual order structure. Certain subgroups of the group A of automorphisms of the unit interval—the group operation being composition of functions—play an important role in the theory, for example the multiplicative group of positive real numbers, which is embedded in A by $r(x) = x^r$. Several standard families of t-norms are in natural one-to-one correspondence with subgroups of A. We examine this phenomenon, and various other group theoretic aspects of t-norm theory.

Operations on fuzzy sets are defined in terms of their membership functions, and as such, are basically operations on the unit interval. The standard operations suggested by Zadeh in 1965 [14] included minimum and maximum for "and" and "or", and the function $1-x$ for "complement." He also suggested the algebraic product xy as an alternative for "and." These are not the only ways to extend classical set theory consistently, and many people have suggested alternatives to these fuzzy set theoretic operations. A general definition for t-norms (and), t-conorms (or), and negations (not) has evolved, and several families of t-norms are now regularly referred to: Hamacher, Dombi, Yager, Schweizer, and Frank, to name a few. Such families are typically one-parameter families that are in one-to-one correspondence with some subset of the real numbers, such as the positive ones. The Hamacher family [7]

$$\left\{\frac{xy}{a + (1 - a)(x + y - xy)} : a > 0\right\}$$

S.E. Rodabaugh and E.P. Klement (eds.), Topological and Algebraic Structures in Fuzzy Sets, 455–467.
© 2003 Kluwer Academic Publishers.

is one example. These and other continuous Archimedean t-norms can be obtained from automorphisms of the lattice $([0,1], \leq)$. For the Hamacher family of t-norms, this family of automorphisms is

$$\left\{\frac{x}{x+a(1-x)} : a > 0\right\}$$

That is, the automorphism defined by $f_a(x) = x/\left(x + a(1-x)\right)$ generates the corresponding Hamacher t-norm in the sense that

$$\frac{xy}{a+(1-a)(x+y-xy)} = f_a^{-1}(f_a(x)f_a(y))$$

This family $\{f_a : a > 0\}$ of automorphisms is closed under composition and forms a subgroup of the group A of all automorphisms of the ordered unit interval. The principal fact that we want to point out is that several of the well known families of t-norms come from just a few simply expressed subgroups of A, or cosets or conjugates of such subgroups. Having generators expressed as compositions of simply expressed automorphisms makes some computations easier, and suggests new families. We provide a number of examples, and consider nilpotent as well as strict t-norms.

1 Definitions and notation

Triangular norms (t-norms) are central items of study in fuzzy set theory. A **t-norm** is a binary operation $\triangle$ on $\mathbb{I} = ([0,1], \leq)$, the unit interval with its usual order relation, such that, for all x, y, $z \in [0,1]$,

1. $1 \triangle x = x$
2. $x \triangle y = y \triangle x$
3. $(x \triangle y) \triangle z = x \triangle (y \triangle z)$
4. The operation $\triangle$ is increasing in each variable—that is, $x \leq x_1$ and $y \leq y_1$ imply that $x \triangle y \leq x_1 \triangle y_1$.

A dual notion to t-norm is that of **t-conorm**, a binary operation $\triangledown$ on $\mathbb{I}$ such that, for all x, y, $z \in [0,1]$,

1. $0 \triangledown x = x$
2. $x \triangledown y = y \triangledown x$
3. $(x \triangledown y) \triangledown z = x \triangledown (y \triangledown z)$
4. The operation $\triangledown$ is increasing in each variable.

A t-norm is **convex** if whenever $x \bigtriangleup y \leq c \leq x_1 \bigtriangleup y_1$, then there is an r between x and x_1 and an s between y and y_1 such that $c = r \bigtriangleup s$. This condition is equivalent to continuity in each variable. We write $a^{[n]}$ for $a \bigtriangleup a \bigtriangleup \cdots \bigtriangleup a$, the t-norm of a with itself n times. A t-norm $\bigtriangleup$ is **Archimedean** if for each $a, b \in (0,1)$, there is a positive integer n such that $a^{[n]} < b$.

A t-norm $\bigtriangleup$ is **nilpotent** if it is convex, Archimedean, and for $a \neq 1$, $a^{[n]} = 0$ for some positive integer n, the n depending on a. Those convex Archimedean t-norms that are not nilpotent are called **strict**. Multiplication xy is the generic example of a strict t-norm, and $(x+y-1) \vee 0$ the generic example of a nilpotent one.

An **automorphism** of $\mathbb{I}$ is a one-to-one mapping of $[0,1]$ onto $[0,1]$ such that $f(a) \leq f(b)$ if and only if $a \leq b$. An **anti-automorphism** of $\mathbb{I}$ is a one-to-one mapping g of $[0,1]$ onto $[0,1]$ such that $g(a) \geq g(b)$ if and only if $a \leq b$. The set M of all automorphisms and anti-automorphisms of $\mathbb{I}$ is a group under composition of maps, and the set A of all automorphisms is a subgroup of index 2 in M. A copy of the multiplicative group of positive real numbers is contained in A by associating a positive real number r with the mapping $[0,1] \to [0,1] : x \to x^r$. This association is a homomorphism mapping the group of positive real numbers isomorphically onto a subgroup of A. This subgroup will be denoted by $\mathbb{R}^+$. If f is any automorphism of $\mathbb{I}$, rf will denote the composition of functions—that is, $rf(x) = (f(x))^r$. Anti-automorphisms in M of order two are called **negations**, or **involutions**. They will play an important role in what we do.

The connection between strict t-norms and the group A is this [8]:

Theorem 1 An Archimedean t-norm $\bigtriangleup$ is strict if and only if there is an automorphism $f \in A$ such that $f(x \bigtriangleup y) = f(x) f(y)$. Another automorphism $g \in A$ satisfies this condition if and only if $f = rg$ for some $r \in R^+$ (where r denotes the function defined by $r(x) = x^r$).

Such an automorphism f is called a (multiplicative) **generator** of the t-norm $\bigtriangleup$ since $x \bigtriangleup y = f^{-1}(f(x)f(y))$, that is, the strict t-norm $\bigtriangleup$ is determined by f and ordinary multiplication.

The connection between nilpotent t-norms and A is the following (see [5, 8]).

Theorem 2 A t-norm $\bigtriangleup$ is nilpotent if and only if there is an automorphism $f \in A$ such that $f(x \bigtriangleup y) = (f(x) + f(y) - 1) \vee 0$. The automorphism f is unique.

The automorphism f is called an **L-generator** of the nilpotent t-norm $\bigtriangleup$ since $x \bigtriangleup y = f^{-1}((f(x) + f(y) - 1) \vee 0)$, that is, the nilpotent t-norm $\bigtriangleup$ is determined by f and the Łukasiewicz t-norm $(x + y - 1) \vee 0$ [6].

In subsequent sections we describe several connections between families of t-norms and groups of generating functions. These connections are generally obscured by the common use of "additive generators" for triangular norms, as the collection of additive generators does not form a group in a natural way. Also, multiplicative and L-generators are isomorphisms between the algebras $(\mathbb{I}, \circ)$ and $(\mathbb{I}, \triangle)$, where $\circ$ represents either multiplication or the Łukasiewicz t-norm, and $\triangle$ a strict or nilpotent t-norm, respectively. Additive generators do not play such a natural algebraic role.

We refer to [4, 5] for additional details.

2 Subgroups of the automorphism group

Theorem 2 puts the set of nilpotent t-norms in one-to-one correspondence with the automorphism group A. This places a group structure on the set of all nilpotent t-norms for which the Łukasiewicz t-norm acts as the identity element.

For strict t-norms, the situation is a bit different. Strict t-norms are in one-to-one correspondence with right cosets of $\mathbb{R}^+$ in A, where the multiplicative group $\mathbb{R}^+$ of positive real numbers is identified with a subgroup of A by $r(x) = x^r$ for each positive r. Multiplication of elements of $\mathbb{R}^+$ corresponds to composition of functions in A. Were $\mathbb{R}^+$ a normal subgroup in A, this would put a natural group structure on the set of all strict t-norms. This is not the case, however. That is, the normalizer $\{f \in A : f^{-1}\mathbb{R}^+ f = \mathbb{R}^+\}$ of $\mathbb{R}^+$ in the group A is not all of A. But the set of t-norms with generators in the normalizer of $\mathbb{R}^+$ in the group A does carry a group structure. This group has been identified and its structure determined [10]. The corresponding set of t-norms is the one-parameter family of Aczél-Alsina t-norms [1]. Details also appear in [9], Section 5.8.1. We describe those results briefly.

2.1 The normalizer of the subgroup of positive reals

Let N be the **normalizer** of $\mathbb{R}^+$ in the group M; that is,

$$N = \{f \in M : f^{-1}\mathbb{R}^+ f = \mathbb{R}^+\}$$

The following hold for the group N.

1. $N = \{f \in M : f(x) = e^{-c(-\ln x)^r},\ c > 0,\ r \neq 0\}$

2. N is isomorphic to the group of pairs of real numbers

$$\{(c, r) : c > 0, r \neq 0\}$$

with multiplication given by

$$(c, r)(d, s) = (cd^r, rs)$$

The subgroup

$$\mathbb{R}^+ = \{e^{-c(-\ln x)} : c > 0\}$$

corresponds to the group $\{(c, 1) : c \in \mathbb{R}^+\}$, and the subgroup

$$S = \{e^{-(-\ln x)^r} : r \neq 0\}$$

to the group $\{(1, r) : r \neq 0\}$. Thus N **splits** over $\mathbb{R}^+$—that is, $\mathbb{R}^+$ is normal in N and any element of N is uniquely a product rf, $r \in \mathbb{R}^+$, $f \in S$. The group S is called the **Richman group**. Notice that elements of the Richman group S fix $\frac{1}{e}$.

3. The t-norms with generators in N are given by

$$x \circ_r y = e^{-((-\ln x)^r + (-\ln y)^r)^{\frac{1}{r}}}$$

with r positive. These are **Aczél-Alsina** t-norms [1]. The t-conorms with generators in N are given by

$$x \diamond_r y = e^{-((-\ln x)^r + (-\ln y)^r)^{\frac{1}{r}}}$$

with r negative. The group S gives the same family of t-norms and t-conorms since an element of N is of the form rf with $r \in \mathbb{R}^+$ and $f \in S$.

4. The negations in N are precisely the elements $\eta_c = e^{-c(-\ln x)^{-1}} = e^{\frac{c}{\ln x}}$—that is, the elements in N with parameter $r = -1$.

5. The t-norm $x \circ_r y = e^{-((-\ln x)^r + (-\ln y)^r)^{\frac{1}{r}}}$ is dual to the t-conorm $x \diamond_s y = e^{-((-\ln x)^{-s} + (-\ln y)^s)^{\frac{1}{-s}}}$ $(r > 0, s < 0)$ if and only if $r = -s$, in which case they are dual with respect to precisely the negations $e^{\frac{c}{\ln x}}$ in N. (A t-norm and negation, together with the dual t-conorm, is called a **De Morgan system** or **triple system**. The normalizer of $\mathbb{R}^+$ gives a two-parameter family of De Morgan systems for $s > 0$, $c > 0$.)

Fixing the parameter c at $c = 1$ gives a one-parameter family of isomorphic De Morgan systems.

Theorem 3 The De Morgan system $(\mathbb{I}, \circ_r, \eta_1)$ is isomorphic to the De Morgan system $(\mathbb{I}, \cdot, \eta_1)$ for all $r > 0, r \neq 0$. In particular, the De Morgan systems $(\mathbb{I}, \circ_r, \eta_1)$ are isomorphic to one another for all $r > 0$.

Proof. Let $h(x) = e^{-(-\ln x)^r}$, $\eta_1(x) = e^{\frac{1}{\ln x}}$, and $x \circ_r y = e^{-((-\ln x)^r + (-\ln y)^r)^{\frac{1}{r}}}$. Then

$$\begin{aligned} h(x \circ_r y) &= \exp\left(-\left[[(-\ln x)^r + (-\ln y)^r]^{\frac{1}{r}}\right]^r\right) \\ &= \exp(-((-\ln x)^r + (-\ln y)^r)) \\ &= h(x)\, h(y) \end{aligned}$$

Also,

$$\begin{aligned} h(\eta_1(x)) &= h\left(e^{\frac{1}{\ln x}}\right) = \exp\left(-\left(-\ln\left(e^{\frac{1}{\ln x}}\right)\right)^r\right) \\ &= \exp\left(-(-\ln x)^{r-1}\right) = \exp\left(\left(\ln\left(e^{-(-\ln x)^r}\right)\right)^{-1}\right) \\ &= \exp\left((-\ln h(x))^{-1}\right) = \eta_1(h(x)) \end{aligned}$$

Thus h is an isomorphism between the two systems. □

2.2 Some special generator functions

Many of the well known families of t-norms (along with some other interesting, but not so well known, families) are very closely connected with the group N and its subgroups $\mathbb{R}^+$ and S. We will express the sets of generators of these families of t-norms as simple combinations of these three subgroups of M, two special functions in M, and one special subset of M. These are the following:

- the subgroup $\mathbb{R}^+$ of A;
- the subgroup $N = \{e^{-c(-\ln x)^r} : c > 0,\ r \neq 0\}$ of M;
- the subgroup $S = \{e^{-(-\ln x)^r} : r \neq 0\}$ of N;
- the function $\alpha(x) = 1 - x$ in M;
- the function $\varepsilon(x) = e^{-\frac{1-x}{x}}$ in A;
- The set $F = \{\varphi_a(x) = \frac{a^x - 1}{a-1} : a > 0,\ a \neq 1\} \subset A$.

These groups and functions are closely connected. It was mentioned earlier that $N = \mathbb{R}^+ \times S$. An easy computation shows that $\varepsilon\alpha\varepsilon^{-1}(x) = e^{\frac{1}{\ln x}}$ so that $\varepsilon\alpha\varepsilon^{-1} \in S$, and in particular, ε and $\varepsilon\alpha$ determine the same right cosets of S and of N. That is, $S\varepsilon = S\varepsilon\alpha$ and $N\varepsilon = N\varepsilon\alpha$. This comes into play in the discussion below of the Dombi t-norms. Also, in some cases, α gives the duality between pairs of t-norms and t-conorms in a family even though α is not a member of the generating group or coset.

Another family of subgroups that come into play are the centralizers of negations, in particular, the centralizer of the negation α:

$$Z(\alpha) = \{z \in A : z\alpha = \alpha z\}$$

You may recognize F as the set of generators of the Frank family of t-norms. The subset F is not a subgroup of A under composition. It is, however, closely related to the subgroups $\varepsilon^{-1}\mathbb{R}^+\varepsilon$ and $Z(\alpha)$ as follows. For each $r \in \mathbb{R}^+$,

$\varepsilon^{-1}r^{-1}\varepsilon\varphi_{r^2} \in Z(\alpha)$. This implies that $\varepsilon^{-1}r\varepsilon$ and φ_{r^2} represent the same left coset of $Z(\alpha)$ as φ_{r^2}—that is, $\varepsilon^{-1}r\varepsilon Z(\alpha) = \varphi_{r^2}Z(\alpha)$. Moreover, $\varepsilon^{-1}\mathbb{R}^+\varepsilon \cap Z(\alpha) = \{1\}$. This follows easily from the fact that $z\left(\frac{1}{2}\right) = \frac{1}{2}$ for all $z \in Z(a)$. The set F is otherwise an anomaly in this setting. In particular, it is not closed under composition.

Now we list some families of t-norms (and t-conorms in some cases), and as promised, express their sets of generators in terms of the subgroups, subsets, and functions mentioned above.

3 Families of strict t-norms

1. The Hamacher family (t-norms and generators) [7]:

$$\left\{\frac{xy}{x+y-xy+a(1-x-y+xy)} : a > 0\right\}$$

$$\left\{\frac{x}{x+a(1-x)} : a > 0\right\} = \varepsilon^{-1}\mathbb{R}^+\varepsilon$$

Comments on this family: The group $\mathbb{R}^+$ gives only one t-norm, namely multiplication. However, the group $\varepsilon^{-1}\mathbb{R}^+\varepsilon$, which is a conjugate of $\mathbb{R}^+$, gives a one-parameter family of t-norms. Since the group $\varepsilon^{-1}\mathbb{R}^+\varepsilon$ contains only automorphisms, it does not generate any t-conorms.

2. The Jane Doe #2 family (t-norms and generators):

$$\left\{xye^{-a\ln x \ln y} : a > 0\right\}$$

$$\left\{\frac{1}{1-\ln x^a} : a > 0\right\} = \varepsilon^{-1}\mathbb{R}^+$$

Comments on this family: The coset $\varepsilon^{-1}\mathbb{R}^+$ gives a one-parameter family of t-norms. Since the coset contains only automorphisms, it does not generate any t-conorms.

3. The Jane Doe #3 family (t-norms and generators):

$$\left\{1-(1-(1-(1-x)^a)(1-(1-y)^a))^{\frac{1}{a}} : a > 0\right\}$$

$$\{1-(1-x)^a : a > 0\} = \alpha\mathbb{R}^+\alpha$$

Comments on this family: The group $\mathbb{R}^+$ gives only one t-norm, namely multiplication. However, the group $\alpha\mathbb{R}^+\alpha$, which is a conjugate of $\mathbb{R}^+$, gives a one-parameter family of t-norms. Since the group $\alpha\mathbb{R}^+\alpha$ contains only automorphisms, it does not generate any t-conorms.

4. The Schweizer family (t-norms and generators) [11]:

$$\left\{(x^{-a}+y^{-a}-1)^{-\frac{1}{a}} : a > 0\right\}$$

$$\left\{e^{-\left(\frac{1-x^a}{x^a}\right)} : a > 0\right\} = \varepsilon\mathbb{R}^+$$

Comments on this family: The coset $\varepsilon\mathbb{R}^+$ gives a one-parameter family of t-norms. Since the coset contains only automorphisms, it does not generate any t-conorms.

5. The Aczél-Alsina family (t-norms, t-conorms and generators) [1]:

$$\left\{e^{-((-\ln x)^r+(-\ln y)^r)^{\frac{1}{r}}} : r > 0\right\}$$

$$\left\{e^{-((-\ln x)^r+(-\ln y)^r)^{\frac{1}{r}}} : r < 0\right\}$$

$$\left\{e^{-(-\ln x)^r} : r \neq 0\right\} = S$$

Comments on this family: The only negation in this group of generators is the generator with $r = -1$, which gives the negation $e^{1/\ln x}$. This group gives a family of De Morgan systems, namely the t-norm with parameter r, $r > 0$, the negation $e^{1/\ln x}$, and the t-conorm with parameter $-r$.

6. The Jane Doe #1 family (t-norms, t-conorms and generators):

$$\left\{\left(1+\left[(\tfrac{1-x}{x})^r+(\tfrac{1-y}{y})^r+(\tfrac{1-x}{x})^r(\tfrac{1-y}{y})^r\right]^{\frac{1}{r}}\right)^{-1} : r > 0\right\}$$

$$\left\{\left(1+\left[(\tfrac{1-x}{x})^r+(\tfrac{1-y}{y})^r+(\tfrac{1-x}{x})^r(\tfrac{1-y}{y})^r\right]^{\frac{1}{r}}\right)^{-1} : r < 0\right\}$$

$$\left\{\frac{x^r}{x^r+(1-x)^r} : r \neq 0\right\} = \varepsilon^{-1}S\varepsilon$$

Comments on this family: The only negation in this group of generators is the generator with $r = -1$, which gives the negation α. This group gives a family of De Morgan systems, namely the t-norm with parameter r, $r > 0$, the negation $\alpha(x) = \varepsilon^{-1}\left(e^{\frac{1}{\ln \varepsilon(x)}}\right) = 1 - x$, and the t-conorm with parameter $-r$. The family of generators in this example is a conjugate of the group S of generators of the Aczél-Alsina family. Any conjugate of S will give a family of De Morgan systems having the same algebraic properties as those of the Aczél-Alsina family.

7. The Jane Doe #1-Hamacher family (t-norms, t-conorms and generators):

$$\left\{\left(1 + [(\tfrac{1-x}{x})^r + (\tfrac{1-y}{y})^r + a((\tfrac{1-x}{x})^r(\tfrac{1-y}{y})^r]^{\frac{1}{r}}\right)^{-1} : a > 0, r > 0\right\}$$

$$\left\{\left(1 + [(\tfrac{1-x}{x})^r + (\tfrac{1-y}{y})^r + a((\tfrac{1-x}{x})^r(\tfrac{1-y}{y})^r]^{\frac{1}{r}}\right)^{-1} : a > 0, r < 0\right\}$$

$$\left\{\frac{x^r}{x^r + a(1-x)^r} : a > 0,\ r \neq 0\right\} = \varepsilon^{-1}N\varepsilon$$

Comments on this family: The negations in this group of generators are those with parameters $a > 0$, $r = -1$, which are the negations $(1-x)/(1+(a-1)x)$. This group gives a family of De Morgan systems, namely the t-norm with parameters $a > 0$, $r > 0$, the negation $(1-x)/(1+(a-1)x)$, and the t-conorm with parameters a, $-r$.

8. The Dombi family (t-norms, t-conorms and generators) [2]:

$$\left\{\left(1 + (\tfrac{1-x}{x})^r + \left(\tfrac{1-y}{y}\right)^r\right)^{-\frac{1}{r}} : r > 0\right\}$$

$$\left\{\left(1 + (\tfrac{1-x}{x})^r + \left(\tfrac{1-y}{y}\right)^r\right)^{-\frac{1}{r}} : r < 0\right\}$$

$$\left\{e^{-(\frac{1-x}{x})^r} : r \neq 0\right\} = S\varepsilon$$

Comments on this family: The coset $S\varepsilon$ has no negation in it. The t-norm with parameter r ($r > 0$), is dual to the t-conorm with parameter $-r$, with respect to the negation α. Denote the element $e^{-(\frac{1-x}{x})^r}$ of $S\varepsilon$ by $t_r\varepsilon$. Then the element $\varepsilon\alpha\varepsilon^{-1} = e^{1/\ln x} = t_{-1} \in S$. For a generator $t_r\varepsilon$ of a t-norm in the Dombi family, $t_r\varepsilon\alpha = t_r\varepsilon\alpha\varepsilon^{-1}\varepsilon = t_r t_{-1}\varepsilon \in S\varepsilon$.

9. The Frank family (t-norms and generators) [3]:

$$\left\{\log_a\left(1+\frac{(a^x-1)(a^y-1)}{a-1}\right) : a>0,\ a\neq 1\right\}$$

$$\left\{\frac{a^x-1}{a-1} : a>0,\ a\neq 1\right\} = F$$

Comments on this family: The set of Frank t-norms is the set of t-norms $\triangle$ satisfying the equation $x \triangle y + x \bigtriangledown y = x + y$, where $\bigtriangledown$ is the t-conorm dual to $\triangle$ with respect to the negation α. This set of t-norms does not fit the pattern of the other well-known families of strict t-norms, as it does not come from a group or a coset of a group of generators. Nevertheless, the negation α plays a critical role in its defining equation. The dual t-conorms for the Frank family relative to α and their generators are

$$\left\{1-\log_a\left(1+\frac{(a^{1-x}-1)(a^{1-y}-1)}{a-1}\right) : a>0, a\neq 1\right\}$$

$$\left\{\frac{a^{1-x}-1}{a-1} : a>0,\ a\neq 1\right\} = F\alpha$$

This family can be reparametrized as follows: For each $a > 1$, $a \neq 1$, write $a = e^{-b}$. Then for $b > 0$, $e^{-b} > 0$ and $e^{-b} \neq 1$. So replacing a by e^{-b} reparametrizes this family with the parameters the positive reals. This makes the parameters into a group under multiplication. However, this operation is not related to the group operation of composition of the generators.

There does not appear to be a direct group theoretic representation of this family, but as pointed out in Section 2.2, it has close ties to the subgroup $Z(\alpha)$, the centralizer of α.

4 Families of nilpotent t-norms

1. The Schweizer-Sklar family (t-norms and L-generators) [11]:

$$\left\{((x^a+y^a-1)\vee 0)^{\frac{1}{a}} : a>0\right\}$$

$$\{x^a : a>0\} = \mathbb{R}^+$$

Comments on this family: The set of generators of this family forms a group, inducing a group structure on the set of Schweizer-Sklar t-norms. The identity of this group is the Łukasiewicz t-norm $(x + y - 1) \vee 0$, corresponding to the parameter $a = 1$.

2. The Yager family (t-norms and L-generators) [13]:

$$\left\{(1-((1-x)^a+(1-y)^a)^{\frac{1}{a}})\vee 0 : a>0\right\}$$

$$\{1-(1-x)^a : a>0\} = \alpha^{-1}\mathbb{R}^+\alpha$$

Comments on this family: The set $\alpha^{-1}\mathbb{R}^+\alpha$ of generators forms a group, being a conjugate of $\mathbb{R}^+$. The Schweizer-Sklar and Yager families have only the Łukasiewicz t-norm in common, corresponding to the fact that $\mathbb{R}^+\cap\alpha^{-1}\mathbb{R}^+\alpha=\{1\}$.

3. The Weber family (t-norms and L-generators) [12]:

$$\{(a(x+y-1)-(a-1)xy)\vee 0 : a>0\}$$

$$\{1-\log_a(1-(1-a)(1-x)) : a>0, a\neq 1\} = \alpha^{-1}F^{-1}\alpha$$

Comments on this family: This set of generators does not form a group and is not a coset of a subgroup of M. The generators of this family, and the next two, are related to the generators of the Frank family of strict t-norms or t-conorms.

4. Jane Doe #4 (t-norms and L-generators):

$$\left\{\left(\tfrac{1}{a}\left(x+y-1+(a-1)xy\right)\right)\vee 0 : a>0,\ a\neq 1\right\}$$

$$\{\log_a(1+(a-1)x) : a>0,\ a\neq 1\} = F^{-1}$$

5. Jane Doe #6 (t-norms and L-generators):

$$\left\{\log_a\left(1+\tfrac{|a-1|}{a-1}\left(\left(\tfrac{|a-1|}{a-1}\left(a^x+a^y-a-1\right)\right)\vee 0\right)\right) : a>0, a\neq 1\right\}$$

$$\left\{\frac{a^x-1}{a-1} : a>0, a\neq 1\right\} = F$$

5 Summary comments

Just why the generators of so many of the well known families of strict and nilpotent t-norms are simply group theoretic combinations—for example, cosets or conjugates of the groups $\mathbb{R}^+$, S, and N—is not too clear. It is of interest to note that the groups S and N come from $\mathbb{R}^+$ rather directly, N being the normalizer of $\mathbb{R}^+$ and S being a complementary factor to $\mathbb{R}^+$ in N, as indicated in Section 2. As mentioned in Section 2.2, the Frank family (and consequently the Weber, Jane Doe #4, and Jane Doe #6 nilpotent families with related generating sets) does not come from a subgroup, but the set of Frank family generators is closely related to the subgroups $\varepsilon^{-1}\mathbb{R}^+\varepsilon$ and $Z(\alpha)$.

References

[1] J. Aczél, C. Alsina, *Characterizations of some classes of quasilinear functions with applications to triangular norms and to synthesizing judgements*, Aequationes Mathematicae **25**(1982), 313–315.

[2] J. Dombi, *A general class of fuzzy operators, the De Morgan class of fuzzy operators and fuzziness measures induced by fuzzy operators*, Fuzzy Sets and Systems **8**(1982), 149–163.

[3] M. J. Frank, *On the simultaneous associativity of $F(x,y)$ and $x+y-F(x,y)$*, Aequationes Math. **19**(1979), 194–226.

[4] M. Gehrke, C. Walker, E. Walker, *De Morgan systems on the unit interval*, International Journal of Intelligent Systems **11**(1996), 733–750.

[5] M. Gehrke, C. Walker, E. Walker, *A note on negations and nilpotent t-norms*, International Journal of Approximate Reasoning **21**(1999), 137–155.

[6] R. Giles, *Łukasiewicz logic and fuzzy theory*, Inter. J. Man-Mach. Stud. **8**(1976), 313–327.

[7] H. Hamacher, *Über logische Aggregationen nicht-binär expliziter Entscheidungskriterien*, Frankfurt/Main (1978).

[8] C. H. Ling, *Representation of associative functions*, Publ. Math. Debrecen **12**(1963), 182–212.

[9] H. Nguyen, E. Walker, A First Course In Fuzzy Logic, Second Edition, Chapman & Hall/CRC Press (Boca Raton), 2000.

[10] F. Richman, E. Walker, *Some group theoretic aspects of t-norms*, International Journal of Uncertainty, Fuzziness, and Knowledge-based Systems **8**(2000), 1–6.

[11] B. Schweizer, A. Sklar, *Associative functions and statistical triangle inequalities*, Publicationes Mathematicae Debrecen **8**(1961), 169–186.

[12] S. Weber, *A general concept of fuzzy connectives, negations, and implications based on t-norms and t-conorms*, Fuzzy Sets Systems **11**(1983), 115–134.

[13] R. R. Yager, *On a general class of fuzzy connectives*, Fuzzy Sets and Systems **4**(1980), 235–242.

[14] Zadeh, L. A., *Fuzzy Sets*, Information and Control **8**(1965), 338–353.

TRENDS IN LOGIC

1. G. Schurz: *The Is-Ought Problem.* An Investigation in Philosophical Logic. 1997 ISBN 0-7923-4410-3
2. E. Ejerhed and S. Lindström (eds.): *Logic, Action and Cognition.* Essays in Philosophical Logic. 1997 ISBN 0-7923-4560-6
3. H. Wansing: *Displaying Modal Logic.* 1998 ISBN 0-7923-5205-X
4. P. Hájek: *Metamathematics of Fuzzy Logic.* 1998 ISBN 0-7923-5238-6
5. H.J. Ohlbach and U. Reyle (eds.): *Logic, Language and Reasoning.* Essays in Honour of Dov Gabbay. 1999 ISBN 0-7923-5687-X
6. K. Došen: *Cut Elimination in Categories.* 2000 ISBN 0-7923-5720-5
7. R.L.O. Cignoli, I.M.L. D'Ottaviano and D. Mundici: *Algebraic Foundations of many-valued Reasoning.* 2000 ISBN 0-7923-6009-5
8. E.P. Klement, R. Mesiar and E. Pap: *Triangular Norms.* 2000 ISBN 0-7923-6416-3
9. V.F. Hendricks: *The Convergence of Scientific Knowledge.* A View From the Limit. 2001 ISBN 0-7923-6929-7
10. J. Czelakowski: *Protoalgebraic Logics.* 2001 ISBN 0-7923-6940-8
11. G. Gerla: *Fuzzy Logic.* Mathematical Tools for Approximate Reasoning. 2001 ISBN 0-7923-6941-6
12. M. Fitting: *Types, Tableaus, and Gödel's God.* 2002 ISBN 1-4020-0604-7
13. F. Paoli: *Substructural Logics: A Primer.* 2002 ISBN 1-4020-0605-5
14. S. Ghilardi and M. Zawadowki: *Sheaves, Games, and Model Completions.* A Categorical Approach to Nonclassical Propositional Logics. 2002 ISBN 1-4020-0660-8
15. G. Coletti and R. Scozzafava: *Probabilistic Logic in a Coherent Setting.* 2002 ISBN 1-4020-0917-8; Pb: 1-4020-0970-4
16. P. Kawalec: *Strucıural Reliabilism.* Inductive Logic as a Theory of Justification. 2002 ISBN 1-4020-1013-3
17. B. Löwe, W. Malzkorn and T. Räsch (eds.): *Foundations of the Formal Sciences II.* Applications of Mathematical Logic in Philosophy and Linguistics, Papers of a conference held in Bonn, November 10-13, 2000. 2003 ISBN 1-4020-1154-7
18. R.J.G.B. de Queiroz (ed.): *Logic for Concurrency and Synchronisation.* 2003 ISBN 1-4020-1270-5
19. A. Marcja and C. Toffalori: *A Guide to Classical and Modern Model Theory.* 2003 ISBN 1-4020-1330-2; Pb 1-4020-1331-0
20. S.E. Rodabaugh and E.P. Klement (eds.): *Topological and Algebraic Structures in Fuzzy Sets.* A Handbook of Recent Developments in the Mathematics of Fuzzy Sets. 2003 ISBN 1-4020-1515-1; Pb 1-4020-1516-X

21. V.F. Hendricks and J. Malinowski: *Trends in Logic*. 50 Years Studia Logica. 2003
ISBN 1-4020-1601-8

KLUWER ACADEMIC PUBLISHERS – DORDRECHT / BOSTON / LONDON

MIX
Papier aus verantwortungsvollen Quellen
Paper from responsible sources
FSC® C105338
FSC
www.fsc.org

If you have any concerns about our products,
you can contact us on
ProductSafety@springernature.com

In case Publisher is established outside the EU,
the EU authorized representative is:
Springer Nature Customer Service Center GmbH
Europaplatz 3, 69115 Heidelberg, Germany

Printed by Libri Plureos GmbH
in Hamburg, Germany